IET ENGINEERING SERIES 250

Interactions of Wind Turbines with Aviation Radio and Radar Systems

Other volumes in this series:

Volume 1	**Power Circuit Breaker Theory and Design** C.H. Flurscheim (Editor)
Volume 4	**Industrial Microwave Heating** A.C. Metaxas and R.J. Meredith
Volume 7	**Insulators for High Voltages** J.S.T. Looms
Volume 8	**Variable Frequency AC Motor Drive Systems** D. Finney
Volume 10	**SF_6 Switchgear** H.M. Ryan and G.R. Jones
Volume 11	**Conduction and Induction Heating** E.J. Davies
Volume 13	**Statistical Techniques for High Voltage Engineering** W. Hauschild and W. Mosch
Volume 14	**Uninterruptible Power Supplies** J. Platts and J.D. St Aubyn (Editors)
Volume 15	**Digital Protection for Power Systems** A.T. Johns and S.K. Salman
Volume 16	**Electricity Economics and Planning** T.W. Berrie
Volume 18	**Vacuum Switchgear** A. Greenwood
Volume 19	**Electrical Safety: A guide to causes and prevention of hazards** J. Maxwell Adams
Volume 21	**Electricity Distribution Network Design, 2nd Edition** E. Lakervi and E.J. Holmes
Volume 22	**Artificial Intelligence Techniques in Power Systems** K. Warwick, A.O. Ekwue and R. Aggarwal (Editors)
Volume 24	**Power System Commissioning and Maintenance Practice** K. Harker
Volume 25	**Engineers' Handbook of Industrial Microwave Heating** R.J. Meredith
Volume 26	**Small Electric Motors** H. Moczala *et al.*
Volume 27	**AC–DC Power System Analysis** J. Arrillaga and B.C. Smith
Volume 29	**High Voltage Direct Current Transmission, 2nd Edition** J. Arrillaga
Volume 30	**Flexible AC Transmission Systems (FACTS)** Y.-H. Song (Editor)
Volume 31	**Embedded Generation** N. Jenkins *et al.*
Volume 32	**High Voltage Engineering and Testing, 2nd Edition** H.M. Ryan (Editor)
Volume 33	**Overvoltage Protection of Low-Voltage Systems, Revised Edition** P. Hasse
Volume 36	**Voltage Quality in Electrical Power Systems** J. Schlabbach *et al.*
Volume 37	**Electrical Steels for Rotating Machines** P. Beckley
Volume 38	**The Electric Car: Development and future of battery, hybrid and fuel-cell cars** M. Westbrook
Volume 39	**Power Systems Electromagnetic Transients Simulation** J. Arrillaga and N. Watson
Volume 40	**Advances in High Voltage Engineering** M. Haddad and D. Warne
Volume 41	**Electrical Operation of Electrostatic Precipitators** K. Parker
Volume 43	**Thermal Power Plant Simulation and Control** D. Flynn
Volume 44	**Economic Evaluation of Projects in the Electricity Supply Industry** H. Khatib
Volume 45	**Propulsion Systems for Hybrid Vehicles** J. Miller
Volume 46	**Distribution Switchgear** S. Stewart
Volume 47	**Protection of Electricity Distribution Networks, 2nd Edition** J. Gers and E. Holmes
Volume 48	**Wood Pole Overhead Lines** B. Wareing
Volume 49	**Electric Fuses, 3rd Edition** A. Wright and G. Newbery
Volume 50	**Wind Power Integration: Connection and system operational aspects** B. Fox *et al.*
Volume 51	**Short Circuit Currents** J. Schlabbach
Volume 52	**Nuclear Power** J. Wood
Volume 53	**Condition Assessment of High Voltage Insulation in Power System Equipment** R.E. James and Q. Su
Volume 55	**Local Energy: Distributed generation of heat and power** J. Wood
Volume 56	**Condition Monitoring of Rotating Electrical Machines** P. Tavner, L. Ran, J. Penman and H. Sedding
Volume 57	**The Control Techniques Drives and Controls Handbook, 2nd Edition** B. Drury
Volume 58	**Lightning Protection** V. Cooray (Editor)
Volume 59	**Ultracapacitor Applications** J.M. Miller
Volume 62	**Lightning Electromagnetics** V. Cooray

Volume 63 **Energy Storage for Power Systems, 2nd Edition** A. Ter-Gazarian
Volume 65 **Protection of Electricity Distribution Networks, 3rd Edition** J. Gers
Volume 66 **High Voltage Engineering Testing, 3rd Edition** H. Ryan (Editor)
Volume 67 **Multicore Simulation of Power System Transients** F.M. Uriate
Volume 68 **Distribution System Analysis and Automation** J. Gers
Volume 69 **The Lightening Flash, 2nd Edition** V. Cooray (Editor)
Volume 70 **Economic Evaluation of Projects in the Electricity Supply Industry, 3rd Edition** H. Khatib
Volume 72 **Control Circuits in Power Electronics: Practical issues in design and implementation** M. Castilla (Editor)
Volume 73 **Wide Area Monitoring, Protection and Control Systems: The enabler for smarter grids** A. Vaccaro and A. Zobaa (Editors)
Volume 74 **Power Electronic Converters and Systems: Frontiers and applications** A.M. Trzynadlowski (Editor)
Volume 75 **Power Distribution Automation** B. Das (Editor)
Volume 76 **Power System Stability: Modelling, analysis and control** A.A. Sallam and B. Om P. Malik
Volume 78 **Numerical Analysis of Power System Transients and Dynamics** A. Ametani (Editor)
Volume 79 **Vehicle-to-Grid: Linking electric vehicles to the smart grid** J. Lu and J. Hossain (Editors)
Volume 81 **Cyber-Physical-Social Systems and Constructs in Electric Power Engineering** S. Suryanarayanan, R. Roche and T.M. Hansen (Editors)
Volume 82 **Periodic Control of Power Electronic Converters** F. Blaabjerg, K. Zhou, D. Wang and Y. Yang
Volume 86 **Advances in Power System Modelling, Control and Stability Analysis** F. Milano (Editor)
Volume 87 **Cogeneration: Technologies, optimisation and implementation** C.A. Frangopoulos (Editor)
Volume 88 **Smarter Energy: From smart metering to the smart grid** H. Sun, N. Hatziargyriou, H.V. Poor, L. Carpanini and M.A. Sánchez Fornié (Editors)
Volume 89 **Hydrogen Production, Separation and Purification for Energy** A. Basile, F. Dalena, J. Tong and T.N. Veziroğlu (Editors)
Volume 90 **Clean Energy Microgrids** S. Obara and J. Morel (Editors)
Volume 91 **Fuzzy Logic Control in Energy Systems with Design Applications in MATLAB®/Simulink®** İ.H. Altaş
Volume 92 **Power Quality in Future Electrical Power Systems** A.F. Zobaa and S.H.E.A. Aleem (Editors)
Volume 93 **Cogeneration and District Energy Systems: Modelling, analysis and optimization** M.A. Rosen and S. Koohi-Fayegh
Volume 94 **Introduction to the Smart Grid: Concepts, technologies and evolution** S.K. Salman
Volume 95 **Communication, Control and Security Challenges for the Smart Grid** S.M. Muyeen and S. Rahman (Editors)
Volume 96 **Industrial Power Systems with Distributed and Embedded Generation** R Belu
Volume 97 **Synchronized Phasor Measurements for Smart Grids** M.J.B. Reddy and D.K. Mohanta (Editors)
Volume 98 **Large Scale Grid Integration of Renewable Energy Sources** A. Moreno-Munoz (Editor)
Volume 100 **Modeling and Dynamic Behaviour of Hydropower Plants** N. Kishor and J. Fraile-Ardanuy (Editors)
Volume 101 **Methane and Hydrogen for Energy Storage** R. Carriveau and D.S.-K. Ting
Volume 104 **Power Transformer Condition Monitoring and Diagnosis** A. Abu-Siada (Editor)
Volume 106 **Surface Passivation of Industrial Crystalline Silicon Solar Cells** J. John (Editor)
Volume 107 **Bifacial Photovoltaics: Technology, applications and economics** J. Libal and R. Kopecek (Editors)

Volume 108 **Fault Diagnosis of Induction Motors** J. Faiz, V. Ghorbanian and G. Joksimović
Volume 109 **Cooling of Rotating Electrical Machines: Fundamentals, modelling, testing and design** D. Staton, E. Chong, S. Pickering and A. Boglietti
Volume 110 **High Voltage Power Network Construction** K. Harker
Volume 111 **Energy Storage at Different Voltage Levels: Technology, integration, and market aspects** A.F. Zobaa, P.F. Ribeiro, S.H.A. Aleem and S.N. Afifi (Editors)
Volume 112 **Wireless Power Transfer: Theory, technology and application** N. Shinohara
Volume 114 **Lightning-Induced Effects in Electrical and Telecommunication Systems** Y. Baba and V.A. Rakov
Volume 115 **DC Distribution Systems and Microgrids** T. Dragičević, F. Blaabjerg and P. Wheeler
Volume 116 **Modelling and Simulation of HVDC Transmission** M. Han (Editor)
Volume 117 **Structural Control and Fault Detection of Wind Turbine Systems** H.R. Karimi
Volume 118 **Modelling and Simulation of Complex Power Systems** A. Monti and A. Benigni
Volume 119 **Thermal Power Plant Control and Instrumentation: The control of boilers and HRSGs, 2nd Edition** D. Lindsley, J. Grist and D. Parker
Volume 120 **Fault Diagnosis for Robust Inverter Power Drives** A. Ginart (Editor)
Volume 121 **Monitoring and Control using Synchrophasors in Power Systems with Renewables** I. Kamwa and C. Lu (Editors)
Volume 123 **Power Systems Electromagnetic Transients Simulation, 2nd Edition** N. Watson and J. Arrillaga
Volume 124 **Power Market Transformation** B. Murray
Volume 125 **Wind Energy Modeling and Simulation, Volume 1: Atmosphere and plant** P. Veers (Editor)
Volume 126 **Diagnosis and Fault Tolerance of Electrical Machines, Power Electronics and Drives** A.J.M. Cardoso
Volume 127 **Lightning Electromagnetics Volume 1 Return stroke modelling and electromagnetic radiation, 2nd Edition** V. Cooray, F. Rachidi, M. Rubinstein (Editors)
Volume 127 **Lightning Electromagnetics Volume 2 Electrical processes and effects, 2nd Edition** V. Cooray, F. Rachidi, M. Rubinstein (Editors)
Volume 128 **Characterization of Wide Bandgap Power Semiconductor Devices** F. Wang, Z. Zhang and E.A. Jones
Volume 129 **Renewable Energy from the Oceans: From wave, tidal and gradient systems to offshore wind and solar** D. Coiro and T. Sant (Editors)
Volume 130 **Wind and Solar Based Energy Systems for Communities** R. Carriveau and D. S.-K. Ting (Editors)
Volume 131 **Metaheuristic Optimization in Power Engineering** J. Radosavljević
Volume 132 **Power Line Communication Systems for Smart Grids** I.R.S. Casella and A. Anpalagan
Volume 134 **Hydrogen Passivation and Laser Doping for Silicon Solar Cells** B. Hallam and C. Chan (Editors)
Volume 139 **Variability, Scalability and Stability of Microgrids** S.M. Muyeen, S.M. Islam and F. Blaabjerg (Editors)
Volume 142 **Wind Turbine System Design, Volume 1: Nacelles, drive trains and verification** J. Wenske (Editor)
Volume 143 **Medium Voltage DC System Architectures** B. Grainger and R.D. Doncker (Editors)
Volume 145 **Condition Monitoring of Rotating Electrical Machines** P. Tavner, L. Ran, and C. Crabtree
Volume 146 **Energy Storage for Power Systems, 3rd Edition** A.G. Ter-Gazarian
Volume 147 **Distribution Systems Analysis and Automation, 2nd Edition** J. Gers
Volume 151 **SiC Power Module Design: Performance, robustness and reliability** A. Castellazzi and A. Irace (Editors)
Volume 152 **Power Electronic Devices: Applications, failure mechanisms and reliability** F Iannuzzo (Editor)

Volume 153 **Signal Processing for Fault Detection and Diagnosis in Electric Machines and Systems** M. Benbouzid (Editor)
Volume 155 **Energy Generation and Efficiency Technologies for Green Residential Buildings** D. Ting and R. Carriveau (Editors)
Volume 156 **Lithium-Ion Batteries Enabled by Silicon Anodes** C. Ban and K. Xu (Editors)
Volume 157 **Electrical Steels, 2 Volumes** A. Moses, K. Jenkins, Philip Anderson and H. Stanbury
Volume 158 **Advanced Dielectric Materials for Electrostatic Capacitors** Q Li (Editor)
Volume 159 **Transforming the Grid Towards Fully Renewable Energy** O. Probst, S. Castellanos and R. Palacios (Editors)
Volume 160 **Microgrids for Rural Areas: Research and case studies** R.K. Chauhan, K. Chauhan and S.N. Singh (Editors)
Volume 161 **Artificial Intelligence for Smarter Power Systems: Fuzzy logic and neural networks** M.G. Simoes
Volume 165 **Digital Protection for Power Systems, 2nd Edition** Salman K. Salman
Volume 166 **Advanced Characterization of Thin Film Solar Cells** N. Haegel and M Al-Jassim (Editors)
Volume 167 **Power Grids with Renewable Energy: Storage, integration and digitalization** A.A. Sallam and B. Om P. Malik
Volume 169 **Small Wind and Hydrokinetic Turbines** P. Clausen, J. Whale and D. Wood (Editors)
Volume 170 **Reliability of Power Electronics Converters for Solar Photovoltaic Applications** F. Blaabjerg, A.l Haque, H. Wang, Z. Abdin Jaffery and Y. Yang (Editors)
Volume 171 **Utility-Scale Wind Turbines and Wind Farms** A. Vasel-Be-Hagh and D.S.-K. Ting
Volume 172 **Lighting Interaction with Power Systems, 2 volumes** A. Piantini (Editor)
Volume 174 **Silicon Solar Cell Metallization and Module Technology** T. Dullweber (Editor)
Volume 175 **n-Type Crystalline Silicon Photovoltaics: Technology, applications and economics** D. Munoz and R. Kopecek (Editors)
Volume 178 **Integrated Motor Drives** Xu Deng and B. Mecrow (Editors)
Volume 180 **Protection of Electricity Distribution Networks, 4th Edition** J. Gers and E. Holmes
Volume 182 **Surge Protection for Low Voltage Systems** A. Rousseau (Editor)
Volume 184 **Compressed Air Energy Storage: Types, systems and applications** D. Ting and J. Stagner
Volume 186 **Synchronous Reluctance Machines: Analysis, optimization and applications** N. Bianchi, C. Babetto and G. Bacco
Volume 190 **Synchrophasor Technology: Real-time operation of power networks** N. Kishor and S.R. Mohanty (Editors)
Volume 191 **Electric Fuses: Fundamentals and new applications, 4th Edition** N. Nurse, A. Wright and P.G. Newbery
Volume 193 **Overhead Electric Power Lines: Theory and practice** S. Chattopadhyay and A. Das
Volume 194 **Offshore Wind Power Reliability, Availability and Maintenance, 2nd edition** P. Tavner
Volume 196 **Cyber Security for Microgrids** S. Sahoo, F. Blaajberg and T. Dragicevic
Volume 198 **Battery Management Systems and Inductive Balancing** A. Van den Bossche and A. Farzan Moghaddam
Volume 199 **Model Predictive Control for Microgrids: From power electronic converters to energy management** J. Hu, J.M. Guerrero and S. Islam
Volume 201 **Design, Control and Monitoring of Tidal Stream Turbine Systems** M. Benbouzid (Editor)
Volume 204 **Electromagnetic Transients in Large HV Cable Networks: Modeling and calculations** Ametani, Xue, Ohno and Khalilnezhad
Volume 208 **Nanogrids and Picogrids and their Integration with Electric Vehicles** S. Chattopadhyay
Volume 210 **Superconducting Magnetic Energy Storage in Power Grids** M.H. Ali

Volume 211 **Blockchain Technology for Smart Grids: Implementation, management and security** H.L. Gururaj, K.V. Ravi, F. Flammini, H. Lin, B. Goutham, K.B.R. Sunil and C. Sivapragash
Volume 212 **Battery State Estimation: Methods and models** S. Wang
Volume 215 **Industrial Demand Response: Methods, best practices, case studies, and applications** H.H. Alhelou, A. Moreno-Muñoz and P. Siano (Editors)
Volume 213 **Wide Area Monitoring of Interconnected Power Systems, 2nd Edition** A.R. Messina
Volume 214 **Distributed Energy Storage in Urban Smart Grids** P.F. Ribeiro and R.S. Salles (Editors)
Volume 217 **Advances in Power System Modelling, Control and Stability Analysis, 2nd Edition** F. Milano (Editor)
Volume 225 **Fusion-Fission Hybrid Nuclear Reactors: For enhanced nuclear fuel utilization and radioactive waste reduction** W.M. Stacey
Volume 227 **Processing and Manufacturing of Electrodes for Lithium-Ion Batteries** J. Li and C. Jin (Editors)
Volume 228 **Digital Technologies for Solar Photovoltaic Systems: From general to rural and remote installations** S. Motahhir (Editor)
Volume 236 **Green HV Switching Technologies for Modern Power Networks,** K. Niayesh (Editor)
Volume 238 **AI for Status Monitoring of Utility Scale Batteries** S. Wang, K. Liu, Y. Wang, D. Stroe, C. Fernandez and J.M. Guerrero
Volume 239 **Intelligent Control of Medium and High Power Converters** M. Bendaoud, Y. Maleh and S. Padmanaban (Editors)
Volume 247 **Power System Strength Evaluation: Methods, best practice, case studies, and applications** H. Alhelou, N. Hosseinzadeh and B. Bahrani (Editors)
Volume 905 **Power System Protection, 4 volumes**

Interactions of Wind Turbines with Aviation Radio and Radar Systems

Alan Collinson

The Institution of Engineering and Technology

Published by The Institution of Engineering and Technology, London, United Kingdom

The Institution of Engineering and Technology is registered as a Charity in England & Wales (no. 211014) and Scotland (no. SC038698).

First published 2023

The Institution of Engineering and Technology
Futures Place
Kings Way, Stevenage
Hertfordshire SG1 2UA, United Kingdom

www.theiet.org

British Library Cataloguing in Publication Data
A catalogue record for this product is available from the British Library

ISBN 978-1-83953-845-2 (hardback)
ISBN 978-1-83953-846-9 (PDF)

Typeset in India by MPS Limited

Cover Image: rusm /E+ via Getty Images

I dedicate this book to my family; to Harry and Mary,
Peter and Val and Valerie

Contents

List of figures

List of tables

About the author

Alan Collinson is the managing director of Collinson Systems Limited (CSL), UK. CSL was formed in 1995 and provides advice on radar systems, phased array and multifunction radar, and radar interference to the military and industry. CSL provides workshops on radar and interference. He taught at Hull University for over 20 years. He has been an expert witness at Public Inquiries into the effects of wind farms on aviation radar systems. He has worked on NATO Industrial Advisory Groups. He is a Registered European Engineer, a Chartered Electrical Engineer, a fellow of the IET, and was appointed OBE for services to the defence industry by her late Majesty the Queen in 1999.

Preface

In 2002, I was asked by a respected United Kingdom engineering consultancy to take part in their due-diligence actions and review a report they had produced on windfarm and aviation system interactions commissioned by the Department of Trade and Industry. My specific task was to consider whether there were any other aspects of the problem that might be worthy of attention that had not already been considered. The aviation systems of concern supported Air Defence and Air Traffic Control missions: missions that are defence and safety critical. A literature search revealed how little material had been published, especially considering the importance of the subject. This was over 20 years ago. Some military investigations had taken place but any information from these was not widely available. More detailed investigations were starting to take place and a few scientific papers were beginning to emerge. Such information that was available in the public domain tended to be concerned with identifying the effects, there was little analysis of causes and less about mitigations. And there were certainly no textbooks written on the subject.

In 2000, the renewable energy sector had been likened to a toddler taking its first tentative steps [1]. But the toddler grew rapidly as the need for alternative energy sources, and in particular as the need for decarbonisation/de-fossilisation, became more and more important. The wind power industry was expanding rapidly. On-shore wind farms could be built relatively quickly and cheaply, compared with off-shore windfarms but, with the benefit of hindsight, in 2002, the storm clouds were gathering. The on-shore sites could only be built provided local authorities were prepared to grant planning permission/zoning. Also, on-shore wind power was unpopular and problematic, especially in small European countries, such as the UK, where proximity to air infrastructure was becoming a serious concern. Notwithstanding the cost, the first off-shore wind farm (Vindeby in Demark) had been in production for ten years and the potential for the off-shore industry was becoming obvious.

At the same time, every sector of the aviation industry was growing and the industry was optimistic [2]; there were plans to build more fleets of aircraft to satisfy demand and airports and air traffic were expanding. The new aircraft were being built using new materials which made them harder to detect by conventional radar systems [3]. Factors including the use of composite materials led to the aviation community foreseeing a time when 'primary' radar would be redundant; superseded by 'secondary' radar and derivative technologies [4]. Replacing methods for primary surveillance became an ambition but it was set back on 11 September 2001 (9/11) when air attacks on Washington, DC and New York killed 2,996 people. Replacing primary surveillance remains an ambition today; although now, if it is to be superseded, it may be by

Automatic Dependent Surveillance-Broadcast (ADSB) and multilateration (SSR derivatives).

Over the course of the next 20 years, I have been asked to investigate a wide range of technical problems, prompted for the most part, by the desire to build new wind turbines or windfarms and concerns about undesirable interactions with aviation radar systems. Each new type of investigation has left me still bemoaning the fact that there was no helpful reference that would address these problems. There comes a time when instead of complaining it is better to do something about it.

I wish to express my gratitude to the following people for their help and support in creating this book. I begin with Nicki Dennis and Christoph von Friedeburg; commissioning editors at the Institution of Engineering and Technology (IET). A chance meeting with Nicki in Edinburgh at the 2022 Radar Conference started the ball rolling. Nicki introduced me to Christoph and it was he who gave the project shape, momentum and a fund of good ideas. The book would not have been completed without the hard work of the following editorial and publishing staff, Olivia Wilkins and Natalie Harper from the IET, and Srinivasan N from MPS Limited; thank you. It is very satisfying to me that the IET should have decided to take this book forward because the IET, formerly the Institution of Electrical Engineers (IEE), has been my professional home for over forty years and I am grateful on many counts.

I am overwhelmed by and want to recognise the invaluable help I received from the following individuals and organisations. The librarians and archivists at the IET, in particular Anne Locker, Yen Le and Aisling O'Malley who provided a lot of practical help with research. Likewise, the librarians and archivists at the Institution of Mechanical Engineers, in particular, Ellen Haggar and Adrian Clement gave valuable support. The librarian and archivist of the Royal Aeronautical Society at National Aerospace Library, Tony Pilmer and Ailean McKillop gave both research assistance and valuable encouragement. Thanks to Greg Smith and Will Boxx of the US 85th Engineering Installation Squadron for their academic assistance with Near-Field/Far-Field boundary evaluation. Thanks to Dr Anne Cameron, Senior Archives Assistant, University of Strathclyde for her assistance with obtaining permission to use images of Professor James Blyth and his work; Peter Bourne of the Cranbrook Windmill Association was kind enough to explain the intricacies of tower windmills and provide a guided tour of one of the finest windmills in the United Kingdon. Bjarke Thomassen Chairman of the Poul La Cour Foundation at Askov in Denmark was kind enough to give his time and help by providing material and images on early wind turbine history and Poul La Cour. Thanks to Margaret Campion of the Legal Affairs Department, the International Telecommunications Union, Geneva, for assistance with and explaining permissions for the reproduction of ITU Recommendations. Thanks to Robert Fairnie, Digital Content Editor Edinburgh Airport, for permission to reproduce an image of the runway at Edinburgh. Special thanks to the Engineering staff at Newquay International Airport, Andy Ormshaw, Marsha Lee, Lee Richardson, Matt Stewart and Richard Philips, who freely gave a lot of their time to provide practical assistance with ground-based aviation systems and for permissions to reproduce images of the aviation systems. Thanks to Euan Cameron of Wind Prospect Ltd Victoria, Australia and Lesley Murray EDF Energy, for assistance with providing information on the collection of meteorological data. Thanks to the staff at Exeter Devon Airport, Alan Freeman and Mark Dulling, for their permission to reproduce images of their radar antenna. Thanks to Wing Commander (Wg Cdr) Andy Calder, Principal Engineer at Air Command, Flt Lt Richie Weeks, media officer RAF Fylingdales, the whole of the media team at RAF Brize Norton and to Simon (Si) Moore of the Media and Communications Office at RAF Boulmer for their help in the complicated process of securing permission to reproduce images of RAF Air Traffic Control and Air Defence radar systems. Thanks to Tracey Newland of VWT Power for information on and images of the Qr6 Vertical Axis Wind Turbine (VAWT) and Chris Newland of Global Partnerships for technical information on the VAWT. Thanks to Cecilia Fikes Jacobson of the US Department of Energy for assistance with reproducing a US National Renewable Energy Laboratory wind atlas image. Thanks to Bill Combatti of the US Federal Aviation Administration (FAA) for his time and trouble explaining the intricacies of using US Government Rights of Use and help in providing a helpful FAA report. Thanks to Daniel Elsey of the UK Copyright Licensing Agency for his assistance in interpreting the UK Government Open Government Licence v 3.0. Thanks to Nichole Gauden of Ysgwydd Gwyn Uchaf Farm Deri for her assistance and permission to use her land for taking photographs of the Brecon Radio Beacon. Thanks to John D. Campbell KC for his helpful review of planning considerations, his encyclopaedic knowledge about the UK energy supply, and his useful and insightful comments. Thanks to Captain Paul Clinton, a commercial airline pilot, for help with aviation communications and navigation.

I would also like to thank my long-term colleagues and friends, Group Captain (Gp Capt) Adrian Parrish (Royal Air Force (RAF) Retd.), WO Mark (Sam) Gallagher (RAF Retd.), Dr Frank Smith (North West Research Associates), Thomas Torgerson (USAF Retd.) and John Wilby (a retired senior civil servant in the UK Ministry of Defence), who have provided encouragement, suggestions, corrections and patience in equal measure coupled, with their vast knowledge of avionics, communications systems and MOD practice.

The following individuals kindly offered their personal opinions and not those of the organisations they now represent, on the themes that needed to be developed in this book: Wg Cdr Andy Calder; Air Commodore (Air Cdre) Malcolm Crayford OBE, former Air Officer Battlespace Management; Gp Capt (RAF Retd.) Maurice Dixon, Command lead on Climate Change & Sustainability, Gp Capt Richard James (RAF Retd.), former head of Air Defence and Air Traffic Delivery Teams for the UK Ministry of Defence, Ms Sam Johnson, Senior Aviation Manager, RES; Gp Capt Chris Knapman (RAF Retd), former O6 lead for Air Battlespace Management Capability Development; Dujon Goncalves-Collins, Senior Strategy Advisor on Aviation and Matt Keal Aviation Radar specialist at Vattenfall; David Jones, Head Of Offshore Development Korea; Dr Steve Leach, Principal Spectrum Technology Analyst Ofcom; Dr Richard Rudd, Plum Consulting; Steve Smith, Windfarm Sector Lead, Thales; Malcolm Spaven, Aviation and Windfarm Consultant at Aviatica; Stephen Speke, Ministry of Defence (MOD) Windfarm specialist; Andrew Yates, responsible for UK Wind and Solar Development, Statkraft, UK.

Special thanks to Dujon Goncalves-Collins, Senior Strategy Advisor on Aviation Vattenfall, London; Patrick Delaney, Vattenfall Pen y Cymoedd; Ben Flett, Vattenfall London, Oziel Garcia Jaimes, Vattenfall Europe, Amsterdam; Matt Keal Aviation Radar specialist at Vattenfall London, Amanda Little, Vattenfall, London, Christian Meyer, Vattenfall Europe Windkraft Gmbh Hamburg; Dan Wills, Vattenfall London and Kathy Wood, Vattenfall Hexham, all of whom have taken this project to heart and provided encouragement, invaluable insights and advice into the wind development industry; to all of them I wish to express my sincere thanks.

Aviation enthusiast and colleague; Mal O'Dell for his valuable suggestions. Dr David Bacon, former Chief Propagation Advisor to Ofcom, has taught me about radio propagation, in particular, diffraction: a subject on which he is one of the World's leading authorities. Cyprien de Cosson has kindly given lots of his time; listening to my ideas and sharing his impressive practical experience, particularly, in the field of Air–Ground–Air communications. In addition to providing advice on themes for this book, Stephen Speke provided guidance on copyright issues on defence material as well as being a long-term source of wise counsel on the interaction of wind turbines and aviation radio and radar systems. And, last but by no means least, I wish to thank Commander (Cdr) John Taylor Royal Navy (RN) (Retd) of Wind Power Aviation Consultants (WPAC) and former Head of Air Traffic Control and Air Traffic Managements for the Royal Navy, who over the last 15 years has taught me all I know about Air Traffic Control operational matters and, more importantly, who has set many, many, technical challenges about Aviation Radio and Radar systems interacting with Wind Turbines.

Alan Collinson OBE
Pickering
June 2023

References

[1] Clubb (2011), Renewable Energy 2000 to 2010—From Toddler to Teen. https://www.eea.europa.eu/downloads/a0e5bb12ea5df7b2c351fa77430df1a8/1620729282/renewable-energy-2000-to-2010.pdf. Retrieved January 2023.

[2] ICAO (2013), *World Aviation and World Economy*, International Civil Aviation Organization (ICAO), 2013. https://www.icao.int/sustainability/pages/facts-figures_worldeconomydata.aspx. Retrieved January 2023.

[3] Quilter (2004), *Composites in Aerospace Applications*. https://www.aviationpros.com/engines-components/aircraft-airframe-accessories/article/10386441/composites-in-aerospace-applications Retrieved January 2023.

[4] ICAO (2007), *Guidance Material on Comparison of Surveillance Technologies (GMST)*, Edition 1. https://www.icao.int/APAC/Documents/edocs/cns/gmst_jtechnology.pdf. Retrieved January 2023.

Chapter 1
Introduction

1.1 Wind turbines and aviation radio and radar systems

Wind turbines are an important means of extracting renewable energy and their importance is growing. Even those not swayed by the arguments of climate change will be aware of the influence of world events on energy prices, the need to achieve energy independence and security, and the need to exploit alternative sources. However, wind turbines are efficient reflectors of radio waves; they have characteristics that can make them difficult to process by radar and radio systems and they are physically large which means they may be visible over long distances. These factors are particularly problematic for the aviation radio and radar systems because they provide safety-of-life services in the civil sector and facilitate national defences. These problems cannot be ignored. This book is about these interactions and how they might be addressed to mitigate the risks to safety and security and help facilitate, in the longer term, cleaner energy and national energy independence.

Before looking at the affected systems and examining why they are affected and what can be done about it, this chapter provides a brief introduction to why there is a problem and why that problem is getting worse.

1.2 Climate change and renewable energy

The United Nations (UN) is unequivocal in linking the burning of fossil fuels, which generate the greenhouse gas (GHG) carbon dioxide, to the rapid increase in global warming. They have recognised the problem for a long time now: the Kyoto Protocol, committing nations to limit and reduce GHG emissions was signed over 25 years ago in 1997 [1][*]. In 2015, there was international recognition that more urgent action was needed to slow global warming. The UN Climate Change Conference of that year, held in Paris, introduced the long-term goal of limiting global warming to less than 2 °C above pre-industrial levels. At the time of writing, 194 parties[†] have signed the Paris Accord [2].

[*]The protocol was not ratified by all 192 participating nations until 2005.
[†]193 countries and the European Union.

To meet this goal, the UN recognises that a reduction in the use of fossil fuels is not enough and there must be a transition to alternative forms of energy; that is to renewable energy [3] which it defines as:

> '... energy derived from natural sources that are replenished at a higher rate than they are consumed'.

The UN provides examples of resources that satisfy this definition [4].

> 'Sunlight and wind, for example, are such sources that are constantly being replenished'.

Extreme weather events, such as flooding, droughts and extreme temperatures, have reinforced in people's minds the need for action. There is the international agreement that climate change, the need for decarbonisation of the world's energy sources and the need to move to the use of renewable energy, are some of the greatest challenges of our time.

Identifying renewable resources is important but being able to exploit them at a sufficient scale requires technology which is mature. Whilst harvesting electrical energy from wind was invented at the end of the nineteenth century, development of the required technology was slow during much of the twentieth century until a series of international events led to rapid advances in its state of readiness for large-scale exploitation.

1.3 International events and energy

The cost of energy is strongly influenced by world events. There were armed conflicts in the Middle East in every decade of the twentieth century. The 1970s, in particular, saw conflict in many countries in the region: Jordan, the Yemen, Iran, Lebanon, Iraq, Saudi Arabia and Syria. However, there were two conflicts in that period that were to have particular significance for energy supply; namely the Yom Kippur War in 1973 and the Iranian Revolution in 1979 both of which led to large increases in the cost of energy. In 1970, crude oil prices were $3 a barrel. After the Yom Kippur War, the Organisation of Petroleum Exporting Countries (OPEC) placed an embargo on oil supplies to nations that had supported Israel and this led to the price rising to $10 a barrel. Following the Iranian Revolution, the price rose to $39 a barrel [5]. Petrol prices were not the only ones affected. In the United Kingdom, over the same period, solid fuel prices increased by 131%, and despite the relatively recent introduction of North Sea gas, gas prices rose by 85% and electricity prices increased by 183% [6]. In the rest of Europe, electricity prices rose by a third between 1970 and 1980 [7]. European gas price increases in the same period were more variable from region to region (even within nations) but typically price increases were 100% [8].

In the face of such large and rapid increases in the costs of energy, Governments were forced to take steps to manage the situation. In the United Kingdom in 1979, the newly elected prime minister, Margaret Thatcher, appointed

David Howell as the energy minister and, in 1980, the British government formed the Parliamentary Group for Energy Studies (PGES). Together, Howell and the PGES reviewed the UK's energy mix. Similar actions occurred across Europe. In France for example, the energy policy was adjusted to focus on three areas: energy conservation; replacing the use of oil with nuclear energy, coal and renewables; and development of a more coherent oil and gas policy. The oil and gas plans consisted of developing relations with supplying nations, encouraging technical expertise and diversification of supply [9]. In Germany in 1980, Energie-Wende, literally Energy Transition, was introduced setting out a policy of migrating from oil and nuclear power to renewable energy [10]. In the United States, in 1978, even before the Iranian Revolution, President Carter signed into law the very prescient Public Utilities Regulatory Policies Act, which mandated that utility companies should buy a proportion of their electricity from smaller suppliers (less than 80 MW) thus fostering the development and use of renewable energy including wind power, biomass and geothermal sources [11].

Thus, it was the costs of energy and the associated policy changes that led to the rapid scaling up of electricity generation using wind turbines. The first large-scale wind farms were built in the United States, where they are known as 'utility scale' wind farms. The first was built at Crotched Mountain in New Hampshire in 1980. It had 20 turbines each of which could produce 30 kW [12]. Europe followed soon after in 1983 when its first wind farm was built at Kythnos (a Greek island in the Western Cyclades); it comprised five 20 kW turbines [13]. The United Kingdom followed suit a little later when its first commercial wind farm was built in 1991 at Delabole in Cornwall; a farm comprising 10 turbines each capable of generating 1.2 MW [14].

Those first wind farms proved successful and during the early 1990s more wind farms followed. In the United Kingdom, for example, between 1991 and 1997 (when the Kyoto Protocol was signed), 19 wind farms became operational. Almost all of these were modest in terms of the number of turbines, the average being 22. There were exceptions: for example, the Penrhyddlan and Llidiartywaun (P&L) wind farm at Llandinam in mid-Wales had (and still has) 103 turbines. Some of the technology appears a little different from what we might expect today. For example, the Chelker Reservoir development, shown in Figure 1.1, built near Skipton for Yorkshire Water in 1992, used two bladed turbines. The overwhelming majority of wind farms were built using the more usual three bladed turbines such as the Carland Cross development built near Newquay in 1992 and shown in Figure 1.2.

Therefore, by the time of the Kyoto Protocol and, later the Paris accord, wind turbine technology was tried and tested and sufficiently mature for much wider exploitation.

In 2022, Russia invaded Ukraine and energy price rises comparable with those of the 1990s have taken place affecting markets worldwide. How the consequences of the invasion will translate into wider changes in policy have yet to be determined. It seems likely that nations will place much greater emphasis on the need for energy independence. However, some forecasters predict that it will lead to a drastic reduction in the amount of fossil fuels consumed and that, instead of a

Figure 1.1 Chelker Reservoir wind farm

Figure 1.2 Carland Cross wind farm

gradual decline, the use of oil and gas will plummet. Whether and how these two consequences can be reconciled is outside the scope of this book but it is clear that ever greater dependence on renewable energy will be essential [15].

1.4 On-shore wind farm siting problems

Prior to the US Public Utilities Regulatory Policies Act being enacted, the US Department of Energy was asked to conduct investigations to determine if any problems might arise from the introduction of wind farms. The investigations revealed the potential risk of interference with television services, radio services

[16] and fixed telecommunications links [17]. These concerns were soon being echoed by studies in the United Kingdom and The Netherlands [18]. Later on, with the development of commercial (utility) scale wind farms, it also became clear there was also a potential risk of interference with radar systems [19].

As this potential risk of interference with aviation systems became better understood, it was not long before it began to cause problems for developers trying to secure permission to build wind farms. This gave rise to efforts to develop mitigations of wind farm effects. In the United Kingdom, for example, Cambridge Consultants developed a 3-dimensional (holographic) in-fill radar‡ with built-in wind farm tolerance. In 2011, when the company (now called Aveillant) set out the case for a new type of radar, they estimated that approximately 65% of planning applications for wind farms were held up because of aviation objections [20]. Auld, McHenry and Whale reported a similar situation in the United States: in 2009, 10 GW of capacity was installed, a further 8 GW was delayed in planning and 2 GW of capacity had been abandoned. In 2010, they reported that 5 GW was installed but the capacity being delayed had increased to 10 GW [21]. Ten years later, in 2021, the UK Government estimated that more than half of the planning applications for wind farms receive objections on aviation grounds [22]. Moreover, interference with radio-based systems is not the only aviation concern, restricting low flying and the effects of wake turbulence on gliders and General Aviation (GA) (light aircraft) are other causes for objections. However, there is no doubt that interference with radio and radar systems is the major concern and the problem may be significantly worse than even Cambridge Consultants and the British government have estimated.

1.5 Pre-feasibility

When a developer considers a new wind farm site, it is standard practice to assess the site incrementally. The first stage of assessment is a relatively simple project review or pre-feasibility study which considers the viability of the site and considers those factors that would prevent construction going ahead. For example, wind farms are often in remote locations and, even though they are built on-site, the components are very large. The site may not be viable if road access for heavy vehicles is cost prohibitive; Figure 1.3 illustrates the challenges of moving wind turbine components. Moreover, access to the electricity grid in remote locations may be problematic. Anderson points out that if a site is remote, it is likely to mean there are few inhabitants, little industry and no requirement for large amounts of electricity. The provision of site access and grid access will be a significant cost factor for the project [23]. This element of the considerations is called the **balance of plant**.

Many companies recognise that if aviation problems cannot be overcome in a cost-effective manner, a site may not be viable. And if the site is not viable, for

‡The radar, which has a range of 5 Nautical Miles, is now marketed as the Theia radar.

Figure 1.3 Delivering turbine blades (image reproduced by kind permission of Vattenfall)

any reason, then it will not be worth investing in a planning application or even a pre-planning screening (discussed in Chapter 4). Thus, many sites may be excluded from assessment and the UK Government's 2021 estimate of more than half of sites being affected by aviation concerns may, in fact, be a larger proportion of sites.

1.6 Off-shore

If finding suitable sites on-shore is difficult, then it might be imagined that the problem would be solved by using off-shore sites. However, everything about constructing and maintaining an off-shore wind farm is more difficult than an on-shore site: the environment is far harsher; providing grid connection is more complex; the construction of a wind farm and servicing the site when completed are all more complex and a lot more costly. As a result of the additional complexity, the first off-shore farm was not completed until more than ten years after the first large-scale on-shore site. The first offshore wind farm was the 11-turbine site developed by Orsted[§] at Vindeby off the coast of the island of Lolland in Denmark. Commissioning of the site was completed in 1991[¶]. It is worth noting that the site was decommissioned after a 25-year life because the cost of maintenance became unsustainable [24].

Choosing the location of a site off-shore and associated cable route still poses problems, albeit different ones from those on-shore. One obvious shortcoming is

§Originally known as Danish Oil and Natural Gas (DONG).

¶The site was named after the Danish scientist Christian Orsted who discovered that electricity and magnetism were related.

that some nations, even if not landlocked, may have limited coastline. Off-shore, there are also shipping lanes, cables and existing oil and gas infrastructure that need to be avoided. Marine Protected Areas (MPAs), which include a number of protected areas (discussed in Chapter 4), must also be taken into account; there are over 200 MPAs in English waters [25] and, by comparison, Norway has 150 [26].

Nor are off-shore sites immune from the problems of the interaction between wind turbines and aviation radio and radar systems; arguably the problems are more complex than on-shore sites. A simple example illustrates the additional complexity. On-shore, the ideal mitigation for radar interaction problems would be to make the turbines invisible to radar. But off-shore, this would not be acceptable because the turbines would then pose a hazard for shipping which is also dependent on radar for the Safety Of Life At Sea (SOLAS).

However, the most difficult problem to solve for off-shore radar interactions results from the interactions with military radars:

- Off-shore sites are generally larger than on-shore sites and overflying them takes longer so this might lead to reduced warning times of hostile aircraft, missiles and drones if the effects of the wind farm corrupt their tracks.
- Although on-shore sites are starting to use larger turbines, off-shore sites have always tended to be larger and have greater visibility and a potential for interactions at longer ranges.
- The wind farms may be long distances from the shore:
  - Usually, the radars have slower update rates.
  - Mitigations that involve margins around a wind farm site, based on angular information, will involve larger volumes of airspace.
  - Attempts to repair track information will take longer.
- It might reasonably be assumed that attacks from foreign adversaries will arrive from off-shore, which would make wind farms off the coast a great concern for air defenders.

1.7 Increasing wind turbine footprint

A further complication of on-shore site viability is the increasing scale of the turbines. The following case study, kindly provided by Vattenfall UK is illustrative.

Figure 1.4 shows the South Kyle wind farm in East Ayrshire. The South Kyle development is a large wind farm that consists of 50 wind turbines, spread over an area of 2,402 ha. Construction began in 2020, turbine installation started in 2022 and the site has just been completed. The turbines are Delta 4000 N133/4.8 models provided by the German company Nordex. Each turbine will have a maximum tip height of 149.5 m and will be capable of generating 4.8 MW of electricity. The whole project will reduce Carbon Dioxide emissions in the United Kingdom by 300,000 tonnes every year. The economics of such large farms provides greater scope for mitigation measures and the site has been developed by Vattenfall in cooperation with the UK National Air Traffic Services (NATS).

Figure 1.4 South Kyle Wind farm (image reproduced by kind permission of Vattenfall UK)

Although the wind farm at South Kyle has only just been completed, Vattenfall is already planning another wind farm, effectively an extension of South Kyle which will be called South Kyle II. The proposed site is smaller and will only support nine turbines. Different turbines will be used, each capable of generating 7.4 MW of electricity. The extra yield is possible because the turbine tip height is being increased to 200 m (the reason why this is beneficial is discussed in Chapter 4).

As the size of wind turbines increases, they become visible over a wider area. A simple method of calculating the visibility based on the tip height (the derivation is included in Chapter 3) shows that the increase from 150 m to 200 m increases the area of visibility by 33% (corresponding to 2,670 square km)[1].

Although these examples are drawn from the United Kingdom, it is a worldwide phenomenon. The US Government Office of Energy Efficiency and Renewable Energy has tracked the growth in sizes of wind turbines, reporting the findings on their website in an article entitled: ‘Wind Turbines: The Bigger the Better’. Their study found that increases in both hub height and blade lengths were taking place. Their findings are summarised in the graphic shown in Figure 1.5 [27].

The US Study reported in August 2022 and even these findings appear conservative. In January 2023, the China State Shipbuilding Corporation announced the construction of a wind turbine with an 853 feet (260 m) rotor diameter capable of producing 18 MW.

Not only are wind turbines large structures, but successive generations of turbines are getting larger and more visible.

[1]Based on a bare earth: distance to the horizon (the line of sight) = $\sqrt{(2\,k\,R\,h)}$, see Chapter 3.

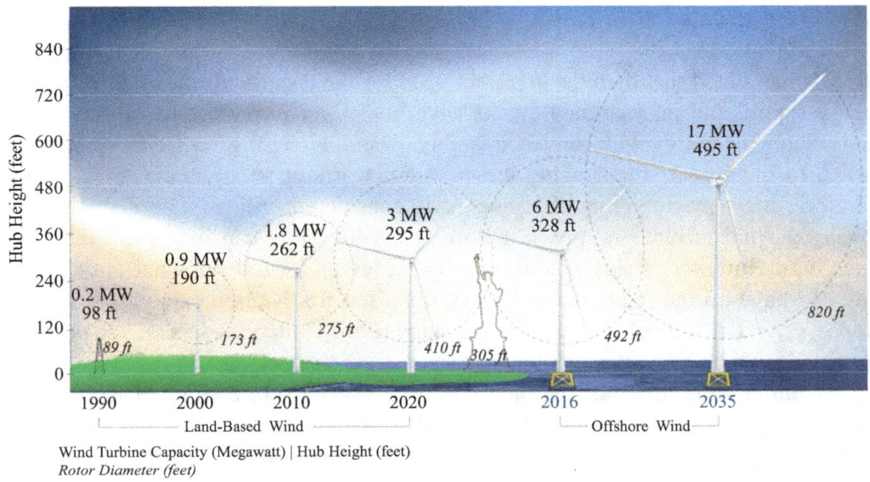

Figure 1.5 The growth of wind turbine sizes (image US government)

1.8 Aviation

There is ample evidence to show that flying is the safest mode of transport [28]. This is because the aviation industry is well regulated, safety focussed, and it has promptly remedied observed safety-related events. It is standard practice to investigate every single safety-related event and react to events promptly. A good example is the Tenerife Airport disaster in 1977 which rapidly led to changes in communications protocols. Moreover, the aviation industry sees opportunities in expanding; the vision of the International Civil Aviation Organization is to 'achieve the sustainable growth of the global civil aviation system' [29].

Of particular importance to this book in the context of aviation is the reaction to the terrorist attacks of 11 September 2001 (9/11). Before 9/11, the trend in civil air surveillance was to migrate from the use of primary radar to what is termed cooperative technology. Primary radars transmit a high-powered signal and then detect the echoes from the hard bodies of aircraft; the aircraft do not have to do anything to be detected. The alternative to primary radar, called secondary radar, requires aircraft to be equipped with transponder systems. The transponder detects coded messages transmitted from the ground and replies with a message containing information about the aircraft, as a minimum its identity and height. Hence, the aircraft cooperates in its detection. The terrorists on board the airliners that struck the World Trade Center in New York and the Pentagon in Washington switched off or modified the responses from the transponders thus preventing them from being tracked. Hence, although it may change in the future, the civil use of primary radar for the detection of aircraft is likely to be required for some time into the future. For the military, the expectation is that all aggressor aircraft will be uncooperative, therefore, military primary radar is a critical defence asset.

1.9 The aim of this book

The UN has stated that there is an urgent need for the decarbonisation of the world's energy supply and international events have both forced up the cost of energy and made national energy independence highly desirable. Using less energy will not suffice to meet goals for slowing down global warming or decrease energy costs; therefore, the need for renewable energy is considered critical. The generation of power by wind turbines is now a mature technology that can help meet the UN's objectives. However, a significant brake to achieving its full potential is the effect they can have on the radio and radar systems used for aviation safety and defence. Moreover, as the technology matures and wind turbines increase in size, the potential to cause interference over a greater area also increases.

Enabling wind turbines and radio and radar systems to operate efficiently and safely within competing geographical areas will require the concerted effort of a number of disparate communities. Included in the wider wind farm community are the developers, their technical consultants, their lawyers who provide support during the complex planning process and the financial institutions that provide the funds for developments; the planning inspectors who need to reach conclusions about the suitability of developments; the civil aviation and defence communities who will have to coexist with wind turbine development; and the radio and radar developers who must facilitate any technical changes. To solve the problems, the academic and research communities will also need to engage, as will Governments.

Each of these disciplines is complex in its own right and, while the members of these disciplines are experts in their own domains, none of them are necessarily experts in any of the others. Even communicating is difficult because of the diverse nature of the problem and the domain-specific language in use.

Therefore, the aim of this book is to explain the nature of these interference problems and do so in such a way that the different communities can interact, solve the problems and make a contribution to solving perhaps the most challenging issues of our time.

1.10 The composition of the book

Before introducing the contents of this book, it is appropriate to say what the book does not contain. This is a technical book; it is **not** a book about operating civil aviation systems safely, nor of defending nations using radio and radar systems, nor about planning law. These distinctions are important. Understanding the technical complexities of an aviation system, either civil or military, and how it works requires quite different skill sets and scopes of knowledge from those needed to operate those systems safely and effectively. Should the reader ever be asked to provide evidence in a court of law or a public inquiry, this distinction will not be lost on a cross-examining barrister! For those readers with military experience, the distinction being made is equivalent to the operational community being able to produce a Concept of Operations, whereas an engineer may be able to propose a Concept of Employment.

There are six chapters following this introduction.

Chapter 2 provides a brief history of wind turbines, electricity generation and radar. The history is interesting in its own right but the chapter also serves as an introduction to the concepts and terminology of disciplines that may be unfamiliar. The development of the wind turbine from its ancient beginnings is described. The story of the first electricity-generating wind turbine is told and taken forward to the development of the first wind farms. The early history of electricity generation is shared with the early history of radio. The evolution of the radar (as an exemplar of the other radio-based systems) is described. The brief history concludes in 2001 when the first reports of interference with radar by wind turbines were published.

Chapter 3 provides an introduction to aviation and to the radio-based systems used to maintain flight safety that might be susceptible to adverse interactions with wind turbines. It is assumed that if a system is to be understood it is helpful to know its purpose. Therefore, the aim of this chapter is to introduce readers to the purpose and design principles of the aviation systems. Other subjects covered include matters such as the radio footprint of the systems including the frequencies used and the antenna patterns. The following systems are considered: air–ground–air communications, instrument landing systems, navigation aids, primary surveillance radar systems, secondary surveillance systems and its derivatives. Military systems are also covered but, for example, the military primary surveillance radars have much in common with civil systems and only the differences are discussed here.

Chapter 4, in some ways the key part of the book, describes wind, how it shapes the development of wind turbines and how the systems discussed in Chapter 3 are affected by wind turbines. Wind shear and turbulence are described as drivers of the scale of wind turbines. The physical and radio footprint of turbines; the radar cross section (RCS) and the differences in RCS with frequency, polarisation and time. The Doppler signature of turbines is described. It will be seen that there are two dominant effects on systems. The continuous carrier systems are subject to corruption by fast fading, and primary radar systems are prone to time-domain effects. Other effects are also described including saturation and shadowing. A way of looking at this chapter is as a set of problems to be addressed in subsequent chapters.

Chapter 5 deals with analytical techniques that may be used to investigate whether interactions will occur and what might be their severity. Locating the various elements of the problem (turbines and radio/radar systems) may seem an obvious and simple thing to do but experience shows this is perhaps the most frequent source of error. Earth curvature effects are described. Analysis techniques supporting all the common interactions are described including methods for predicting impacts on systems. Diffraction methods are described starting with Huygens' construction and building up to the Delta Bullington method.

Chapter 6 describes mitigation options. The chapter starts by considering what is meant by the term mitigation. Three broad strategies for mitigation are considered. The solution must be drawn from modification of the plans for a wind farm to remove the cause of the interference; modifying the aviation service provided, for example by changing the use of airspace to cope with the interference; or by changes to the equipment used to provide the service. Whichever method is chosen, it must not place air safety at risk. Technical options, clutter removal, clutter removal with augmentation and clutter discrimination, are described with their advantages and disadvantages. A frequent question is, when is a solution good enough? The chapter addresses what makes a good mitigation solution and offers suggestions as to how the technical maturity of solutions might be addressed. Some technical solutions require new equipment but, what is critical, is delivering new capability not just new equipment. A systematic process for delivering capability is discussed. The chapter concludes by looking at the factors that need to be considered in a project to deliver mitigation.

Chapter 7 picks up some of the points raised in earlier sections and identifies where more work is necessary, in particular where research and development are required.

A selected bibliography is provided with a list of publications that readers may want to consult to find more information.

The book concludes with a glossary.

References

[1] United Nations (2023), *What Is the Kyoto Protocol?*, https://unfccc.int/kyoto_protocol. Retrieved 28 January 2023.

[2] United Nations (2023), The UN Nationally Determined Contributions Registry, https://unfccc.int/NDCREG?gclid=Cj0KCQiAic6eBhCoARIsANlox86rV3Gsm6x8mpa0vVvm8_k2BLTi29LekSmN7b8hNe6xlu2sLWNW_UQaAjbhEALw_wcB. Retrieved 27 January 2023.

[3] United Nations (2023), *What Is Climate Change?* https://www.un.org/sites/un2.un.org/files/fastfacts-what-is-climate-change.pdf. Retrieved January 2023.

[4] United Nations (2023), *What Is Renewable Energy?* https://www.un.org/en/climatechange/what-is-renewable-energy?gclid=Cj0KCQiAic6eBhCoARIsANlox86k8UklZL7z1KrakgA8DCwl0M1WhClGP4Md-lRftr-or3AVc1eJrpIaAgokEALw_wcB. Retrieved on 27 January 2023.

[5] Macrotrends (2023), *Crude Oil 70 Year Historical Chart*, https://www.macrotrends.net/1369/crude-oil-price-history-chart. Retrieved 28 January 2023.

[6] Hansard (1980), *Statement by Mr John Cartwright MP for Woolwich East*, https://api.parliament.uk/historic-hansard/commons/1980/may/12/fuel-prices. Retrieved 30 January 2023.

[7] European Environment Agency (2009), *Past and Projected Prices of Fossil Fuels and Electricity 1970–2050 in the Baseline and LCEP Scenarios*, https://www.eea.europa.eu/data-and-maps/figures/past-and-projected-prices-of-fossil-fuels-and-electricity-1970-2050-in-the-baseline-and-lcep-scenarios-1. Retrieved 30 January 2023.

[8] Eurostat (1979), *Gas Prices 1976–1978*, Office for Official Publications of the European Communities, 1979, http://publications.europa.eu/resource/cellar/ddb05239-10dc-45ef-b4db-e359c8b23372.0001.02/DOC_1. Retrieved on 30 January 2023.

[9] Giraud, A.L. (1983), *Energy in France*, In the Annual Review of Energy August 1983, pp. 165–191.

[10] Krause, F., Bossel, H., and Muller-Reibmann, K.-F. (1980), *Energie Wende: Wachstum und Wohlstand ohne Erdöl und Uran: Ein Alternativ-Bericht* (*Energy Transition: Growth and Prosperity Without Petroleum and Uranium: An Alternative Report*), Amwalz.

[11] US Congress (1977), *President Carter's Energy Proposals: A Perspective*, US Congressional Budget Office, Staff Working Paper, June 1977.

[12] Kessler, A. (2017), *Gone with the Wind: When Crotched Mountain Had a Wind Farm*, Monadnock Ledger Transcript, 9 February 2017, https://www.ledgertranscript.com/When-Crotched-Mountain-had-a-wind-farm-7910099. Retrieved 20 April 2023.

[13] Wind Europe (2023), *The First European Wind Farm*, Wind Europe Archives, https://windeurope.org/about-wind/history/timeline/the-first-european-wind-farm. Retrieved 20 April 2023.

[14] Powersystems (2021), *The UK's First Wind Farm Turns 30 Meet the Family Who Started It All*, 15 December 2021, https://www.powersystemsuk.co.uk/uks-first-wind-farm-turns-30-meet-family-who-started-it-all. Retrieved 20 April 2023.

[15] Bond, K. (2022), *How Putin's War Marks the End of the Fossil Fuel Era*, Rocky Mountain Institute, https://rmi.org/how-putins-war-marks-the-end-of-the-fossil-fuel-era. Retrieved February 2023.

[16] Senior, T.B.A, Sengupta, D.L., and Ferris, J.E. (1977), *TV and FM Interference by Windmills*, The University of Michigan Radiation Laboratory Ann Arbor, Michigan 48109 Final Report 1 January 1976–31 December 1976, Publication Date: February 1977.

[17] Sengupta, D.L. and Senior, T.B.A. (1978), "Electromagnetic Interference by Wind Turbine Generators" Final Report No. 2, January 1977–March 1978, University of Michigan. Department of Electrical and Computer Engineering. Radiation Laboratory, United States Department of Energy, Wind Systems Branch.

[18] Van Kats, P.J. and De Jager, O.P.E. (1984), *Reflections of Electromagnetic Waves by Large Wind Turbines and Their Impact on UHF Broadcast Reception*, Dr. Neher Laboratorium, Report 511 TM/84.

[19] GPIA (2001), *The Operational Effects of Wind farm Developments on ATC Procedures for Glasgow Prestwick International Airport*, 30 January 2001.

[20] Cambridge Consultants (2011), *Wind Farms, ATC Headaches and 3-D Radar*, Published by Cambridge Consultants, the original news article was published as: http://www.cambridgeconsultants.com/news_pr305.html. However, the article is no longer available on this website as the radar company was bought out by Thales Air Systems. A record of the original

publication can be found at http://www.capdallas.org/aerospace-education/aeo-s-bulletin-board/wind farmsatcheadachesand3dradar.
[21] Auld, T., McHenry, M.P., and Whale, J. (2013), "US military, airspace, and meteorological radar system impacts from utility class wind turbines: Implications for renewable energy targets and the wind industry", *Renewable Energy*, vol. 55, pp. 24–30.
[22] HMG (2021), https://www.gov.uk/government/publications/wind farm-mitigation-for-uk-air-defence/competition-document-wind farm-mitigation-for-uk-air-defence. Retrieved 21 April 2023.
[23] Anderson, C.G. (2020), *Wind Turbines: Theory and Practice*, Cambridge: Cambridge University Press, 2020.
[24] Orsted (2019), *Making Green Energy Affordable: How the Offshore Wind Energy Industry Matured – and What We Can Learn from It*, https://orsted.com/-/media/WWW/Docs/Corp/COM/explore/Making-green-energy-affordable-June-2019.pdf. Retrieved 21 April 2023.
[25] HMG (2023), *Marine Conservation Zone Designations in England*, https://www.gov.uk/government/collections/marine-conservation-zone-designations-in-england. Retrieved 15 February 2023.
[26] Norwegian Directorate of Fisheries (2023), *Marine Protected Areas*, https://www.fiskeridir.no/English/Coastal-management/Marine-protected-areas#:~:text=Areas%20covered%20may%20vary%20from,management%20measures%20have%20been%20introduced. Retrieved 15 February 2023.
[27] US Office of Energy Efficiency and Renewable Energy (2022), *Wind Turbines: The Bigger, the Better*, https://www.energy.gov/eere/articles/wind-turbines-bigger-better#:~:text=The%20average%20rotor%20diameter%20in,wind%2C%20and%20produce%20more%20electricity. Retrieved 15 February 2023.
[28] IATA (2023), *Aviation Safety*, https://www.iata.org/en/youandiata/travelers/aviation-safety/#:~:text=Is%20flying%20safer%20than%20taking,the%20world%20has%20ever%20known. Retrieved 21 April 2023.
[29] ICAO (2023), *Vision and Mission*, International Civil Aviation Organization, https://www.icao.int/about-icao/Council/Pages/vision-and-mission.aspx. Retrieved 8 May 2023.

Chapter 2
A brief history of windmills, electricity generation and radar

2.1 Problems are reported

In 2001, Glasgow Prestwick International Airport (GPIA) [1] and Spaven [2] reported problems with aviation radar systems caused by interference from wind turbines. The UK Department of Trade and Industry (DTI)* set up a Wind Energy, Defence and Civil Aviation Interests Working Group to conduct investigations into potential problems which were reported in 2002 and 2003 [3–5]. These reports were quickly followed by numerous others from diverse agencies including, the Royal Air Force (RAF) [6,7], US Congress [8] and the International Energy Agency (IEA) [9] which described several classes of interference, all of which remain concerns today. Currently, the majority of radar systems are not capable of filtering out this type of interference, and this begs the question, 'why have radars not been developed to coexist with wind turbines?'

2.2 Approach

It might be assumed that the development and use of wind turbines post-dated the development of radar and, had it been the other way around, the problem would, through necessity, have been solved. What has actually taken place is more complex. A brief historical overview of the development of three basic technologies of energy collection using windmills; electricity generation; and radar; addresses the question of precedence. This historical approach also illustrates that the development of innovative systems tends to require stages of development to be observed and this is equally true of innovations to mitigate the problems of windfarm and infrastructure. These stages are examined in Chapter 6. There is a further reason for considering the history; it is interesting.

Wind technology has evolved in many different parts of the world, has been affected by some of the most momentous events in history and has involved some of the greatest scientists and engineers. To introduce the topic, this brief history has been divided into three parts. The first part deals with the development of machines

*The DTI was replaced by the Department of Innovation, Universities and Skills and the Department of Business, Enterprise and Regulatory Reform in 2007.

for extracting energy from wind and covers the period from approximately 10,000 BCE (Before the Current Era) to the end of the 19th century, when the first electricity-generating wind turbine was built. This history shows that many of the principles of wind turbine development were established a long time ago. The second part discusses the development of electricity generation, starting with the discovery that electricity and magnetism were related in the early nineteenth century, up to the start of the 21st century when the impact of wind turbines on radars was first documented. The final part deals with a brief history of radar that, it will be seen, shares a common heritage with electricity generation. The development of radar technology is considered up to the initial identification of issues caused by wind turbines.

2.3 Machines for extracting energy from wind

2.3.1 Sails

The first machines to extract energy from wind were sailing craft built by the Mesopotamians living in the region surrounding the Tigris and Euphrates rivers and their hinterlands, that is modern-day Iraq, parts of Iran, Eastern Turkey, Syria and Kuwait [10]. The Mesopotamian era lasted from the end of the Neolithic Period, approximately 12,000 years ago, until 539 BCE when Babylon, its capital, was conquered by the neighbouring Achaemenid (Persian) Empire. The date of the invention of the sail predates written records and is not accurately known. However, images of sail-bearing vessels found on shards of pottery found at an archaeological site at Subiyah, in Kuwait, have been dated, implying the invention is at least 7,500 years old [11]. Why the invention took place in this region is disputed. The Shamal winds, called the si-sá in Babylonian times, blow from north west which is auspicious for river craft moving downstream on the Tigris and Euphrates. However, other authorities point out that the rivers themselves assist that process and the sail would be of greater value in providing flexibility of travel for sea-going vessels in the Persian Gulf [12].

The invention of the sail was a significant technological milestone. A sail has aerodynamic properties extracting energy using, what is termed, ‘lift’; a force which allows the sail to move faster than the wind driving it and with higher efficiency. Whereas, the earliest windmill designs extracted energy using flat plates which can only exploit drag forces. It follows that these windmills could never exceed the speed of the wind and had limited efficiency. All modern wind turbines use aerodynamic blades but it was not until the end of the nineteenth century that the advantages of such blades were better understood and the two technologies started to merge.

2.3.2 Rotating machines

The origin of the first rotating wind-powered machines is even less certain than the invention of the sail. Emperor Hammurabi (c1850–c1750 BCE) of Babylonia (in the south eastern part of Mesopotamia) planned to irrigate land using wind-powered machines in the 17th century BCE. A Sanskrit document refers to wind-powered

Figure 2.1 Nashtifan Windmills of Iran image by Morvaridi Meraj (Creative Commons Licence)

machines for lifting water in the 4th century BCE which would place the origin in what is now India. However, there is no conclusive evidence that either of these systems was actually built. The first conclusive evidence of a wind-powered machine comes from the 2nd century BCE and this was used by the Persians to grind corn. Such wind turbines were used in the Sistan-Baluchestan, South Khorasan and Khorasan Razavi regions. Examples of these early machines can still be found in Nashtifan (in Khorasan Razavi) and there are plans for these to be granted United Nations Educational, Scientific and Cultural Organization (UNESCO) World Heritage status; see Figure 2.1.

Some texts refer to this class of windmills as Sistan mills [13]. The sails of these mills were made from reed bundles or wood; they were between 5 m and 9 m long and were suspended from horizontal arms on a vertical shaft [14]. The sails of the Persian windmill moved in the horizontal plane (pushing the vertical shaft around). Today, such machines are usually called vertical-axis machines. But it is common, especially in older literature, for these machines to be called horizontal machines after the plane in which the sails moved [15]. Figure 2.2 illustrates the difference between the two types of machines [16–18].

2.3.3 The influence of wind direction

Whereas modern vertical axis machines have blades that can operate with the wind blowing from any direction, early machines were built with fixed structures (windscreens and supporting walls) to ensure wind could only reach sails on one side of the axis of rotation. Figure 2.3 shows an analysis of wind directions, known

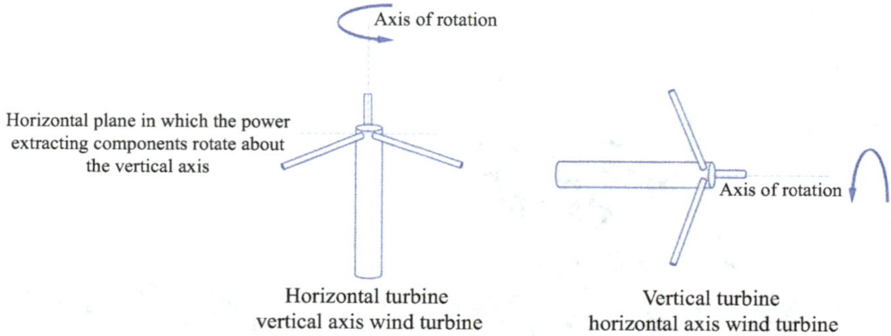

Figure 2.2 Horizontal and vertical wind turbines

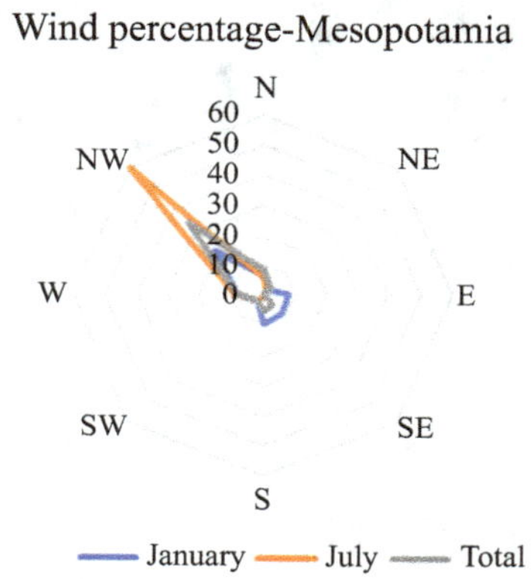

Figure 2.3 Mesopotamia wind percentage versus direction (Wind Rose) (after Neumann)

today as a wind rose (explained in Chapter 4), in Mesopotamia based on the work of Neuman [12]. It shows that a fixed structure would have imposed fewer limitations on operational use and would have been much simpler in construction.

2.3.4 The first horizontal axis machine?

The first horizontal axis wind turbine (HAWT), using the term in the modern-day sense, has been credited to Heron (or Hero), of Alexandria (10–75 CE)[†]. Heron was a lecturer at the Museum of Alexandria where he taught a wide range of technology-related subjects, including control engineering (he is credited with

[†]The date of Heron's birth and death have been disputed. This problem has been exacerbated because Heron was a common name at that time. A major factor in identifying the dates quoted was a translation made by Otto Neugebaur (1899–1990) in 1938 of Heron's writing in which he described a solar eclipse. This eclipse is known to have taken place in Alexandria on 13 March 62 CE.

inventing feedback control), calculating machines, steam engines and optics. Heron also wrote a book about children's toys. In separate papers, Heron described a musical organ driven by a 'wind wheel'. But, as with the Babylonian and Indian machines, it is uncertain whether Heron's design was ever implemented and it may only ever have been intended as a toy. The wind wheel consisted of a multiple-bladed, horizontal rotor. The rotor's motion was translated into a reciprocating action to drive a pump to provide air to the organ pipes [19].

2.3.5 Increasing complexity

Although windmill development in the remainder of the first Millennium was limited, there was one significant innovation in this period. In the seventh century, in Eastern Peria, that is modern-day Sistan and Baluchestan, close to Iran's Afghanistan border; the Panemone windmill incorporated a new invention; the variable pitch blade. The Panemone windmill was a vertical axis machine; the blades on the one side of the axis were fully exposed to the wind whereas on the opposing side of the axis the blades were rotated (feathered) to minimise their resistance to the wind [20,21]. The difference between the Sistan and Panemone constructions is illustrated in Figure 2.4 (note that the screening applied to one side of the windmill can also be seen in the image of the Nashtifan windmill shown in Figure 2.1).

2.3.6 Windmill proliferation

By the middle of the first millennium, the use of windmills was widespread in Persia, to the extent that millwright became a recognised occupation. For example, several authors report stories about Abū Lu'lu'a, a Persian millwright. Most of these stories focus on Lu'lu'a's reaction to the fall of Persia to the Rashidun Caliphate in AD 642 led by Caliph Omar (AD 584–644)[‡] and his refusal to pay taxes. Thus, not only do the Lu'lu'a' stories recognise the millwright's position in

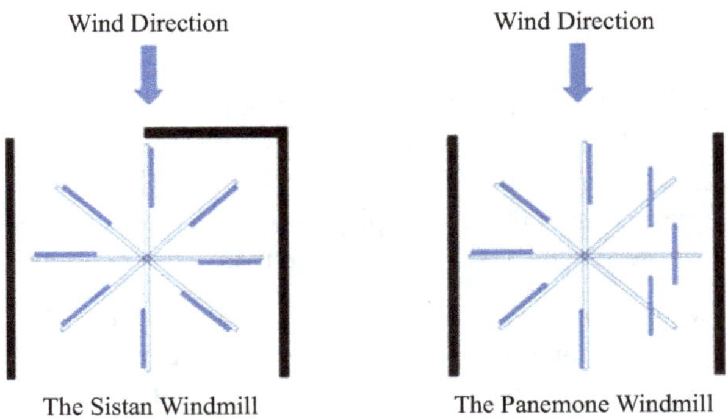

Figure 2.4 Comparison of Sistan and Panemone Windmills

[‡]In some texts, Omar is called Umar.

society but they also allude to some of the large geo-political events of the period which were to provide vectors for the proliferation of technology [21].

In addition to the rise of the Rashidun Caliphate, over the next thousand years, there were many changes in the world order, many of which contributed to transfer of windmill and other forms of technology. Empires rose and fell. The Mongol Empire spread throughout what is now called Asia; one of the emperors, Genghis Khan, also called the Great Khan (c1162–1227), ordered Persian millwrights to move to China where their skills were used to improve drainage [22–24]. The rise of the Moorish Empire led to windmill technology being introduced to the Iberian Peninsula from North Africa [25]. In addition, there were wars, some long running like the Crusades, which started in 1096 and went on for four hundred years [26]. All of these events caused displacement of peoples, some trying to escape the effects of war and some trying to avoid persecution. Even treaties, such as the Byzantium-Venetian Treaty of 1277, led to mass movements of people. All contributed to the spread of ideas. A key socio-economic change also took place in this era; it became possible for individuals to move more freely. Diplomats, and spies, moved between regions. Trade expanded, particularly between Europe and Asia. It was in this era that the famous journeys of Niccolò, Maffeo and Marco Polo took place between 1260 and 1294 [27]. Arguably more important, the scholarship also spread; illustrated by the mathematician Fibonacci who was born in Algeria and spread knowledge of mathematics, travelling throughout the Mediterranean [20].

2.3.7 European winds of change

Driven by these events, windmill technology spread beyond the Middle East into North Africa, China and Europe. However, from the outset, on entering Europe, horizontal axis machines dominated, although no one is certain whether horizontal technology was first developed in the Middle East and exported to Europe or whether Europeans developed these systems independently. Several ideas have been put forward for the preference for horizontal axis machines. These theories include:

- The extensive use in Europe of water wheels which have a horizontal axis configuration.
- A recognition of the greater efficiency available from the horizontal axis (it does not require half the wind collection system to be screened).
- The wind patterns in Europe are less consistent than those of the Middle East, favouring a design that allowed the plane of the windmill blades to be changed to ensure that the blades always extract the maximum energy from the wind.

However, whilst horizontal axis machines may have become the dominant technology in use in Europe, vertical axis technology continued to evolve and is still exploited today. Further reading on the development of vertical axis technology can be found in Wailes (1968), which also includes a comprehensive collection of designs [28,29].

Although little is known about them, it is believed that the first European windmills were built in France around 1105. Perhaps surprisingly, the country in Europe most associated with windmills, the Netherlands, did not start building until

1200. However, despite their later start, the Dutch went on to use windmills very extensively reaching a peak of 9,000 in service by the end of the 19th century. The Netherlands still has windmills in operation which can be visited and the region around Dordrecht has been awarded World Heritage status because of the large number of windmills in service as part of a wider water management scheme [30].

In the United Kingdom, the Domesday Book, compiled in 1086, contained records of over five thousand watermills but no windmills [31]. What is believed to be the first English windmill was built, a little later than those in France, in 1191; the date is known so precisely because the construction was surrounded by controversy. The mill was of wooden construction, built by Herbert, the Dean of the Abbey of St Edmunds. It was built on glebe land, that is land allocated to support parish priests, at Haberdon in what is now Suffolk. However, at that time, watermills were used for the important job of grinding corn for flour and they required complicated infrastructure projects and a lot of resources for their construction. Hence, Manorial watermills were protected by law from competition. For a servant of the lord of the manor, even having corn ground by a mill not owned by the lord was a punishable offence. On learning of the construction of Herbert's mill, the Abbot [Adam] Samson of Tottington (1135–1211) ordered the windmill to be dismantled by the servants of the sacristan. Before they were able to carry out the order, Herbert and his son dismantled the windmill themselves [32–34].

All of the earliest European mills were of a design called the 'Post Mill'. The Post Mill comprises a trestle base, made of stout timbers. In the earliest mills these timbers were sunk into the ground. There is an obvious disadvantage to this approach and, to reduce the risk of the trestle rotting, later designs introduced masonry foundations and pillars to support the trestle. A fine example of a post mill can be found at Great Chishill, near Royston, in Cambridgeshire; Figure 2.5 clearly shows how exposed the mill timbers are to the elements. A further evolution of the design surrounded the base with a weatherproof masonry shroud or roundhouse. Sitting on top of the wooden base, the whole of the upper structure, including the sails (which are referred to by millers as sweeps), the millstones and the coupling mechanisms had to be rotated into the wind. Moving this large structure required a lot of physical effort applied by means of a long 'tailpole'. Rolvenden Mill in the County of Kent, shown in Figure 2.6, is typical of this later post mill design. The roundhouse is dated 1772 but the mill itself was built in either the late 17th or early 18th century and there has been a mill on the site since 1596 [35].

2.3.8 Cap Mills

The problem of the effort required to turn a post mill into the 'eye of the wind' led to a design innovation in the late 14th century; namely the 'Cap Mill'. In such a mill, the 'windshaft' that carries the sails is mounted in a dome-like structure at the very top part of the mill; the cap. Only the cap needs to be rotated into the wind and this allows the heavy millstones, the connecting equipment and all the other parts of the mill to remain stationary. Even though the cap mill is lighter and easier to move into the wind, it still has to be moved. The earliest forms used a tailpole similar to

Figure 2.5 Great Chishill Post Mill

Figure 2.6 Rolvenden Post Mill

that used by the post mill but this was soon replaced by a manually operated system of ropes and gears. In 1745, Edmund Lee, a blacksmith from Wigan, invented the fantail; a wind-driven device that automatically points the sails into the wind.

There are two basic types of cap mills: the tower mill and the smock mill. As the name implies, the tower mill consists of a masonry tower with a number of different floors which contain all the milling machinery. Figure 2.7 shows the Holgate Mill which is a good example of a tower mill. The land for the mill was bought in December 1768 and it became operational in October 1770. Holgate, at the time of the mill's construction was a village that has now become part of the City of York. As can be seen in the photograph, Holgate Windmill has five blades instead of the normal four. Figure 2.8 shows Lee's great invention, the fantail, as implemented on the Holgate Windmill.

A variation on the tower mill is the smock mill; which usually has an octagonal wooden tower with a pronounced bulge around the girth where a walkway is provided. The term smock mill is derived from the appearance of a farmer's smock [36]. Figure 2.9 shows an example of a smock mill, the Union Mill located in the town of Cranbrook in Kent and built in 1814. Rex Wailes, an authority on windmills, considered this to be the finest smock mill in England [36]. The Union Mill is of an unusual design because it was built on land which was already surrounded by a built environment. To cope with the wind blockage created by adjacent buildings, the mill was built with an exceptionally high tower; it is 72 ft or 22 m to the ridge of

Figure 2.7 Holgate Tower Mill

Figure 2.8 Holgate's Fantail

Figure 2.9 Cranbrook Union Mill

Figure 2.10 Upminster Smock Mill

the cap. Having to build higher turbines to exploit more free-flowing air remains a feature of modern on-shore wind turbines. Figure 2.10 shows the Upminster Mill in the London Borough of Havering. Unlike the Union Mill, the scale and proportion of the Upminster Mill is more like others of its type.

The distribution of the different types of windmills was not uniform across the United Kingdom. For example, whereas almost all the windmills built in Scotland were tower mills, the windmills of Orkney were an exception being of the post type. In the south east corner of England around the Weald, the smock mill dominated because of the ready availability of wood and the relative scarcity of clay for bricks [37].

2.3.9 Evolution towards the modern wind turbine design

The tower mill and the smock mill both illustrate the key characteristics of a modern wind turbine: they consist of a tower surmounted by a nacelle that rotates into the wind to extract energy and there are few blades (sweeps), typically only four or five. There is another characteristic that all these mills have in common with a modern wind turbine. All have a windshaft that is tilted back from the horizontal thus tilting the plane of the sails back from the vertical. This feature offered several advantages that remain valid today. It is most convenient to taper the supporting towers and by tilting the blades it prevents them from fouling the tower. But there is another technical reason for the tilt. The windshaft has to be supported in a way that allows it to turn with the sails. Two bearings facilitate this: one at the front and another at the rear of the cap. By tilting the windshaft, the rear bearing carries some of the weight of the sails, distributing the weight across the cap and making it simpler to build and move.

Windmills became ubiquitous in England and, at their peak, there were more windmills in England than in the Netherlands; it is estimated there were ten thousand. But as their numbers peaked other alternative energy sources were being developed that would prove, at least in the short term, more cost-effective than the windmill.

2.3.10 Options and understanding

In 1712, the English inventor Thomas Newcomen (1664–1729), invented a new form of power generation, the Atmospheric Steam Engine. The Newcomen engine burned coal, to heat a vessel of water and produce steam. On the opening of a valve, the pressure in the boiler forced steam into a cylinder pushing a piston up the cylinder as it did so. When the piston reached its maximum travel, the steam pressure valve was closed and cold water sprayed into the cylinder. The cold water caused the steam to condense, creating a vacuum whereupon air pressure on the other side of the piston forced the piston to retract into the cylinder (hence the term the atmospheric steam engine). As the cycle repeated, the engine created a reciprocating action that was intended to be used to drive a pump capable of lifting water out of mines. Tin mines took up the engine but it found more applications in coal mines which had the same flooding problems as tin mines but also a ready source of fuel to fire the boilers [38].

Newcomen's invention meant that, for the first time, there was a real alternative to the use of wind or water power to produce mechanical movement. Paradoxically, while coal power began to compete with wind (and water) power, and continues to do so to the present day, this competition led to the first scientific assessment of wind and water power. In 1759, following a series of innovative experiments using the forerunner of a wind tunnel, John Smeaton (1724–1792), an English civil engineer, published findings that proved to be the benchmark in the understanding of aerodynamic force for a hundred and fifty years. They held sway until they were refined by the Wright Brothers when they started to develop the Wright Flyer, world's first heavier-than-air flying machine, in 1899 [39,40].

Steam engines gradually reduced the wide-scale use of water mills and large-scale windmills as the Industrial Revolution took hold. However, windmills remained a common sight in Northern Europe and the United States of America until the start of the 20th century, especially in rural areas, where they continued to be used for pumping water and griding corn. These windmills have the attributes of a modern wind turbine: a tower and a rotating cap (nacelle) which is used to rotate the sails to face the wind. Yet, despite the recognition of the benefits of the horizontal axis machine when the first electricity-generating wind turbine was built, its designer turned to a vertical axis design to extract wind power. But before discussing this, the origins of machines for generating electricity will be described.

2.4 Machines for generating electricity

2.4.1 Electricity and magnetism: Oersted

In 1820, the Danish physicist, Hans Christian Oersted (1777–1857), see Figure 2.11, published the results of an experiment that demonstrated that if a current is passed through a conductor, it caused an adjacent magnetic needle to be deflected (to align with the magnetic field being created by the current). Figure 2.12 illustrates how the magnetic field was set up in the experiment. This was the first conclusive evidence that there was a relationship between a current flow and a magnetic field [41].

It is often reported that this discovery was accidental but this is not the case; Oersted had been working on this idea for 2 years before he was successful. His early experiments used low currents and much of Oersted's report to the *Annals of Philosophy* was given over to the description of the 'galvanic apparatus' used to

Figure 2.11 Hans Christian Oersted (Creative Commons Licence)

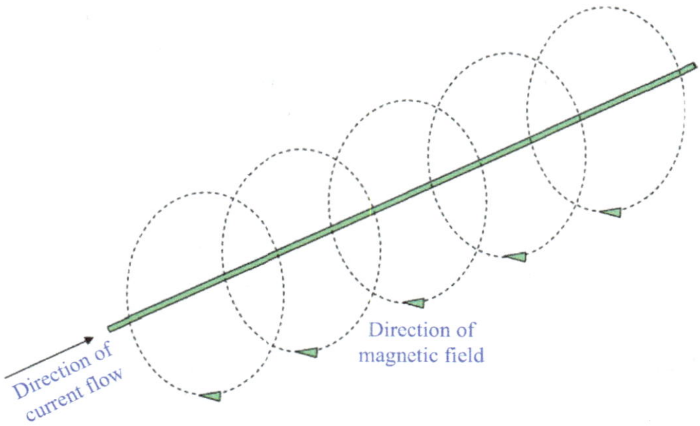

Figure 2.12 Oersted's magnetic needle experiment

create a sufficiently high current to provide a reliable demonstration. In practice, the current created by his new apparatus was so great that the conductor glowed red hot. Oersted explained his observations based on what he called the 'conflict of electricity' which appeared to envisage the electrical current battling to pass through the conductor giving rise to the heat and light. Oersted believed that the force causing the deflection of the needle was another consequence of this conflict. However, this explanation was flawed, because whereas the heat and light caused by the conflict are created as the electricity passes along the conductor, the force produced acts at right angles to the conductor.

These problems were not completely resolved for another 10 years; the first steps being made by the French mathematician and scientist, Andre Marie Ampere (1775–1836). Ampere showed that the lines of the magnetic force acting on Oersted's needle were annular (circular centred on the conductor). This finding was confirmed in 1821 by a friend of Ampere's with whom he regularly corresponded; the English engineer Michael Faraday (1791–1867), see Figure 2.13 [42,43].

2.4.2 Faraday, motors and generators

The insight into Oersted's experiment provided by Ampere prompted Faraday to design an experiment in which a magnet was set in wax at the bottom of a glass container. The container was then filled with mercury so that only the tip of the magnet was visible. A wire was then suspended above the container with its tip dipping into the mercury. The final step of the experiment was to connect the fixed end of the suspended wire to one terminal of a battery and the other battery terminal to the mercury. In effect, this arrangement reversed that of Oersted's experiment in which the conductor was fixed and the magnetic needle could move. When the electrical circuit was completed, the wire moved in rapid circles around the magnet. This experiment was the first demonstration of the ability to

Figure 2.13 Michael Faraday (Creative Commons Licence)

convert electricity to mechanical movement; Faraday had created the first electric motor [44].

Despite the success of the motor demonstration, the publication of his findings in 1821 was to prove a problem for Faraday. It has been suggested that, in a rush to publish, Faraday did not acknowledge collaborations that were influential in his success, which led to accusations of plagiarism. Later in his career, these accusations were refuted but it is possible that this was a factor in his not returning to his interests in electromagnetism until 1831. In the intervening years, other investigators worked on electromagnetism and two in particular made useful progress. In 1823, Ampere discovered that a relatively weak magnetic field could be intensified by winding the conductor into a helical coil, a solenoid. The magnetic field created by a solenoid is directly proportional to the number of turns. Then, in 1824, the French physicist and polymath (who later went on to become the Prime Minister of France), Francois Arago (1786–1853) showed that if a copper disk is spun rapidly, a magnetised needle (such as a compass needle) suspended above it is dragged around by the disk's rotation even if it is not touching the disk. It is now known that the magnetisation of the needle causes currents, called 'eddy currents' to flow in the disk causing magnetisation that interacts with the magnetic needle. The reason was not understood until Faraday provided an explanation but the experimental apparatus was to prove instrumental in subsequent investigations.

When Faraday restarted his work on electromagnetism, he first investigated the effect of coupling two magnetic 'circuits' together. His goal was to establish how the second circuit behaved when exposed to the magnetic field being created by the first. The tool he used to carry out the investigation was Ampere's solenoids. Faraday wound the solenoids onto a toroidal iron ring. To the first solenoid, he connected a battery and to the second a galvanometer to measure the signal. The results of this experiment came as a surprise. When the primary circuit was connected, the galvanometer showed a current generated in the secondary circuit, but once the current in the primary coil stabilised, the signal in the secondary coil disappeared. When the primary circuit was disconnected, a current was once again detected in the secondary coil, but the current flow was in the opposite direction to the original. Investigating this further in October 1831, Faraday discovered that similar effects were observed when a bar magnet was introduced into a coil of wire [45]. As the magnet (and its field) moved into the arrangement, a current was produced in the coil. A current flowed in the opposite direction when the motion was reversed. This was the first time that mechanical energy had been converted to electricity. Faraday had demonstrated the principle of the electricity generator [46].

Although the principle of generation had been demonstrated, the method was unsuitable for practical exploitation. A more practical process might be provided by a system using smooth continuous motion and Faraday's next experiment was based on an apparatus that used a copper disk similar to that used by Arago. In Faraday's disk experiment, the copper disk was spun rapidly using a hand crank. The disk was mounted so that it spun in a fixed magnetic field. Direct current (DC) induced in the disk could be extracted between the centre and the edge of the disk. The device was inefficient and produced only small amounts of electrical energy but it did show that smooth and continuous motion could produce electrical energy.

Faraday published his findings about electromagnetic induction in November 1832. Within a year of the publication, two events took place that paved the way for the practical exploitation of electricity generation. Hyppolyte Pixii (1808–1835), an instrument maker from Paris, created a prototype alternating current (AC) generating the machine in 1832 and a British engineer William Richie invented the commutator. This latter device allows the current to be reversed within a machine as it rotates. The commutator is used extensively in modern DC motors and generators. Ampere suggested to Pixii that a commutator be introduced into his AC machine which earned Pixii the accolade of having invented the world's first Dynamo [47].

2.4.3 Commercialisation of power generation

Over the course of the next 40 years, numerous companies based in different countries commercialised power generation. The years 1866 and 1867 were particularly important in this period. A shortcoming of designs up to that point had been that machines relied upon permanent magnets. On 24 December 1866, Samuel Alfred Varley (1832–1921), an English Engineer, patented a design for an improved dynamo. Then on 17 January 1867, German Ernst Wernher Siemens (1816–1892) and Englishman Charles Wheatstone (1802–1875) published, on the same day, designs for similar improved dynamo designs. Common to all the designs was the use of electromagnets in place of permanent magnets, a design feature called self-excitation. [48] Thus, by the start of the 1880s, electrical power generation based on self-exciting dynamos was starting to become commonplace, generally for the purpose of artificial lighting. The power to drive these generators relied on steam created by burning coal, but an alternative means of powering the generators was about to be developed.

2.4.4 James Blyth – the first electricity-generating wind turbine

In 1887, following 2 years of experimentation, the first electricity-generating wind turbine was installed by James Blyth (1839–1906) at his holiday cottage in his home village of Marykirk in Kincardineshire, approximately 20 miles south of Aberdeen in Scotland; see Figures 2.14 and 2.15.

Blyth's wind turbine used a tripod to support a 33 ft (10 m) long vertical 'wind shaft' from which four 13 ft (4m) long canvas sails were hung. A flywheel on the wind shaft was connected to a Burgin dynamo by rope. The dynamo charged lead acid batteries and the system was capable of powering ten 25-volt lamps. It is worth noting that vertical axis wind turbines (VAWT) have a distinct advantage over the, now more common, HAWT because they do not require a yawing mechanism to turn them into the wind. Blyth patented his invention in 1891, see Figure 2.16 [49]. An interesting feature of the patent is a reference to the aerodynamic properties of the design which protected it from the effects of overspeed in high winds now a design factor in modern wind turbines.

Blyth devised a number of improvements to his turbine and in 1895 licensed a Glasgow engineering company, Mavor and Coulson, to build a larger version for the Royal Asylum of Montrose. One of the principal modifications was to replace the canvas sails with eight, solid, semi-cylindrical vessels to harvest the wind. The

Figure 2.14 James Blyth c1900 (image reproduced courtesy of the University of Strathclyde)

Figure 2.15 Wind turbines at Blyth's Home in Maryhill, Aberdeenshire 1891 (image reproduced courtesy of the University of Strathclyde)

new design was based on the invention in 1846 of a cup-anemometer by Rev. John Thomas Robinson (1792–1882). Blyth's turbine remained in service until 1914 when the 4in. (10 cm) vertical shaft fractured [15].

Blyth, who also contributed to the development of microphones for telephones, was Freeland Professor of Natural Philosophy at Anderson's College in Glasgow, which became the Glasgow and West of Scotland Technical College and is now the University of Strathclyde. He is regarded by many as a pioneer of renewable energy, which he championed because he believed wind-powered generators were cheaper to manufacture than fossil-fuelled generators. Blyth was awarded the Makdougal Brisbane Gold Medal from the Royal Society of Edinburgh in 1892 for his work on wind power [50]. It seems fitting that Marykirk, now part of the larger community of Laurencekirk, is overlooked by the Twinshiels Wind Farm on the Hill of Garvock which became operational in 2014.

2.4.5 US wind turbines – the first horizontal axis machines

Blyth's wind turbine predated the first US-built wind turbine, developed by Charles Brush (1849–1929) by only 7 months. Brush was an inventor and industrialist; his company Brush Electric became part of the General Electric (GE) Company in 1892. His turbine was a horizontal axis machine, unlike Blyth's, and it was also

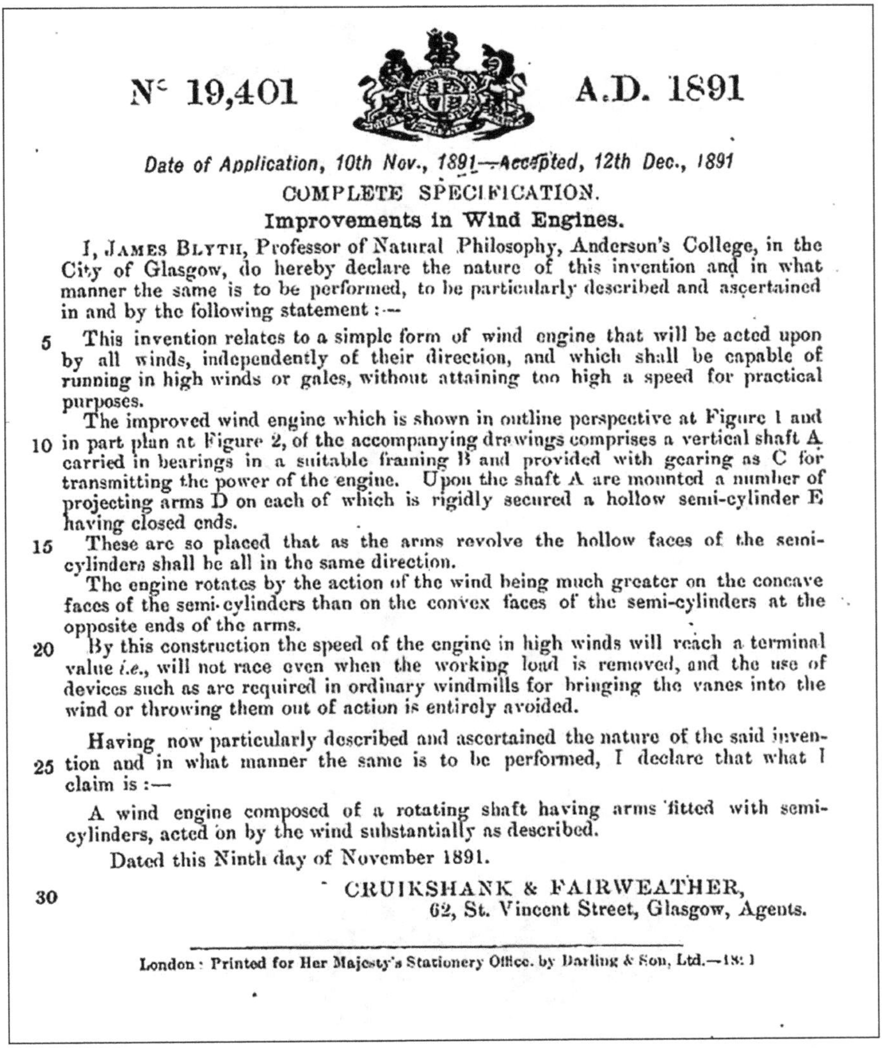

N° 19,401 A.D. 1891

Date of Application, 10th Nov., 1891—Accepted, 12th Dec., 1891

COMPLETE SPECIFICATION.

Improvements in Wind Engines.

I, James Blyth, Professor of Natural Philosophy, Anderson's College, in the City of Glasgow, do hereby declare the nature of this invention and in what manner the same is to be performed, to be particularly described and ascertained in and by the following statement :—

This invention relates to a simple form of wind engine that will be acted upon by all winds, independently of their direction, and which shall be capable of running in high winds or gales, without attaining too high a speed for practical purposes.

The improved wind engine which is shown in outline perspective at Figure 1 and in part plan at Figure 2, of the accompanying drawings comprises a vertical shaft A carried in bearings in a suitable framing B and provided with gearing as C for transmitting the power of the engine. Upon the shaft A are mounted a number of projecting arms D on each of which is rigidly secured a hollow semi-cylinder E having closed ends.

These are so placed that as the arms revolve the hollow faces of the semi-cylinders shall be all in the same direction.

The engine rotates by the action of the wind being much greater on the concave faces of the semi-cylinders than on the convex faces of the semi-cylinders at the opposite ends of the arms.

By this construction the speed of the engine in high winds will reach a terminal value *i.e.*, will not race even when the working load is removed, and the use of devices such as are required in ordinary windmills for bringing the vanes into the wind or throwing them out of action is entirely avoided.

Having now particularly described and ascertained the nature of the said invention and in what manner the same is to be performed, I declare that what I claim is :—

A wind engine composed of a rotating shaft having arms fitted with semi-cylinders, acted on by the wind substantially as described.

Dated this Ninth day of November 1891.

CRUIKSHANK & FAIRWEATHER,
62, St. Vincent Street, Glasgow, Agents.

London: Printed for Her Majesty's Stationery Office. by Darling & Son, Ltd.—1891

Figure 2.16 Patent for the world's first electricity-generating wind turbine (Crown Copyright)

much larger than either of Blyth's; it had 144 cedar rotor blades with a 56 ft [17 m] rotor diameter. The turbine was capable of producing 12 kW which was used to charge lead acid batteries to light Brush's Ohio mansion. The turbine was operational from 1888 until 1908. Brush's turbine was also much more sophisticated than Blyth's with automatic control features including the capability to shut down in high winds, see Figure 2.17 [51].

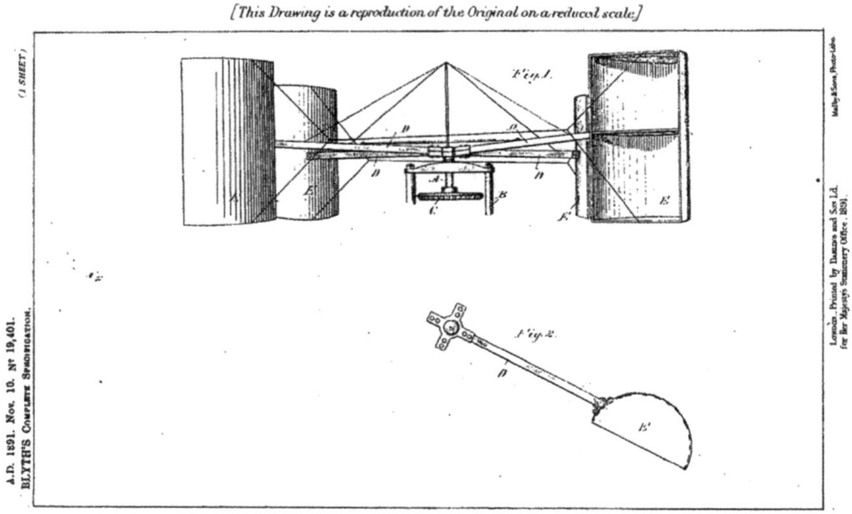

Figure 2.16 (Continued)

Figure 2.17 Brush's wind turbine (courtesy of the Poul La Cour Fonden, Denmark)

2.4.6 Danish and German wind turbines – appliances of science

Blyth and Brush built their turbines by adding electricity generation to adaptations of existing design concepts for windmills. At the time, there was no scientific basis for the design of windmills, instead they were based upon generations of experience in their construction. This situation was about to change. In 1896, a Danish scientist, Poul la Cour (1846–1908) built a wind tunnel and started carrying out experiments on scale models of complete windmills (Figure 2.18).

Useful results followed in 1899 with additional investigations to extend Smeaton's work to investigate the aerodynamics of different blade sections considering the relative merits of flat and curved blades. La Cour's findings were published in 1903 and may be summarised as follows:

- There should be few blades on a turbine.
- The pitch of the blades should be small.
- Turbines worked better at high speed rather than low speed.

La Cour validated his findings by building full-scale turbines at his home in Askov, Denmark, approximately 200 km West of Copenhagen. In 1903, he established the Danish Society of Wind Electricity, one of the aims of which was to train electrical technicians and publish technical material on wind turbines. His work was fundamental in establishing Denmark as a leader in wind turbine technology, a position that it still enjoys today [52].

The understanding of wind turbine aerodynamics was extended early in the 20th century when the theoretical limit for the power that can be extracted from wind was discovered independently by three scientists: namely, the British Frederick Lanchester (1868–1946), Russian Nikolay Joukowsky (1847–1921) and German Albert Betz (1885–1968). The limit, 59.3%, has become associated with Betz and the value as the 'Betz limit', or 'Betz's Law', was published in 1919. Turbine efficiency should be viewed with respect to the Betz limit; typical values for HAWT are between 35% and 45% [53].

Figure 2.18 Poul La Cour (image reproduced courtesy of the Poul La Cour Fonden, Denmark)

2.4.7 Turbine development during the 20th century

Even with Smeaton's and La Cour's findings, La Cour's Askov turbines still had the appearance of a 19th-century smock mill. But as the 20th century progressed turbines evolved through a number of steps to become what we recognise today as a typical wind turbine. Alternative milestones may be found but the following turbines and farms illustrate that evolution.

Despite the success of La Cour's HAWT Askov turbines, some of the earliest developments of the twentieth Century were new designs of VAWT. In 1922, the Finish engineer Sigurd Johannes Savonius (1884–1931) invented a simplified drag turbine and in 1926, the French aeronautical engineer, Georges Jean Marie Darrieus (1888–1979) registered a patent for an aerodynamically bladed VAWT. The principal advantages of these machines are their relative simplicity and they do not need to be pointed into the wind. The efficiency of the Darrieus wind turbine is similar to an HAWT but they are difficult to start turning from being stationary. The example in Figure 2.19 (reproduced courtesy of Quiet Revolution) is a helically bladed Darrieus-type turbine [54]. The

Figure 2.19 A helical-bladed Darrieus type wind turbine (image courtesy Quiet Revolution, St Neots)

principal disadvantage of the Savonius wind turbine is its low efficiency, typically 15%. The example shown in Figure 2.20 is located at London's O2 arena.

In 1931, the first 100 kW wind turbine, what some authorities have referred to as the first 'utility-scale' wind turbine, was built at Balaclava, near Yalta then part of the Union of Soviet Socialist Republics (USSR). The turbine was a HAWT, with a 30-m steel lattice tower with three blades. It is widely reported that it achieved a 32% load factor (the proportion of time it generated electricity). The turbine provided power for the Balaclava to Sevastopol tram line and it operated from 1931 to 1942 when it was destroyed during the Second World War [55].

The first wind turbine to exceed 1 MW capacity was designed by Palmer Cosslett Putnam (1900–1984) and built by S. Morgan Smith Company. The 1.5 MW turbine was installed in 1941 at Grandpa's Knob near Castleton, in Rutland County, VT in the USA. It was a HAWT, with a 37-m steel lattice tower with two blades and a rotor diameter of 53 m, each was 2.4 m wide and weighing seven and a half tons. The date of its construction is important because it was during the Second World War and shortages of raw materials led to a known weak point in one of the blades failing after only 1,100 h of operation [56].

The first multi-turbine site was built by the US National Aeronautics and Space Administration (NASA) at Plum Brook, New York. The Plum Brook wind farm had only two turbines but it was capable of powering over 4,000 homes and it went operational in 1975. In 1980, a windfarm of twenty turbines was installed at Crotched Mountain, between Bennington and Francestown in New Hampshire. Some authorities have argued that a wind farm should have more than five turbines and the Crotched Mountain installation was the World's first real wind farm.

Figure 2.20 A Savonious Wind Turbine O2 Arena London

In the United Kingdom, the first commercial wind farm was built at Delabole, 5 km North East of Port Isaac, in Cornwall. Construction was completed in December 1991 and it consisted of ten turbines with a total capacity of 4 MW. The original turbines were replaced in 2010 with fewer more powerful turbines; at the time of writing, the site remains operational. By 1999, when the Hare Hill Wind farm was completed [57], another 31 wind farms became operational, ranging in capacity from single turbines to the 103 turbines in the P+L Wind Farm in Wales. Such was the rate of growth at the time that a further 12 wind farms had been built before a complaint from Glasgow Prestwick International Airport (GPIA) was raised in 2001, identifying Hare Hill as the source of the interference on their radar [58].

2.5 Radar

2.5.1 Getting over the influence of Aether

In the middle of the 19th century, it was well known in scientific circles that light could propagate through a vacuum, in other words, empty space. But if the space was empty, it was reasoned, a light would have no propagation medium; therefore, there must be some medium present that was not yet detectable. The underlying principle had been proposed by the French philosopher, René Descartes (1596–1650) in 1644 when he was working on a theory of gravity. But the name given to this medium was ancient, it had been proposed by Plato (428 or 427 BCE–348 or 347 BCE) and his student Aristotle (384 BCE–322 BCE); the name was Aether. In this philosophy, Aether was the fifth element, after Earth, Wind, Water, and Fire; in Latin, 'quintessence'. If proof were needed that Aether existed, it was the fact that light could propagate in a vacuum which it would not be able to do unless Aether existed. So apparently obvious was this reasoning that it influenced science for over two hundred years and even Isaac Newton built models which attempted to define the role of Aether [59,60]. And, of course, if Newton believed in Aether, then this only added weight to the argument for its existence.

2.5.2 Faraday and Maxwell

Faraday's report to the Royal Society on 24 November 1831 about his discovery of electromagnetic induction was a simple factual, understated, account of his findings [43]. In the paper, he introduced a new term 'electro-tonic state'; a state in which a current could flow remotely from a source of energy by the action of lines of force. Although Faraday did not say that Aether was not required for this propagation, he certainly did not say that it was. Fifteen years later, he removed any doubts. Faraday routinely organised speakers to deliver Friday evening lectures at the Royal Institution and for the evening of 11 April 1846, he had engaged the distinguished physicist Charles Wheatstone to speak. As he was about to be introduced to the audience, Wheatstone suffered stage fright and ran off leaving Faraday to provide an impromptu alternative lecture. In an unprepared speech, he set out his belief that light was a manifestation of vibrating lines of force. It was now clear that Faraday's position was that Aether was not required for the propagation of light (or any other form of electromagnetic wave). This speech effectively questioned Newton's opinion on this subject and it was too radical a concept for the scientific community of the time. When his earliest biographers recorded Faraday's life it was deemed kind not to refer to these speculations.

However, the idea was accepted by at least one man, a Scottish physicist James Clerk Maxwell (1831–1879) who was born at Glenlair in Dumfries and Galloway, only 65 km from GPIA; see Figure 2.21. Maxwell developed the series of equations that now bear his name; he expressed Faraday's lines of force as vector quantities, that is possessing both magnitude and direction. Maxwell set out his work in a series of papers, the first of which was published in 1855, and it is fitting that he entitled it 'On Faraday's Lines of Force'. What came out of this work was more far-reaching than simply accounting for electromagnetic induction. Maxwell's equations describe how electromagnetic waves propagate and how they interact with the environment. Moreover, he describes how radio waves are reflected off surfaces (initially assumed only to be metallic surfaces) and their refraction passing through di-electric (non-conducting) media.

Taken together, the work of Faraday and Maxwell marked a profound change in the understanding of physics; Einstein is said to have revered Faraday and Maxwell and had paintings of them hanging in his flat in Berlin [61].

Figure 2.21 James Clerk Maxwell (Creative Commons Licence)

2.5.3 Proving Maxwell's theories

Whilst Maxwell established the theory explaining the behaviour of electromagnetic waves, he did not know how to prove that the theory was correct because there were no tools for creating electromagnetic waves. Moreover, there was no obvious method for creating such a tool. It was not until 1886 that a means of creating electromagnetic waves was discovered, accidentally. The discovery was made by the Professor of Physics at Karlsruhe Polytechnic, Heinrich Hertz (1857–1894) (Figure 2.22). Hertz was experimenting with a device called a Riess, or Knochenhauer, spiral; a device which in operation could have been a prototype for a vehicle ignition, that is, a circuit deliberately intended to produce a spark.

A spark occurs when high voltage causes the normally insulating air, or other forms of gas to break down in the gap between two electrodes. The resulting current is a rich source of electromagnetic waves that covers virtually the whole electromagnetic spectrum of frequencies including what we now call radio waves, heat and light. The fact that light was produced was instrumental in the discovery of the phenomenon of electromagnetic waves [62]. Hertz observed that when the spark was created, a spark was also created in a second Riess spiral not connected to the first. The only way the second device could spark was if an electromagnetic wave produced by the first Riess spiral was propagated to the second. Now there was the basis of a tool for proving Maxwell's equations. Hertz refined the apparatus to create a spark transmitter and in 1887 he proved that Maxwell's equations accurately predicted the performance of electromagnetic waves. Hertz demonstrated the following:

- The speed of the electromagnetic wave corresponds with the speed of light, also an electromagnetic wave.
- Reflection.

Figure 2.22 Heinrich Hertz (Creative Commons Licence)

- Refraction.
- Polarisation, that is, the orientation of the electric and magnetic components of the electromagnetic wave.
- Interference, that is, the constructive and destructive addition of waves from different sources.

Hertz was a popular lecturer at his university, although his own personal notes suggest that he did not like lecturing. An oft, and perhaps unfairly, quoted anecdote concerning his lectures, was that he repeatedly stated that electromagnetic waves had no practical value [63,64].

> 'I do not think that the wireless waves I have discovered will have any practical application'.
>
> Heinrich Hertz 1890

2.5.4 Propagation and Marconi

Hertz's conclusion that electromagnetic waves had no value arose because there was one aspect of their behaviour that was not fully understood. All of Hertz's work had been carried out in the confines of a university laboratory. All the preceding work by Faraday on electromagnetic induction was carried out within the confines of his laboratory at the Royal Institution and, similarly, Oersted worked in a laboratory environment at the University of Copenhagen. Moreover, electromagnetism is an effect that is applied within machines (motors, dynamos and generators). None of this work considered the possibility that what was to be called Hertzian waves could propagate over long distances, but that was to change.

Hertz's proof of the existence of electromagnetic waves prompted work at other universities, including the University of Bologna where the Professor of Physics Augusto Righi (1850–1920) made useful contributions to extending the understanding of Hertzian waves including, for example, generation of microwaves. Bologna was also home to the Marconi family; an old family and part of Italy's middle-ranking nobility originally from Venice. Guglielmo Marconi (1874–1937), Figure 2.23, was the family's second son and, consistent with his background, was educated privately. His parents hired appropriate tutors right into his teenage secondary-education years, when he was given a basic education in the then-new field of electrical engineering. At the age of 18 (1892), Marconi met Righi and was allowed to participate in lectures and have access to other university facilities including the library and laboratories. By the age of 20 (1894), Marconi had built his own transmitting and receiving equipment and was able to transmit from one side of a room to the other. In 1895, by making improvements to the equipment, he increased the range, first to 1 km and then to 3 km. In 1897, he achieved a range of 6 km. In 1899, with further refinements, he was able to transmit a signal across the English Channel. Two years later (1901), he made a transmission across the North Atlantic proving that Hertzian waves could be transmitted over great ranges.

Figure 2.23 Guglielmo Marconi (Creative Commons Licence)

2.5.5 Hülsmeyer and the Telemobiloskop

Thus, by 1900, there had been multiple demonstrations (first by Hertz and later by others including Righi) that Hertzian (radio) waves are reflected off metallic surfaces and could propagate over long distances. These are the two critical requirements for a radar system. A commonly held view is that radar was not developed until the Second World War, at least another 30 years away. But this is not true. As early as 1903, a German inventor, Christian Hülsmeyer (1881–1957) invented a radar system he called the Telemobiloskop. It is widely accepted that Hülsmeyer's rationale for developing the Telemobiloskop came about because of an event in his childhood. Hülsmeyer was born in a village called Eydelstedt in the Lower Saxony region of Germany close to the River Weser. The Weser is the longest river that runs entirely through Germany and it can be dangerous for shipping due to the high frequency of fogs especially in the lower reaches [65]. Someone from Eydelstedt was killed in a ship collision on the Weser and Hülsmeyer witnessed first-hand the grief this caused. At School, he proved to be adept at physics and went on to a teacher training college in Bremen which was equipped with a Hertzian wave experimental system. Hűlsmeyer left the college before completing the course to become a trainee at Siemens and Halske, a highly reputable electrical engineering company. He then went on to form his own company which was to develop the Telemobiloskop, with the express purpose of improving marine safety. After an initial failure to secure a patent for the device, modifications were made to the submission and a patent was granted on 30 April 1904. The patented system comprised: a spark transmitter, a coherer receiver, a wide beamwidth transmitting antenna and a narrow beamwidth receive antenna. These allowed the direction of an approaching vessel to be localised. The detection of a ship caused a bell to ring. The system was demonstrated on numerous occasions on shore, and some important organisations showed an interest in the device including the Holland America Lijn shipping company. Encouraged by this success and interest, Hülsmeyer wanted to trial the equipment on board ship and an initial trial was conducted in Rotterdam Harbour on 9 June 1904 onboard the Steam Ship (SS) Columbus; a 70-tonne riverboat built by Bonn and Mees of Rotterdam in 1893. This first sea-trial was considered successful but limited in scope. In particular, the detection range of ships appears to have been low and a second trial was conducted in the Autumn of 1904 again on board the SS Columbus. The later trial was conducted at the Hook of Holland where ships would be observed at longer range. Sadly, this trial was to prove disastrous for the Telemobiloskop and Hülsmeyer's ambitions. As soon as the device was trialled in this more representative environment it failed to work. The most likely cause of the failure was the presence of Marconi shore stations and ships carrying Marconi equipment. The Marconi station, whether they were on-shore or on-board ship used similar spark transmitters to the Telemobiloskop and it had no way of distinguishing its own transmissions from those being made by the Marconi stations [66].

It is worth noting that when Hülsmeyer's prototypical radar device was failing its second sea trial off the Hook of Holland, the first electricity-generating wind turbine had been in service for 17 years.

2.5.6 *Improving the early systems*

If Faraday's discoveries and inventions ushered in the age of electrical engineering, the next 2 years would usher in the age of electronic engineering. In 1904, the British inventor Dr John Ambrose Fleming (1849–1945), who had been a student of James Clerk Maxwell at Cambridge University, invented a device that could detect the presence of radio signals. The device had two electrodes, hence it is called a diode, in an evacuated glass envelope known in the United Kingdom as a valve and in the United States as a tube. One of the electrodes was heated to create a stream of electrons that can only move through the device in one direction. Thus, the diode could rectify the alternating current of the radio signal changing it to the direct current. Two years later, in 1906, another form of rectifier was invented by American Dr Greenleaf W Pickard; this became known colloquially as 'the cat's whisker'. The device consisted of a crystal of silicon which when touched with a small filament (the cat's whisker) proved to be more sensitive than the first diodes. Also in 1906, the American inventor Lee de Forest (1873–1961), added a third electrode between the electrodes in the diode configuration which allowed the flow of current in the device to be modulated. A small change in the voltage applied to the third electrode, called the grid, causes a larger change in the current flowing between the other two electrodes. The triode was the first device capable of increasing the power of a signal and was developed over the next 6 years to create the first amplifiers [67].

The introduction of amplification revolutionised many different branches of technology particularly in the fields of telegraphy and telecommunications.

However, in the context of radar and taken together with the improvements in signal detection offered by diodes, they facilitated more powerful transmitters and receivers that were able to detect weaker radio signals.

Radar echoes can be weak, and they become increasingly weak as the distance of the target from the radar grows. To detect such weak signals, multiple stages of amplification are required but detection can easily be disrupted if there are any sources of interference nearby (as Hülsmeyer found at the Hook of Holland trials). To counter interference, it is useful to be able to select from a number of different frequencies of transmission. However, having several stages of amplification in the receiver, each of which should be tuned to the same frequency, becomes problematic if that frequency has to change. In 1918, American, Edwin Howard Armstrong (1890–1954) solved this problem when he invented a system for converting different received frequencies to a single frequency used by each stage of amplification. The process is called super-heterodyning and a super-heterodyne receiver is capable of detecting very weak signals which was to lead to an important finding about the reflection of radio waves.

2.5.7 Non-metallic objects and the Naval Research Laboratory

In 1922, two scientists at the US Naval Research Laboratory (NRL) in Washington, DC, Dr Albert Hoyt Taylor (1879–1961) and Leo Crawford Young (1891–1981), were experimenting with a superheterodyne receiver. The NRL is located on the Potomac River, where they set up a transmitter and on the other side of the river, they set up a receiver. During the course of their measurements, it was observed that the experiment was upset when a ship sailed between the transmitter and the receiver [68]. The cause of the disruption was shown to be reflections of radio waves off the ship, the SS Dorchester which, depending on the precise location of the ship, either added to or subtracted from the signals received directly from the transmitter. Hoyt and Young recognised the importance of this phenomenon and with an associate worker, Lawrence A. Hyland (1897–1987), they went on to develop the system and they submitted a patent application for 'A System for Detecting Objects by Radio' on 13 June 1933. The patent was granted in November of the following year [69].

However, there is another important feature of the NRL work that is important to the discussion here. The SS Dorchester was a wooden ship, demonstrating that it was not only metallic objects that reflect radio waves. The sensitivity of the NRL equipment showed that, for all practical purposes, all objects reflect radio waves, it is simply a matter of degree [70].

2.5.8 Other nations

With the Second World War anticipated, several nations rapidly developed their own radar systems which included:

- In June 1934, Rudolph Kuhnold (1903–1992) demonstrated the detection of ships in Kiel Harbour at a distance of two km and in October of the same year the system detected an aircraft.

- In February 1935, British Scientist Robert Watson Watt (1892–1973) with Arnold Wilkins (1907–1985) carried out the Daventry Experiment in which a Heyford Bomber was detected at a range of 13 km [71].
- Pierre David (1887–1987), a French engineer, started experimenting with the detection of distant aircraft in 1927, using a method similar to that of NRL. The results of the early experiments were mixed but sufficient to continue work. The years 1934 and 1935 were spent trying to 'operationalise' the system and make it more reliable. In 1935 there were two sets of trials, the one in July 1935 proving particularly successful by demonstrating the ability to detect aircraft at heights of over 26,000 feet [72,73].
- In the next 2 years, Russia, Italy and Japan also demonstrated similar capabilities.

All the above-mentioned systems were based upon the NRL method, but in June 1935, Edward George (Taffy) Bowen (1911–1991) introduced a modification to use pulses of energy in place of a continuous signal. This change had two effects, it meant that the transmitters could operate at higher power but also the pulse provided a time reference. If the time when the echo was received was compared with the time when the transmission was made, the distance of the aircraft from the transmitter could be calculated. These changes were rapidly adopted by many nations.

In the United Kingdom, using Bowen's pulsed system, by 1937, Watson Watt (Figure 2.24) was able to build a system that was able to provide early warning of attacks on London. These radars became operational in 1938 before the start of the Second World War. In the United States, the introduction of this capability was given a name, and, from 1939 onwards, equipment sets with this capability were referred to as Radio Detection and Ranging (RADAR) systems [74].

Figure 2.24 Sir Robert Watson Watt (Creative Commons Licence)

2.5.9 Secondary surveillance radar

The radar systems described above were designed to warn of the approach of enemy aircraft. The early warning provides the defending forces with options; take shelter and/or attempt to intercept the enemy. But if both sides in a conflict have aircraft, it becomes necessary to distinguish between an enemy aircraft attacking and your own aircraft returning from an attack. The problem is not easily resolved. A simple solution might be procedural, with your own aircraft returning via air corridors that are known in advance. But if the enemy accidentally used the same air corridor, it would still be impossible to identify if they were enemy aircraft. Pilots might report their positions using Air–Ground–Air communications but that has intrinsic dangers if the enemy was able to decode these communications or discern the position based on locating the source of the transmissions.

To investigate these scenarios, the UK Royal Air Force arranged a series of experiments, named after the airfield used to provide the trials aircraft; Biggin Hill. The Biggin Hill experiments started in 1936 and they determined that it was critical to know the identity of the aircraft as well as its location [75]. To overcome this problem a second type of radar system was required; one which can provide the identity of the aircraft. The solution to the problem, called Identify Friend or Foe (IFF), consists of elements on the ground and on-board the aircraft that work together as follows.

The ground-based element has a transmitter and a rotating antenna. Typically, the antenna completes a full scan in 4 s, but it can be longer (this is discussed in Chapter 3). As the antenna rotates, sequences of precisely timed pulses are transmitted. When an aircraft is illuminated by the scanning antenna, equipment on board the aircraft detects the message and responds with a message which includes information that can identify the aircraft. After the Second World War, IFF was adopted for both civilian and military users and in civil use alternative names were used. In the United Kingdom, this became known as Secondary Surveillance Radar (SSR) and in the United States as Air Traffic Control Radar Beacon System (ATCRBS). The interaction of wind turbines and SSR/ATCRBS is problematic and this will be discussed in more detail in Chapter 3. To assist in the distinction between the two types of radar, the original is known as primary surveillance radar (PSR).

2.5.10 Development of PSR

War and the threat of war, always lead to improvements in technology, both to create new weapons and, in response, to sensor systems to detect the new technology earlier and to provide more time to react. The Second World War was no exception. In 1943, Dwight Oliver North (1909–1998) invented a process for extracting the greatest amount of energy from a received signal, **the matched filter**, and another system for enhancing the sensitivity of radar which could diminish the effects of noise, **multiple pulse integration**. As the sensitivity of radars increased, more and more unwanted returns (clutter) could present to operators and complicate the detection of the targets of interest, in this case, aircraft. A simple strategy to simplify the task is to filter out all those targets that are not moving. The echo returns from static targets do not contain any Doppler shift and recognising the absence of Doppler shift is the basis for removing them from the display. The first systems implementing **Moving Target Indication** (MTI) appeared towards the end of the Second World War [76]. However, these early systems were only partially successful in removing static returns. If the effects of Doppler are to be detected, then it is important that the radar transmitter has a very stable frequency of transmission because any changes in the frequency of the transmitter can be misinterpreted in the receiver as changes caused by target motion. The problem was particularly acute for strong returns from the ground and the built environment close to the radar antenna. Ingenious solutions were designed to mitigate the problems; for example, the Airfield Control Radar (ACR) 430 radar had an adjustable antenna that allowed the operator to increase the elevation of the antenna to track arriving and departing aircraft more accurately and which also offered a measure of control over ground clutter [77,78].

The radar in use at Glasgow Prestwick when the first public reference to wind turbine interference was made, was the EN 4000 which was from this era [79][§].

2.5.11 Computerisation

Another technology developed during the Second World War, which was designed to help solve encryption problems, was digital computing. After the war, digital computing contributed to radar in two ways. First, it enabled manually intensive problems to be solved rapidly and accurately. For example, in the 1950s, the first computer programming languages were invented, such as Fortran (Formula Translation) in 1954 and Algol (the Algorithmic Language) in 1958. These languages allowed non-computer specialists to write code to solve complex mathematical problems and this revolutionised the understanding, for example, of how targets reflected radio waves. By the 1970s and 1980s, it was commonplace for radars to have integrated digital technology to reduce the effects of clutter.

2.5.12 Networking

Another early application of digital technology which started the development in the 1950s and 1960s allowed radar systems to be networked together to provide operators with an integrated air picture. For example, the Semi-Automatic Ground Environment (SAGE) system, an American Air Defence radar network, was conceived in 1951 and was fully deployed in 1963[¶]. The United Kingdom developed a similar system named Linesman Mediator; Linesman was the military component of the network and was the forerunner of the United Kingdom Air Defence Ground Environment (UKADGE) while Mediator was the civil component which evolved into the National Air Traffic Services (NATS).

[§]The radar consisted of an upgraded version of an earlier radar called an ACR-6 modified to include what were in the 1980 state-of-the-art electronics from a new radar called the S511.

[¶]https://www.ll.mit.edu/about/history/sage-semi-automatic-ground-environment-air-defense-system. Retrieved 17 March 2023.

2.6 Summary

Extracting power from wind has been known since antiquity but its use to generate electricity, following the work of men like Faraday, dates back to the end of the nineteenth century. The first electricity-generating wind turbine, created by Blyth, was operational ten years before the equivalent prototypical radar set created by Hülsmeyer. By the time the first radar systems were being used in anger during the Second World War, a 1.5 MW electricity-generating wind turbine was in service. It might be argued that wind turbine technology was more mature than radar at this time.

The identification of problems associated with wind turbines and technical infrastructure were first noted as multi-turbine, commercial, wind farms were entering service. The Hare Hill Wind Farm and the GPIA radar are good examples. Addressing these problems required, and still requires, the development of new technologies so it is useful to be able to understand how technical innovation works as a benchmark for assessing the new remedial techniques.

Although the technologies of wind harvesting, electricity generation and radar are quite dissimilar in nature, the processes used for their development have much in common and demonstrate the kinds of benchmarks required to judge the maturity of new remedial technology. These concepts are discussed in Chapter 6.

References

[1] GPIA (2001), *The Operational Effects of Windfarm Developments on ATC Procedures for Glasgow Prestwick International Airport*, 30 January 2001.

[2] Spaven, M. (2001), 'Wind turbines and radar: operational experience and mitigation measures'.

[3] Jago, P. and Taylor, N. (2002), *Wind Turbines and Aviation Interests: European Experience and Practice*, Stasys Ltd. ETSU W/14/00624/REP DTI PUB URN No. 03/515, Crown Copyright, 2002.

[4] Butler, M. and Johnson, D. (2003), *Feasibility of Mitigating the Effects of Windfarms on Primary Radar*, Alenia Marconi Systems, ETSU W/14/00623/REP, DTI PUB URN No. 03/976, Crown Copyright, 2003.

[5] Poupart, G. (2003), *Wind Farms Impact on Radar Aviation Interests – Final Report*, Qinetiq, FES W/14/00614/00/REP, DTI PUB URN 03/1294, Crown Copyright, 2003.

[6] AWC (2005), *The Effects of Wind Turbines on [Air Traffic Control] ATC Radar*, Report AWC/WAD/72/665/TRIALS, 10 May 2005.

[7] AWC (2005), *The Effects of Wind Turbines on Air Defence Radar*, AWC/WAD/72/652/TRIALS, 6 Jan 2005.

[8] Congress (2006), *The Effect of Windmill Farms On Military Readiness*, Report to the Congressional Defense Committees, 2006.

[9] IEA (2007), 'Radar, radio and wind turbines', In *53rd IEA Topical Expert Meeting*, Oxford, UK, March 2007.

[10] Golding, E.W. (1976), *The Generation of Electricity by Wind Power*, London: E. and F.N. Spon Ltd., 2nd ed., Chapter 2.

[11] Carter, R.A. (2006), 'Boat remains and maritime trade in the Persian Gulf during sixth and fifth millennia BC'. *Antiquity*, 80 (307):52–63.

[12] Neumann, J. (1977), 'The winds in the worlds of Ancient Mesopotamian Civilisations', *Bulletin of the American Meteorological Society*, 58 (10):1050–1055.

[13] Majlesi, A. (2021), 'The story of how ancient Iranians harnessed the power of wind', *Teheran Times*, 11 September 2021. https://www.tehrantimes.com/news/464934/The-story-of-how-ancient-Iranians-harnessed-the-power-of-wind. Retrieved February 2023.

[14] Drachmann, A.G. (1961), *Heron's Windmill*, Centaurus, vol. 7, pp. 145–151.

[15] O'Connor, J.J. and Robertson, E.F. (1999), *Heron of Alexandria*, University of St Andrews. http://www-history.mcs.st-and.ac.uk/Biographies/Heron.html. Retrieved 3 June 2023.

[16] Mathew, S. (2006), *Wind Energy: Fundamentals, Resource Analysis and Economics*, Springer, Chapter 1.1.

[17] Golding, E.W. (1976), *The Generation of Electricity by Wind Power*, London: E. and F.N. Spon Ltd., 2nd ed., Chapter 12.

[18] Wailes, R. (1967), 'Horizontal windmills', *Transactions of the Newcomen Society*, 40(1):125–145, doi:0.1179/tns.1967.007.

[19] Keysor, P. (1992), 'A new look at Heron's "Steam Engine"', *Archive for History of Exact Sciences*, 44(2):107–124.

[20] Hassan, A.Y. and Hill, D.R. (1992), *Islamic Technology: An Illustrated History*, Cambridge University Press. Chapter 2.

[21] Shepherd, D.G. (1992), *Historical Development of the Windmill*, Technical Report produced for the US Department of Energy. https://www.osti.gov/biblio/6342767. Retrieved February 2023.

[22] Hildinger, E. (1997), *Warriors of the Steppe: A Military History of Central Asia 500 B.C. to 1700 A.D.*, De Capo Press.

[23] Saunders, J.J. (1971), *The History of the Mongol Conquests*, University of Pennsylvania Press.

[24] Letcher, T.M. (ed.) (2017), *Wind Energy Engineering: A Handbook for Onshore and Offshore Wind Turbines*, Academic Press.

[25] Al-Hassan, A.Y. (2006), *Transfer of Islamic Science to the West*, Foundation for Science, Technology and Civilisation, December 2006, Publication ID 625.

[26] Jones, D. (2019), *Crusaders: An Epic History of the Wars for the Holy Lands*, Head of Zeus Ltd.

[27] Humble, R. (1975), *Marco Polo*, Weidenfeld and Nicolson.

[28] Wailes, R. (1968), *Horizontal Windmills*, Paper read at the Science Museum, London, April 1968.

[29] Dodge, D. (1996), *Illustrated History of Wind Power Development*. https://web.archive.org/web/20181002082917/http://www.telosnet.com/wind/early.html. Retrieved 3 June 2023.

[30] Chen, L. (2020), *Windmills in the Netherlands*, Department of Civil and Environmental University of Wisconsin – Madison. https://ancientengrtech.wisc.edu/the-netherlands-windmill/#:~:text=Dutch%20started%20building%20windmills%20as,are%20being%20used%20and%20maintained. Retrieved February 2023.

[31] Tremenheere, W. (2017), *A Guide To Cranbrook Union Mill*, Cranbrook Windmill Association, 6th ed.

[32] Skilton, C.P. (1947), *British Windmills and Watermills*, Collins, London.

[33] de Brakelond, J. (2011), 'The chronicle of Jocelin of Brakelond: a picture of monastic life in the days of Abbot Samson', Edited by Sir Ernest Clarke. https://www.gutenberg.org/files/37780/37780-h/37780-h.htm Retrieved February 2023.

[34] Wailes, R. (1976), *Windmills in England: A Study of Their Origin, Development and Future*, Charles Skilton Ltd., London.

[35] HE (2023), *Rolvenden Windmill*, Historical England Listed Building Directory. https://historiengland.org.uk/listing/the-list/list-entry/1116206?section=official-list-entry. Retrieved February 2023.

[36] Wailes, R. (1976), *Windmills in England: A Study of their Origin, Development and Future*, Charles Skilton Ltd; New edition.

[37] Bourne (2022), Personal conversation between the author and Mr Peter Bourne of Cranbrook Union Mill, August 2022.

[38] BBC (2014), 'Thomas Newcomen (1663–1729)', *BBC History*. https://www.bbc.co.uk/history/historic_figures/newcomen_thomas.shtml. Retrieved February 2023.

[39] ICE (2023), *John Smeaton*, Institution of Civil Engineers, Who are Civil Engineers. https://www.ice.org.uk/what-is-civil-engineering/who-are-civil-engineers/john-smeaton. Retrieved February 2023.

[40] Anderson, C.G. (2020), *Wind Turbines: Theory and Practice*, Cambridge University Press.

[41] Oersted, C. (1820), 'Experiments on the effect of a current of electricity on the magnetic needle', *Annals of Philosophy*, XVI(4):273–276.

[42] Faraday, M. (1821), 'On some new electro-magnetic motions and on the theory of magnetism', *Quarterly Journal of Science*, 12:74–96.

[43] Forbes, N. and Mahon, B. (2014), *Faraday, Maxwell and the Electromagnetic Field: How Two Men Revolutionised Physics*, Prometheus Book.

[44] RI (2023), *Michael Faraday's Electric Magnetic Rotation Apparatus*, Royal Institution Collection Description. https://www.rigb.org/explore-science/explore/collection/michael-faradays-electric-magnetic-rotation-apparatus-motor. Retrieved February 2023.

[45] RI (2023), *Michael Faraday's Generator*, Royal Institution Collection Description. https://www.rigb.org/explore-science/explore/collection/michael-faradays-generator. Retrieved February 2023.

[46] Faraday, M. (1832), *Experimental Researches in Electricity*. https://royalsocietypublishing.org/doi/10.1098/rstl.1832.0006. Retrieved February 2023.

[47] NMAH (2023), *Pixii Magneto Machine*, The Smithsonian National Museum of American History, Metadata link http://n2t.net/ark:/65665/ng49ca746ab-e432-704b-e053-15f76fa0b4fa. Retrieved February 2023.

[48] Harvey, A., Larson, A., and Patel, S. (2020), *History of Power: The Evolution of the Electric Generation Industry*, *Power*. https://www.power-mag.com/history-of-power-the-evolution-of-the-electric-generation-industry/. Retrieved February 2023.

[49] Blyth (1891), *Improvements in Wind Engines*, Blyth patent. GB19401 of 1891. The British Library.

[50] Hardy, C. (2010), *Renewable Energy and Role of Marykirk's James Blyth*, Dundee Courier, Available on-line at Courier.co.uk.

[51] DWIA (2023), Danish Wind Industry Association Archive. http://web.archive.org/web/20090218230815/http://www.windpower.org/en/tour.htm/. Retrieved February 2023.

[52] Warne, K. (2023), *Poul La Cour Pioneered Wind Power in Denmark*. https://windowstoworldhistory.weebly.com/poul-la-cour-pioneered-wind-power-in-denmark.html. Retrieved February 2023.

[53] See for example, REUK (2010), *Betz Limit*, REUK with a commentary by Professor Richard M. Andres. http://www.reuk.co.uk/wordpress/wind/betz-limit/. Retrieved 23 February 2023.

[54] Archer (2006), *Wind Power Calculators for Various Wind Turbines—HAWT/VAWT*. http://smartservo.org/en/wind-turbine-efficiency-compen/#:~:text=The%20maximum%20efficiency%20of%20a%20typical%20Darrieus%20lift%20wind%20turbine,of%20the%20drag%20wind%20turbine. Retrieved 23 February 2023.

[55] Fuergy (2019), *First Blows of Energy Independence – Wind Power II*, Fuergy Industries, Bratislava, https://fuergy.com/blog/first-blows-of-energy-independence-wind-power-ii. Retrieved February 2023.

[56] Voaden (1982), 'Smith-Putnam wind turbine: a step forward in aero-electric power research', in *Large Horizontal-Axis Wind Turbines*, NASA Conference Publication 2230, 28–30 July 1981.

[57] Windpower (2022), *Hare Hill*. https://www.thewindpower.net/windfarm_en_1469_hare-hill.php. Retrieved February 2023.

[58] Renewables UK (2023). https://www.renewableuk.com.

[59] Pilkington, M. (2004), *Filling Space: The Aether*. https://www.theguardian.com/science/2004/feb/12/research.science#:~:text=In%20the%20fourth%20century%20BC,describe%20the%20medium%20of%20space. Retrieved February 2023.

[60] Baird, E. (2000), *Newton's Aether Model*. https://arxiv.org/ftp/physics/papers/0011/0011003.pdf. Retrieved February 2023.

[61] Robinson, A. (2019), *How British Scientists Inspired and Ensured Einstein's Place in History*. https://www.sciencefocus.com/science/how-british-scientists-inspired-and-ensured-einsteins-place-in-history/. Retrieved February 2023.

[62] Li, T., Tong, Z., Tang, S., *et al.* (2018), 'Identification of the electric spark electromagnetic waveform based on SVM', *Advances in Engineering Research*, 127.
[63] Conor and Robertson (2007), *Heinrich Hertz*. https://mathshistory.st-andrews.ac.uk/Biographies/Hertz_Heinrich/. Retrieved February 2023.
[64] IOP (2023), *Hertz's Useless Discovery*. https://spark.iop.org/hertzs-useless-discovery. Retrieved February 2023.
[65] US Weather (1942), *Preliminary Report on Climate and Weather of Northwestern Europe*, United States Weather Bureau, 1 January 1942.
[66] Bauer, A.O. (2005), *Christian Hülsmeyer and About the Early Days of Radar Inventions: A Survey*. https://www.cdvandt.org/Huelspart1def.pdf. Retrieved February 2023.
[67] Watson, R.C. (2009), *Radar Origins Worldwide: History of Its Evolution in 13 Nations Through World War II*, Trafford Publishing.
[68] Naval History (2018), *Leo C. Young, Radar Pioneer (1891–1981)*. https://www.history.navy.mil/content/history/nhhc/research/library/manuscripts/u-z/leo-c-young-radar-pioneer.html. Retrieved February 2023.
[69] US Patent (1933), Patent Number 1981884.
[70] Parry, D. (2010), *NRL History – Radar*. https://www.nrl.navy.mil/Media/News/Article/2577147/nrl-history-radar/. Retrieved February 2023.
[71] Kendal, B. (2001), *The Birth of Radar*. Serialised in the Rad Com Journal of the Radio Society of Great Britain, October–December 2001.
[72] Blanchard, Y. (2016), 'A French pre-WW II attempt at air-warning radar: Pierre David's "Electromagnetic Barrier"', *Radio Science Bulletin*, 358: 18–34.
[73] David, P. (1969), *Le Radar*, Presses Universitaires de France, 1969.
[74] OCS (2013), *RADAR Blooms*, US Army Signal Corps, Office Candidate School. http://www.armysignalocs.com/index_oct_13.html. Retrieved 3 June 2023.
[75] Gough, J. (1993), *Watching the Skies: The History of Ground Radar in the Air Defence of the United Kingdom*, HMSO.
[76] Selove, W. (1947), 'MTI receivers', in L.N. Ridenour (ed.), *Radar System Engineering*, pp. 579–612 (Chapter 23).
[77] ATCEU (1985), *Evaluation of the Area Moving Target Indicator as fitted to the Plessey ACR 430 radar at Bristol Filton Aerodrome*, National Air Traffic Services, Air Traffic Control Evaluation Unit Memorandum 120, Dated 1985. National Archive, DR 14/565.
[78] Plessey (1977), *ACR 430 Airfield Control Radar*, Equipment Manual.
[79] EN4000 radar Manchester Airport, UK National Archive Record. DR 14/567.

Chapter 3

Aviation and aviation radio systems

3.1 Introduction

The aim of this chapter is to identify and describe the civil and military aviation systems that may be affected by wind turbines. The systems can be categorised by the nature of the effects they might experience.

The first category consists of systems that use continuous (not pulsed) radio transmissions, for example, the instrument landing systems (ILSs) that help aircraft land safely in bad weather. The systems in this category transmit signals from the ground to aircraft. Some systems also require transmissions from aircraft to the ground, for example, air–ground–air (AGA) communications. Consider what happens when a signal from a transmitter on the ground is reflected off a structure such as a wind turbine. The antenna on the aircraft will now receive the reflection as well as the direct signal. Sometimes these signals will interfere constructively, increasing the amplitude received and sometimes destructively, with the opposite effect. This effect is called **fading**.

The second category of systems uses pulses of radio frequency energy. The various types of radar systems and their derivatives belong to this category. These pulses may be affected by the Doppler effect (described later in the chapter); however, the principal concerns are the production of **false targets**, **clutter returns**, and **loss of genuine targets**.

Before considering the systems themselves, it is useful to understand something about the environment in which these systems operate, for example, where they are installed, and why. This information is essential for analysis of performance. Furthermore, it is useful to understand the terminology used in aviation.

In the previous chapter, it was stated that aviation is well regulated. Some of the regulators publish useful guidance on the interaction between wind turbines and aviation radio and radar systems. Therefore, the chapter begins by describing the aviation regulatory framework, it introduces some of the regulators and some of the most important publications that deal with the interaction of wind turbines and aviation radio and radar systems.

3.2 Regulation

Aviation is organised and regulated nationally but it is, by its very nature, an international enterprise with aircraft routinely travelling between different nations.

There are clear advantages when flying internationally for standardisation; for example, it is important to know that an aircraft from a different nation will be airworthy. The principal organisation that harmonises the many national regulators, as well as harmonising civil and military aviation, is the International Civil Aviation Organization (ICAO). ICAO was set up in December 1944 and it is part of the United Nations. One hundred ninety-three states are members of ICAO (only the Vatican and Liechtenstein are not members) and 36 nations are represented on its council. ICAO publishes regulations in annexes to the 1944 Chicago Convention on Civil Aviation. ICAO regulations and standards cover all aspects of aviation, for example, ICAO sets recommendations for obstacle lighting (including wind turbines), runway design, and navigation systems such as VHF Omnidirectional Range (VOR) and precision approach radar (PAR) [1]. ICAO publishes an important document concerning wind turbines and radio system safeguarding; its European Guidance Material on Managing Building in Restricted Areas [2]. This document sets out exclusion zone dimensions for buildings in a restricted area (BRA) which includes the recommended safeguarding distances of wind turbines from navigational aids.

The Comet, the Boeing 707 and the Douglas DC-8 were the World's first passenger jet airliners; they entered service in the 1950s. In Europe, it was correctly predicted that air travel would grow dramatically as a result. However, this growth presented challenges in Europe where a single flight may fly over many countries. In 1960, **EUROCONTROL** was founded to foster improved aviation interoperability between the nations, to make international flight over Europe as seamless as possible. EUROCONTROL still fulfils that role today. It has 41 full members which include all the nations in the European Union (EU), the UK and Armenia and nations such as Turkey that are applying for EU membership. Although a membership criterion is that members should be nations in Europe, EUROCONTROL also has comprehensive agreements with non-members Morocco and Israel. EUROCONTROL publishes an important guideline document entitled, 'How to Assess the Potential Impact of Wind Turbines Surveillance Sensors' [3].

Notwithstanding its importance, EUROCONTROL is not a regulator. Within Europe, the supranational regulator for the EU nations is the European Union Aviation Safety Agency (EASA). Founded in 2002, EASA has aviation regulatory authority over the 27 EU member states as well as Switzerland, Norway, Iceland and Liechtenstein. EASA works closely with EUROCONTROL.

The organisation of aviation regulation in many countries is similar to that in the United Kingdom. There is a national civil regulator, the Civil Aviation Authority (**CAA**) and a military regulator, the Military Aviation Authority (**MAA**); the latter being an autonomous part of the Ministry of Defence. Thus, both provide regulation, monitoring, inspection of and assurance of both operating and technical domains. The CAA promulgates its regulations via Civil Aviation Publications (CAP). Among the large canon of regulations, a number are directly relevant to the subject of this book: CAP 670, Air Traffic Services Safety Requirements, of particular relevance is Part 3, Section C, SUR 13: Requirements for Implementation of Wind Turbine Interference Mitigation Techniques and CAP 764, Policy and

Guidelines on Wind Turbines. Another very important regulatory document published by the CAA includes CAP 493, Manual of Air Traffic Services Part 1. Part 1 is a generic document; Part 2 is specific to each aerodrome. The MAA promulgates its regulations via MAA Regulatory Publications (MRP).

The scope of regulations published by ICAO, EASA, the CAA and the other national regulators covers Aerodromes, Air Operations, Air Traffic Management, Aircraft and Products (including those systems affected by wind turbines), Aircrew and Medical, General Aviation, Civil Drones and the Environment. In the context of Air Traffic Management, the regulations apply to all Air Navigation Service Providers (ANSP). The ANSP includes aerodromes (defined below) and national providers who staff the area control centres. For example, in the United Kingdom, the area control centres (London and Scottish) are staffed by a commercial entity, National Air Traffic Services En Route Limited (NERL); although the control centres also have military staff providing liaison and control for military services. Similar situations are found worldwide.

There is one exception to this model of national regulator and commercial entities providing the services; the US regulator, the Federal Aviation Administration (**FAA**) is both a regulator and a service provider. The FAA was formed in 1958 in part in response to a collision between a Super-Constellation and a DC-7 over the Grand Canyon in 1956 which killed everyone on-board the two aircraft, 128 people. Associated with changes in the US Department of Transport, the Agency became the FAA in 1967 [4]. FAA regulations are promulgated as Federal Aviation Regulations (FAR).

Other organisations referred to in Figure 3.1 and not mentioned in the text above are: The French Government civil service provider Direction des Services de la Navigation Aérienne (DSNA) and Centre de Détection et de Contrôle militaire (CDC) the military equivalent of DSNA, The German commercial service provider Deutsche Flugsicherung (DFS) which provides both civil and military Air Traffic Management (ATM) services.

3.3 Aviation's ground environment

3.3.1 Introduction

The placement of the precision landing radio aids that are described later are determined by features on a runway and, therefore, it is useful to know where those features are located and it is informative to know why those locations are chosen. Understanding more about runways also provides some useful geographic reference points for analysis.

Although the principal interest of this section is runways, to assist the non-aviation specialist, the discussion begins with an introduction to some aviation terminology associated with airfields (known as aerodromes). It is stressed that this is an introduction, for a more detailed and an authoritative discussion, the reader is referred to the glossaries provided by regulators such as the Civil Aviation Authority (CAA), the Federal Aviation Agency (FAA) and EUROCONTROL.

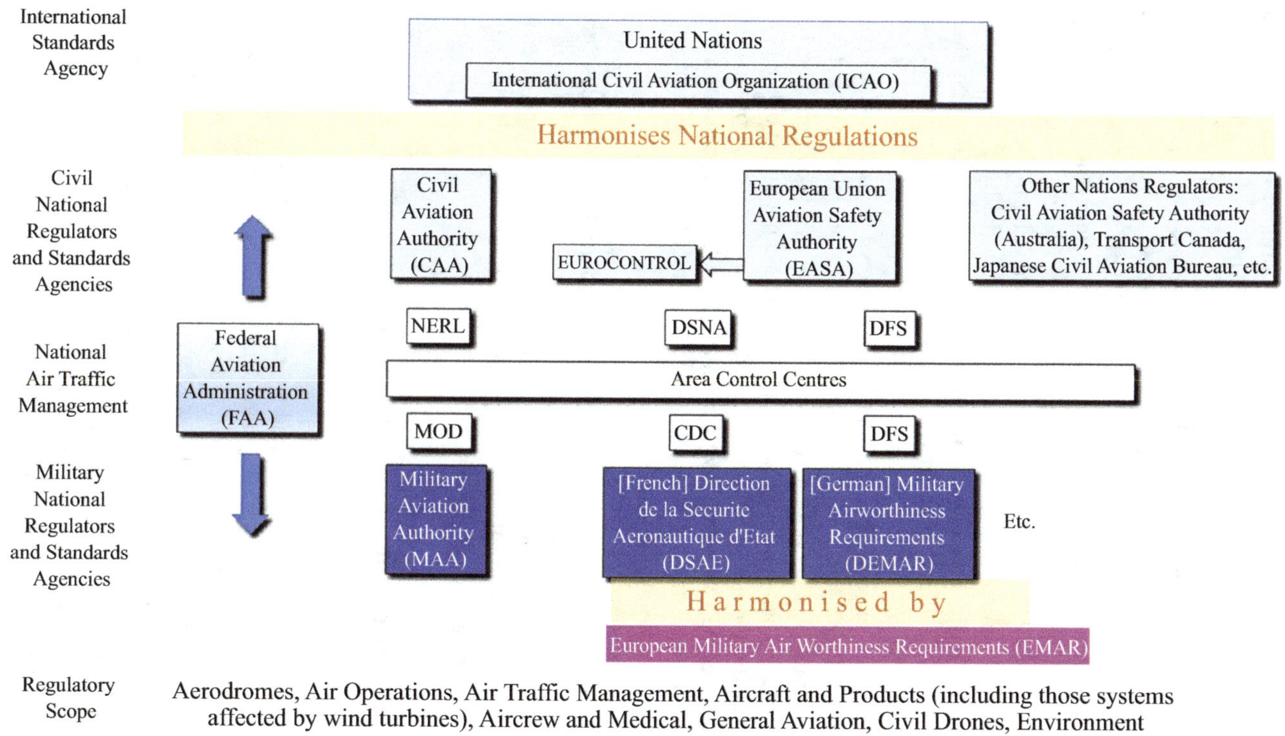

Figure 3.1 ATM regulation

3.3.2 Ground terminology

For anyone whose background is outside aviation, the terms Airfield and Airport are probably used interchangeably and the term Aerodrome may even be thought archaic. Whereas in aviation circles, the three terms have specific meanings.

Aerodrome, or in the US an airdrome, is the name given to a place, which must have a defined border and at least part of the area within that border must be used for flying activities, that is the takeoff, landings and surface movement of aircraft. The area can be on land, or on a floating or fixed platform at sea or on water. The infrastructure supporting the flying activities, the buildings, installations and equipment are all part of the aerodrome. The formal definitions of an aerodrome make no distinction about the size and nature of aircraft using the aerodrome; they may be light aircraft [which also has a specific name in aviation circles; General Aviation (GA)], civil/commercial or military [5].

The principal national aerodromes are allocated a four-character code by International Civil Aviation Organisation (ICAO) [6].

Licensed Aerodrome, or in some countries, such as the United States and Canada, a **Certified Aerodrome** or, simply an **Airport** is the term applied to an aerodrome licensed for the purpose of the use by civil/commercial aircraft. Notwithstanding the use of the term airport, in all technical aviation documentation associated with the airport, it will be referred to as an aerodrome.

Airfield is best understood by its literal meaning it is the field on which, and from which, air operations take place.

The types of equipment that might be affected by wind turbines are usually associated with larger aerodromes.

3.3.3 Aerodrome or airport location

An aerodrome location is specified by its Aerodrome Reference Point or the Airport Reference Point (ARP). This is a location that is the geometric middle of the air operations at the aerodrome; it is specified in latitude and longitude in World Geodetic System (WGS) 84 coordinates. Thus, if an aerodrome has a single runway, then the reference point is usually at the centre of the runway. For example, the ARP for Leeds Bradford Airport (ICAO code, EGNM) is:

Latitude: 53° 51′ 58.00″ North	Longitude: 001° 39′ 39.00″ West [7]

The ARP in this case is in the middle of runway 14/32 which is the only active runway at Leeds Bradford Airport. There will be more discussion about the naming of runways later on in this chapter.

Taking another example, at Charles De Gaulle airport (ICAO code LFPG) [8] at Paris, there are four active runways and the ARP is:

Latitude: 49° 00′ 46.00″ North	Longitude: 002° 33′ 00.00″ East

This location is on the edge of a taxiway midway between the four runways.

The ARP and its whereabouts on the airfield are specified in the documentation for the aerodrome lodged with and promulgated by the national regulator. The ARP is an important reference point for protecting operations at an aerodrome. For example, the ARP forms the origin of the Aerodrome Traffic Zone (ATZ) which is the airspace surrounding the aerodrome that is protected to ensure the safety of aircraft within the zone. For more discussion on the ATZ, refer to Section 3.3.8. The centre of the ATZ also forms the origin of the horizontal surfaces that are the basis for the physical protection of the airfield [5].

The ARP is used loosely when referring to an aerodrome which can create potential for confusion. A common question when performing analysis is, where is the location of a particular piece of equipment? It would be unusual for equipment to be placed on the actual ARP.

3.3.4 Runways

3.3.4.1 A formal definition

The ICAO definition of a runway is:

> A defined rectangular area, on a land aerodrome selected or prepared for the landing and take-off run of aircraft along its length [9].

International regulators such as the Civil Aviation Authority [10] in the United Kingdom and the Federal Aviation Authority [11] in the United States of America have similar definitions.

3.3.4.2 Number of runways

If possible, aircraft take off and land into the wind to maximise lift from the wings with the benefit of reducing take off and landing distances and also to be able to exploit the wind resistance to aid in slowing down. Therefore, it might be expected that guidance on the number and orientation of runways at an airport might be couched in terms of the prevailing wind; this is not the case. Rather, ICAO's guidance is based on the consideration that cross-winds should not prevent landing or take off; specifically, at any given airport at least one runway should be available 95% of the time. Thus, there are obvious benefits to aligning a runway with the prevailing wind but the guidance also illustrates why at some airports and airfields there are three runways arranged in a triangular pattern.

3.3.4.3 Runway dimensions

The length of runways is determined by the nature of the traffic which is expected to use them; larger and heavier aircraft require longer runways than smaller and lighter aircraft. This consideration will determine what is called the **Basic Length** of the runway. However, air pressure is another consideration and the Basic Length is calculated assuming the runway is at the sea level. If the runway is higher than sea level and the air pressure correspondingly lower then, at least the main runway should be extended to allow a greater distance to accelerate to the speed where take off is possible or landing is safe. The guidance on lengthening runways to account for air pressure advises that their length should increase by five percent for every

Table 3.1 Runway width

Category	Width	Marking stripes
A, B	60 m (200 ft)	16
C, D, E	45 m (150 ft)	12
F, G	30 m (100 ft)	8

1,000 feet (300 m) of altitude. Also factored into this calculation must be expected annual temperature and pressure variations and whether or not the runway is inclined. For example, in general, longer runways are needed in warmer climates and if the runway is on sloping ground.

The widths of runways are categorised A to G, as set out in Table 3.1; the number of stripes on the threshold indicates the width in a way that is visible from the air.

3.3.4.4 Runway designation

A runway designation is taken from its magnetic (compass) bearing to the closest ten degrees. Therefore, a runway could have any bearing from 10° to 360°. However, there are conventions for how this information is presented. Only two digits are used and the third digit which would always be 0 is dropped. An example will be helpful; a runway on a heading of 90° (due East) would be referred to as runway 09. But it will be obvious that runways can usually be used in two directions. Thus, using the same example, a runway on a heading of due East might also be used in the opposite direction with a heading that is due West. Hence, the full title of this runway would be runway 09/27. In general, at least one of these bearings will be aligned with the prevailing wind. The runway designation is painted on the runway. The lettering, size and location conform to the relevant regulations [12].

3.3.4.5 Parallel runways

Busy aerodromes may benefit from having multiple runways and, if they are parallel, to avoid confusion they are given a third character which is either an L for left (as viewed from the air approaching the runway), an R for right and if there is a third a C may be used for centre. For example, the runways at London's Heathrow airport are parallel runways and their full names are 09L/27R and 09R/27L. There are a few aerodromes that have even more parallel runways; for example, Chicago's O'Hare airport has six parallel runways, these are named 09L/27R, 09C/27C, 09R/27L, 10L/28R, 10C/28C and 10R/28L, see Figure 3.2. If there are multiple parallel runways then, some will be used for take off and some for landing but it will be usual for these purposes to be rotated to share the wear and tear on the runways, and also provide some mitigation of noise for those living in the vicinity of the airport.

3.3.4.6 A note for analysts

It is often necessary to calculate the bearing of, say, a wind turbine from an aerodrome or the distance of a wind turbine from the extended centreline of a runway.

Figure 3.2 Multiple runway nomenclature

Checking that the calculation is correct is always prudent and knowing the runway bearings can provide a useful cross-check in such calculations.

3.3.4.7 Runway longitudinal features and their markings

The features of a runway and its markings are determined by the type of aircraft that use them and, associated with the aircraft type, shorter runways are marked differently from longer ones.

A key runway-distance metric is the Landing Distance Available (LDA). Runways with LDA less than 1,200 m (4,000 ft) are regarded as short. These runways would be used by slower moving GA (light aircraft). In these circumstances, there is more time for pilots to assess the runway area and the requirement for special features, and their associated markings, is minimal. When the LDA exceeds 1,200 m and, by implication, faster and/or heavier aircraft are also using the runway, or, if there is a local requirement to emphasize the runway thresholds (the term is described below), then marking is increased. Increased levels of marking may also be required if temporary maintenance work is being carried out on runways that are still active.

However, the focus of this discussion is on runways where radio approach aids are available to facilitate the precision approach. Such runways are used in situations where large, faster moving, aircraft are operating and where visibility may be limited. In these circumstances, the need to distinguish the features of the runway is greater and the markings must be made as clear as possible to help the pilot land the aircraft. The features of such runways are described below.

In all cases, the features of the runway are painted in a fashion that renders them visible from the air and to maximise visibility, all runway markings are painted white. Starting with a feature that is common to all types of tarmac or concrete runway, lines drawn along the central axis of the runway mark the **Runway Centreline**. Another term that is encountered frequently is the **Extended Centreline of the Runway**, which as the name implies is the runway centreline extended beyond the runway and beyond the aerodrome.

All runways used for the precision approach are required to have **Runway Edge** marking to assist pilots as they land in poor visibility.

Stating the obvious, the function of the runway is to allow aircraft to take off and land but often it is only part of the overall length of the runway is given over to this purpose. The bounds of the region where take off and landing should begin are marked by the **Thresholds**. The threshold is marked by a series of parallel stripes aligned with the centreline and depicting the width of the runway, see Figure 3.3.

The **runway designation** is marked inside the threshold marking.

Although in principle the thresholds mark the edge of the part of the runway used for landing (and take off) other factors must be taken into account. Aircraft approaching a runway may experience a sudden loss of altitude, for example, because of turbulence, mechanical failure or pilot error. To allow for these possibilities, aircraft will aim to touch down some distance within the thresholds. Moreover, when an aircraft touches down the wheels are not rotating and at the point of touch down rubber from the tyres is deposited on the runway. Over time, this rubber can obscure markings and the continued use of the same aim point would lead to runway wear and tear. Taking all these factors into account, the **Aim Point** and several **Touch Down Zones** (TDZ) will be marked upon the runway. Anticipating the discussion later in this chapter on ILSs, the ILS radio beam is intended to help the pilot find the Aim Point.

At some aerodromes, the threshold may mark the end of the runway; at others, there may be taxiways leading to the threshold that extends the length of the runways. Such features are marked to ensure that it is clear to pilots that these are part of the taxiway and are not available for take off and landing.

The features and their markings are illustrated in Figure 3.3.

3.3.5 Take off, take-off or departure?

The term 'take off' has been used so far because the term is in common usage for one of the principal purposes of a runway. However, the term is now only used in aviation operations in one specific context, that is when Air Traffic Services grant a pilot permission for an aircraft to take off. The term is no longer in general operational usage since 27 July 1977 when two airliners collided at Tenerife airport with 583 fatalities. At the time of writing, this remains the worst disaster in aviation history. The cause of the accident was attributed to a misunderstanding about a manoeuvre being requested by ATC following take off. The crew of a KLM Boeing 747 mistakenly believed they had been given permission to take off before a second Boeing 747 aircraft, operated by Pan Am, had cleared the runway. Unable to see the Pan Am aircraft until it was too late, because of fog, the KLM crew commenced take off and collided with the Pan Am aircraft. Since that time any discussion about events surrounding take off uses the term **departure** [13].

Readers consulting other documents will notice that there is no consistency in the spelling of the term 'Take Off'. Take Off and Take-Off are both in common usage and are sometimes even mixed up in the same documents. It will be clear that the former style has been adopted here.

3.4 The air environment

Airspace is categorised in a number of different ways. The first category to be considered is the flight information regions (FIR).

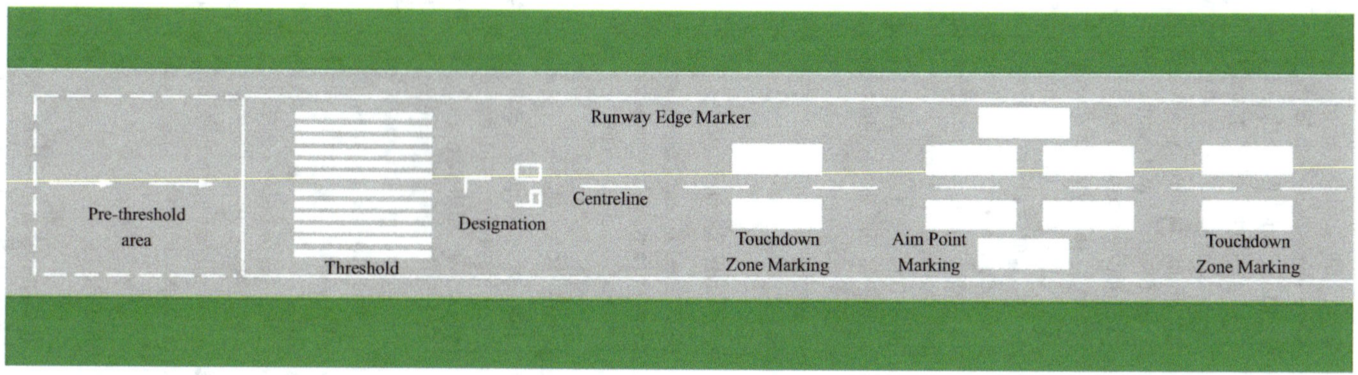

Figure 3.3 Runway markings

3.4.1 FIRs

The International Civil Aviation Organisation (ICAO) is a global forum that develops policies and sets standards for civil aviation; these standards include those that form the basis for the management of airspace [14]. Although ICAO sets the standards, it delegates regional control of airspace to **Controlling Authorities**.

As their name implies, controlling authorities are responsible (inter alia) for providing a **Flight Information Service** (FIS), that is, in simple terms, air traffic control (ATC), within a **FIR**. FIR are not necessarily aligned with national boundaries although the majority are. If the airspace is complex, or the country is large, a controlling authority may be responsible for more than one FIR. Whereas the airspace over small countries may be aggregated into a single FIR.

An example of complex airspace is that over the British Isles which is divided into two FIRs: the London FIR and the Scottish FIR. The London FIR covers the airspace over the South of England and the whole of Wales. The Scottish FIR covers the airspace over Scotland, the North of England, the Isle of Man and Northern Ireland. The dividing line between the two regions runs from the North Sea, through Newcastle to the Lake District, just below the Isle of Man and then onto Northern Ireland's border with the Republic of Ireland, and it is completed in the North Atlantic. The Controlling Authority for these British FIRs is the Civil Aviation Authority (CAA) which is also responsible for the Shanwick Oceanic FIR which extends out into the North Atlantic.

An example of a large country with multiple FIRs is Canada which has seven, namely, Edmonton, Gandar (Domestic and Oceanic), Moncton, Montreal, Toronto, Vancouver and Winnipeg. The Controlling Authority for the Canadian FIRs is NAV Canada.

Belgium and Luxemburg share a single FIR. The Controlling Authority is Skeyes (also known as Belgocontrol).

FIRs may be divided into lower airspace, still referred to as the FIR and upper air space which is referred to as the **upper information region** (UIR). The boundary between the FIR and the UIR is, nominally, 25,000 ft [15]. Altitudes and the method of measuring them are discussed later in this chapter.

For the practical purposes of providing FIS, FIR may also be divided geographically into **sectors**. For example, there are eight high-level sectors in the Scottish FIR; Rathlin, Dean Cross, Montrose, Humber, Tyne, Central, Moray and Hebrides.

3.4.2 Airspace class

Every part of the airspace within the FIR is assigned a **Class** according to the rules that must be followed by the aircraft using that airspace. The classes are named with the letters A to G; Class A airspace rules are the most stringent and Class G rules are the least stringent. Classes A to E are all controlled airspace, that is, operations within these classes must be controlled by ATC services. Classes F and G are uncontrolled (and ATC services are not mandated). Controlling authorities do not have to use every class of airspace within their FIR. For example, in the UK FIRs, the CAA does not use either Class B or Class F and the Republic of Ireland (Eire) uses only Class A, C and G airspace.

The ICAO encourages some general principles concerning the way airspace is allocated. For example, one principle is that there should be minimal regulation which in practical terms, means maximising the amount of uncontrolled airspace and allocating as much Class G airspace as possible. Notwithstanding this guidance, there are many variations in the allocation of Class G airspace by different Controlling Authorities, often based on height restrictions as the following examples illustrate:

- In the United Kingdom, the majority of the airspace below a height of nominally 19,500 ft is Class G.
- In the Danish FIR where, with the exceptions of airspace over water and controlled airspace around aerodromes, all airspace below 3,500 ft is Class G [16].
- In Eire, all airspace above 66,000 ft is Class G. Airspace below this level is either Class A or Class C. With the exception of aerodromes and restricted military airspace, Class G airspace extends from the surface to 7,500 ft underneath Class C airspace and from the surface to 5,500 ft under Class A airspace [17–19].

3.4.3 Airspace types

There are other ways of classifying airspace. In the immediate vicinity of all licensed civil aerodromes and military aerodromes, the airspace is referred to as the **Aerodrome Traffic Zone** (ATZ). The purpose of this zone is to protect aircraft when they are departing, arriving and flying in the immediate vicinity of an aerodrome; all of which are regarded as critical stages of flight. This protection is afforded by enforcing rules to prevent the movement of aircraft without permission. The volume of airspace depends on local circumstances either 2 Nautical Miles (Nm) (3.7 km) or 2.5 Nm (4.6 km) in diameter and it extends from the ground usually to 2,000 ft [20,21].

In the case of civil aerodromes, the ATZ may, in effect, be replaced with a zone tailored to meet the specific local requirements. Such airspace is called the **Control Zone** (CTZ). The military equivalent of the CTZ is the **Military Air Traffic Zone** (MATZ). Both CTZ and MATZ are cylinders of airspace which are 5 Nm (9.3 km) in diameter with a maximum altitude of 3,000 ft. Sometimes MATZ is extended along the extended centreline of the runway in what is called a stub. Sitting above the CTZ is the **Control Area** (CTA); a funnel-shaped piece of airspace wider at the top than at the minimum altitude. To avoid confusion between a CTZ and a CTA, the minimum altitude of a zone (ATZ, CTZ or MATZ) is ground level (so the **z**one corresponds to **z**ero feet), whereas CTA does not reach the ground.

The CTA is linked together using another type of airspace: the **Airway**. Airways are in lateral width (8Nm in the US) and have a vertical extent from between 5,000 ft and 7,000 ft up to 24,500 feet. Above the Airways are **Air Routes**, sometimes referred to as the **Upper Air Routes** (UAR); these extend from 25,000 ft to 46,000 ft. In the United Kingdom, all airspace above 24,500 ft

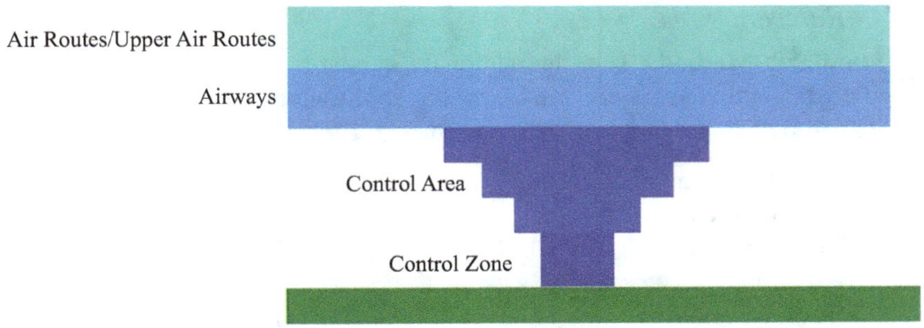

Figure 3.4 Airspace types

is Class C (that is controlled airspace) and aircraft are mandated to use ATC services. These airspace types are illustrated in Figure 3.4.

3.4.4 The role of AGA communications

There is an implicit requirement arising from the foregoing discussion. To provide a FIS (ATC), AGA communications are essential. There is only one class of airspace in which carrying a radio receiver may not be mandated, that is, Class G. However, even in this class of airspace, it is recommended that aircraft should carry a radio to allow pilots to communicate with controllers on the ground.

AGA radio can be subject to interference by wind turbines (and many other structures and terrain).

3.4.5 Altitude measurement

Knowing the altitude of aircraft is critical for ensuring safe vertical separations. Aircraft measure their altitude using a form of barometer called an altimeter. Atmospheric pressure is created by the weight of air pushing down from above. As altitude increases, the amount of air above decreases, and therefore, the pressure decreases (and vice versa). It also follows that the difference in pressure between two points is proportional to the difference in their heights.

But a problem arises, local atmospheric pressure is also affected by temperature and the effects of wind, both of which vary continuously. The key to successful altitude measurement, and the safe separation of aircraft from other aircraft and the terrain, is the selection of appropriate atmospheric pressure reference that can be used as a comparison with the measured pressure and a protocol for switching references as required to take account of the temperature/wind variations.

Before discussing the protocol, the three types of atmospheric pressure reference that are used in aviation are described along with their relevant characteristics.

3.4.6 Reference pressures

The Standard Method. The International Standards Organization (ISO) has defined a 'standard atmospheric pressure' that applies at mean sea level (msl); this pressure can be expressed using different units:

- 14.7 pounds per square inch,
- 760 mm of mercury,
- 29.92 in. of mercury.

Or the ones most commonly used in aviation

- 1,013.2 millibars (mb) which is expressed as an International System of Units (SI)
- 1,013.2 hecto Pascal (hPA) (hecto is a factor of 100)

This unit has been adopted as the standard by aviation regulators such as the Civil Aviation Authority (CAA) [22], EUROCONTROL [23] and the Federal Aviation Authority [24]. When this reference is provided to the altimeter, the height measurements it provides with be with respect to the mean sea level.

In referring to this and the other measurement methods, it is convenient to use three-letter 'Q' codes which were developed early in the twentieth century to facilitate multilingual communications. When the altimeter is set to standard pressure, the Q code is **QNE**, sometimes referred to as the En-route pressure setting.

Aerodrome (or Field) Elevation Method. This is the surface pressure at the aerodrome and, when this reference is provided to the altimeter, it will provide height measurements with respect to the airfield. In practice, meteorological measurements (not just pressure but including other useful information such as temperature and wind speed) will be made at some practical location on the aerodrome. If the aerodrome is not level, which is defined as a variation in surface height of 7 m or greater, then the measurement must be adjusted to account for the difference between the runway threshold and the measurement location. The purpose of this correction is to ensure precision approaches work to the correct datum (and prevent the aircraft from 'landing' above or below the runway!). The aerodrome elevation method is more frequently referred to by the Q code, **QFE**.

Measured Sea Level Method. This method is similar to the standard method but the sea level pressure is measured. Altimeter readings will, therefore, be with respect to the true sea level. This method is more frequently referred to by the Q code, **QNH**. However, there are two QNH variants. Ideally, an aerodrome will be able to make pressure measurements: **Local QNH** is usually updated every hour but some aerodromes update QNH every half hour. If Local QNH is not available, an [wide] **Area QNH** can be used. The Controlling Authorities or their designated service providers decide on the appropriate value of Area QNH which remains valid for 3 h.

3.4.7 Accommodating variations in air pressure

It will be clear from the foregoing discussion that protocols must be observed when referring to the height of an aircraft to avoid confusion and ensure safety. For example, addressing such questions as the height above the airfield being referred to or the height above sea level. To eliminate confusion, the following definitions are important.

Flight Level. A method is necessary to ensure aircraft departing from aerodromes, where local surface air pressures may be different, have a consistent approach to determining height, to ensure safe vertical separations can be maintained. The method universally adopted is the use of Flight Levels, a form of **pressure altitude**, where aircraft height is measured in values corresponding to hundreds of feet, for example, Flight Level 200 corresponds to 20,000 ft. To ensure aircraft in the same airspace are at the correct altitude (Flight Level), the height of all aircraft is calculated using the Standard Method's air pressure (1,013.2 mb) as a reference. This is a safe method for flying at high altitudes, but it would not be equally safe at low altitudes where flying at a height above the ground is safer. Therefore, two height measurement methods have to coexist and there has to be a well-understood process for switching between the two.

Transition Altitude. The Transition Altitude is the altitude at which measurements change from feet (at lower altitudes) to Flight Levels. The Transition Altitude varies between FIR and within FIR. In the United States and Canada, the transition altitude is 18,000 ft; authorities have considered adopting this as an international standard. Currently the transition altitude in the UK FIR is 3,000 ft but in the airspace around London the transition altitude is 6,000 ft. Transition altitudes are published for aerodromes.

Transition Level. The Transition level is the first Flight Level above the transition altitude.

Transition Layer. The transition layer is the region of airspace (the layer) between the transition altitude and the first Flight Level available for use.

These concepts are illustrated in Figure 3.5.

3.5 The rules of flight

As might be expected, there are high degrees of international harmonisation of the rules of flight and many of the rules set out, for example, by the United Kingdom

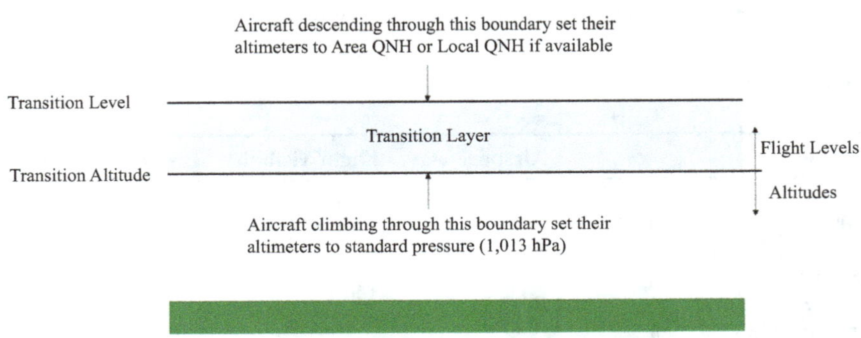

Figure 3.5 Height measuring protocols

[25], Europe [26] and the ICAO [27] are identical. As a generalisation, these rules identify two categories of flight rules: **Visual Flight Rules** (VFR) and **Instrument Flight Rules** (IFR). Overviews of these rules are discussed in turn.

3.5.1 Visual flight rules

As the name implies, VFRs apply when the pilot is able to see where the aircraft is going without the need for instruments. In simple terms, these are rules that might apply to aircraft flying at low levels such as GA (light aviation), some helicopter operations and, in some circumstances, aircraft during take-off and landing.

The first element of VFR provides objective criteria to quantify the statement about being able to see where the aircraft is going to prevent situations where the visibility might be impaired (i.e., proximity to clouds). These criteria are closely associated with the class of airspace and VFR cannot be used at all in some classes of airspace.

The term used in aviation circles for visibility is **Visual Meteorological Conditions** (VMC) and the limiting VMC are referred to as **Visual Flight Minima** (VFM) [28]. The criteria, which are commonly adopted by countries, are summarised in Table 3.2, note that it is custom and practice to measure horizontal distances in kilometres or nautical miles and altitudes in feet [26].

The rules of VFR are what might be expected. Aircraft flying under VFR are not to enter airspace where VFR flying is not allowed, for example, VFR flying is not allowed in controlled airspace, unless under exceptional circumstances. There are separate speed limits for fixed-wing and rotary-wing (helicopters) aircraft. And, aircraft under VFR unless landing or taking off, must observe a minimum allowable altitude of aircraft; generally, 500 ft or 1,000 ft over built-up areas. In the United Kingdom, a dispensation is allowed in more remote areas where flying at lower altitudes is permitted providing that an aircraft maintains a minimum separation distance of 500 ft from any person, vehicle or structure.

3.5.2 IFRs

IFRs apply when VFM cannot be met and IFRs must also be followed in certain classes of airspace. They were developed to maintain safety when visibility is impaired, a situation which is sometimes referred to as 'blind flying'. There are two aspects to maintain safety: control of the aircraft and navigation.

Table 3.2 VFR VMC

Altitude	Airspace class	Flight visibility	Distance from cloud
Altitude $\geq$ 10,000 ft	B, C, D, E,	8 km (4.3 Nm)	1,500 m horizontally
3,000 ft $\geq$ Altitude < 10,000 ft or 1,000 ft above terrain (whichever is greater)	F, G	5 km (2.7 Nm)	300 m 1,000 ft vertically
Altitude $\geq$ 3,000 ft or 1,000 feet above terrain (whichever is greater)	B, C, D, E F, G		Clear of cloud with the surface in site

3.5.3 Control of flight

Pilots can quickly become disoriented when flying in poor visibility and it is very easy for them to lose control of the aircraft. This problem was particularly serious in the very early days of flying because aircraft were not capable of flying above fog or low cloud. Even experienced pilots can get into trouble. When John Alcock (1892–1919) and Arthur Brown (1896–1948) made the first transatlantic crossing in June 1919 (in under 72 h) they flew into thick fog and Alcock lost control of the aircraft twice. On both occasions, the aircraft spiralled out of control only surviving because the cloud base was high enough to allow recovery before the aircraft hit the sea [29].

3.5.4 Navigation

The second problem is that without being able to see the ground, it is not possible to see landmarks that would assist navigation. This is a particular problem when linked to the aforementioned problem of poor visibility close to the ground. While, in general, over-flying bad weather may be acceptable, it is not acceptable when attempting to land. In his book describing the first solo flight across the Atlantic, Charles Lindbergh described his experience delivering mail. In 1926, the year before his own Trans-Atlantic flight, Lindbergh got lost in thick fog and could not find anywhere to land. When his fuel was running low, he flew his aircraft as high as he could and parachuted to the ground. He recovered the lost mail the following day from the wreckage of the crashed aircraft. This happened to him twice that year [30]. This situation was not unusual. In one year starting in July 1924, the US Postal Service recorded 554 forced landings because of bad visibility, and deaths in these circumstances were commonplace.

3.5.5 Flying using instruments

Although there was a clear requirement for instruments to facilitate blind flying, an additional incentive was provided by the First World War when there was a drive to create pilotless aircraft, cruise missiles.

The first flight using instruments is credited to James H. Doolittle (1986–1993). On 24 September 1929, he flew a 15-min circuit from Mitchel Air Force Base in Long Island. To prove that he was not cheating, Doolittle had his cockpit covered with a hood during the flight. The instruments which made this flight possible were a gyroscope to provide the aircraft heading, an attitude indicator and an altimeter [31]. Doolittle went on to become famous for leading the first air attack against Tokyo in the Second World War in 1942.

Instruments to assist the pilot flying blind were also developed; for example, the Bank and Turn indicator invented by Carl J. Crane (1900–1982) and William C. Ocker (1880–1942) in 1930 [32] and the Artificial Horizon/Attitude Indicator (AI) invented by Leslie F. Carter of Sperry Gyroscope in 1938 [33]. These instruments, which also allow pilots to fly at night, are based on the gyroscope, they do not rely on radio and are outside the scope of this discussion.

There are two sets of requirements to fly using instruments: the aircraft must be suitably equipped and the pilots must be suitably qualified. Pilots must also be able to demonstrate that they have regular annual experience in IFR flying (i.e., a minimum number of flight hours each year).

3.6 AGA communications

The earliest communications to aircraft from ground controllers used visual aids, including large signs painted on the ground and coloured flags. At closer quarters, other methods could be used; hand signals, paddles (also known as bats) and, more recently, illuminated wands have been used; the latter are still in use today for marshalling aircraft on the ground [34].

The earliest forms of communication from the aircraft occurred when a pilot wanted to signal the intent to land the aircraft. The intention was announced by flying close to the control tower and wingtip-dipping, a practice that went on for many years until radio messaging was widely available.

Binns [35] reports that the first radio communication between an aircraft and the ground was a Morse code message sent in August 1910 by a Canadian pilot, James McCurdy, when flying over Brooklyn, New York. Shortly after that, a successful longer-range experiment carried out in England led to a research programme between the Marconi Company and the Royal Flying Corps (RFC), which allowed aircraft to guide artillery fire during the First World War. The resultant equipment was heavy, the spark transmitter weighing more than 36 kg but, arguably more problematic, was the antenna a 76 m (250 ft) long wire that had to be fed in and out from a spool in the cockpit.

In the late 1920s [36], Radio Telephony (R/T) voice communications were introduced. The equipment of the day was only capable of using frequencies in the high frequency (HF) (3–30 MHz) and low frequency (LF) (in this context below 3 MHz) bands. A significant event at the end of that decade, in 1929, was the formation of the Aeronautical Radio Incorporated (ARINC) organisation, set up to develop and promulgate standards for aeronautical radio services. ARINC, in association with the Airlines Electronic Engineering Committee (AEEC), remains an important source of standards today. The reader may wish to use these standards as a reference source [37].

The use of LF and HF was highly problematic. These frequency bands suffer from atmospheric interference and in the late 1930s and 1940s, as the technology became available, AGA voice communications moved to the very high frequency (VHF) spectrum and this band remains in use today.

The VHF spectrum is now also used to provide digital communications between the ground and the cockpit, for example, for flight planning. However, the principal concern and the focus of the following discussion are analogue voice communications.

3.6.1 The importance of AGA communications

AGA communications have become a critical element of flight safety requiring appropriate safeguarding measures. The following list of AGA message classes illustrates why [38]:

- Distress calls, distress messages and distress traffic, including May Day and Pan Pan messages*.

*The terms May Day and Pan Pan originate from the French expressions M'aidez (meaning help me) and Panne (meaning breakdown). The May Day message indicates there is a risk to life. The Pan Pan, message indicates an urgent problem but no immediate risk to life.

- Urgency messages, including messages preceded by the medical transports signal.
- Communications relating to direction finding.
- Flight safety messages.
- Meteorological messages.
- Flight regularity messages including messages that deal with aircraft arrival and departure times and servicing; in particular, when parts are required as soon as possible to prevent aircraft from becoming unserviceable.
- Messages relating to the application of the United Nations Charter.
- Government messages for which priority has been expressly requested.
- Service communications relating to the working of the telecommunication service or to communications previously exchanged.
- Other aeronautical communications.

3.6.2 Spectrum use

The frequency band between 117.975 MHz and 137.000 MHz has been allocated internationally† for 'aeronautical mobile'‡ services, that is, AGA communications [39]. Telecommunication regulators have allocated this spectrum on what is termed 'a primary basis'; in other words, other services are not allowed to cause interference to users. Any infringement of these frequencies is taken very seriously internationally; for example, in the United Kingdom, Ofcom has enforcement powers against interferers, and in France, Autorité de Régulation des Communications Électroniques et des Postes (ARCEP) and, in the United States, the Federal Communications Commission (FCC) have similar powers. Frequencies in this part of the electromagnetic spectrum are traditionally called VHF but it is becoming increasingly common to use North Atlantic Treaty Organisation (NATO) terms to describe different parts of the spectrum and some readers may be more familiar with these frequencies being described as A-Band.

The spectrum is utilised as discrete channels with each channel providing two-way communications between the ground controller and an aircraft. Over time, the growth in air travel has led to the demand for more and more channels, particularly to service an increase in the number of en-route sectors required. The reasons for this are discussed later. The demand has been satisfied by improvements in radio technology which have allowed the spacing between channels to be reduced, accommodating more channels in the same block of spectrum. When channels were introduced, in 1947, the channel spacing was 200 kHz. In 1958, this was reduced to 100 kHz, in 1964 to 50 kHz, in 1972, it was further reduced to 25 kHz to provide 760 channels. The most recent change was carried out in 2019, tripling the number of channels by reducing the spacing to 8.33 kHz [40].

Some channels have specific purposes. For example, the International Civil Aviation Organisation (ICAO) mandates 121.500 MHz as the emergency channel.

†The International Telecommunications Union (ITU) divides the World into three regions. All three regions recognise this allocation.

‡The reference to 'mobile' acknowledging that the services are used by aircraft as well as ground facilities.

Other channels have specific purposes such as 122.480 MHz which is dedicated to ballooning [41]. But, in general, channels are assigned to airfields for specific ATC purposes. For example, the current assignments at London's Heathrow Airport (ICAO airport code EGLL) are [42]:

• 118.505 MHz	Tower Control of the Southern Runway (09R/27L)
• 118.705 MHz	Tower Control of the Northern Runway (09L/27R)
• 121.905 MHz	Ground Movement Controller 1
• 121.705 MHz	Ground Movement Controller 2
• 121.855 MHz	Ground Movement Controller 3
• 124.475 MHz	Standby Frequency

3.6.3 Additional military spectrum use

The military also requires AGA communications and in addition to using VHF they use ultra high-frequency (UHF) spectrum between 225 and 400 MHz [39,43]. Bandwidths used are higher than VHF and are currently set to 25 kHz. For example, the frequencies currently assigned to RAF Coningsby (ICAO airport code EGXC) are:

• 124.675 MHz	Tower
• 362.300 MHz	Approach
• 255.950 MHz	Departures
• 379.950 MHz	Director
• 121.850 MHz	Ground movements
• 357.125 MHz	Ground movements
• 279.325 MHz	41 Sqn air to air
• 379.350 MHz	Ops
• 338.025 MHz	HAVEQUICK (a jamming resistant frequency hopping system) Ops
• 234.575 MHz	Talkdown

The principles discussed below are the same for UHF as VHF and in the examples VHF frequencies have been assumed.

3.6.4 Modulation method

For historical reasons (it has never changed since voice communications were first introduced), AGA communications use amplitude modulation. This is one of the earliest methods used to modulate a radio frequency carrier and, as the name implies, the amplitude of the carrier is modulated in sympathy with the volume of the audio signal that it is passing. The frequency (pitch) components of the voice signal appear in the modulated signal as sidebands, one greater than the carrier frequency (the upper sideband) and one lower than the carrier frequency (the lower sideband); AGA communications retain both sidebands. Modulation types are

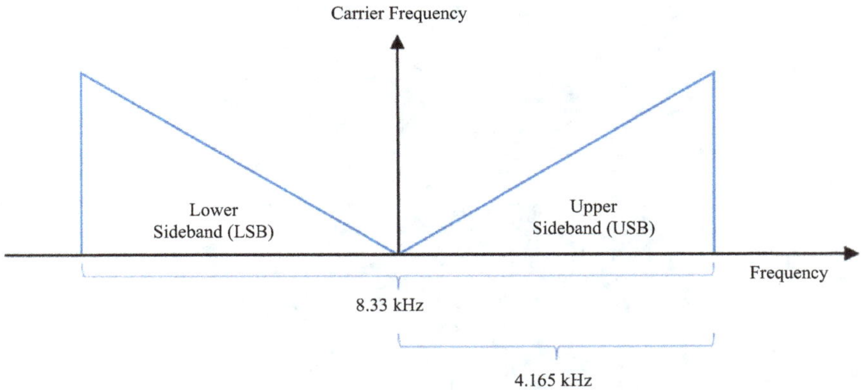

Figure 3.6 A3E modulation

described by an ITU emissions code; this type of modulation is being referred to as A3E modulation, specifically Double Sideband, Full Carrier, modulation. This arrangement is illustrated in Figure 3.6. A feature of this arrangement is that the maximum frequency excursion of the voice is limited to 4.165 kHz which, although the human voice has a much greater span of frequencies is quite adequate for the purpose; by comparison, an analogue telephone signal is limited to 3.4 kHz.

However, unlike some other methods, A3E modulation provides no error detection and correction and cannot easily mitigate the potential degradation experienced in the presence of wind turbines or a wind farm.

3.6.5 AGA protocols

The AGA channels are used in 'half duplex' mode, that is, the same channel is used for communications in each direction, ground to air and air to ground in turn. Mandated by ICAO, messages are highly organised with standard use terminology. Aircraft are selected for communications by use of call signs. As a matter of standard procedure, messages received by aircrew are acknowledged by repeating the direction received [44]. Corrupted messages have been likened to the breaking up of a mobile phone signal. If a message is corrupted or unacknowledged, it must be repeated adding to the workload of controllers and reducing the ATC tempo.

3.6.6 Equipment considerations

Ground-based segments of the AGA link may need to communicate with aircraft in any part of their area of responsibility (AOR). In other words, throughout a full 360° coverage. Aircraft also must be able to communicate with ground stations in any direction. Omni-directional antennas are the simplest way to meet these requirements. On the ground, the antennas are, typically, centre fed, half wave dipole or folded dipole which to be consistent with the VHF frequencies being used, will be a little over a metre in length. For ATC purposes, these antennas will be located on a high structure,

Figure 3.7 Control Tower Jersey International Airport (EGJJ)

typically at an airport on the control tower. Company channels may be supported from other locations. Figure 3.7 shows the control tower at Jersey International Airport (ICAO code EGJJ), note the antennas located around the roof of the tower. The gain of a dipole or folded dipole is 2.14 dBi [45]. dBi and other dB values are discussed in Chapter 6. An example of AGA antennas being mounted on other structures are shown in Figures 3.8 and 3.9 which show the antenna systems in use at Newquay International Airport/Cornwall Airport (ICAO Code EGHQ).

Aircraft antennas for AGA communications are typically of the blade type. The blade shape, most are trapezoidal, is to make the assembly aerodynamic, that is to reduce drag. These antennas provide vertical polarisation and are omni-directional in the azimuth plane and they are broader bandwidth than a thin wire vertical antenna. A minimum of two antennas are required on all airliners and these might typically be mounted above and below the fuselage to minimise the possibility of screening, see Figure 3.8. The gain of a blade antenna is typically between −1 and −3 dBi (Figure 3.10).

3.6.7 Calculating the distance to the radio horizon (constraining mitigation of effects)

It is useful to be able to calculate a rough approximation of the distance to the radio horizon. In this instance, the purpose is to establish a rule of thumb for

Figure 3.8 VHF communications towers (transmitter left) (receiver right) at Newquay International Airport

Figure 3.9 Close-up of the AGA antennas at Newquay International Airport

determining the potential for interference to radio services based on the height of the transmitting antenna. However, the method is useful for other purposes as well. If the height of the transmitting antenna is replaced by the height of the tip of a wind turbine, then a rough estimate can be provided for the distance from which it will be visible. The method set out below is only an approximation because it takes no account of the effects of terrain, sometimes referred to in the literature as 'a bald Earth' or a homogeneous surface. For detailed analysis, it will be necessary to account for terrain and this is discussed in Chapter 5. However, the approximate value is calculated thus [46]:

$$\text{Distance to the horizon (the line of sight)} = \sqrt{(2\,k\,R\,h)} \qquad (3.1)$$

where R is the radius of the Earth in metres (6,371,000), k is a refraction correction factor (discussed below) = 1.333 and h is the height of the transmitting antenna in metres.

The method used to arrive at this equation is a simple application of Pythagoras's theorem. The geometry is illustrated in Figure 3.11 [47].

Figure 3.10 Aircraft blade antennas

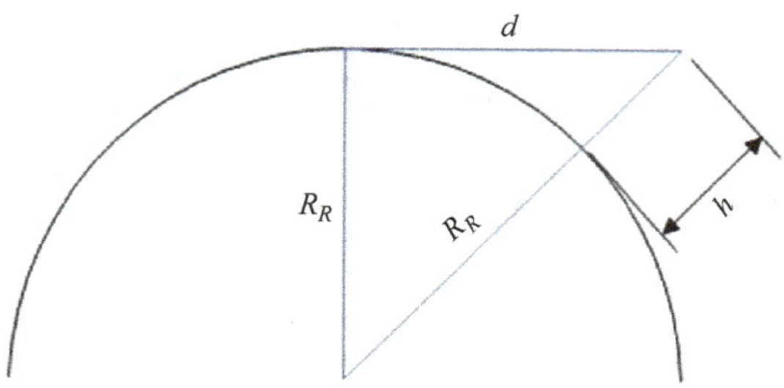

Figure 3.11 Calculation of the radio horizon

Note the use of R_R, the radio radius of the Earth. This is discussed below.

$$\begin{aligned} & (R_R + h)^2 = d^2 + {R_R}^2 \\ \text{Rearranging} \quad & d^2 = (R_R + h)^2 - {R_R}^2 \\ \text{Expanding} \quad & d^2 = (R_R + h)(R_R + h) - {R_R}^2 \\ & d^2 = {R_R}^2 + 2R_R h + h^2 - {R_R}^2 \\ & d^2 = 2R_R h + h^2 \\ \text{Taking the square root} \quad & d = \sqrt{(2R_R h + h^2)} \end{aligned} \tag{3.2}$$

But $2R_Rh$ term is very much larger than h^2 and h^2 is usually ignored, reducing the equation to:

$$d = \sqrt{(2R_R h)} \tag{3.3}$$

The Troposphere is one of the five[§] main layers within the Earth's atmosphere and is the layer closest to the Earth's surface[¶]. This layer is 10–15 km deep and contains 99% of all the water content of the atmosphere, and is also the region where the air pressure is the highest. The combination of humidity (water content), pressure and also temperature causes radio waves to refract within this part of the atmosphere. Because the degree of refractivity is affected by temperature, it will be clear that refraction varies with time of day, season, etc. and this leads to a concept of a proportionate atmosphere usually either a 50% atmosphere or a 90% atmosphere, which will be discussed later. The effect of refraction is to bend the radio beam downward, extending the radio horizon farther than geometry would predict. Simply put, it is as if the radio wave clings to the surface of the Earth. A simple correction factor can be applied to represent refraction, treating the Earth's radius as greater than it really is. A factor of 4/3 is a good correction factor, it is usually given the symbol k. Hence the above equation can be modified thus:

$$d = \sqrt{(2\,k\,R\,h)} \tag{3.4}$$

where R is the radius of the Earth in metres (6,371,000), k is the Earth radius correction factor = 1.333 and h is the height of the transmitting antenna in metres.

3.6.8 Radio horizon implications

If VHF signals operate reliably when the transmitting and receiving antennas are within line of sight, it follows that those signals from beyond line of sight will either not be received or will be received unreliably. Propagation is discussed in Chapter 5, but assuming for the time being that the radio horizon represents the limit of propagation, then applying the formula derived above gives useful insights into how the frequency channels can (and should not) be used.

The distance to the radio horizon from a ground-station antenna at a height of 20 m is a little over 18 km. Assuming its antenna was the same height as the first, or

[§]The five main layers of the Earth's atmosphere are the Troposphere, Stratosphere, Mesosphere, Thermosphere and Exosphere.

[¶]The literal translation of the term is the region of the atmosphere where the weather takes place.

lower, a second ground station could use the same radio channel as the first and neither would interfere with the other, provided the two stations are separated by at least double the distance to the radio horizon. Planning frequency allocations using this simple rule might be quite satisfactory for communications with aircraft on the ground. However, communicating with aircraft in the air is a different matter. Applying the formula again for aircraft flying in an airway, the vertical limits of which are 5,000–24,500 ft [48], the corresponding radio horizons are 60–350 km, respectively. Moreover, it is necessary for en-route Air Navigation Service Providers (ANSP) to communicate with aircraft flying even higher.

It follows that frequency planning to avoid interference is a complex process. The following corollaries arise:

- As previously stated, the need for additional channel capacity to deal with increased air traffic is driven by the en-route environment (not necessarily the ground environment, although this too will still require careful frequency planning).
- It also follows that if there were to be interference caused by a wind turbine or wind farm (or any other source), there is no simple mitigation method provided by selecting another channel. To do so would risk causing interference to other AGA communications and there would also be practical problem of how to inform the aircraft of a change in channel if communications have already been lost.

3.6.9 Long-range communications

Thus far, all the discussion has been about relatively short-range AGA communications. It is also necessary to consider longer ranges.

A fundamental limitation of most radio frequencies, including VHF and UHF, is that they can only reliably operate in line of sight (most radio wave frequencies are extremely limited in their ability to operate beyond the horizon). But aircraft operate internationally and travel long distances across the oceans where ground, line of sight and contact are not possible. However, an exception to the general rule about the limited capability to propagate beyond the horizon is HF frequencies (sometimes referred to as shortwave) in the range of 2–30 MHz which correspond to wavelengths between 150 m and 10 m. HF signals can operate beyond line of sight (BLOS) because these frequencies are reflected by the Ionosphere; a complex structure in the upper atmosphere. Thus, HF communications are widely used for long-range, particularly transoceanic, purposes. Further, today, HF is backed up by satellite communications [49]. However, relatively little work has been carried out to investigate the impact of wind turbines on HF communications.

Concerns have been expressed about the impact of wind turbines on Over-The-Horizon-Radar (OTHR) and there are some similarities between OTHR and transoceanic communications. Like transoceanic communications, OTHR relies upon the Ionosphere to provide propagation beyond the horizon and both use HF frequencies. However, the purposes are quite different. OTHR is intended to detect moving, usually military, platforms (vehicles, vessels, missiles, etc.) at long ranges on or above the surface of the Earth. Detecting wind turbines is problematic

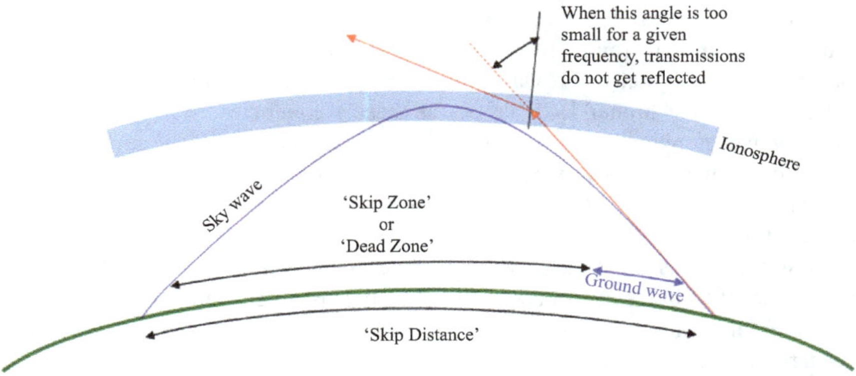

Figure 3.12 Skip distance

because the movement of the blades creates a Doppler signal that can be confused with these platforms. In other words, wind turbines produce clutter for OTHR [50].

However, with regard to HF radio communications, although the span of frequencies is only 28 MHz, this covers almost four octaves[1], and the distance over which signals can propagate differs and, at any given frequency within the range, the propagation varies with time. A simple rule of thumb is the higher the frequency, the longer the propagation distance that is possible and very long ranges are possible. Paradoxically, because HF signals are bounced off the Ionosphere, propagation over shorter distances is unreliable. Figure 3.12 explains the processes involved. In the HF spectrum, **ground waves** propagate over relatively short ranges before they are absorbed by the lowest levels of the atmosphere. However, **sky waves** can reach the Ionosphere where, depending on their frequency, the state of the ionosphere and the angle of incidence, they may pass through or they may be reflected back to Earth. Under the influence of Solar weather, the ionosphere can contain fewer or more free electrons. The higher the count of electrons, the better able the ionosphere to reflect radio waves. For any given state of the ionosphere if the frequency used exceeds a **Maximum Usable Frequency** (MUF), then the signal will pass through the Ionosphere and escape the Earth. At lower frequencies, provided, the angle of incidence with the Ionosphere is large enough, the signals can be reflected back towards the Earth. However, this creates a **skip zone** or **dead zone**, too distant for ground waves to propagate but too close to receive a signal reflected off the ionosphere to be received. The length of the skip zone depends on many factors but may be hundreds to a thousand kilometres long.

Taking these propagation factors into consideration, it may be concluded that the potential for interference between, say, large off-shore wind farms and very long-range communications is small. However, as the depth at which off-shore wind farms can operate increases, their distance from the shore will increase and

[1]Within this range the frequency doubles almost four times.

there may be potential for interference. Moreover, the wavelengths of the higher frequencies in the HF range are comparable to the scale of wind turbine components: a 28 MHz signal has a wavelength of approximately 11 m. Consequently, wind turbines have higher RCS values at these frequencies. Whereas, the lower frequency wavelengths are more likely to be electrically short compared to the scale of components. These would tend to the lower RCS values associated with Rayleigh scattering, discussed later, significantly reducing the potential for interference with communications.

In summary, as the size of wind turbines increases and the depth at which off-shore wind turbines can operate, the potential for interference with long-range air–ground–air communications may need to be monitored. This is not widely considered a cause for concern at present.

3.7 Aeronautical Navigation Aids (Navaids)

Most aeronautical Navigational Aids (Navaids) perform their roles using radio and may be subject to radio interference from many sources including, potentially, wind turbines and wind farms. The need for navaids is as old as flight itself; from the earliest flights, it was known that the pilot might lose sight of the ground and of landmarks for determining location and direction. These navaids comprise:

- Non-Directional Beacon (NDB) [and automatic direction finding (ADF)], a system for finding the bearing from a known point on the ground.
- VOR, a system for assisting with setting a course based on fixed airways or an aerodrome.
- Distance measuring equipment (DME), a system for determining the distance of the aircraft from a known location on the ground.
- ILS, a system to assist landing in inclement weather or the dark.
- Tactical Air Navigation (TACAN), a military system providing a similar role to VOR and DME.

These systems are old in design and, more importantly, they are old in concept. They were conceived when there were relatively few aircraft in the air and there were not the current broader ecological concerns. The requirements for air navigation have had to evolve further to improve efficiency, cost-effectiveness and be more ecologically sound. These goals are achieved by, for example, reducing fuel consumption and emissions, and organising airspace to reduce noise levels on the ground.

The new navigation paradigm is performance-based navigation (PBN). In simple terms, the requirement has shifted from following airways as precisely as possible to knowing precisely where the aircraft is at any moment, whether inside or outside an airway. Satellite navigation combined with other techniques such as multilateration provides the means whereby these modern requirements can be met.

Thus, many of the systems listed above are being actively decommissioned by aviation regulators worldwide. This begs the question, why discuss them here if they are being withdrawn from service? The answer to this is that safety is paramount and the changing face of aviation navigation must proceed with caution. Plans for rationalising navaid provision have long timescales. Moreover, these systems provide useful backup for satellite navigation, which has its own problems, and they also provide a valuable source of collateral information to check satellite information [51].

For the time being, it is not possible simply to ignore these systems. The problem for the analyst investigating a proposed wind farm that might cause interference is to establish the status of any given system and establish whether there are long-term plans for it to remain in service. This section will describe the various legacy navigation aids that may be raised as concerns and potential interference from wind turbines and wind farms.

3.7.1 Non-Directional Beacon

The NDB was developed by Captain Albert Francis Hegenberger (1895–1983), US Army Air Corps. Hegenberger demonstrated the effectiveness of the beacon system on 9 May 1932. In an impressive feat of flying, he landed an aircraft, equipped with NDB receiver equipment, with the cockpit blacked-out [52]. Hegenberger went on to have a distinguished career retiring with the rank of Major General in 1949 [53].

3.7.1.1 NDB standards

Definitions and standards for NDBs are set out by ICAO [54]. The standards to be maintained are:

- Field strength of the radiated signal.
- Centre frequency: for example, Newquay airport's NDB operates on 357 kHz.
- Modulation depth, which should be 95%.
- The Morse code identification: The Ident is a one, two or three letter code. Canada is an exception to this rule using a one letter and one number code. For example, Newquay airport's Ident is SM (. . . - -). Before becoming a civil airport, Newquay was a military aerodrome called RAF St Mawgan. The Ident is repeated seven times a minute.
- The modulation frequency, that is, the tone of the morse that is either 400 Hz or 1,020 Hz [55].

As the title implies, NDBs do not transmit any information that conveys information about direction and navigational equipment on-board the aircraft is needed to exploit the beacon transmissions.

3.7.1.2 NDB status

As previously stated, in the introduction to this section, NDB is slowly being withdrawn from service, partly as cost-saving measures but principally because they are not as important due to the introduction of PBN and satellite navigation. However, the withdrawal is slow and it is anticipated that some will remain in service for some years to come. The rate of withdrawal depends on the NDB system type and this is discussed below.

3.7.1.3 NDB types

There are four types of beacons. The first two of the four beacon classes are for the purpose of longer-range navigation, essentially for determining the correct heading to follow, called **homing**:

- Enroute: these beacons mark airways. The NDBs are in fixed locations on the ground marking the associated airway. These beacons are the least compatible with PBN and are being withdrawn from service first. For example, no enroute NDB remain in the United Kingdom.
- Terminal, beacons that are located on aerodromes. Many aerodrome beacons remain in service.

and the remaining two types of beacons are associated with ILS:

- Localiser: Beacons used as part of an instrument landing procedure.
- Locater: these beacons are used to identify where an ILS approach should start. They usually have a range of between 18.5 and 46 km (10 and 25 Nm).

3.7.1.4 NDB emissions

An NDB transmits a vertically polarised, continuous wave (CW), signal. ICAO requirements state this signal should be in the band 190–1750 kHz, although in some countries, such as the United States, the band is restricted to 190–535 kHz. These frequency bands propagate via ground waves. The use of higher frequencies was avoided because of the potential for ambiguity caused by sky wave propagation. The wavelength of NDBs may be greater than 1.5 km and practical-sized antenna cannot be resonant at such a low frequency. They are typically 20 m high towers with some form of top loading, such as the one shown in Figures 3.13 and 3.14 reproduced courtesy of Newquay Airport.

Figure 3.13 NDB antenna (courtesy Newquay Airport)

Figure 3.14 Close-up of the top loading of the NDB antenna (courtesy Newquay Airport)

NDB is also classified by their transmission power:

- Low power: 50 W.
- Medium power: between 50 W and 2 kW.
- High power: greater than 2 kW.

The signals from NDB stations are radiated omnidirectionally. Each beacon has a radius of coverage. This figure is the coverage over which the bearing measurements are known to be accurate and they take into account the effects of the local terrain.

3.7.1.5 Shortcomings

NDBs suffer from a number of disadvantages related to their low-frequency of operation:

- Against the background of the introduction of PBN, one of the principal disadvantages of NDB/ADF is its accuracy, ±5%.
- Lightning causes wideband noise generation that peaks in the low-frequency bands [56]. Binns points out that lightning discharges may produce a more powerful signal than the NDB beacon which may cause ADF equipment in the cockpit to point to a lightning storm instead of the beacon [35].
- Normally, the lowest layer of the ionosphere (the D Layer) absorbs low-frequency radio waves. However, after the Sun has gone down, the electron density in the ionosphere decreases. In the D layer, this reduces absorption which can lead to NDB signals being interfered with by distant stations.

- In mountainous regions, reflections can cause interference with the NDB signal causing bearing errors.
- In coastal regions, anomalous propagation of NDB signals can cause bearing errors.
- Finally, reflection off the airframes when the orientation of the aircraft changes can cause bearing **banking errors**.

3.7.2 VOR/DME

In the 1930s and 1940s, if no visual reference was available, pilots navigated using a system of low-frequency, directional, beacons. The beacons transmitted two radio beams slightly offset from each other in the azimuth (lateral) dimension. In technical literature, this arrangement is sometimes referred to as 'squint' beams. One beam transmitted the letter A in Morse code (a dot followed by a dash) and the second beam transmitted the letter N (a dash followed by a dot). Where the beams intersected, a pilot listening to a receiver tuned to the beacon frequency would hear a continuous tone. The pathway thus created could be followed to the range at which the signals could no longer be detected. The system was called **A–N Range** or, just, Range. This system suffered from the following operational drawbacks: to maintain a steady course, the pilot needed to listen to the transmissions continuously, and the system did not support a deviation from predefined airways. A–N Range also had technical limitations: if the antenna on the aircraft was not mounted correctly, or if the signals were reflected off adjacent terrain, it was possible to follow the wrong course. Furthermore, because of the low frequencies employed, 190–565 kHz, the receivers in the aircraft were prone to interference from lightning causing radio interference (often called 'static') which prevented the pilot from hearing the A–N signals [57].

In 1937, the US Department of Commerce demonstrated an improved version of the A–N Range system, called **Visual Aural Range (VAR)**. VAR used VHF radio frequencies to eliminate the problem of atmospheric conditions and to mitigate the problems of reflections off terrain. VAR also had design improvements to allow it to handle crossing airways. However, VAR still did not overcome the problem of the pilot having to listen continuously to the beacon transmissions [57].

In the same year that VAR was first demonstrated, David Luck and Lowell Norton of the Radio Corporation of America (RCA) submitted a patent application, awarded a year later, for an 'apparatus for the timing of periodic events', a system which, when developed, allowed the accurate measurement of the phase of a signal [58]. This system was to form the basis of a navigation system that was superior to VAR, called **VHF Omnidirectional Range (VOR)**, and in 1949, ICAO chose VOR to be the replacement for A–N Range. For consistency with its predecessor, the term 'Range' was retained in the system name; but causing confusion today because none of the systems, A–N Range, VAR nor VOR, measured the distance of the aircraft from the beacons. However, two of the advantages of VOR over A–N Range and VAR are evident from its name. The system operates at VHF which, as was the case with VAR, eliminated the problems of operating at low frequency, and

VOR is omnidirectional allowing the pilot to determine the position of the aircraft with respect to a beacon instead of being obliged to follow predetermined airways. Moreover, VOR does not require the pilot to listen continuously to transmissions. Instead, the navigation information is presented to the pilot on a cockpit display. Hence, VOR offered not only greater flexibility but also improved performance compared with VAR and A–N Range.

3.7.2.1 VOR beacon use and network density

VOR beacons were used to mark the centre point of the start and end of airways/air routes and, depending on the airway length, beacons may also be needed at intermediate waypoints. The spacing of beacons in a long airway is determined by two factors: VOR accuracy requirements and the potential for mutual interference. Early VOR systems were assumed to provide a measurement accuracy of $\pm 5°$. For the purposes of calculating how many VOR stations were required in a network, a safety margin was added and the accuracy was assumed to be $\pm 7.5°$. Assume the width of an airway is 10 Nautical Miles (Nm); therefore, in long airways, to ensure that aircraft always remained within the airway, VOR beacons were required every 80 Nm. There are also criteria specifying how close together VOR beacons may be, based on their potential for mutual interference, these criteria are set out in ICAO (2006). Similarly, ICAO (2006) sets out criteria for the minimum signal levels for a VOR to deliver an acceptable level of service (90 μV per metre or -107 dBW/m^2). VOR can also be located at aerodromes to assist with arrival navigation.

In the literature, it is often stated that VOR beacons also mark convenient locations for hold points that can assist in the regulation of air traffic flow. For example, London's Heathrow Airport used four 'stacks' for arriving aircraft, at Bovingdon, Lambourne, Biggin and Ockham and these were all originally marked by VOR beacons. However, the modernisation of airspace is reducing the importance of holding stacks and this application of VOR is reducing [59].

Considering their applications, technical limitations and the proliferation of airways that resulted from the expansion of civil aviation after the Second World War, VOR requires a large number of beacons to provide the necessary coverage. At their peak, there were, for example, 967 VOR stations in the United States, 44 in the United Kingdom, France had 96 and Germany 58.

A corollary of this large number of stations, some of which may be close together, is the need for VOR stations to identify themselves so pilots can confirm they are using the correct beacon. The identification, or ident, is provided by having the beacon transmit a unique three-letter code in Morse code, for example, in the United Kingdom, the VOR at Ottringham shown in Figure 3.25, is OTR (--- - .-.), Kennedy airport VOR is PKE (.--. -.- .) and the VOR at Charles De Gaulle Airport in Paris is CGN (-.-. --. -.). Some beacons can also interleave the coded ident with pre-recorded voice messages.

A characteristic of VHF propagation is that it only works reliably when the transmitting and receiving antennas are within line-of-sight (LOS), but an aircraft at height, particularly in upper airspace may be within coverage of many VOR beacons. This situation is useful because it gives the pilot flexibility to select the best airway to reach the desired destination. However, it is important that the beacons marking the different routes do not cause mutual radio interference with each other so deciding on the allocation of spectrum to each beacon is an important task. Spectrum considerations influencing the characteristics and use of VOR are discussed below. However, before addressing the technical characteristics of VOR, the current usage of VOR must be discussed.

3.7.2.2 Current status of VOR

From the 1950s until the early part of this century, VOR was the staple method of short-range air navigation. However, PBN concepts are slowly taking over. King stated in a NATS Blog in 2015 [60], '[Commercial Aviation] now almost exclusively relies on satellite navigation'. However, VOR is still used by General Aviation (GA) (light aircraft) but private pilots are now trained in satellite navigation. Indeed, King's 2015 Blog was an encouragement to private pilots to migrate to satellite navigation.

As part of the drive to PBN, VOR is being relegated to a secondary navigation system which can provide a useful cross-check for satellite systems and as a backup for satellite navigation. Many nations are reducing their inventory of VOR stations, but the reduction is deliberately slow. For example, in the United States, the FAA has been working since 2016 to reduce the number of VOR stations from 967 in 2014 to create a Minimum Operational Network (MON) of 308 stations in a project due to be completed in 2030 [61]. The United Kingdom is also reducing its complement of VOR, for example, the Brecon VOR is no longer active, see Figure 3.15.

Figure 3.15 Brecon VOR (courtesy Nichole Gauden Ysgwydd Gwyn Uchaf Farm Deri)

Notwithstanding the reduction in its importance, there is still widespread availability of VOR and it is, therefore, appropriate to consider the effect that wind turbines could have on VOR systems.

The specifications for VOR are included in ICAO Document ICAO Annex 10 Volume 1 Chapter 2: General Provisions for Radio Navigation Aids [62].

3.7.2.3 VOR system types descriptions

Anyone analysing the effect of wind turbines on VOR will find it helpful to understand the different types of VOR beacons and the scope and limitations of their use:

- Conventional VOR (CVOR): the standard VOR system.
- Doppler VOR (DVOR): a more modern version of VOR which uses modulation in a different way from CVOR and is less susceptible to errors.
- Standard Service Volumes (SSV): VOR beacons have different service volumes. The SSV for each type of VOR are set out below. However, it is common for there to be local restrictions in place caused by the local terrain or built environment affecting signal propagation and the performance of individual beacons. These restrictions are published in national aviation documentation and Notice to Air Men/Notice to Air Missions (NOTAMS). An analyst should be aware of any long-term restrictions before predicting the impact of wind turbines.
- En-route VOR: These beacons typically use a transmit power of 200 W. However, the analyst should confirm the figure for a given beacon before reaching any conclusions about performance and degradation of performance. There are two service classes of en-route VOR: low and high.
    - Low altitude VOR (LVOR): The benefit of a low altitude VOR is that, by limiting the highest usable altitude, it is possible to be confident that aircraft will not experience interference from other VOR beacons sharing the same frequency. The traditional LVOR SSV is illustrated in Figure 3.16. However, within the United States, as VOR are withdrawn and satellite-based PBN is

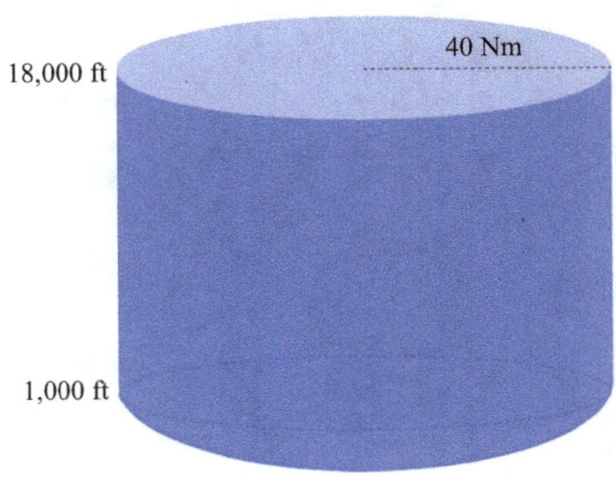

Figure 3.16 LVOR SSV

introduced, there has been a reduced pressure on the RF spectrum and the potential for mutual interference. Hence, the FAA are introducing a modernised SSV as shown in Figure 3.17.

- High altitude VOR (HVOR): The HVOR SSV is illustrated in Figure 3.18. As with the LVOR, a modified SSV is being introduced in the United States consistent with the MON for HVOR; the modified SSV is shown in Figure 3.19.

- Terminal VOR (TVOR): Usually, these beacons have 50 W transmitter power. The SSV of TVOR is illustrated in Figure 3.20.

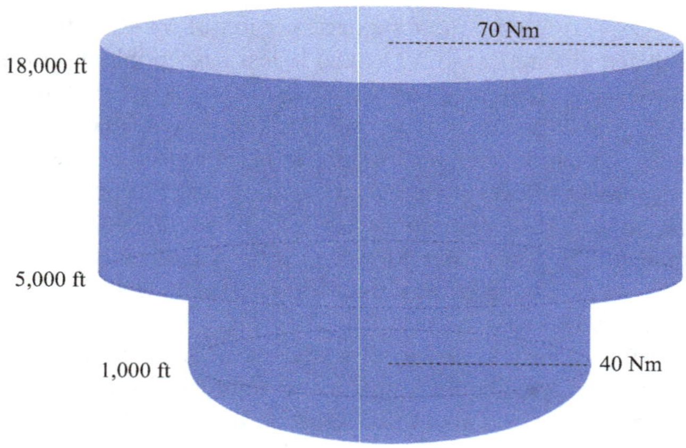

Figure 3.17 Modified LVOR SSV

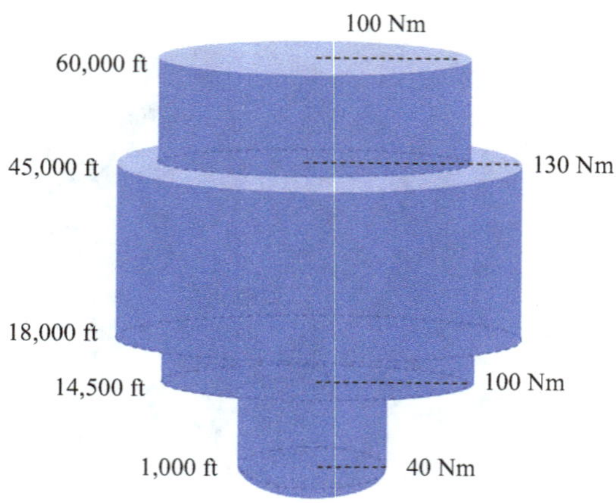

Figure 3.18 HVOR SSV

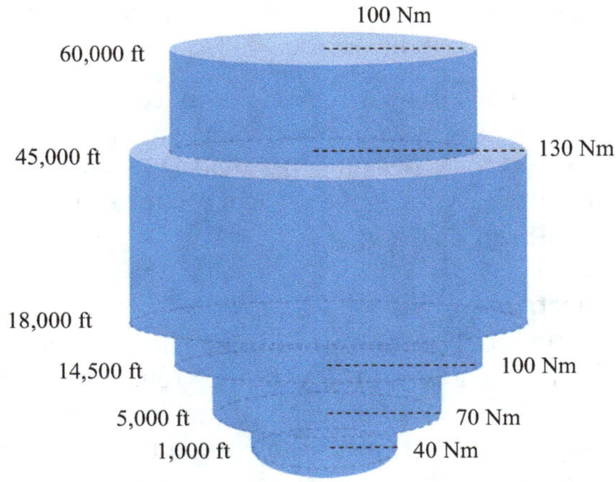

Figure 3.19 Modified HVOR SSV

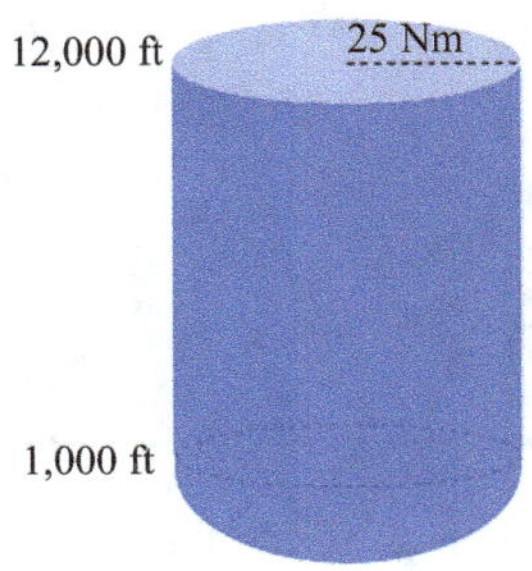

Figure 3.20 TVIOR SSV

- Broadcast VOR (BVOR): A BVOR is usually a type of TVOR that incorporates the Automatic Terminal Information Service (ATIS). ATIS provides information on local conditions at an aerodrome such as weather and weather-related information including temperature, visibility, cloud over, QNH** and transition altitude; active runways available and whether ILS is available. Any warning information such as maintenance work being carried out on aprons and taxiways is also included. This information is interleaved with VOR identification information.
- VOR/DME: VOR can only provide bearing information. To know the location of an aircraft, it is also necessary to know the distance to the beacon. Often a DME system is included which provides an aircraft's distance from a DME beacon.
- VOR/Tactical Air Navigation System (VORTAC): A collocated VOR and (military) Tactical Air Navigation (TACAN) facility.

**Air pressure.

- VOT: A low-powered test VOR system used at some aerodromes to allow aircraft equipment serviceability checks.

Not every type of VOR system is found in every country. For example, in the United Kingdom, all the VOR systems are en-route providing both high-level and low-level service with just one exception, the VOR at Inverness is an LVOR, and there are no TVOR. Whereas, in France, for example, all types of VOR are in service (HVOR, LVOR and TVOR).

3.7.2.4 VOR spectrum

The spectrum from 108 to 117.95 MHz is divided into 200 channels. Within that spectrum, the channel spacing is 50 kHz. VOR uses 160 of these channels: 120 are allocated to en-route VOR and 40 channels are allocated to TVOR. Generally, TVOR channels occupy the lower part of this spectrum, from 108 to 112 which they share with ILS localisers (discussed later in this chapter). Here, the frequencies are allocated as follows [63]:

108.000 MHz	TVOR
108.050 MHz	TVOR
108.100 MHz	ILS localiser
108.150 MHz	ILS localiser
108.200 MHz	TVOR
108.250 MHz	TVOR
108.300 MHz	ILS localiser
108.350 MHz	ILS localiser
Etc.	

En-route VOR can use every channel in the frequency range from 112.050 MHz to 117.950 MHz. These general rules are adapted on a national basis with some countries using what are, nominally, TVOR frequencies for en-route purposes.

3.7.2.5 Beacon identification

The three-letter beacon identification in Morse code is amplitude modulated as a 1020 Hz tone onto the carrier. In some beacons, the Morse code signal is interleaved with a voice recording of the identification and, in others, this is also interleaved with ATIS.

3.7.2.6 VOR theory of operation

VOR beacons transmit signals modulated with 30 Hz tones. One tone, called the **reference,** is modulated in such a way that the phase of the tone is independent of the aircraft bearing. The second tone, called the **variable**, has a phase difference compared with the reference that is dependent on the bearing of the receiver (the aircraft) with respect to the beacon. The two tones are in phase (that is aligned) when the bearing of the receiver from the beacon is magnetic north. This principle is illustrated in Figure 3.21. The symbol in the middle of the figure is that used in aviation charts to identify VOR locations.

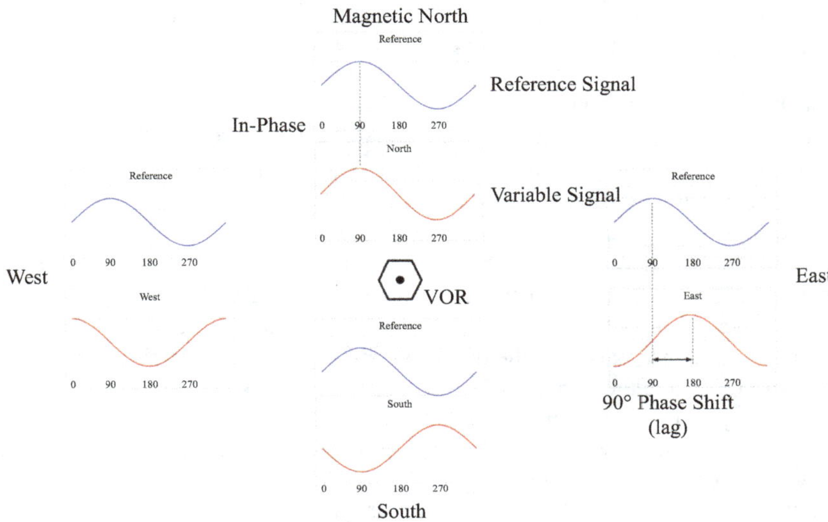

Figure 3.21 VOR principle

Figure 3.21 shows the relationship between the reference and variable signal on four bearings, which are usually referred to in this context as **radials**. The phase shift (in degrees) between the reference and the variable phase signals corresponds to the radial (in degrees from magnetic north) of the receiver in the aircraft that would detect that phase shift. For example, the pilot tunes the VOR receiver to the frequency of the desired beacon, if the aircraft is on a radial due-East of that VOR, the phase difference between the reference and variable components would be 90°. The process for achieving this phase shift is described below. One minor technical problem then has to be resolved. On-board the aircraft, it is the bearing of the beacon from the aircraft that is of interest, not the bearing of the aircraft as seen from the VOR. The correction is simple, 180° is added to the measured phase shift and the resulting number is the bearing of the beacon from the aircraft. Using the above example, the measured phase shift is 90°, adding 180° to this measurement provides the bearing of the VOR beacon from the aircraft, that is 270°.

These characteristics are required to be maintained up to elevations of 40° from the horizontal.

The VOR systems of the 1940s and 1950s used thermionic devices (valves/ tubes) to produce the necessary signals and had to use a mechanically rotated dipole antenna to create the variable signal. When solid-state devices became widely available in the 1960s, what is now called the CVOR was produced and DVOR followed as sophistication of electronics was developed further. Although the underlying principle remains the same, CVOR and DVOR present the phase shift to receivers in different ways. Although DVOR was the later development, it is easier to understand than the CVOR and will be described first. However, before considering either type of VOR system, it is useful to describe a common component of both: the Alford Loop Antenna.

3.7.2.7 The Alford Loop

The Alford Loop antenna was patented by Andrew Alford (1904–1992) in 1942 [64]. An Alford Loop consists of a pair of modified dipole antennas mounted back-to-back. To understand the construction, refer to Figure 3.22. The process starts with a conventional half-wavelength dipole (A in Figure 3.22). The current distribution in the dipole when operational is illustrated by the blue line; there is a current maximum at the centre of the dipole and current is a minimum at the ends. The ends of the dipole are folded (shown in B) and then the centre feed points of the dipole are folded similarly (C). The two halves of the dipole are arranged to form a right-angled dipole (as shown in D); note the current when the dipole is arranged in this manner.

When a pair of dipoles are arranged back-to-back and cross-fed they form an Alford Loop. A plan view of the completed structure is shown in Figure 3.23. Note that each side of the loop has a high current in each of the sides and these are all in-phase creating what is effectively a circular current and creating a circular radiation pattern.

Although the design shown below is a pair of folded 'open' dipoles, the design can be realised with folded dipoles in, what is the more conventional sense of the term, closed folded dipoles.

In summary, the antenna has very useful characteristics for VOR beacons:

- The antenna produces horizontally polarised signals.
- An individual Alford loop produces a circular radiation pattern with constant gain.

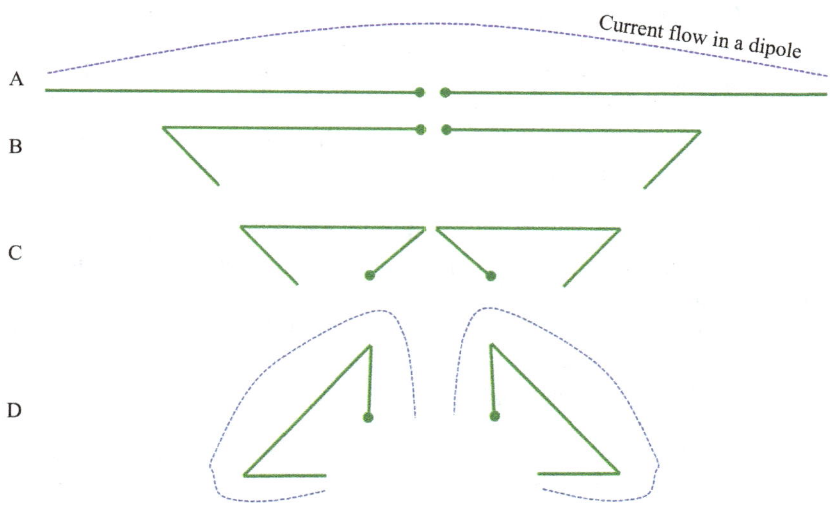

Figure 3.22 Construction of an Alford Loop antenna

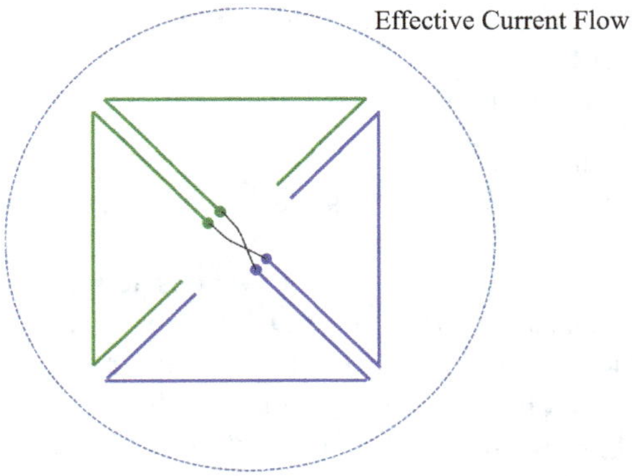

Figure 3.23 The Alford Loop antenna plan view

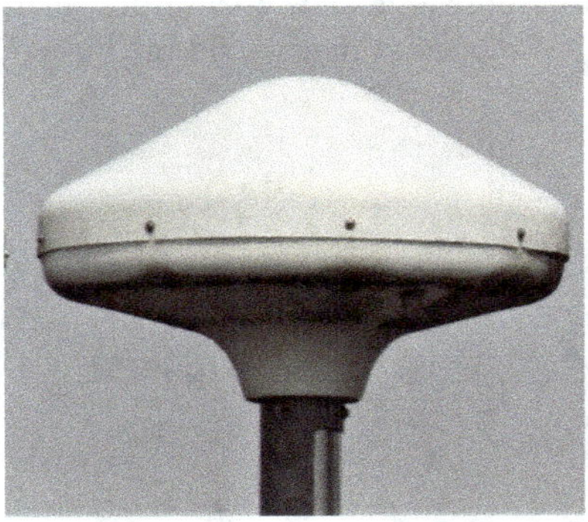

Figure 3.24 Alford Loop antenna radome

- Alford Loop antennas can be manufactured to provide broad bandwidth by making each of the elements from sheet metal, typically aluminium or brass. In addition to broadening the bandwidth, this method of construction makes the antennas physically strong.

Typically, the gain of an Alford Loop antenna is 0.25 dBi. In VOR use, Alford Loop antennas are used as part of an array. It is usual to place the antenna in a radome. The radomes may be individual, which is normal for DVOR antennas. Alternatively, a radome may cover a group of antennas, which is normal for CVOR antennas. The radomes usually have domed or pointed covers to prevent rain from collecting on top of the antenna, see Figure 3.24 [65].

3.7.2.8 DVOR

Figure 3.25 is a photograph of the DVOR beacon site located at Ottringham in East Yorkshire. The beacon marks the centre point of airway Lima 602 where it turns to the South East to head outwards over the North Sea towards the Netherlands. This site is typical of DVOR sites.

The principal components of a DVOR site are:

- The transmitter building.
- On top of the transmitter building is the ground plane or counterpoise. There are two sizes of counterpoise depending on the type of signal transmitted (described below): 45 m diameter, for what is termed the single sideband (SSB) method, and, in the majority of DVOR systems, 30 m diameter for the double sideband method (DSB). The counterpoise fulfils two roles. Sometimes, it is useful operationally to locate the beacons at sites where the ground has a high electrical resistance and does not provide good electrical earth; such as hilltops and mountainous areas, for example, see Figure 3.15. In these locations, the counterpoise provides a good electrical earth beneath the antenna elements. The second role of the counterpoise is to reduce reflections from objects such as trees and buildings or just the terrain close to the antenna, see the illustration in Figure 3.26. The area of the counterpoise is large; nearly 730 m^2 for the DSB DVOR and over 1,600 m^2 for a SSB DVOR. Such large

Figure 3.25 Ottringham DVOR/DME site

Figure 3.26 DVOR ground plane

Figure 3.27 Central reference antenna in the DVOR

expanses of gridwork weigh a great deal and they require a substantial framework to support them to minimise deformation.

- There is a circle of Alford Loop antennas, each with its own radome mounted 1.4 m above the counterpoise. Typically, there are 48 individual antennas in the circle but some manufacturers produce systems with 50 or 51. The diameter of the circle of antennas is approximately 13.5 m.
- In the centre of the circle of antennas is another Alford Loop antenna, this can be seen a little more clearly in Figure 3.27.
- Also visible in Figure 3.27 is the base of a vertical antenna mounted above the central DVOR antenna. This is the base of the DME antenna. DME are often collocated with VOR, though they are separate navigation systems described later in this chapter.

3.7.2.9 DVOR operation

As stated above, the VOR relies on an aircraft on a particular radial being able to work out its bearing from the beacon by comparing the phase of a reference signal and a variable signal. In the DVOR, the reference signal is amplitude modulated by a 30 Hz signal and is transmitted from the centre antenna exploiting the circular radiation pattern of the Alford Loop antenna. The variable signal is produced by the ring of antennas. Each of these variable signal antennas is enabled in turn by a commutator circuit, a circuit that switches the transmitter output from one antenna, to the next and the next in an anti-clockwise direction completing thirty revolutions a second. The arrangement is shown in Figure 3.28.

Viewed from a distance, the signal transmitted by the variable antennas appears to come from a moving source. Given the diameter of the circle of antennas of 13.5 m, and the rotation rate of 30 times a second, the signal appears to be

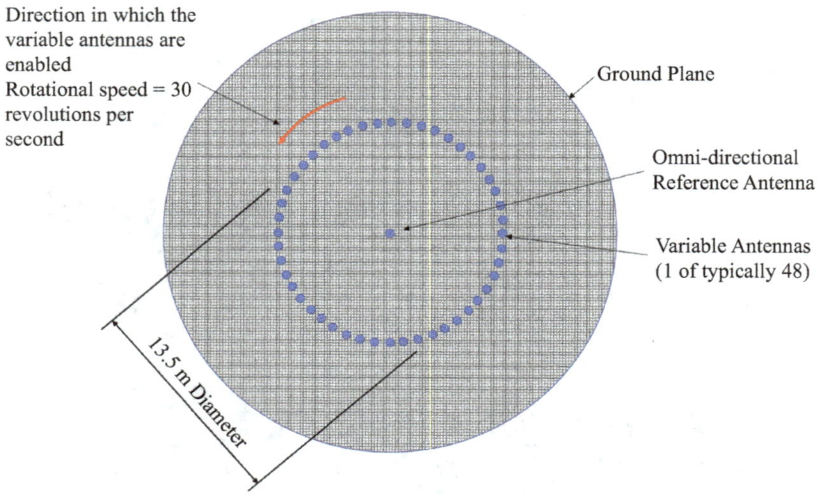

Figure 3.28 DVOR principle of operation

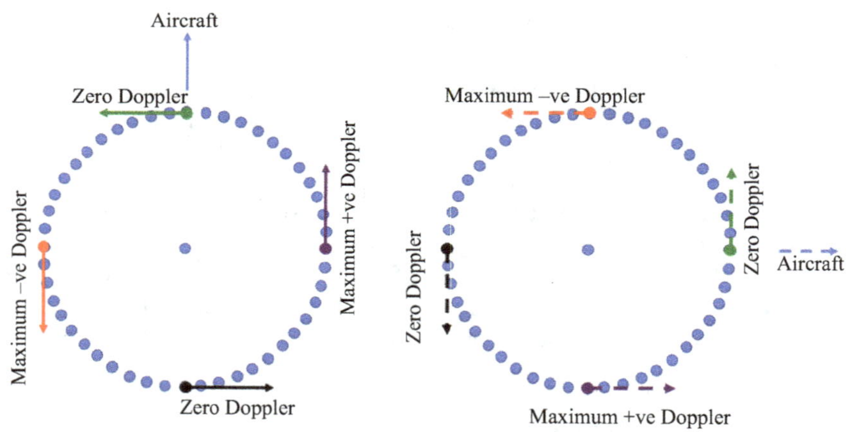

Figure 3.29 DVOR Doppler/frequency modulation

moving around the perimeter of the circle at an equivalent linear speed of 1,297 m/s. At times in the cycle, the signal source appears to be approaching a remote observer (a receiver in an aircraft), for part of the time it is moving away and for part of the time appears to be stationary. This situation is illustrated in Figure 3.29. The motion translates into a Doppler signal with a maximum spread of ±480 Hz. To the remote observer this appears to be a frequency-modulated signal with a maximum deviation of ±480 Hz.

Consider the left-hand side of Figure 3.29. An aircraft is on the radial that is pointing directly towards magnetic north. Assume the green-coloured antenna is the first to transmit (that is the antenna that is in the 12 o'clock position). At this point in the cycle, the signal has no motion towards or away from the aircraft. Therefore, it presents no Doppler shift to the observer and, from the perspective of frequency modulation, there is no deviation from the centre frequency of the transmission. As the subsequent antennas transmit, after a quarter of a cycle, the signal appears to be moving at its maximum speed away from the observer. This situation applies when the red antenna in the diagram, located at the 9 o'clock position, is transmitting. At this point, the signal will present the maximum negative Doppler. As the cycle evolves, another zero Doppler position is encountered when the black-coloured antenna, at the 6 o'clock position, transmits. And so, the cycle continues until it is completed. Figure 3.30 shows the FM variable signal and the AM reference signal that an aircraft on the radial of magnetic north with respect to the DVOR beacon will receive. Note that both signals are in phase indicating that the aircraft is on the radial of magnetic north.

In the right-hand side of Figure 3.29, a second case is illustrated. This time the aircraft is due East of the beacon. The waveforms presented to the receiver in the aircraft are shown in Figure 3.31. In the general case, there will have been cycles

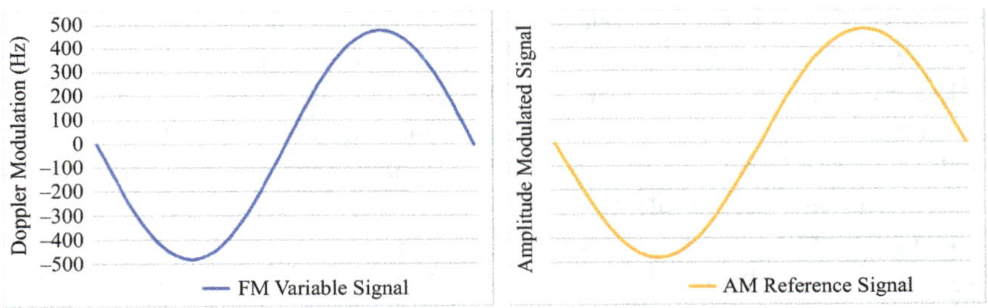

Figure 3.30 FM variable signal and AM reference signal when the aircraft is on the magnetic north bearing

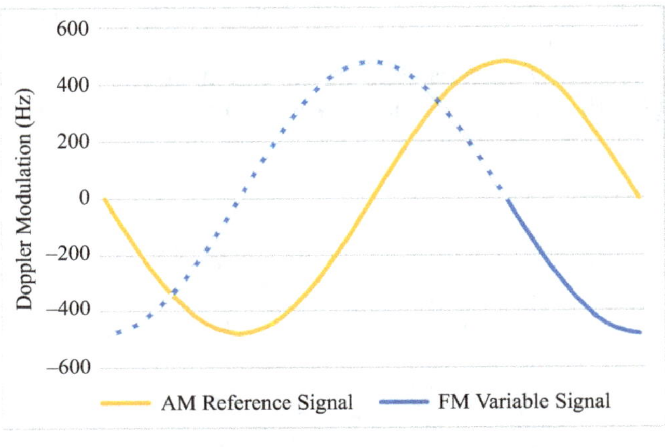

Figure 3.31 DVOR waveforms for an aircraft due East of the beacon

before the time of interest and these will have been detected by the aircraft's VOR receiver. In Figure 3.31, the variable signal waveform produced by earlier transmissions is shown in Figure 3.31 as a dotted line, later the solid blue line starts three quarters of a cycle behind the AM reference (a phase lag of 270°). Ostensibly, this is incorrect. Why this works will be explained following a discussion on CVOR.

In summary, a DVOR uses an AM reference waveform and an FM variable signal and it is organised to provide a variable signal from which it is possible to calculate the radial of a receiving aircraft.

3.7.2.10 CVOR

CVOR systems use four Alford Loop antennas mounted on a counterpoise, similar but smaller than the DVOR's, usually between 6.5 m and 16 m in diameter. The four antennas are mounted in a square formation as shown in Figure 3.32 and, as labelled are usually referred to as the North West (NW), North East (NE), South East (SE) or South West (SW) elements.

3.7.2.11 CVOR operations

The simplest element of CVOR operations to understand is the reference phase signal. In CVOR, this signal is electronically frequency modulated. Like DVOR, the frequency modulation has a maximum deviation of ±480 Hz. The modulated signal is provided, in phase, to all four Alford Loop antennas (NW, NE, SE and SW). The effect is to create an antenna radiation pattern that is circular and horizontally polarised. To prevent any vertically polarised signals being radiated caused by the close proximity of the four antennas, which can cause navigation errors when aircraft bank to change direction, it is usual to mount vertical elements above the Alford Loops. These 'parasitic elements' cancel any spurious vertical radiation.

The CVOR's 30 Hz variable phase signal is amplitude modulated. The modulation is organised so that an aircraft receiver tuned to the beacon frequency sees the signal amplitude peak when the phase difference between the reference and the variable is at the same angle as the aircraft's bearing from the beacon. This situation is achieved as follows.

The modulating (30 Hz variable) signal is split into two components separated in phase by 90° [an in-phase and quadrature (I&Q) pair]. Each component is then

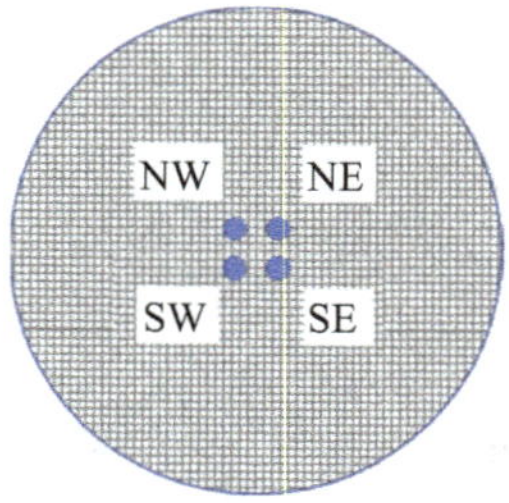

Figure 3.32 CVOR plan view

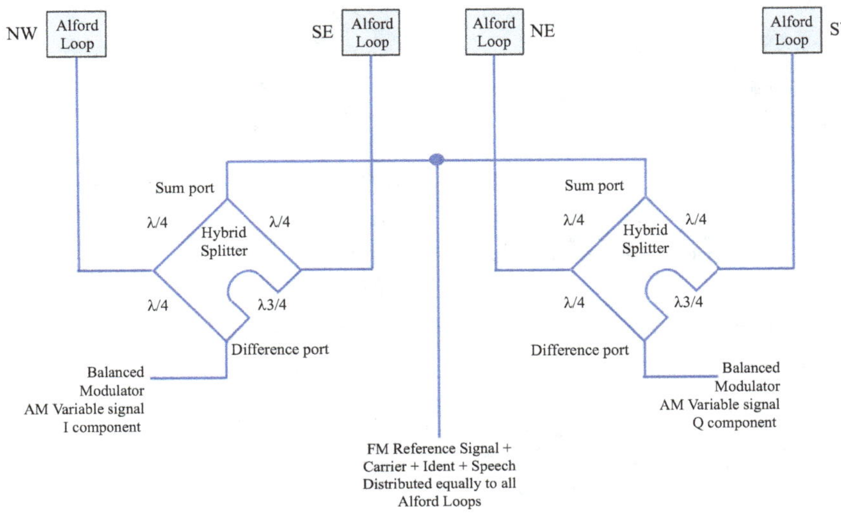

Figure 3.33 CVOR antenna feeds

multiplied by the carrier frequency in a balanced modulator which removes the carrier frequency and leaves an upper sideband of the carrier plus the 30 Hz and lower sideband of the carrier minus 30 Hz. The resulting signals are fed to pairs of Alford Loops via a hybrid splitter. The hybrid splitter is a four-port device with two inputs, a sum port and a difference port, and two output ports. A signal applied to the sum port is distributed to both output ports via a path that is quarter of a wavelength long. A signal applied to the difference ports is distributed to the output ports via one path that is a quarter wavelength and the other port via a path that is three quarters of a wavelength long. The sketch, illustrated in Figure 3.33, was adapted from Lee (2016). A sum port signal is split in two and shared equally between the two output ports and, therefore the two associated Alford Loops. However, because the path lengths to the difference port are different lengths, one half a wavelength longer than the other, the signals on the two paths cancel each other out at the difference port, thus providing isolation between the two input ports.

An individual Alford Loop has a circular radiation pattern. Driving the antennas as a pair creates a figure-of-eight radiation pattern. When driven with an I&Q pair, the pattern rotates. The radiation patterns associated with the different components of the signals are illustrated in Figure 3.34.

The resulting antenna pattern, shown in Figure 3.34, is called a Limacon Curve. A receiver tuned to the beacon frequency would detect a maximum signal strength when the phase of the variable signal was on the radial of the aircraft with respect to magnetic north.

3.7.2.12 VOR spectrum

The spectrum used by VOR is illustrated in Figure 3.35. The figure shows the difference in use by CVOR and DVOR. CVOR was devised first and aircraft

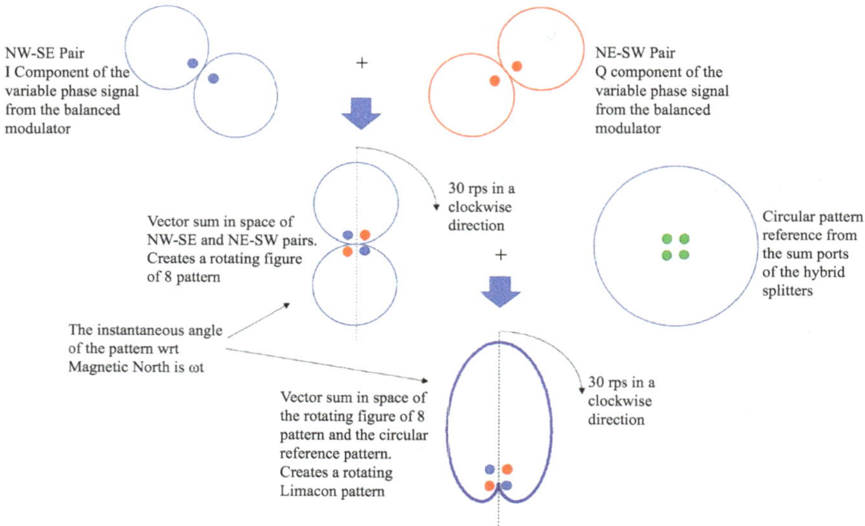

Figure 3.34 CVOR radiation patterns

navigation systems were designed to calculate their bearing by subtracting the variable signal phase from the reference phase. When DVOR was introduced, the use of the signals was reversed, that is the am signal became the reference and the fm signal became the variable signal. To retain compatibility, the DVOR systems were made to rotate in the opposite direction to CVOR.

3.7.2.13 Technical shortcomings of the VOR beacons

Multipath

The radiation from a CVOR is the vector sum of the signals from all four Alford Loops. However, the overall Limacon pattern is formed in space, not within the antennas themselves. If the signal from one of the antennas is corrupted in some way, then the bearing information can be affected with the potential for a navigation error. For example, **multipath fading** can cause this problem. This is the situation where two paths can form between ground and aircraft, one a direct path and the other a reflection of terrain. Multipath fading is described in Chapter 4. CVOR are prone to such problems and siting the beacons to minimise the possibility of this eventuality is important.

Cone of confusion

The antennas of both CVOR and DVOR provide peak radiation close to the horizontal plane. An aircraft approaching the beacon at a constant height appears from the ground to increase in elevation. As the elevation angle increases, the gain of the VOR ground antenna decreases. ICAO (2006) sets a requirement that system performance must remain usable until the elevation angle reaches 40°. Beyond 40°, the antenna gain can decrease to the point where the VOR receiver on the aircraft does not receive a strong enough signal to provide a reliable measurement in the cockpit.

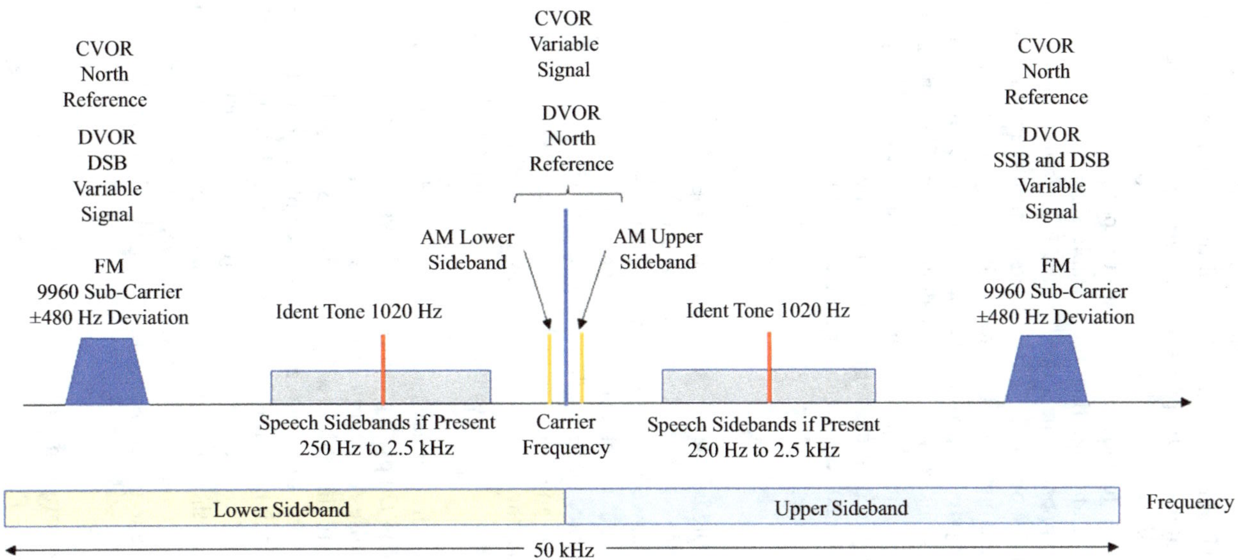

Figure 3.35 VOR spectrum usage

An additional limitation applies to DVOR. The diameter of the DVOR antenna is chosen so that as the different elements transmit the signal, it appears to the aircraft to come from a moving source and, therefore, it possesses a Doppler shift. The Doppler shift translates in the aircraft receiver as frequency modulation. The greater the Doppler shift, the greater the apparent amplitude of the modulating tone. However, as the elevation angle of the aircraft increases, the apparent motion of the signal decreases until the point where if the aircraft were immediately above the antenna, then the variable signals would have no Doppler shift at all. It would appear to the aircraft that the variable signal had diminished to the point where it disappeared. This problem adds to the reduction in the signal level compounding the unreliability of the VOR at high elevations. The region of this degradation of system performance is called the **Cone of Confusion**. The lateral dimensions of this region can be calculated for different aircraft heights. For example, assuming that degradation occurs at an elevation angle of 45° then at 60,000 ft altitude, the diameter of the region of uncertainty (where navigation is unreliable) would be 20 Nm. At an altitude of 18,000 ft, the diameter decreases to 6 Nm.

3.7.3 DME

A shortcoming of VOR is that it is unable to provide the distance of the aircraft from the beacon (either NDB or VOR). DME addresses this problem by providing the slant distance between the aircraft antenna and the ground equipment. Therefore, it is usual, but not essential, to collocate DME with some other form of navigation. For example, DME might be collocated with a VOR system, note in Figure 3.25, the DME antenna is mounted above the DVOR reference antenna. Thus, in this example, the aircraft would be able to establish both its bearing and range from Ottringham. Requirements for DME systems are set by ICAO and other regulators.

DME was derived from a navigation system called Gee that was developed in 1942 by Robert James Dippy for use by the Royal Air Force (RAF). The Gee system transmitted pulses of energy from three ground transmitters, a master and two slaves. The pulse timings allowed a receiver on board the aircraft to triangulate its position by comparing the relative times of receipt of the pulses. Gee had two shortcomings: it could only be used if the transmitter and the receiver was in line-of-sight and it was not very accurate. Buderi [66] reports that it could locate an aircraft 350 miles from the transmitters within an ellipse with a major axis 6 miles long and a minor axis 1 mile wide. However, Gee successfully demonstrated the principle of being able to use pulse timings for navigations and was the forerunner of DME.

3.7.3.1 The slant range problem

It was stressed that the DME provides a slant range, see Figure 3.36. The slant range, problem is discussed in Chapter 6 in the context of integrating information from two, or more, 2-dimensional radars, where no height information is available.

In the context of DME, height information is available from the aircraft's altimeter. Below the transition altitude, described earlier in this chapter, the height above the ground is known accurately, above the transition altitude the uncertainty is greater. The system-level requirement for accuracy is not to exceed an error of 0.6 Nm [67].

Figure 3.37 shows the ground antenna system for a DME transponder (reproduced courtesy of Newquay Airport). To provide the necessary accuracy ICAO (2006) requires the antenna to be 2.5 m above ground level.

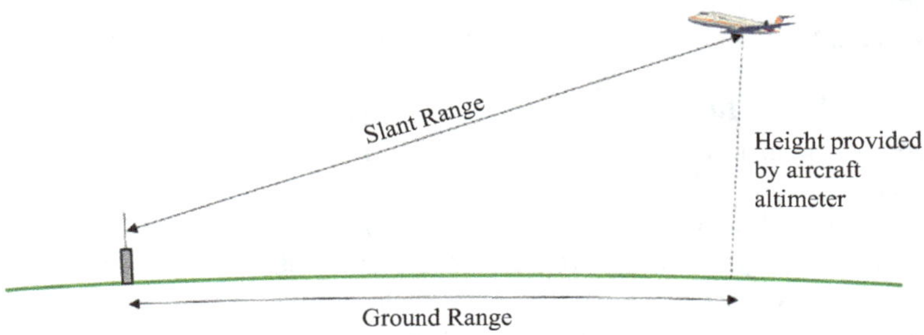

Figure 3.36 Slant range versus ground range

Figure 3.37 DME ground equipment (courtesy Newquay Airport)

3.7.3.2 DME classes

There are three classes of DME:

- Terminal DME: usually located at the aerodromes for which they are intended to be used. These are only required to provide a service up to an altitude of 12,000 ft and a range of 25 Nm.
- Medium altitude DME: these provide a service up to 18,000 ft and a range of 40 Nm.
- High altitude DME: these provide a service up to at least 45,000 ft and a range of 130 Nm.

These classes are served by low-power DMEs (LPDME) 100 W power and high-power DMEs (HPDME) 1 kW power.

3.7.3.3 DME status

DME complements navigation aids that provide a bearing, such as VOR, and these are sometimes collocated, the Ottringham DVOR, Figure 3.25, and the Brecon DVOR, Figure 3.15, are both examples. Ottringham is currently not due to be decommissioned in the near future but Brecon has already been withdrawn. More generally, DVORs are slowly being withdrawn from service and this begs the question, what will happen to the DME systems collocated with a DVOR that is no longer required? In the United Kingdom, CAP 1781 sets out the UK view on the future of DME. It is possible that the network of DME will be rationalised to ensure that a network is available for backup navigation that would allow a minimum of two DME sites to be accessible to aircraft to allow a process of triangulation to be employed [68]. In the United States, the FAA are following a similar process introducing what they call Next Gen DME to support PBN [69]. Implementing these plans may involve the movement of some existing DME stations and the analyst should consider the status on a case-by-case basis.

In summary, DME sites are likely to be retained for the foreseeable future.

3.7.3.4 DME spectrum

DME uses UHF spectrum between 960 and 1,215 MHz. Two operating modes are used: X and Y. If the interrogation is on an X-channel, the transponder response will be 63 MHz lower than the interrogation and if the interrogation is on a Y-channel, the transponder will reply with a signal 63 MHz higher than the interrogation. Airnav.EU contains channel pairings [70].

3.7.3.5 DME operation

The requirements for DME are set out in ICAO (2006). There are two components to DME: the interrogator system on-board the aircraft and the ground-based transponder.

The aircraft interrogator periodically transmits pairs of pulses. Pairs of pulses are used to prevent false triggering if a spurious single pulse is detected. When the pulses are detected by the transponder, replies are transmitted on a frequency of 63 MHz and the aircraft equipment calculates the slant range distance from the ground system using the known propagation time and the assumed transponder reply time.

3.7.4 TACAN

In effect, Tactical Navigation (TACAN) is a militarised version of VOR and DME. Like the civilian systems, TACAN is giving way to satellite navigation. However, because TACAN can be used by both civil and military systems, withdrawing TACAN from use is inevitably made more complex. An overview of the signals used by TACAN can be found in Specification Mil Std 291-C [71]. The main operating principles are the same as DME and CVOR but an overview is provided below.

3.7.4.1 TACAN distance measurement

TACAN's distance-providing capability and DME are directly compatible and TACAN may be used by civilian as well as military aircraft. Hence the interrogation frequency depends on the TACAN system selected and the frequency used by the transponder which has an offset selected in the same manner as a civilian DME system.

3.7.4.2 TACAN Azimuth measurement

The Azimuth-navigation capability of TACAN can only be used by military aircraft. TACAN is more accurate than VOR/DME and is capable of being used in a maritime environment because it is tolerant to being used on a wave-affected platform. TACAN operates in the same UHF spectrum as DME (960–1,215 MHz) which is less susceptible to (but not immune from) atmospheric interference than the VHF frequencies used by VOR. TACAN transmission powers are 10 kW for older systems and 400 W more modern systems (Figure 3.38).

Figure 3.38 High-power and low-power TACAN transponders (images courtesy of John Hogan, Thales, UK)

3.7.4.3 TACAN Azimuth operation

TACAN uses two tones, 15 Hz and 135 Hz, to amplitude modulate the carrier. The 15 Hz tone signal is transmitted using the same principle as CVOR but it rotates clockwise at 15 rps, not 30 rps. Like CVOR, the early TACAN systems used a mechanically rotating antenna; these have now been replaced by an omni-directional antenna surrounded by other, parasitic, elements that can be controlled electronically to switch on an off. This process creates a Limacon antenna pattern that acts in the same way the variable signal works in CVOR. The 15 Hz tone superimposed on 135 Hz, modulation tone, illustrated in Figure 3.39. The amplitude-modulated signals are applied to the pulse pair returns from the range-providing (DME equivalent) element of TACAN.

Thus far, the operating principle is the same as CVOR, the major difference between TACAN and CVOR is that the reference signals are sent out as bursts. Nine bursts are used in each 15 rps revolution of the signal. There is a North reference burst and that is followed by eight bursts, called **auxiliary reference bursts** that allow the receiving system on-board the aircraft to use the 135 Hz references. Each burst uses the DME-like pulse pairs as discussed above.

In other regards, TACAN fulfils the same requirements as its civilian counterparts. A disadvantage of TACAN is that it is an active system that relies on the aircraft making transmissions: this is undesirable if an adversary is able to detect the transmissions.

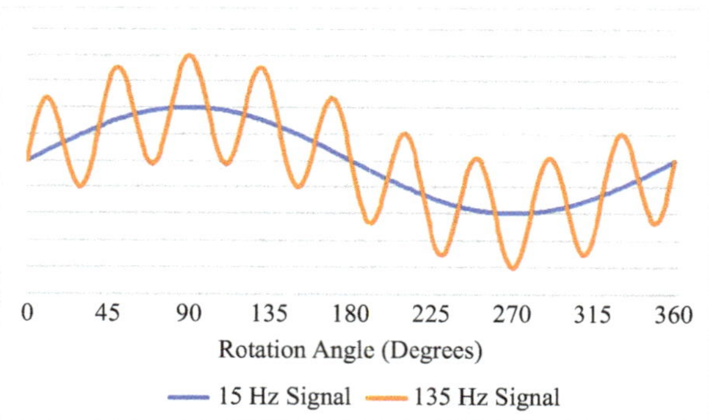

Figure 3.39 TACAN amplitude-modulated waves

3.8 Precision landing aids

There are two types of radio-based precision landing aids, sometimes called blind-landing aids, which facilitate the safe landing of an aircraft in adverse weather conditions:

- Controllers on the ground can use radio aids to determine the precise location of an aircraft so they are able to provide voice instructions to the pilot to land; or the pilots must be given direction information in the cockpit so they can precisely locate the runway. The former strategy is called the ground-controlled approach (GCA) and this is dealt with later.

- The alternative strategy uses an equipment set called the ILS; arguably ILS is actually a system of interdependent systems. This section describes the importance of ILS, using its development as an illustration of this importance; ILS component systems are described and how ILS might be affected by wind turbines.

3.8.1 The development of ILS

From the invention of aircraft until the end of the First World War, the principal focus of aviation development was improvement of aircraft and their performance; after the First World War, the focus until the Second World War was making it safer to fly aircraft [72]. Unusually, the development of ILS spanned both these periods even though the major effort took place after the First World War. The importance of improving the safety of landing may be judged by the number of different technologies that were considered before radio-based systems became the de facto standard. A brief review of the technologies and the reasons for their rejection shows how the requirements for ILS evolved.

3.8.1.1 Lighting

Assuming that the pilot has navigated the aircraft to the vicinity of the aerodrome, the precise location of the runway must be found and the wind direction determined so the aircraft can be landed in the wind. One of the earliest methods to achieve these aims was to locate a powerful floodlight at the end of the runway and point it in the direction from which the wind was blowing so that when landing into the wind the floodlight is not shining into the pilot's eyes. It was found that fog obscures white light but by swapping white lights for neon gas tube lights, their 'rose red' glow could be seen several miles away even in dense fog. By the mid-1920s, the majority of European airports were equipped with neon-based lighting [73]. An alternative solution was to illuminate the airfield and to provide illuminated wind direction indicators, a process that remains in use to this day (Figure 3.40).

3.8.1.2 Leader cables

An alternative all-weather landing aid technology was based on a naval system called 'Leader Cables'. This was first conceived by the Scottish scientist Charles Stephens in 1893 and then developed and patented in 1901 and 1903 by the American engineer Robert Bowie Owens (1870–1940). Conceptually, the leader cable system is similar to the induction sound loop systems installed in public buildings to assist the deaf. In the marine version, the leader cable was laid on the sea floor, the harbour end was fed with alternating current from one terminal of a generator and the remote end of the cable and the other generator terminal were earthed. The resulting current in the leader cable and the surrounding seawater was detectable using telephone-like receivers on-board the vessel which indicated when the vessel was over the cable and when to steer left or right. Cables were installed in Portsmouth (UK), New York and Brest Harbours. This was no small undertaking, the leader cable at Portsmouth harbour, for example, was 18 miles (29 km)

Figure 3.40 Illuminated wind direction indicator (courtesy of Cornwall Airport Newquay)

long. The reason that these aids were dropped was because they were difficult to maintain and tended to get fouled by ship's anchors [74]. In 1922, William Loth, an engineering officer in the French Navy investigated the use of leader cables for aviation; subsequent references to the systems called them 'Loth leader cables'. The leading European proponent of these systems, France, proposed constructing a system for navigation between London and Paris although there is no evidence this was ever completed [75]. The UK Royal Aircraft Establishment (RAE) and US government also investigated variations on the Loth system. The idea appears to have been dropped, in part because of the same maintenance problems found in the maritime environment, and in part because more promising alternatives were coming to the fore; namely, balloons and the use of radio beams [76,77].

3.8.1.3 Balloons

When the RAE abandoned leader cables in 1930, it briefly experimented with the use of high-altitude balloons. The concept, originally tried during the First World War, was to fly the balloon at an altitude that was above fog. The pilot had to fly as close as possible to the balloon and then fly at a constant rate of descent on a predetermined compass bearing until the aircraft reached the ground level. A refinement of this technique proposed by RAE was to use a ground 'proximeter' with the specific intention of making landings less scary for passengers. This device comprised a weight hanging by a thread below the undercarriage. When the weight touched the ground the tension in the thread dropped and a light came on in the cockpit signalling the pilot to flare the aircraft (pull the nose up). This reduces the stress on the undercarriage and made landing more comfortable for passengers.

The value of balloons was easily discounted. The depth of fog changes over time, but knowing the depth is difficult from the ground: the balloons were all too easily obscured rendering them useless. The second premise, a constant rate of descent, was also problematic in the nineteen twenties. At the time, height could only be determined using barometric altimeters, described earlier in this chapter.[††] However, as air pressure does not change consistently with height, especially in fog, a constant rate of descent is impossible and, in tests of the system, aircraft had landed outside the limits of airfields. Moreover, a barometric altimeter is not accurate enough for precision landing; even in good circumstances where it might only be accurate to 6 m. A more subtle but equally important, problem with the balloon system was that some of the limitations could only be experienced when pilot and aircraft were in fog. It follows that pilots could not routinely practice and train on the system and were never able to achieve confidence in the method [77].

3.8.1.4 Radio beams

In 1907, Otto Scheller (1876–1948), a German engineer patented a method for producing two overlapping, equal powered, beams of radio energy expressly for the purpose of guiding aircraft. Nine years later, he improved on the original design and was awarded another patent in 1916. By this time, Scheller had moved to work for the C Lorenz Company in Berlin. The new design was developed by another employee of the Lorenz Company, Dr Ernst Ludwig Kramar (1902–1978), who successfully produced a multiple beam landing system in 1930. The first system, which came to be known as a Lorenz Beam, was installed at Berlin's Templehof Airport in 1932. By 1938, approximately 35 systems had been installed worldwide. Marketing ceased during the Second World War [78].

The principle of the Lorenz Beam system is illustrated in Figure 3.41. Two beams were transmitted, one modulated with Morse code dots, the other with dashes. Both beams were of equal amplitude and operate on the same frequency. A pilot need only have a receiver tuned to the frequency being broadcast. Depending on the position of the aircraft within the beam a stream of dots or dashes will be detected indicating the pilot should fly right or left respectively. When a continuous tone was detected, the aircraft was on the extended centreline of the runway.

Supplementing the direction indicating beams were two marker beams, the first 3 km from the end of the runway, the second 300 m. The beams were pointed

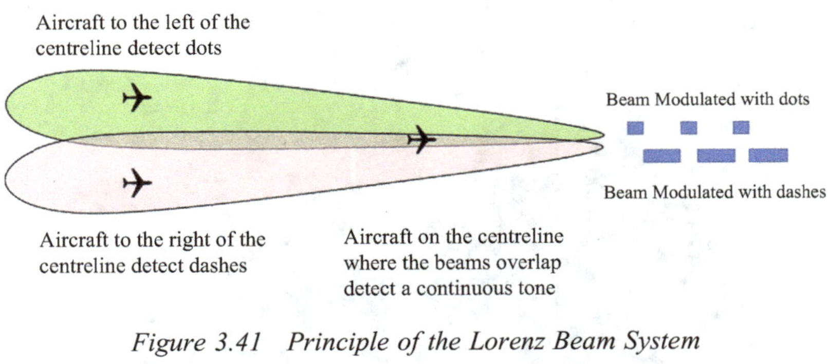

Figure 3.41 Principle of the Lorenz Beam System

[††]Radio altimeters were invented in 1924 by the American engineer Lloyd Espenschied (1889–1986), but it took Bell Labs until 1938 to make the system suitable for use in flight.

directly up in the air and were modulated with a 1,700 Hz and 700 Hz tone, respectively. The beams provided a distance reference for the pilot [79]. The disadvantage of the Lorenz System was that it only provided a bearing telling the pilot the direction of the runway but it provided no vertical information.

3.8.2 ILS

In the United States, a variant of the Lorenz System called SCS-51 was developed and, in 1947, it was accepted by ICAO as the international standard ILS. The benefit of the US system is that it is more accurate than the Lorenz System and offers vertical as well as horizontal guidance.

The principal components of ILS are as follows:

- The Localiser (LLZ) that provides the horizontal guidance (similar to the Lorenz System).
- The Glide Slope (GS) that provides the vertical guidance.
- Marker Beacons or DME; discussed earlier in this chapter.
- Runway visual range (RVR) sensors.

3.8.2.1 Localiser

The critical component of the localiser system is the antenna array. A typical localiser antenna array is shown in Figure 3.42. The antennas are mounted at the far end of the runway as viewed from a landing aircraft as shown in Figure 3.43 and they are normally approximately 300 m from the end of the runway. To prevent them from being an obstacle, the antenna height is usually 3 m, and in some circumstances may be lower.

Localiser arrays use directional antennas; typically, either a Yagi antenna or, as shown in Figure 3.42, log periodic antennas. The use of a directional antenna is an

Figure 3.42 ILS localiser antenna (courtesy Newquay Airport)

important safety feature that reduces the signal transmitted in the opposite direction as part of the wider scheme to ensure approaches are made on the desired runway.

3.8.2.2 Glide slope

There are two types of glide slope antenna: either they have two groups of antenna elements as shown in Figure 3.44, or three groups of antennas as shown in Figure 3.45. The choice of whether to use two or three antenna elements depends on the local terrain.

The glide slope antennas are mounted on towers and, to minimise risks, they are usually mounted 120 m from the edge of the runway as shown in Figure 3.46.

Figure 3.43 Localiser arrays location

Figure 3.44 ILS glide slope antenna (disabled at the time the photograph was taken) (courtesy Newquay Airport)

Figure 3.45 M-Array Newquay Airport (courtesy Cornwall Newquay Airport)

The glide slope antenna is aligned adjacent to the runway aimpoint markings as illustrated in Figure 3.47.

Note that when a pilot requests ILS landing from ATC, the controller informs the pilot of the runway visibility; this information is provided by RVR sensors as shown in Figure 3.48. However, these sensors are not subject to wind turbine interference and so are not analysed further here.

3.8.2.3 Marker beacons

Marker beacons are giving way to satellite navigation alternatives and because of their nature are unlikely to be problematic because of their interactions with wind turbines and wind farms. They are described briefly below for completeness.

The localiser and glide slope systems provide the pilot with bearing information to ensure that the aircraft is on the glide slope and on the extended centreline of the runway. However, these systems do not provide the pilot with any information

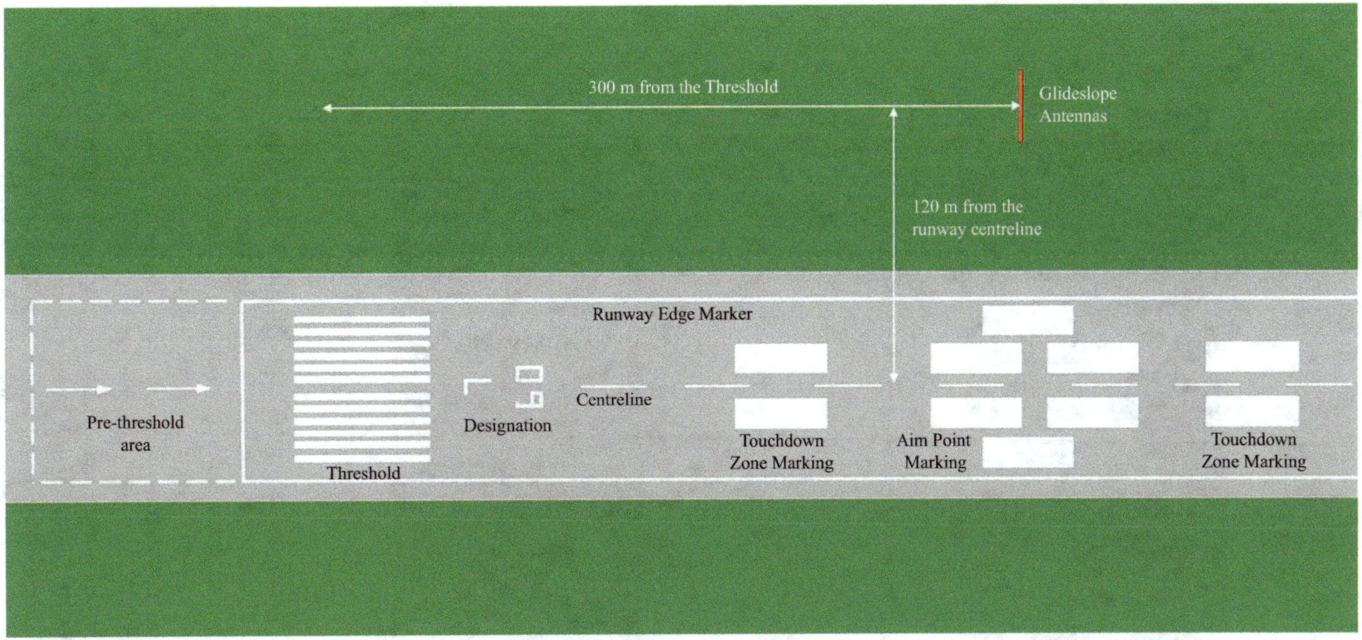

Figure 3.46   Glide slope antenna location

Figure 3.47 Glide slope antenna adjacent to the landing marker (reproduced courtesy of Edinburgh International Airport)

Figure 3.48 RVR sensors (courtesy Cornwall Airport Newquay)

on the distance from the runway. The traditional way to obtain range information is from dedicated beacon transmitters. These beacons point vertically up from the ground and they inform a receiver onboard the aircraft when the beacon has been crossed. Conventionally, there are three markers: outer, middle and inner.

The outer marker was usually placed approximately 5 Nm (9 km) from the runway threshold. The outer marker identifies the final approach fix (FAF) and it may be collocated or replaced with a locator NDB which was discussed earlier. When the aircraft flies over the outer marker, onboard equipment illuminates a flashing blue light in the cockpit and sounds Morse code dashes in a 400 Hz tone. The aircraft should normally be 1,000 ft in altitude above the marker.

The middle marker was usually placed 1 Nm (2 km) from the threshold. When the aircraft crosses the middle marker, an amber light flashes in the cockpit accompanied by the sound of alternate dots and dashes in a 1,300 Hz tone. The aircraft should normally be 200 ft above the marker.

The inner marker, historically known as the Fan Marker and also as the Z marker for reasons that will be explained, has no set location with respect to the threshold. It marks the location where, in certain categories of approach, the pilot must decide whether or not to abort the landing. When the aircraft crosses the marker, a white light illuminates accompanied by either a Morse code s (. . .) or z (- -..), hence the term Z marker, in a 3 kHz tone. The aircraft may still be at 200 ft above the beacon at the inner marker.

3.8.2.4 ILS spectrum

The LLZ and GS systems operate in different frequency bands. The LLZ system uses channels in the VHF spectrum from 108 to 112 MHz. ICAO requirements for the field strength produced by the LLZ transmitter depend on the approach category. These are summarised in Table 3.3.

The GS system uses channels in the UHF spectrum from 329.15 to 335 MHz. ICAO requires that the field strength available throughout the entire coverage should be not less than 400 μV/m. The marker beacons operate at 75 MHz.

3.8.2.5 ILS operation

The LLZ and GS antennas of ILS are both forms of phased array: a network of individual antennas working in concert to produce radio beams. These principles are described later in this chapter when the radar is discussed. What is important to consider is where the beams originate and their information content.

Both LLZ and GS antennas produce a pair of squinted beams, similar in concept to the Lorenz Beam system. For historical reasons, the beams are modulated with 90 Hz and 150 Hz tones on their respective carrier frequencies. The earliest ILS used electromechanical modulation systems that used a type of electric motor, which when running on US mains frequency (60 Hz), rotated at 30 rps. The two tones are 3 × 30 rps and 5 times 30 rps, respectively.

When received on board, the two tones must be maintained at equal levels to keep the aircraft on the glide slope and the extended centreline of the runway (Figure 3.49).

Table 3.3 Localiser transmits field strengths

Condition	Requirement	Notes
Whole of coverage	40 μV/m	
Cat I approaches	90 μV/m	From 10 Nm and 60 m above the threshold
Cat II approaches	100 μV/m	From 10 Nm and 200 μV/m 15 m above the threshold
Cat III approaches	100 μV/m	From 10 Nm and 200 μV/m 6 m above the threshold

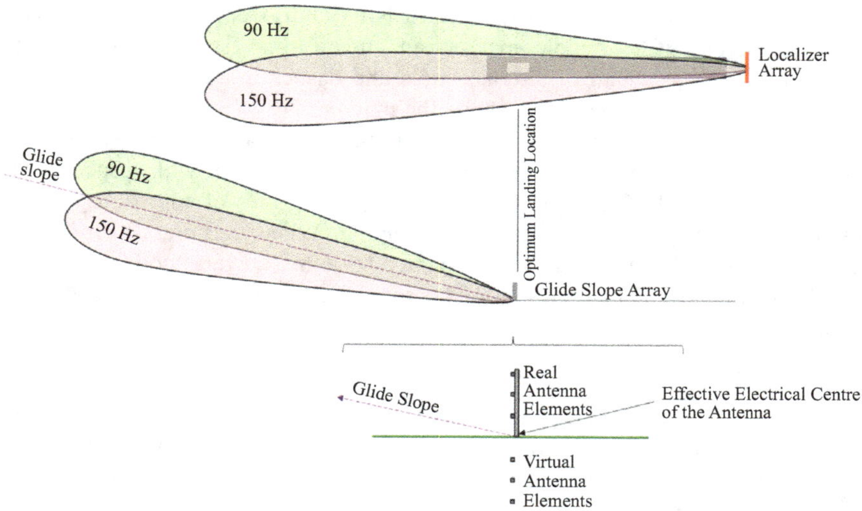

Figure 3.49 Antenna arrangements

A critical component of the GS signal is ensuring that the reflected component of the antenna is effective. The region in front of the antenna is called the **reflection plane**. It is important that this region is free from debris or anything that would degrade the reflection plane. Most GS antennas are located to one side of the runway on a grass field. Even ensuring that the grass does not get too high is important. A routine preventative maintenance procedure of airfield engineers is checking the grass length and ensuring there are no extraneous items present, see Figure 3.50.

3.8.2.6 ILS modulation scheme

Both LLZ and GS systems work in a similar fashion and only the localiser will be described. Holm provides a detailed description of the process for both localiser and glide slope systems [80].

Critical to understanding the operation of ILS is a knowledge of the modulation scheme of the signals transmitted by the antenna array and how the signals are presented to the array. The left and right transmissions observed from a receiver in an aircraft are associated with different modulation tones: 90 Hz to the left of the centreline and 150 Hz to the right. It might be expected that these are created by two separate beams. In practice, the modulation scheme is more complex. Both tones are modulated onto the same carrier frequency. The modulation process creates two separate signals: the first is a conventional amplitude modulation scheme which results in a carrier with two sidebands, one 90 Hz and the other 150 Hz. In the case of the localiser signals, the modulation depth is 20% and it is 40% in the glide slope transmissions. Note that both sidebands are in-phase. This signal is called the carrier and sideband (CSB) signal. The second signal is created in a balanced modulator which has the effect of eliminating the carrier frequency.

Figure 3.50 Glide slope vicinity checks. Courtesy Newquay Airport and Airfield Engineer Marsha Lee

The resulting waveform is called the sideband only (SBO) signal. A key difference between the two modulation schemes is that in the SBO signal, the two sidebands are out of phase. The navigation information is derived from comparing the modulation levels. If the CSB was transmitted without the SBO signal, it would appear to the aircraft that they were on the centreline of the runway anywhere within the coverage of the ILS transmitter/antenna because the modulation depth is constant. The navigation information comes from the CSB and the SBO signals being summed (in space) in the direction the antenna is pointing.

The simplest ILS localiser antenna would be a three-element array, centre, left and right antennas. The centre element is fed with the CSB signal and the left and right antennas are fed with SBO signal. Note that one of the SBO signals is phase shifted by 180°: this is readily achieved using a 180° hybrid device discussed earlier. The results are illustrated in Figure 3.51.

In practice, ILS locator antennas use many more elements than the three illustrated, as shown in Figure 3.44. The use of a larger array improves performance by:

- Increasing gain and reducing beamwidths.
- Greater ability to steer the beam.
- Facilitating better sidelobe control using the process of adding a **taper** is described later in this chapter.

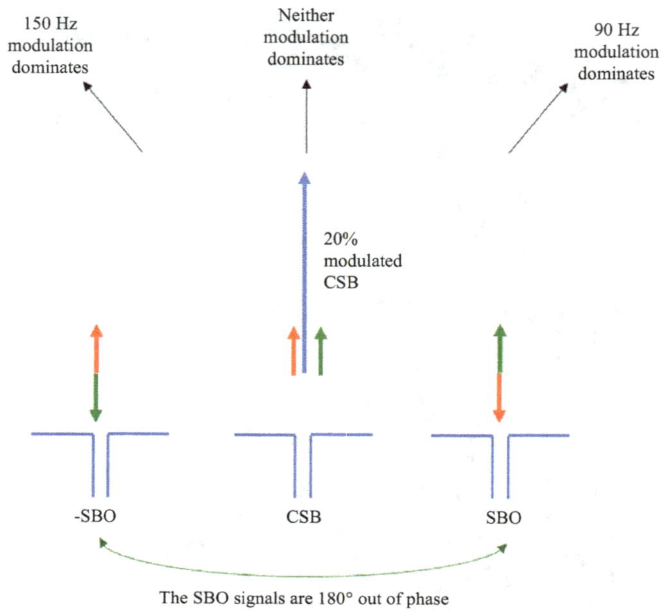

Figure 3.51 ILS operations

3.9 Primary radar

For the purposes of the discussion here, there are two broad classes of radar: primary and secondary. **Primary radars** transmit, generally pulsed, signals and detect echoes from the things illuminated by those signals. Secondary radars, or **secondary surveillance radars** (SSR), transmit a coded signal that prompts a response from transponder equipment on-board an aircraft to reply with information about the aircraft. Depending on the code in the transmission from the ground, the information in the reply may contain information about the flight or just the aircraft identification. The military term for SSR conveys well the original intent and concept, namely **identify friend or foe** (IFF). SSR/IFF is described later in this chapter, this section deals with ground-based primary radar.

3.9.1 ATC

There are three types of air traffic controller: area controllers, terminal controllers and aerodrome controllers [81]. Different types of primary radar are available to support these three activities:

- **Air route surveillance radar** (ARSR) supports area (or en-route) controllers working in ATC centres and provides control of aircraft in airways, air routes and upper air routes. An example of an ARSR site is shown in Figure 3.52.

- **Airfield surveillance radar** (ASR) supports terminal controllers, traditionally working at the aerodromes where the radar is located, who control aircraft leaving an airway and arriving at an aerodrome or departing from an aerodrome prior to entering an airway. In recent years, in the interests of economy, terminal control of some of the smaller regional airports and some military aerodromes has been consolidated in remote locations and, in some cases, one ASR may support control of multiple aerodromes that are close together. Another term that is widely encountered for this type of radar is **primary surveillance radar** (PSR). An example of an ASR is shown in Figure 3.53.

Figure 3.52 ARSR Claxby, Lincolnshire

Figure 3.53 Air Surveillance Radar Humberside Airport, North Lincolnshire

- **Airport surface detection equipment** (ASDE) supports aerodrome controllers generally working in ATC towers located on an aerodrome and, as the name suggests, control aircraft movements on the ground; the radar being useful in fog. An example of an ASDE radar is shown in Figure 3.54.

The military also uses primary radars:

- **Air defence** (AD) radars. These radars provide long-range air surveillance which allows fighter controllers to detect enemy and unknown aircraft entering national airspace and take the appropriate actions. An example of an AD radar is shown in Figure 3.55.

Figure 3.54 Airfield surface detection equipment, Heathrow Airport, London

Figure 3.55 Air defence radar (image courtesy of RAF Boulmer)

Figure 3.56 Military ASR (image courtesy of RAF Brize Norton)

- **Airfield radars.** Military air traffic controllers need to provide terminal ATC services for military aerodromes in the same fashion as their civilian counterparts for airports. The radars used to perform these roles are similar to the civilian radar systems. An example of a military airfield radar is shown in Figure 3.56.
- **Ground-controlled approach** (GCA) radar, also commonly called **precision approach radar** (PAR), provides an alternative form of blind landing aid.
- **Specialised radars**. The military also uses radar for specialised purposes, for example, to engage enemy aircraft with surface to air missiles (SAM). It is also necessary for military personnel to learn how to operate these radars and for pilots to recognise when their aircraft are being illuminated with these '**threat radars**'.

Table 3.4 illustrates some useful heuristic rules; the longer the maximum range of the radar, the greater the time taken to complete a scan. Civil radars tend to be 2-D, providing a range and a bearing but not capable of providing height information. Height information on aircraft is provided by SSR. Whereas air defence radars tend to be 3-D and do not rely (but may still exploit) IFF to determine the height of an aircraft.

It is not proposed to say anything more about ASDE radars. Although these radars are as technically prone to wind turbine interference as the other radars, they have a relatively short range, because they provide coverage of aircraft on the ground at aerodromes and, although they may have some coverage outside the boundaries of the aerodrome it is unlikely that wind turbines/wind farms would be built so close to aerodromes because of other safeguarding measures.

3.9.2 Nomenclature

One aspect of radar nomenclature may be confusing, there is no standardised use of the terms, surveillance and search. What may be referred to as a surveillance radar

Table 3.4 Typical radar parameters [82]

Radar type	ARSR	ASR	ASDE	AD	GCA/PAR
Target information provided	2-Dimensional (2-D)	2-D	2-D	3-Dimensional	3-D
Maximum range	250 Nm (370 km)	60 or 80 Nm (111 or 148 km)	4 Nm (7 km)	220 Nm (410 km)	30 Nm (56 km)
Frequency of operation	L-Band 1–2 GHz	S-Band 2.7–2.9 GHz Military versions 2.7–3.3 GHz	X-Band 8–12 GHz higher resolution models operate in 92–96 GHz	L-Band 1–2 GHz	X-Band 8–12 GHz
Time to complete a scan	12 s	4 s	1 s	10 s	0.5–4 s
Pulse repetition frequency	370 Hz	1,000 Hz	10,000 Hz	Variable depending on height and range	1,200–4,000 Hz
Antenna Azimuth beamwidth	1.5°	1.5°	0.4°	1.5°	1°
Antenna elevation beamwidth	Cosecant squared fan beam, may be modified to enhance higher altitude coverage	Cosecant squared fan beam	1.9°	2° Note modern radars use electronic steering and stacked beams to create the surveillance volume	2°
Peak power	1,000 kW	60 kW	25 kW	50 kW	100 kW

in one forum may be called a search radar elsewhere. Throughout this book, the term surveillance is used for all radars used to detect the presence of aircraft[‡‡].

3.9.3 Primary radar characteristics

This is not a book about designing radars; many fine books have been written for that purpose. There are, however, a number of elements of primary radar design that make them susceptible to undesirable interactions with wind turbines. This

[‡‡]There are types of radar called multifunction radars that can adapt their behaviour as the needs of the operator change. In these radars, the 'search' function looks for objects in a constrained volume of airspace and 'surveillance' functions look for objects in a wide area volume of airspace.

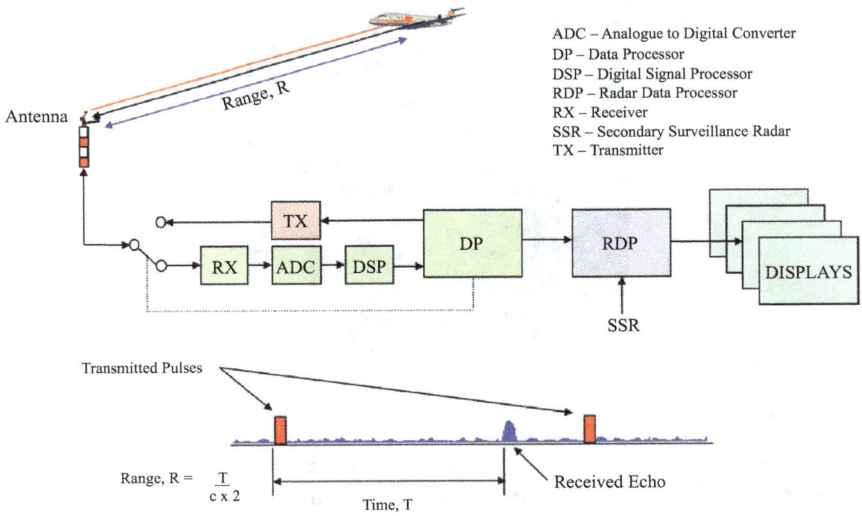

Figure 3.57 Radar principle

section will provide a brief overview of the sub-systems of a radar but it will focus on those aspects of the design that give rise to the problematic performance.

The principal components of a radar are illustrated in Figure 3.57. The radar works as follows. On the command of a timer, a switch is thrown and the transmitter (TX) is commanded to produce a pulse of rf energy that is transmitted by the antenna. In a modern radar, the timing is likely to be part of the radar's Data Processor (DP) components. When the transmission is completed, the switch is returned to the receive position. When the transmitted pulse reaches an aircraft, an echo is radiated and a part of the echo travels back in the direction of the antenna. When the echo arrives at the antenna, it is very weak. From the antenna, it passes to the receiver where it receives several stages of amplification and filtering. At some point in the receiver processing, the signal is converted from an analogue to a digital signal. As radars have developed, digitisation is performed earlier and earlier in the receiver and today processes are carried out using computing techniques that once had to be performed using electronic circuits. Modern radars use computer techniques to extract information about the echo, a process called **plot extraction**. The plot extractor creates a description of each return: its magnitude, range, azimuth, and if available its elevation.

A further processing step in the radar is to work out if the echo is stationary or moving, this is described below.

Reports on aircraft and any clutter (spurious) signals that could not be removed are provided to a radar data processor (RDP) where information from the PSR is merged with returns from secondary radar as illustrated in Figure 3.58. The RDP also adds map data and other geographic data. Some manufacturers have hosted

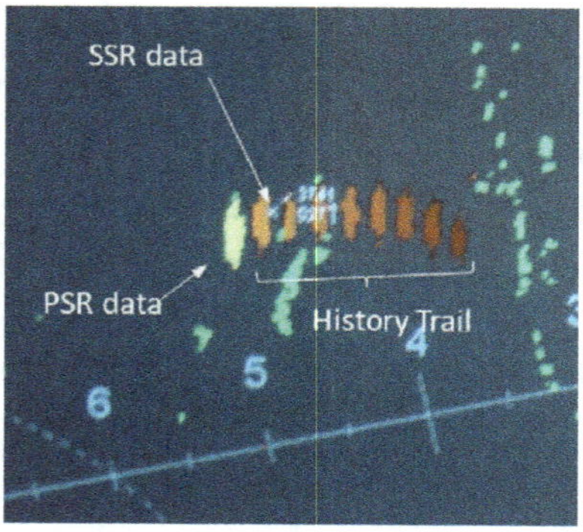

Figure 3.58 Merging of PSR and SSR data

wind farm mitigation in the radar post-processing systems, discussed in Chapter 6. The information from the RDP is made available for display on operators' consoles.

The information on aircraft in the display depends on the radar. If the radar does not use plot extraction, then does not only consist of raw radar returns. If the radar is plot extracted then the only processed target returns are displayed, although sometimes there is an option to display the raw radar returns as well.

Descriptions of those parts of the radar design and performance that are relevant to the interaction with wind turbines and wind farms are split between this chapter and Chapters 4 and 6 depending on which context makes the concept most easily understood.

The first thing to be discussed is the transmitter and, arguably more importantly the waveform that is transmitted.

3.9.3.1 The transmitter and transmitted waveforms

It is easier to understand how a system works (and may be disrupted from working) if the purpose is understood. The purpose of an ATC radar, civil or military, is to provide information to controllers to ensure safe separation distances between aircraft can be maintained. The transmitter has to produce enough power to be able to detect aircraft at the maximum desired range of the radar; this requirement will be discussed in more detail later. But the radar must also be able to resolve aircraft that are close together. There is, therefore, an implicit requirement to transmit pulses of lengths that can provide the resolution. The means of achieving the necessary resolution is affected by wind turbines and, hence, requires some discussion.

Underlying principle

Figure 3.59 illustrates the relationship between pulse length and range resolution; the figure illustrates a pulse, of pulse length τ, at two instants in time, T and $T+\delta T$ (a small increment in time later) and the events associated with the interaction with two aircraft targets. For the purposes of this illustration, it is assumed the aircraft are small in physical length compared with the pulse length.

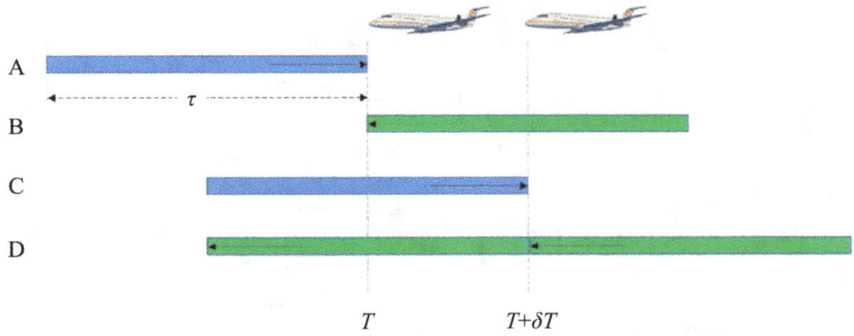

Figure 3.59 Range resolution

Event A occurs at time T when the leading edge (start) of the outgoing pulse arrives at the first aircraft which causes event B which is the target echo starting its return to the radar. The outgoing pulse continues its outward journey and the echo continues on its return path to the radar. Event C occurs when at time $T + \delta T$ it arrives at the second aircraft. Event D is the start of the echo from the second aircraft returning to the radar. If time δT is less than half the pulse length τ, then the return echoes will overlap and the two targets cannot be resolved. Only if δT is greater than half the pulse length, in theory, might two similar sized (RCS) aircraft be resolved.

The minimum allowable horizontal separation of two aircraft is 3 Nm (5.6 km) in the terminal area rising to 5 Nm (9.3 km) outside the terminal area [83]. Consider, for example, the terminal case, and the separation is in range. The time taken for a radio wave to travel 5.6 km at the speed of light = 18.5 μs. From this figure, it is possible to calculate the maximum theoretical length of pulse that can be transmitted and be able to resolve the two aircraft. To meet this criterion, the maximum pulse width can be no greater than, $\tau = 2 \times 18.5$ μs = 37 μs.

This theoretical figure is affected by a number of factors including noise level, target sizes, and the effects of filtering in the receiver. Skolnik [84] points out that instead of $0.5 \times \tau$, a more realistic figure for resolving two equally sized targets is $0.8 \times \tau$. In the above example, to maintain the minimum range separation, this would decrease the maximum possible pulse length to a little over 23 μs.

The figure is also affected by the method used to determine if a target is present and, in some circumstances, described later in this chapter, the maximum length of pulse reduces to 7 μs.

Early radar systems used vacuum device transmitters (such as klystrons, magnetrons and travelling wave tubes) which were capable of producing short pulses, typically of the order of 1 μs wide, at very high power, up to Mega Watts; this combination is ideally suited for ASR and ARSR requirements. However, vacuum devices are expensive and difficult to maintain and there has been a gradual migration to the use of cheaper and more reliable solid-state (semiconductor) transmitting devices. Unfortunately, solid-state devices are more suited to low-power generation and long pulse lengths, what are known as high-duty cycles. The

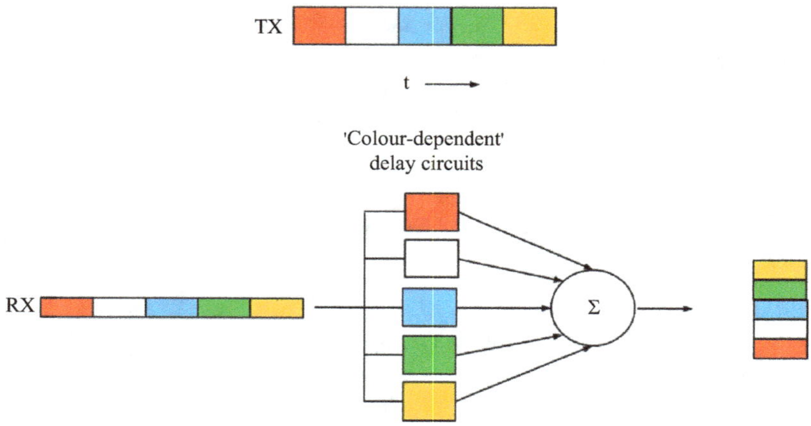

Figure 3.60 Pulse compression explanation

design solution is based on the idea that the energy in the transmission can be maintained using the longer transmitted pulse but encoding it in such a way that the receiver is able to collect energy from the long echo produced but use the coding to aggregate the energy as a short pulse, a process called **Pulse Compression.**

To explain the concept, consider the diagram in Figure 3.60 and the associated 'thought experiment'. Suppose it is possible to encode a pulse of radio waves in different colours. A multi-coloured pulse is transmitted and a little while later an echo from the pulse arrives back at the antenna and is passed to the receiver. In the receiver, the pulse reaches a bank of delay circuits. Each colour is delayed by a different amount and the output from the delay circuits is summed. The result is a pulse that is compressed in time but increased in amplitude. This is the pulse compression concept.

The commonest coding method used today is to change the frequency of the pulse with time, producing what are called **chirped pulses**. There are a number of ways of reconstituting a shorter pulse, for example, some electronic devices conduct signals at different velocities depending on frequency[§§] but it is increasingly common to use computer algorithms to perform the task. Irrespective of the method, all have a disadvantage as illustrated in Figure 3.61. Figure 3.61 shows the waveform produced by compressing a pulse of chirp bandwidth B and pulse length τ. The amplitude of the compressed pulse = $\sqrt{(B\tau)}$ and its width between minima is $2/B$. However, an artefact of the compression process is that time (range) when sidelobes are produced. Without additional processing, these sidelobes will be 13.2 dB lower than the peak return. To the receiver these will appear to be additional targets at fixed offsets in range from the real target, ghost returns. It is possible to reduce the magnitude of these sidelobes (ghost targets) but this incurs a penalty of a lower amplitude return from the genuine target which has a broader pulse width

[§§]Surface Acoustic Wave (SAW) devices

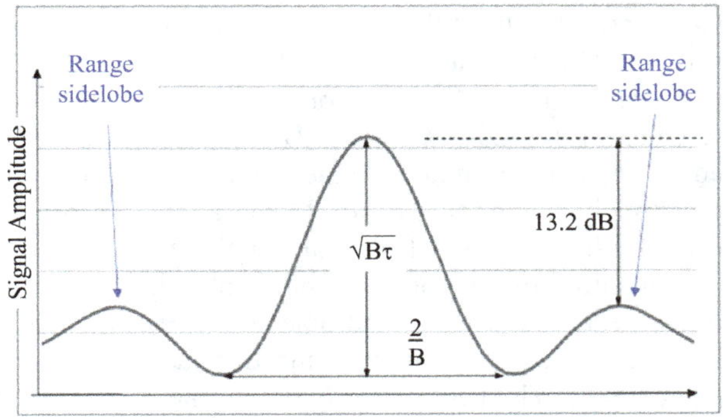

Figure 3.61 Compressed pulse waveform

and, therefore, lower range resolution. The range sidelobes may be tolerable when the target being observed is an aircraft because the target amplitude will not be high and the sidelobe amplitude will tend to the noise level. Chapter 4 will deal with the effect of these artefacts in the presence of wind turbines.

3.9.3.2 The antenna

One of the mitigations that might be considered for reducing the impact of wind turbines and wind farms is the mechanical tilting of the radar antenna [85]. The following discussion outlines the design principles of antennas in the context of the description of radar systems to set the scene for understanding the advantages and disadvantages of tilting which are discussed in Chapter 6.

The discussion starts with the technical requirements for the ASR and the cosecant squared antenna, which is almost ubiquitous in this context.

Aircraft at range from the aerodrome

Figure 3.62 shows a cartoon of an aircraft flying at constant altitude towards a radar. Although the aircraft remains at a constant altitude, at long range from the radar, it appears low in the sky (a low elevation angle) and as the aircraft gets closer to the aerodrome, the elevation angle increases.

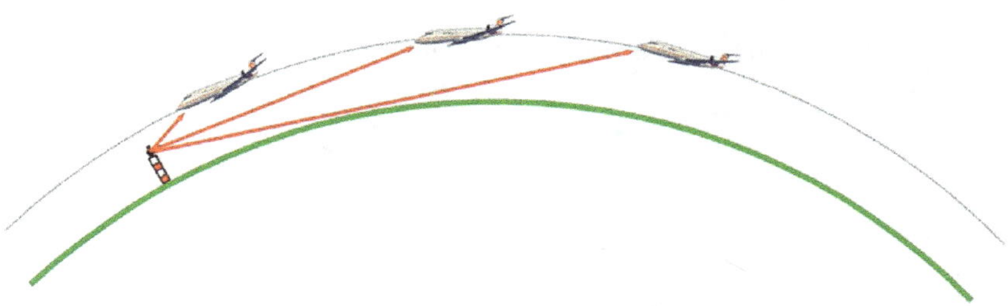

Figure 3.62 An aircraft approaching a radar

As an aircraft approaches the radar, a realistic scenario if the radar is located on an aerodrome that the aircraft is approaching, the range (distance) of the aircraft from the radar decreases and the propagation losses decrease, it follows that the echo power presented to the radar increases. However, it would be helpful if the power presented to the radar was related to the size of the aircraft and its proximity to the radar was not such an important factor. Means are incorporated into the radar receiver to assist with this process but the antenna also plays a role.

The coverage required by the radar is important. Although aircraft will generally approach or depart from an aerodrome along pre-defined corridors, the operator needs wider situational awareness and this mandates a requirement for 360° of azimuth coverage. The geographic extent of the ATC operator's (ATCO) task, the area of responsibility (AOR), is reflected in the maximum range required by ASR radars; indicative values were provided above. The vertical extent of coverage is a more complex matter. Aircraft may approach from and depart to airways but coverage of upper airspace is generally less important. In particular, coverage of upper airspace close to the aerodrome is less important because the controllers at the aerodrome will not, generally, be providing a service to aircraft as they pass overhead. Hence, coverage at height over the aerodrome is not essential. Moreover, while it is important for the operator to be provided with precise height information, this can be provided by SSR and, hence, 2-D coverage is adequate.

The above requirements are satisfied by the cosecant squared antenna.

Cosecant squared antenna principles

Consider first the meaning of the term 'cosecant squared'. A useful starting point is the geometry set out in Figure 3.63. An aircraft is at height H and slant range R which corresponds to an elevation angle ε from a radar. The radar antenna has a gain, G, and the power of the signal received by the antenna is S.

The aircraft height,

$$H = R \sin \varepsilon \tag{3.5}$$

Making R the subject of the equation

$$R = \frac{H}{\text{Sin } (\varepsilon)} \tag{3.6}$$

$$R = H \text{ cosec } (\varepsilon) \tag{3.7}$$

Using the radar equation, described in Chapter 4,

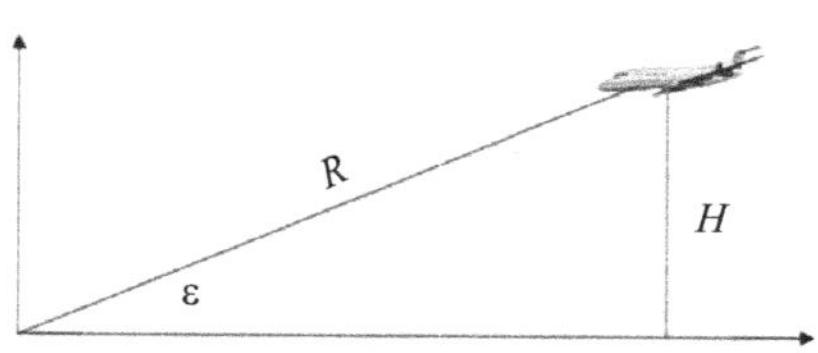

Figure 3.63 Cosecant squared antenna geometry

The received signal,

$$S \propto \frac{G^2}{R^4} \tag{3.8}$$

If the signal power, S, is held constant, then,

$$G^2 \propto R^4 \tag{3.9}$$

$$G \propto R^2 \tag{3.10}$$

Assuming height, H, is constant, and substituting into R for the expression at (3.3)

Then

$$G \propto \mathrm{cosec}^2\varepsilon \tag{3.11}$$

Hence, a cosecant squared antenna is one in which the gain is proportional to the cosecant squared of the elevation angle.

Practical implementations of cosecant squared antennas

In practice, a cosecant squared antenna pattern is created using a double curvature reflector. The antenna has two segments: an upper, hyperbolic-shape, and a lower, more curved segment. Signals from the antenna horn are reflected by the upper half of the radar beam to form the fan-shaped beam, see Figure 3.64. Signals reflected by the lower segment of the antenna are thrown up to higher elevations in such a way that the antenna gain in these directions is flattened to provide a gain that is approximately constant for an aircraft flying at a constant altitude, that is it is a function of the cosecant squared of the elevation angle, see Figure 3.65.

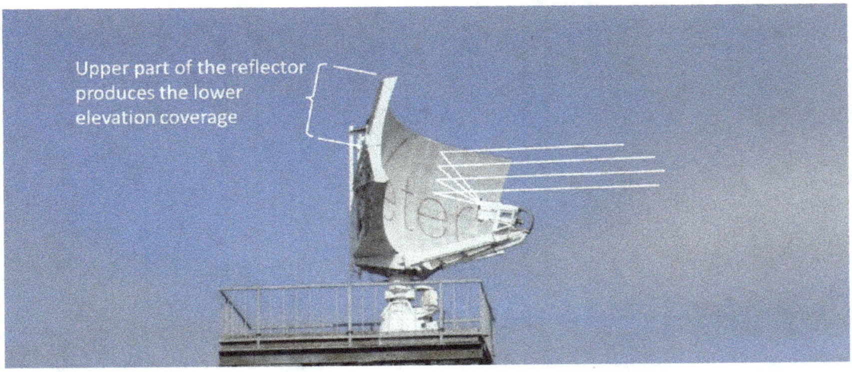

Figure 3.64 Fan beam formation (image courtesy of Exeter and Devon Airport)

Performance

Practical implementations of cosecant squared antennas give rise to a performance envelope that has three performance bands. The boundaries between these bands are the elevations ε_1 and ε_2. The bands are illustrated in Figure 3.66.

- At elevations less than ε_1: The performance is determined by the upper segment of the antenna. This part of the antenna is usually twice the physical size of the lower segment. Therefore, this is the region where the gain is the highest. Maximum gain, which is typically 34 dB_i, occurs between 2° and 3° in elevation. On either side of the peak, the gain falls off. At higher elevations, the fall in gain is compensated for by the lower part of the antenna, of which

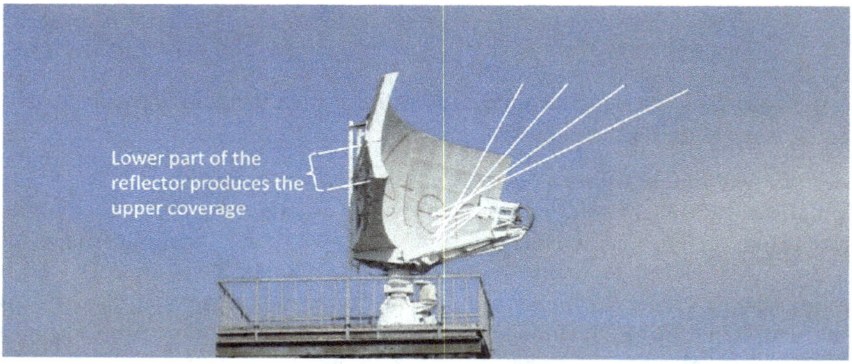

Figure 3.65 Cosecant beam formation (image courtesy of Exeter and Devon Airport)

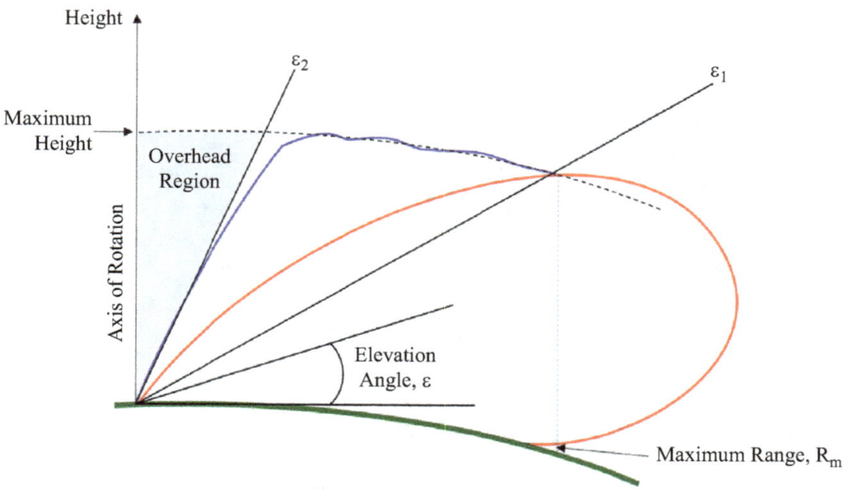

Figure 3.66 Antenna gain of a cosecant squared antenna

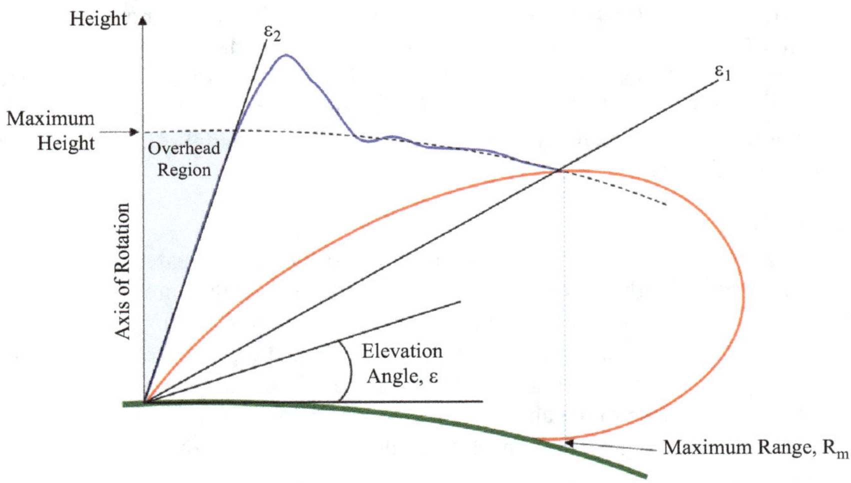

Figure 3.67 Reduced overhead

more in a moment. However, at low elevations that are less than 1°, the antenna gain falls off rapidly. This is important because it is usually this part of the antenna pattern that illuminates objects on the ground such as wind turbines. Chapter 6 discusses this in more detail.

- At elevations between ε_1 and ε_2, the performance is determined by the smaller, more curved, lower segment of the antenna. In this region, the contribution to the pattern from the higher antenna segment decreases but the lower segment of the fan compensates for this loss causing the gain to flatten, which is required to be consistent with the performance required of a cosecant squared antenna.
- At elevations greater than $\varepsilon_{2,}$ that is at elevations greater than 40°, the gain would diminish rapidly which leads to a phenomenon called the **overhead** (or sometimes the **cone of silence)**; the region of airspace above the antenna where gain is reduced and detection is limited.

However, it is commonplace in practical cosecant squared antenna design for manufacturers to modify the pattern deliberately to reduce the overhead; achieved by adding a lobe at high elevations. The effect of this modification is illustrated in Figure 3.67 [86,87].

The double-beam antenna

It is the region of performance close to the horizon that is of interest here. Even though the gain is relatively low, compared with the peak, there can still be useful gain close to the horizon, leading to ground, and objects on the ground, close to the antenna to be illuminated and echoes from that region being presented to the radar receiver. Barton's formula for calculating the distance to the radio horizon is included in the section of this chapter dealing with AGA communications and it is not repeated here. Barton assumes that clutter is uniform in character and uniformly distributed from the antenna to the radio horizon; hence, the name given, **homogenous clutter** [88]. Using this formula, an estimate of the distance to the radio horizon can readily be calculated. There are a number of strategies for dealing with this **ground clutter**: sensitivity time control (STC) and multiple beam antennas. STC is discussed first.

3.9.3.3 STC

The ground terrain and built environment close to the radar create strong echoes that are subject to relatively low transmission losses because they are so close to the radar. It is desirable to minimise the effects of this clutter and one technique is called STC, also known as **swept gain**. STC reduces the gain of the radar receiver close to the antenna countering the effect, between predetermined limits, of the decrease in losses close to the antenna [89]. Note that losses are a function of range from the antenna raised to the power 4, as described in the power equation, discussed later.

STC can be disabled for maintenance purposes and clutter from the ground can be observed without the mitigating effect of STC. This is illustrated in Figure 3.68. Moving out from the centre of the image, there are two collections of ground returns that almost form solid annular rings. Like many ASR, the radar responsible for the picture in Figure 3.68 uses a short pulse, for detecting objects at close range, and a long pulse for detecting objects to the edge of the radar coverage. The two rings of returns correspond to the receiver 'opening' for the echoes detected by the two pulses.

3.9.3.4 Multiple beams

For example, if the antenna is mounted on a tower and the assumed radiation point of the antenna, referred to as the **electrical centre of the antenna** is lifted to 15 m (as a point of reference, this is approximately the average of ASR antenna heights

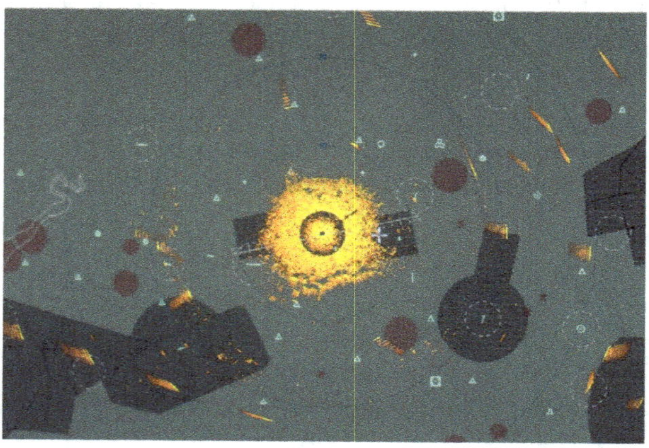

Figure 3.68 Ground cutter without STC mitigation (courtesy RAF Brize Norton)

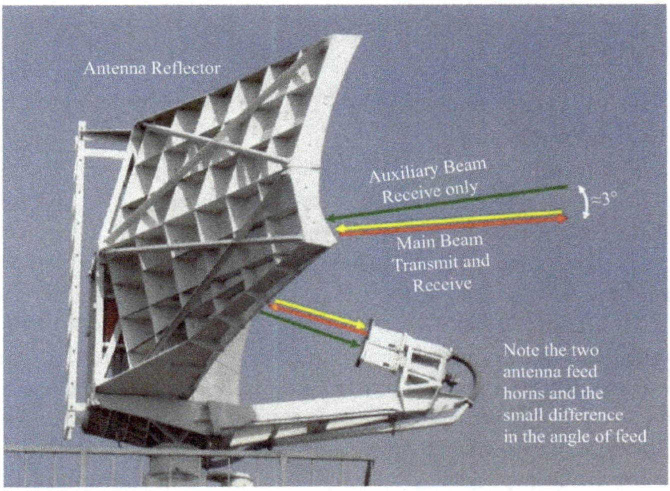

Figure 3.69 Dual-beam antenna (courtesy Exeter and Devon Airport)

in the United Kingdom), then the radio horizon would be located approximately 16 km from the antenna.

Ground clutter echoes may be sufficiently dense that they could obscure the presence of aircraft echoes, degrading the technical performance of the radar and causing potential problems for the operational performance. A means of mitigating this problem is to use two beams. This concept is illustrated in Figure 3.69. In front of the reflector, there is an assembly supporting two components, called **horns**, one above the other. The higher horn and the reflector create a low beam that is used in transmit and in receive. The lower of the two horns is used only in receive and forms a higher beam with the reflector. Following a transmission using the low beam (from the higher of the two horns), the receiver input is taken from the higher beam (lower horn). After a brief time, which corresponds to returns coming from further away from the radar and beyond the radio horizon, the input to the receiver is switched to the lower beam.

Antenna/beam tilt

Barton's model assumed a spherical surface with no terrain. In practice, this may not be the case and, in general, terrain must be taken into account, and the ground clutter model is only a starting point for setting up a radar. It is possible to adjust the angle of the antenna mechanically and tilt the whole antenna up or down. Antenna tilt is discussed in Chapter 6.

3.9.3.5 The receiver

The receiver must amplify signals to prepare them for extracting information about aircraft. The amplification process can be prone to a condition called saturation discussed in Chapter 6. The next stage of processing is to determine whether the echo is stationary or moving.

3.9.3.6 Moving target detection (MTD) and moving target indication (MTI)

An echo from a moving target is Doppler shifted, this property is explained in Chapter 4. The change in frequency can be small and it is convenient to detect the frequency difference via the phase of return signals compared with the signal that was transmitted. This concept is illustrated in Figure 3.70.

A train of pulses (e.g. four) is transmitted in succession. A highly stable 'local oscillator' (LO) is used by both the transmitter and the receiver to ensure that the received signal can be accurately compared with the transmitted signal. After amplification and filtering, the signal is processed in a phase detector and the information is stored until receipt of the next pulse in the train. If the phase has not changed, the report is deemed to have come from a stationary target and it is filtered out. Provided the oscillator is stable, this is an efficient way of filtering out static clutter. The process does have a drawback. If the phase shifts in the echoes between pulses are exactly 360°, or multiples thereof, then the aircraft will not be detected and the radar is said to have **blind speeds**. A second drawback is that aircraft moving perpendicular to the radar's field of view also appear to present no Doppler signal and such targets suffer **tangential fading** [89].

An alternative to MTI is the MTD system. The concept is illustrated in Figure 3.71. In MTD, signals are presented to a **Doppler Filter Bank** via a process called a matched filter that optimises the signal available for further processing. Each filter is tuned to a different Doppler frequency shift. To cater to signals which are not moving and would otherwise suffer from tangential fading, a zero-Doppler filter is also provided.

It is possible to filter out some moving objects from the display. Motor vehicles, ships and trains may all be detected and a moving clutter filter can be applied to the data. Such filters can remove slow-moving vehicular traffic and could, in principle, wind turbine returns but this would be at the expense of removing slower moving aircraft from the display.

Butler investigated the impacts of wind turbines on radars using MTD. A typical MTD filter response is shown in Figure 3.72 [90].

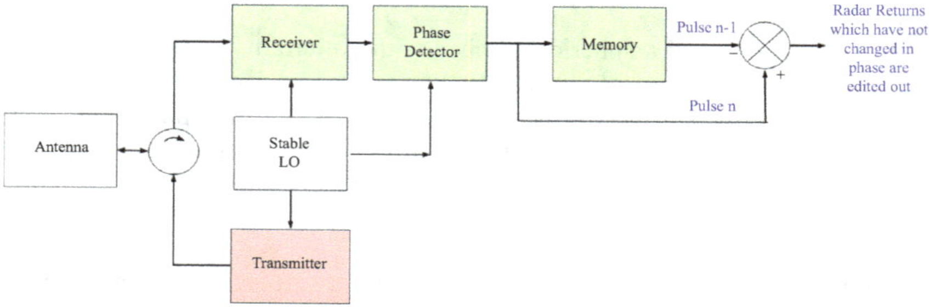

Figure 3.70 Moving target indication

Following the extraction of Doppler information, the next stage is to determine the range of the aircraft and separate it from noise. This is carried out using a Constant False Alarm Rate (CFAR) processor which is described below.

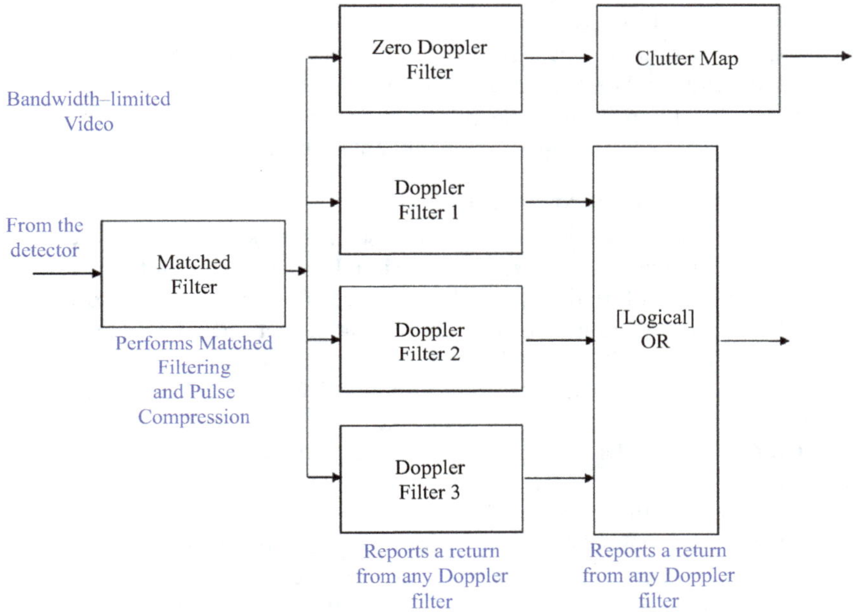

Figure 3.71 MTD

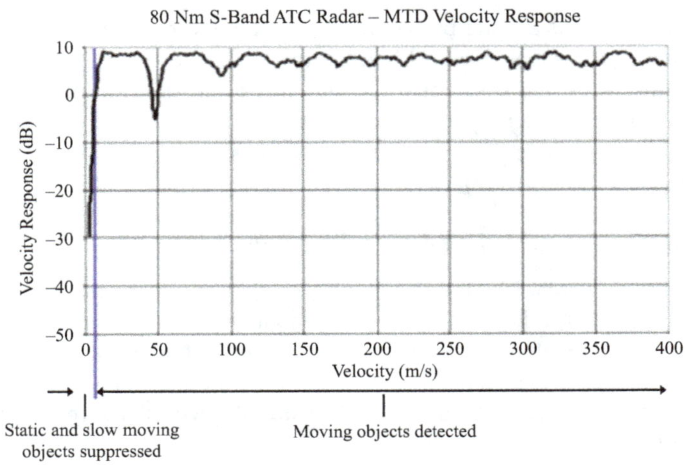

Figure 3.72 MTD filter response after Butler for the Department of Trade and Industry (Crown Copyright)

3.9.4 Detecting the presence of targets in noise

3.9.4.1 Introduction

Most aviation radars in service today, both civil and military, use computer algorithms to determine if target echoes (aircraft) are present in the measurements of the received signal. The performance of these algorithms is inherently problematic in the presence of wind turbines. This section describes the design principles of the detection algorithms. For the purposes of the following discussion, it will be assumed that the only measurements presented to the algorithms are noise or noise (which is always present) plus the echo from an aircraft.

3.9.4.2 Binary hypothesis testing

Whether a measurement value is a result of noise or that there is also a target signal present, is an example of a problem in **Decision Theory** which, in turn, is part of the wider discipline of statistics. In the parlance of statisticians, there are two possible hypotheses; there may be a target present or there may be just noise. The decision to be made is binary and the outcomes are mutually exclusive, that is, there is either a target present or there is not and the process is called **Binary Hypothesis Testing**. Although radar is not the only field in which this problem is encountered, it is considered by many as a good example for a reason that will be described below.

Intuitively, it might be expected that measurements would have higher values if a target is present than when not. However, this may not be the case; sometimes, aircraft may present weak returns and, hence, low measurement values and noise signal levels and measurements may be high or low. It is possible for noise to present a measurement value greater than the measurement value of an aircraft echo plus noise; in which case it is possible for an aircraft echo to be classified as noise and vice versa.

Given that there is a finite probability that the answer may be incorrect, it was widely believed that it was impossible to define a reliable test to prove or disprove binary hypotheses. A few statisticians posited that a reliable test might be possible but only if the measurement quantity on which the decision was made was in some way special. For example, this was the view expressed by the French mathematician Emile Borel (1871–1956).

3.9.5 The Neyman and Pearson Theorem

In 1933, a Polish mathematician called Jerzy Neyman (1894–1981) and a British statistician Egon Sharpe Pearson (1895–1980) jointly published a theorem that provided the basis of a reliable detection algorithm and other binary hypothesis tests. The Neyman–Pearson theorem, sometimes referred to as the Neyman–Pearson Observer [91], is based on a construct called the **Joint Probability Density Function** [92]. This is described later.

In a large number of noise measurements, some measured values will occur more frequently than others (i.e., there is a higher probability that they will occur). And the same applies to target plus noise measurements. The frequency of occurrence of different values is called the probability density. Different classes of signals have different probability densities and the (mathematical) description of the probability density is called the probability density function (PDF). For example, noise detected by most aviation radar receivers will present as a Gaussian¶¶ PDF. Figure 3.73 illustrates an accumulation of measurements presented to the detection algorithm. The abscissa (x-axis) is the measurement value, generally considered to be a voltage measurement, and the ordinate (y-axis) is the frequency of occurrence of the value on the abscissa. The total area under a PDF, as illustrated in Figure 3.74, is the sum of all the probabilities and it is equal to one.

Figure 3.75 illustrates the target plus noise PDF. For convenience, the axes are the same as the noise PDF and the noise PDF is also shown as a blue dashed line. Although both PDF are shown on the same graph, it must be remembered that any individual measurement is a member of either the noise PDF or the target plus noise PDF and the noise PDF. For the purposes of the following discussion, it is assumed that the PDF of a target echo plus noise is also a Gaussian PDF; this

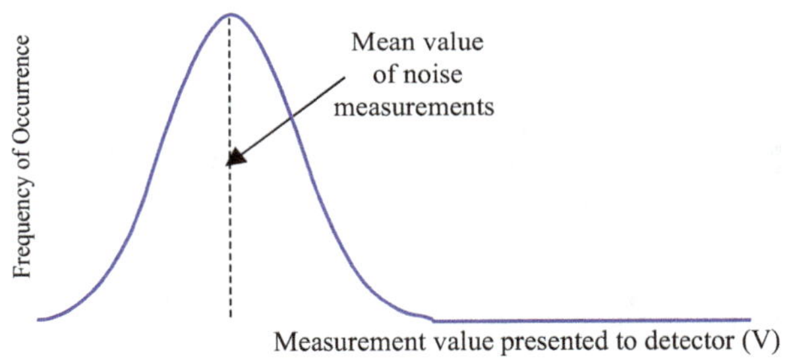

Figure 3.73 Noise PDF

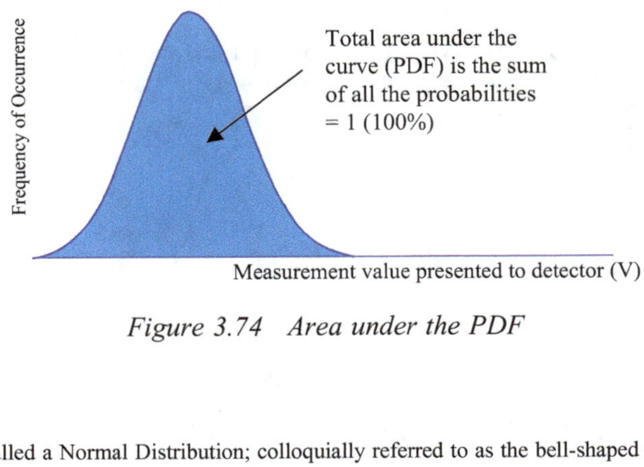

Figure 3.74 Area under the PDF

¶¶Sometimes called a Normal Distribution; colloquially referred to as the bell-shaped curve.

is a simplification and it will be discussed later. The offset in the x-direction of the target plus noise PDF compared with the noise PDF indicates the difference between the mean target plus noise measurement value and the mean value of the noise.

To decide whether a measurement belongs to the noise or the target plus noise, each measurement value is compared to a threshold (T). This threshold setting is illustrated in Figure 3.76.

Before explaining how the threshold value is derived, it is convenient to consider the effect that applying a threshold test will have on a range of measurement values starting with the assumption that the measurements are caused by noise as illustrated in Figure 3.77. Applying a threshold, measurement values less than the threshold will be correctly classified as noise and will not be made known to the operator. For those measurement values that are greater than the threshold, they would be incorrectly classified as target measurements and would give rise to a **False Alarms** which would require additional processing and could be displayed to the operator; in the language of statistics, this would be referred to as the **first error case**. It is important that the proportion of measurements that are incorrectly classified in this manner is very low and that proportion is measured by the **Probability**

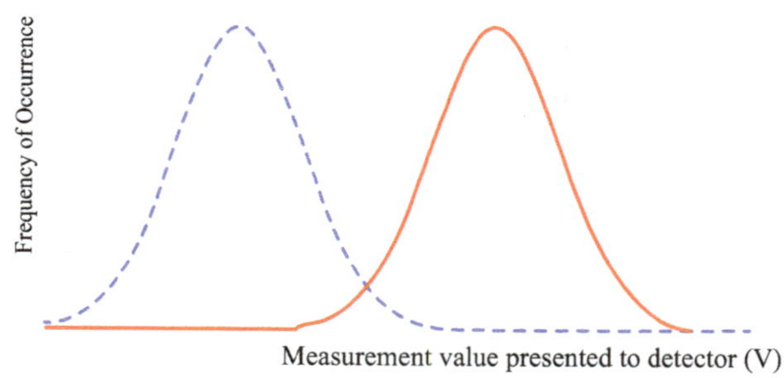

Figure 3.75 Noise PDF and target + noise PDF

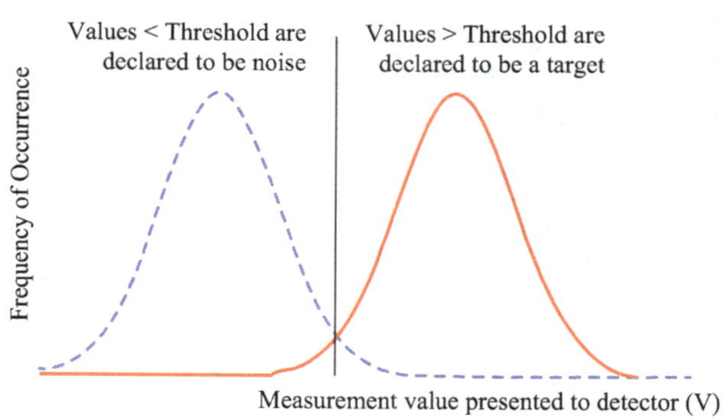

Figure 3.76 Noise PDF, target + noise PDF and threshold T

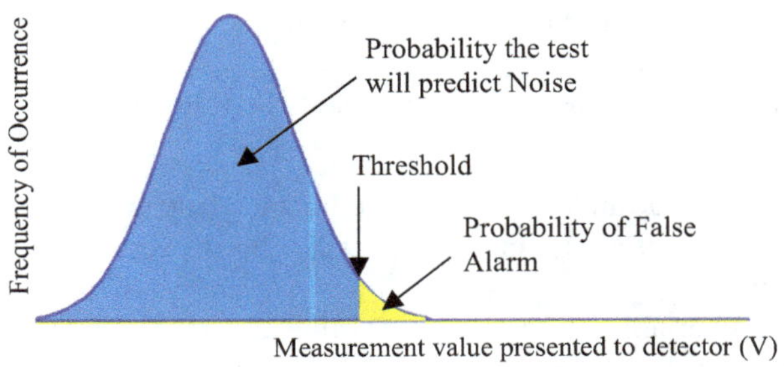

Figure 3.77 Noise measurement test applied

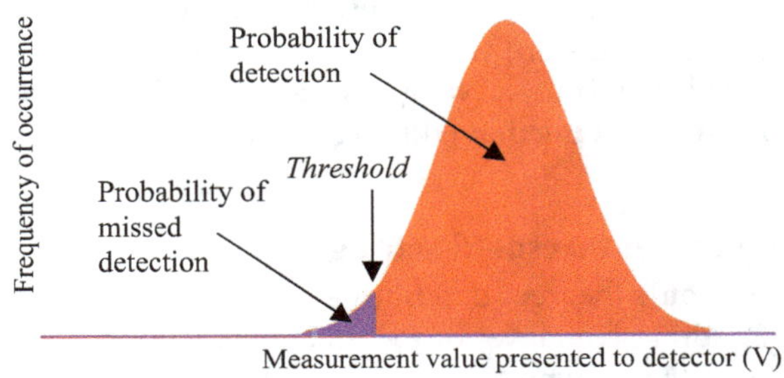

Figure 3.78 Target + noise measurement test applied

of false alarm (P_{fa}). Such is the importance of P_{fa} and its impact on the overall performance of the radar that it is normal for a value to be set as one of the cardinal performance specifications of the radar; a typical value of P_{fa} would be one false alarm in every hundred thousand or every million measurements.

Considering the genuine target measurements, the effect of applying a threshold is shown in Figure 3.78. Measurement values that are lower than the threshold will be incorrectly classified as noise and will be excluded from being presented to the operator. In the terminology of statistics, this is the **second error case**. Measurement values greater than the threshold will be correctly classified as targets; the proportion passing the test constitutes the **Probability of Detection** (P_d). P_d is another cardinal specification for radar systems; the specification takes the form that the radar must be able to detect targets of a specified size and behaviour and range with a 90% (or 95%) probability of detection. A more detailed discussion of P_d is included later in this chapter.

3.9.5.1 More on probability density functions

Before discussing how to choose where the threshold setting should be set, it is helpful to say more about PDF in general and consider the simplifications used so far.

Definition. A PDF maps the frequency of occurrence for a random, continuous, univariate, variable.

Each term in this complex definition means the following:

- The concept of the frequency of occurrence has already been described.
- Measurements of a random variable are independent; which, in the statistical sense, is to say that a new measurement has no dependence on any previous measurement.
- A continuous variable can assume any value as opposed to only being able to take on specific discrete values (e.g., 0, 1, 2, etc.).
- A univariate variable is one determined by a single physical factor.

Even noise, although it is often treated as originating from only a single source, thermal noise, is actually the summed effect of a number of different, independent, sources and in the statistical sense is not univariate||. Therefore, it is clear that the PDF of the target plus noise value cannot be univariate.

For bivariate and multivariate variables, the equivalent function is called a **Joint Probability Density Function** (JPDF) and the Neyman–Pearson Theorem is defined for JPDF [93].

3.9.5.2 The optimum threshold setting

Neyman's and Pearson's theorem describes a means of setting the test threshold to optimise the P_d for a given level of P_{fa}. In the theorem, Neyman and Pearson define a **Likelihood** that a hypothesis is correct which is a function of the PDF of the measurement quantity (noise and target plus noise). The hypotheses are assigned the symbol H_0 and H_1 representing noise and target plus noise, respectively.

Neyman and Pearson's likelihood ratio:

$$\text{Likelihood Ratio}, L(x_1, ..., x_n), \frac{p(x_1, ..., x_n|H_1)}{p(x_1, ..., x_n|H_0)} \geq T \tag{3.13}$$

where $x_1 \ldots x_n$ are an ensemble of representative data points; $p(x_1,...,x_n|H_1)$ is the JPDF of Hypothesis 1 (target is present) being true; $p(x_1,...,x_n|H_0)$ is the JPDF of Hypothesis 0 (target is not present) being true; and T is the threshold.

The Neyman Pearson theorem established that the optimum threshold setting is for the probability of a false alarm to be the same as the probability of a missed detection, see Figure 3.79.

3.9.5.3 Practical implementation of Neyman–Pearson

A characteristic of this scheme is that if the threshold is modified to decrease the probability of false alarms but it also decreases the probability of detection and vice versa (Figure 3.80).

||Noise is often assumed to be thermal (Johnson-Nyquist) noise. But noise is also caused by a wide range of environmental and atmospheric effects.

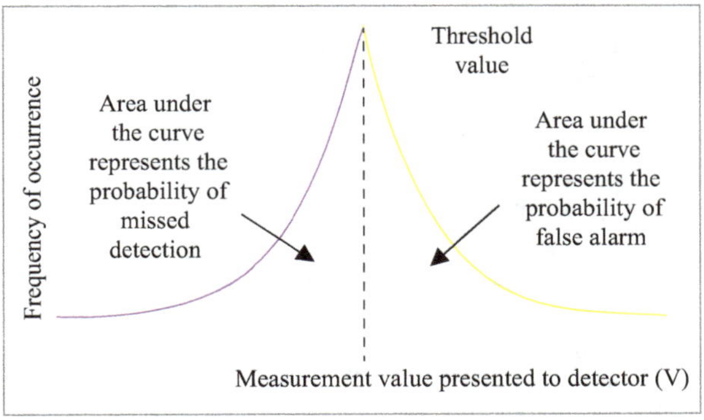

Figure 3.79 The Neyman–Pearson threshold for optimum P_d

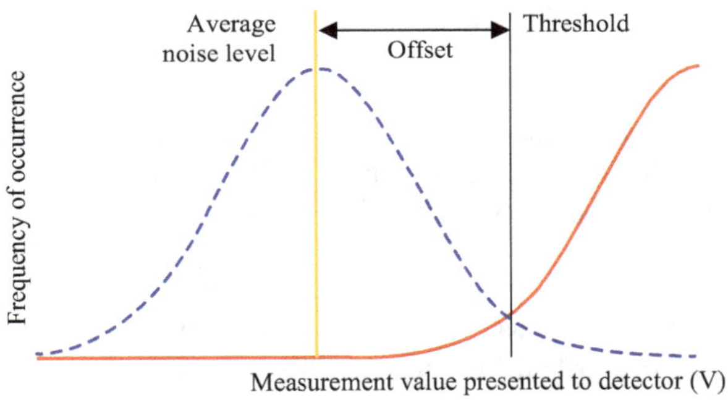

Figure 3.80 Threshold setting

3.9.6 A practical target detector

The Neyman–Pearson detection principle requires an estimate of the average noise level. Finn and Jonson [94] established in 1968 that, without an adaptive threshold, even small increases in the average noise can lead to a large increase in false alarms; specifically, if the noise power density increases by 3 dB, then the false alarm rate increases by four orders of magnitude. Hansen [95] corroborated these findings 5 years later. A practical means of meeting this requirement is the CFAR processor which uses the empirical observation that there are relatively few genuine targets and the majority of the time the radar observes noise.

A CFAR processor works as follows. After the radar pulse has been transmitted, the receiver is opened and the longer the receiver is open echoes from farther away can be detected. The received signal is sampled repetitively using an analogue to digital converter (ADC); each sample is stored and then the next sample is made. The time interval between samples must be constant and, therefore, each sample corresponds to a measurement from a greater distance. It also follows that the time between samples corresponds to an increment in range and

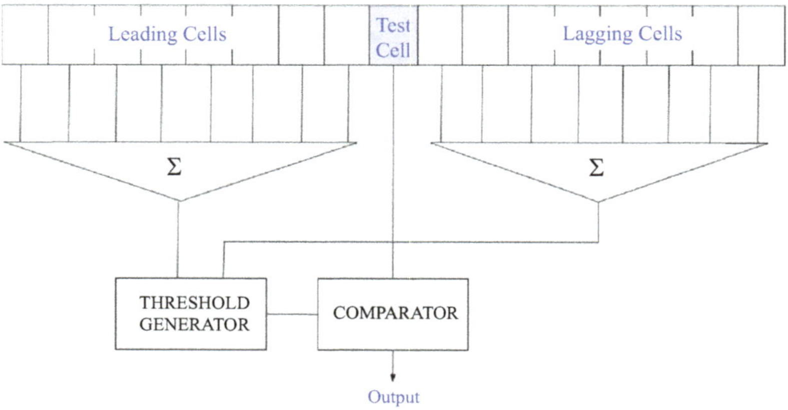

Figure 3.81 A CFAR processor

each sample taken is representative of the signal level within that period. There are a number of ways in which sampling can be performed but it is normal for the sample to represent the maximum signal level that occurred in the period. The name given to each sample is either a **range cell** or a **range bin**. In the majority of cases, the value in the range cell will be a sample of the noise level and if the values in a number of cells are averaged, the resulting value will be the average noise level over the period of time when the range cells were created; this gives rise to an alternative name for this type of processor, viz. the cell averaging CFAR (CACFAR). Using this information, a processor can be configured as shown in Figure 3.81.

Range cells are being filled with the values produced by the ADC; they are shown in Figure 3.81. There is a cell that is being tested to find out if an aircraft might be present. On either side of the test cell, there are range cells from before and after the value in the test cell was received. The values in the leading and lagging CFAR cells are averaged to determine the average noise level around about the time when the test cell was measured. A performance metric for CFAR processors is the probability of detection (P_d) and this is a function of the evaluation of the average noise level. Nitzberg showed that as the number of range cells included in the averaging process is increased, then the better the evaluation of the average noise level, the lower the CFAR loss and the higher the P_d[***]. However, Nitzberg also showed that the benefit is non-linear and the improvements with more than 32 cells in the averaging process are marginal [96]. In Figure 3.81, eight range cells are shown on either side of the gap cells.

It may be the case that if there is an aircraft present it may not be exactly aligned with the range cells as depicted in Figure 3.82.

[***]Subject to the caveat that it is assumed the environment is homogeneous.

To cater to this possibility, the range cells on either side of the test cell are discounted from the averaging process. These cells are called **guard cells** or **gap cells**. Note that in the example shown in Figure 3.82, it has been assumed that the range cells represent a range difference of 50 m. This value is typical of many older ASR in service at the time of writing. The trend is for the range cell size to decrease.

Based on the average noise levels present on either side of the test cell and the gap cells, an offset is calculated and this is used to determine whether an aircraft is present in the test cell range.

3.9.6.1 CFAR operations

To illustrate the operation of a CFAR processor, a simple mathematical model has been built and the results are illustrated in Figure 3.83. The graph in the figure presented shows the abscissa (x-axis) is the range cell number starting 16 cells before the test cell (cell 0) and there are a further 16 cell beyond the test cell.

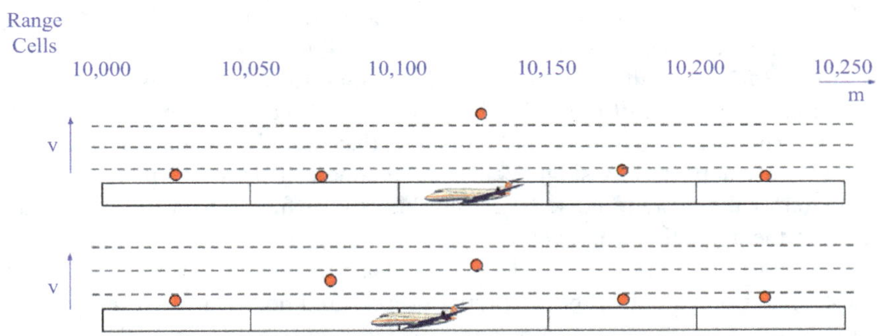

Figure 3.82 Target straddling range cells

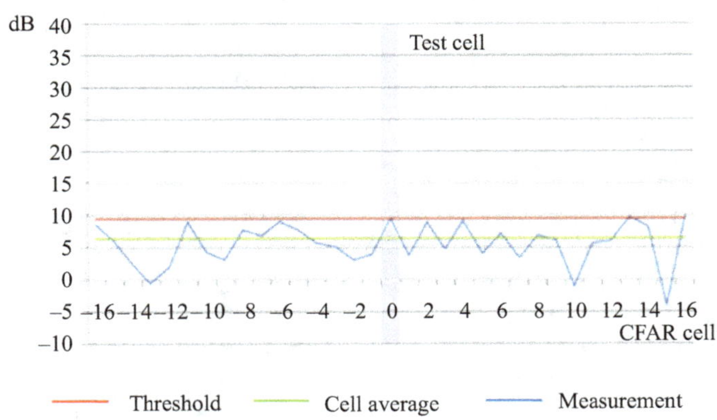

Figure 3.83 CFAR operation

The ordinate axis is the signal level measurement in arbitrary decibel units. The value presented in each cell is a modelled random variable representing a noise measurement. Note that for simplification, the gap cells have not been represented and targets are not represented with any straddling between adjacent cells. The green horizontal line (the lower of the two) is the average noise level computed in the conventional way:

$$\text{Average Noise Level, ANL} = (\sum_{m=1}^{M} q_m)/\text{M} \tag{3.14}$$

where M is the number of CFAR cells; q is the measurement in the mth cell.

The red line (the higher horizontal line) is a simple 3 dB offset from the average noise level which is used as the threshold for detection. Figure 3.83 shows that in this run, the value in the test cell did not exceed the threshold and no target was declared.

In a second example, illustrated in Figure 3.84, the model was modified to increase the noise level. The noise level was doubled. Note how the average noise level has increased by 3 dB and the threshold has also risen by 3 dB. No detection has been announced because the value in the test cell does not exceed the threshold.

Figure 3.85 shows that the noise level has been restored to the levels present in Figure 3.83 but an aircraft return has been introduced; note the higher signal level in cell 12. The feature to note from this result is that even though there is an aircraft present it only appears in a single cell which is one cell out of 32 included in the calculation of the average noise level and this is insufficient to make a large difference to the average noise level.

The next figure, Figure 3.86, illustrates what happens when the target reaches the test cell. It drops out of the averaging process and note that the average noise level is restored to exactly the same value as that shown in Figure 3.83. The value in the test cell now exceeds the threshold and a detection is declared.

Thus far, it has been assumed that the only signals present are either noise or noise plus a target. But it is also possible that clutter will be present. For example, if

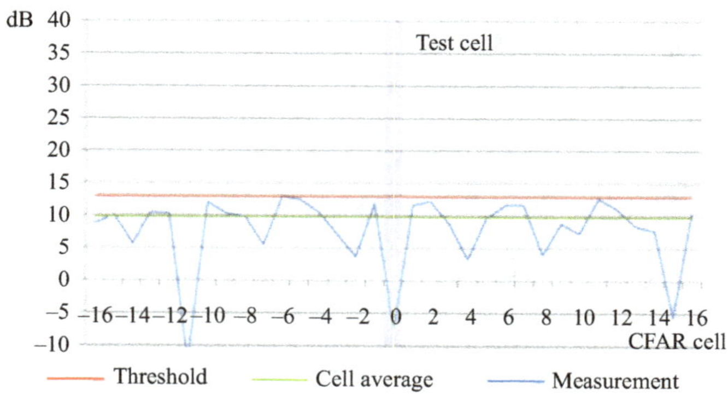

Figure 3.84 CFAR operation increased noise levels

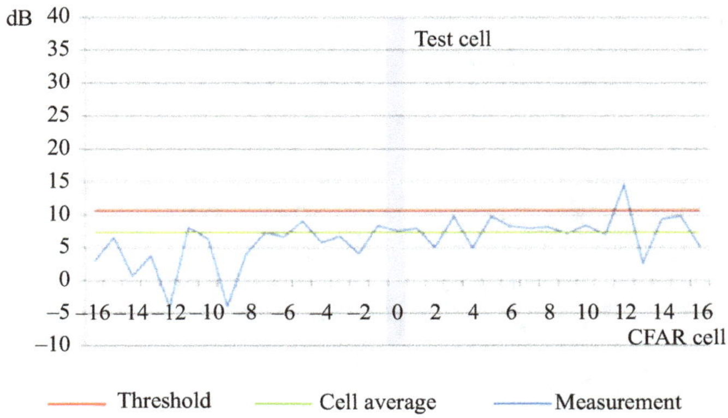

Figure 3.85 CFAR operation in the presence of a target

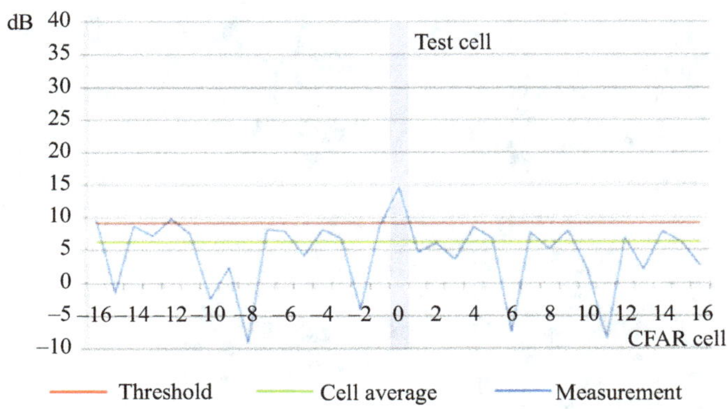

Figure 3.86 Target present in the test cell

the radar beam is grazing the ground, there may be regions of ground clutter present and also heavy cloud can result in high levels of backscatter which increases signal levels observed by the radar. Such features highlight a problem with the standard CACFAR. Regions of clutter such as clouds and ground clutter can be extensive but also well defined, in the sense that there are regions where the clutter is present and regions where it is not. This situation is illustrated in Figure 3.87.

The model was modified so that half of the ensemble of returns measurements are from a region where there is a higher-than-normal level of backscatter, for example, from a cloud. The remaining cells are representative of returns from normal levels of noise. Because a large number of cells are affected by the higher levels of backscatter, then the average noise level is increased but it is weighted by the lower levels in the other half of the ensemble. Thus, when the edge of the clutter region reaches the test cell, a detection is declared. This problem is called the **clutter edge problem**.

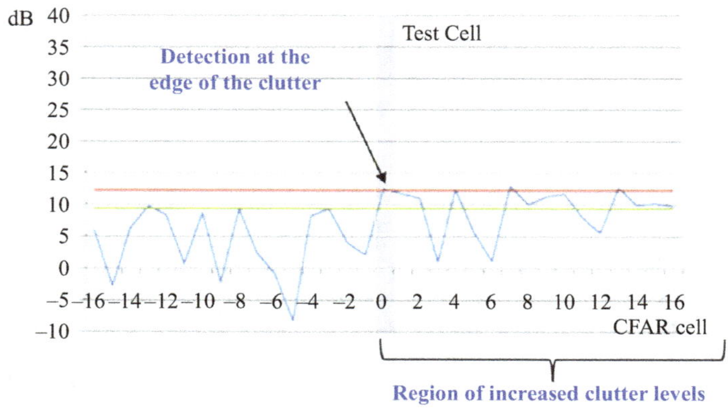

Figure 3.87 Clutter edge

Figure 3.88 Aircraft in formation (image US Air Force)

Solving the clutter edge problem requires a little more work in the threshold generator. The average noise level from the leading and lagging cells is calculated independently and compared. If one set of cells is significantly higher average than the other, the low set is discarded from the threshold calculation. This form of CFAR is called **Greatest Of CFAR** (GOCFAR). The disadvantage of this type of processor is that it reduces the number of cells averaged, reduces the quality of the assessment of the average noise level and, therefore, reduces the P_d.

A further disadvantage of GOCFAR occurs in situations where there are a number of closely spaced returns; for example, aircraft flying in formation. Consider Figure 3.88.

Figure 3.89 illustrates how aircraft in formation might present in a CFAR processor. Two thresholds have been computed; the first is the normal threshold with the

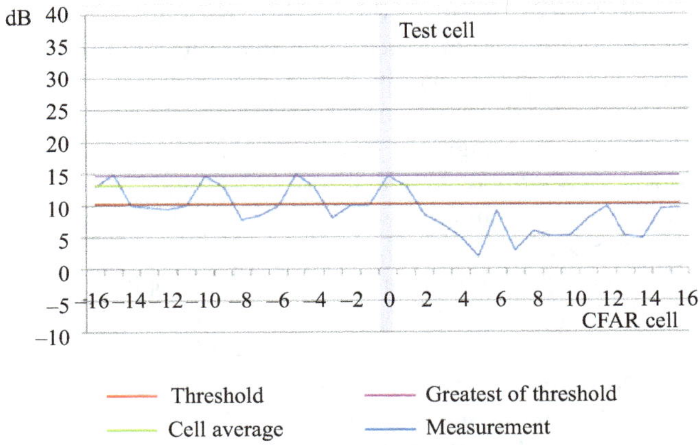

Figure 3.89 CFAR masking

average calculated using both the leading and lagging range cells. The GOCFAR would rise to account for the presence of multiple aircraft but can rise so high that the targets do not cross the detection threshold. This condition is called **masking**. This topic will be taken up in Chapter 4 when the effect of wind turbines is considered.

3.10 Secondary radar

One of the characteristics of primary surveillance radar systems is that they only provide information on the location of objects, they cannot positively identify the objects they observe. Furthermore, measuring the height of an object using a single primary radar is costly. Air defence radar needs to know the height of aircraft to do this and they cost, typically, an order of magnitude more than an ATC radar. Therefore, neither does an ATC radar provide height information (only range and azimuth bearing).

SSR addresses these shortcomings and can supplement the ATC radar data with both aircraft identity and height.

3.10.1 SSR development

In February 1935, an experiment was carried out to determine if it was possible to detect an aircraft at range using radio waves. The BBC's radio transmitter at Daventry was used to provide a source of radio frequency energy and the experiment has come to be called the Daventry experiment. The experiment showed that it was possible to detect a Heyford Bomber at a range of 13 km. The detection system was improved and, as a result, by the Summer of 1935, this range had doubled and, by the end of 1938, it was possible to detect aircraft at a range of 160 km.

This capability was a radical improvement on anything that had existed before and it was necessary for the RAF to find out how best to exploit it. In 1937, a series of trials were conducted known as the Biggin Hill experiments. From the very first trial, it became obvious that it was critical to be able to distinguish their own forces returning home after a mission and opposing forces attacking. Even though own forces would be required to return home in pre-designated air corridors, there was no guarantee that the enemy might not discover these corridors using espionage or simply by accident. Furthermore, although own forces might be intended to use the pre-designated routes, if an aircraft was damaged, or through shortage of fuel, it may have been impractical.

The solution to the problem was arrived at very promptly. Following investigations at the UK Government's research establishment at Bawdsey, the Ferranti company was tasked in 1938 with developing the Mark 1 IFF system and the system went into production in 1939 in time for the start of the war. There were two parts to IFF: equipment on the ground that transmitted 'interrogations' and transponder equipment on board aircraft that transmit a reply back to the ground.

Successive improvements in IFF led to the introduction of the Mark X system in 1950 which was then adopted with minor modifications by ICAO for civilian use. The civilian system is known in the United States as the ATC Radar Beacon System (ATCRBS) and, in the United Kingdom, as SSR. The Mark X remains the basis of SSR/IFF systems that are in use today [97].

3.10.2 Operating concepts

The SSR operating concept is the same now as it was in 1939. A ground-based secondary radar system transmits interrogation pulses using a frequency of 1,030 MHz. These pulses are coded in such a way that when a suitably equipped aircraft detects the interrogation, it responds, providing the requested information. The information is sent to the ground using a frequency of 1,090 MHz. It is decoded in the SSR ground receiver and then provided, possibly over a datalink to a remote location, to a radar data processor that combines primary and secondary data to be displayed to an operator. This concept is illustrated in Figure 3.90. An example of the secondary radar data is shown in Figure 3.91. The information derived from the SSR is kept deliberately concise to minimise congestion of the radar screen. The information is presented in a track data block (TDB).

The TDB contains information such as the aircraft identification, its height expressed as a flight level and derived information such as whether the aircraft is descending or climbing[†††].

To solicit information from the aircraft, SSR/IFF has a number of modes of operation. The modes and the information returned by the aircraft in the different modes are listed in Table 3.5.

Many aircraft are obliged to carry the equipment because of the nature of their roles, for example, airliners and military aircraft. Carrying transponders is also mandatory if an aircraft is to fly in controlled airspace, above 10,000 ft or aircraft wishing to fly in a transponder mandatory zone.

[†††]Aircraft do not ascend, they climb. This terminology is used to avoid the possible confusion of ascending and descending.

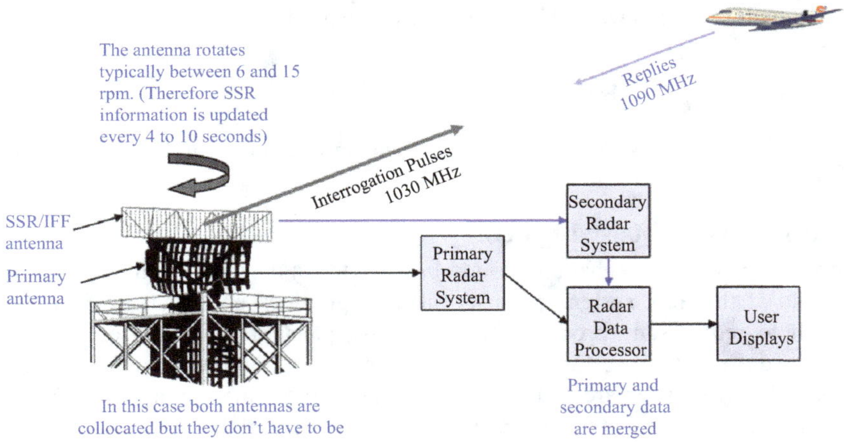

Figure 3.90 SSR/IFF operational concept

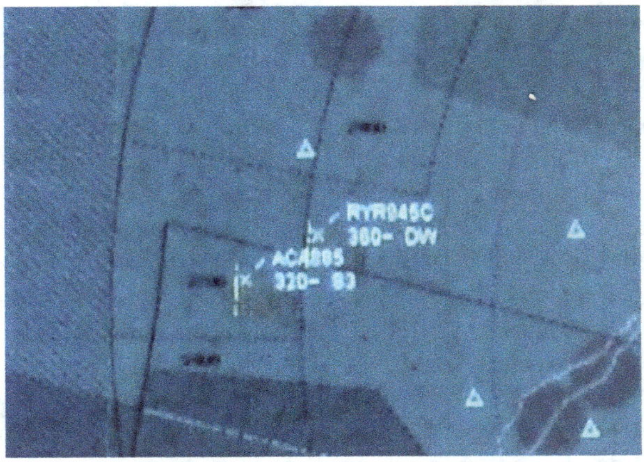

Figure 3.91 Secondary data (courtesy of RAF Brize Norton)

Table 3.5 SSR/IFF modes of operation

Mode	User	Aircraft reply data	Notes
1	Military	Aircraft type of missions	Collectively known as the selective identification feature (SIF) modes
2	Military	Aircraft tail number (specific aircraft ID)	
3/A	Military/civilian	Aircraft squawk code	
4	Military	No longer in use	
5	Military	An encrypted version of Mode S	
C	Civilian/military	Provides pressure height information taken from the aircraft altimeter	
S	Military/civilian	Selective mode allows individual aircraft to be addressed and a variety of information can be returned	

To explain the modes and how they are initiated, it is necessary to know a little more about the equipment used.

3.10.3 Equipment

SSR/IFF consists of a ground-based element and an element located on-board aircraft.

3.10.3.1 Ground-based system

Stephens [98] points out that the ground-based antenna is one of the most important elements of the SSR system because unless it provides good performance, it is difficult to redress any problems through subsequent processing. Both the horizontal and the vertical performance are important. The key aspects of the antenna's horizontal performance are that it must provide accurate measurements of aircraft bearing; discriminate between aircraft that are close together and also to reject replies from aircraft that are not on the intended bearing (e.g., through an antenna sidelobe). The vertical characteristics must minimise ground reflections to reduce multipath-induced fading and false aircraft returns.

Today, the ground antenna is most often a phased array. As the name implies, a phased array consists of a number of individual antenna elements working in concert to produce a beam. The beam produced is vertically polarised, is narrow in the azimuth (horizontal) direction, typically one to two degrees wide, and broad in the elevation direction, typically up to 45°. The gain of the antenna is typically 25dBi. The most common form of antenna is called the large vertical aperture (LVA). The beam may be steered away from the horizon to reduce the possibilities of multipath reflections. Stephens reports that the LVA can reduce the groundward facing, first, sidelobe level by 6 dB [98].

The scan rate is the same as ASR, ARSR or AD radars, that is at either 15 or 6 revolutions per minute (rpm). It follows that aircraft are illuminated either every 4 s or every 10 s. Therefore, the SSR/IFF's ground-based antennas may be mounted either independently or they may be mounted on top of the primary radar antenna and use a common rotating drive system. In the latter case, the antennas tend to be smaller to reduce the loading on the antenna turning gear. The reduction in size decreases the gain of the antenna and increases the beamwidth; for example, the beamwidth may increase to 5° and the gain drop to 19 dBi. Figure 3.92 shows a co-mounted SSR antenna at Newquay Airport and Figure 3.93 shows an independently mounted IFF antenna at RAF Brize Norton.

For more information on SSR/IFF deployment, see CAP 761 [99]: Operation of IFF/SSR interrogators in the UK: Planning principles and procedures.

3.10.3.2 Sidelobe control

Discrimination against sidelobe targets is provided by an additional signal radiated from a collocated antenna called the omnidirectional antenna, or the control antenna. The gain of the control antenna must be lower than the main lobe but higher than the sidelobe level. This is illustrated in Figure 3.94. The sequence starts when the P_1 pulse is transmitted from the main antenna. After a brief delay, a P_2 pulse is

Figure 3.92 SSR antenna collocated with the PSR antenna (image courtesy of Newquay Airport)

Figure 3.93 IFF antenna RAF Brize Norton (reproduced with permission)

transmitted from the control antenna. The aircraft receiver knows whether the P_1 pulse was received from the main lobe or a sidelobe by comparing the relative strength of the P_1 and P_2 pulses, as illustrated in Figure 3.94. It is worth noting that although its primary purpose is sidelobe control, that, in effect, the P_2 pulse provides a means of turning off a reply from an aircraft; this feature is discussed later. The mode being requested is determined by the time when the P_3 pulse is transmitted (from the main antenna), the options are illustrated in Figure 3.95.

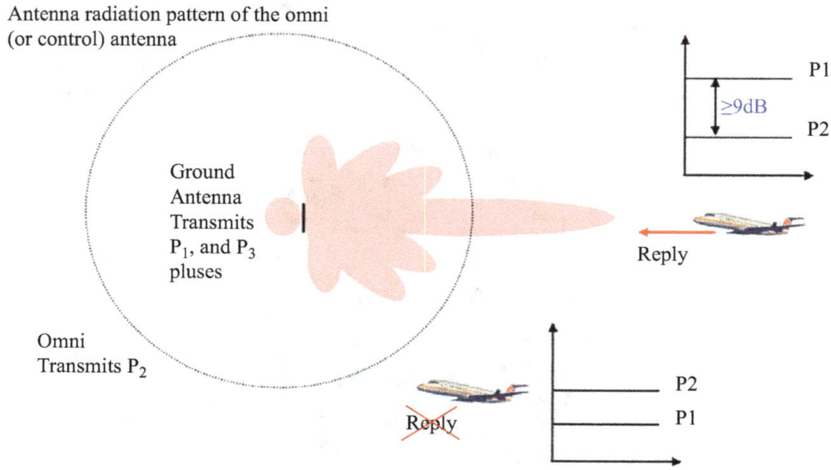

Figure 3.94 SSR antenna and control antenna patterns

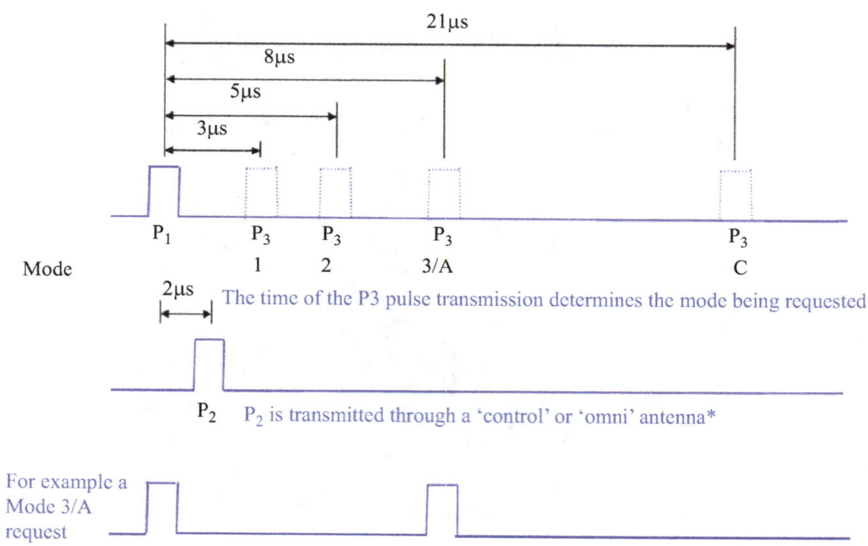

Figure 3.95 Interrogation requests

3.10.3.3 Monopulse SSR

Before describing the airborne equipment and the replies provided, a modern variation of the LVA is the monopulse SSR. A problem with the angular beamwidth of an antenna is that it is not possible to tell whereabouts in the beam a target is located: returns may come from anywhere in the beam. Greater accuracy could be achieved if it was possible to determine the location of the target within the beam. The monopulse technique provides a means of doing this. There are a number of ways of implementing monopulse. The concept originated as a process called **sequential lobing** where to refine the position of a target, a number of transmissions were made around a target. The return providing the largest magnitude was taken to be a better indicator of the target location. The disadvantage of sequential lobing is that multiple transmissions are required. Monopulse achieves the same ends with a single transmission. The antenna is arranged to create synthetic beam positions about the target. Replies from these positions can be compared to provide a refined bearing. Monopulse can provide approximately an order of magnitude improvement in measurement accuracy.

3.10.3.4 Airborne system

When the signal is received by equipment on-board the aircraft, a coded reply is transmitted in accordance with Table 3.5. A typical airborne antenna system is illustrated in Figure 3.96. When redundant systems are required, for example on civil airliners, antennas are located on different parts of the airframe.

The reply coding sequence is shown in Figure 3.97. The normal method of operation for these SSR modes is to alternate between Mode 3/A and Mode C, providing on every other scan the aircraft Ident and height.

In Table 3.5, there is a reference to a squawk code. The blocks of codes available to each country are assigned by ICAO [100]. Squawk codes, which are returned in Mode 3/A, are 4-digit octal codes (each digit can be from 0 to 7), are normally assigned by the air traffic controller and the pilot will select the code directed on the SSR/IFF control panel in the cockpit. Some codes are assigned a specific purpose; for example, in the United Kingdom, the code 0006 is allocated to British Transport Police and 0037 is allocated to helicopters in use by the Royal

Figure 3.96 Typical airborne antennas

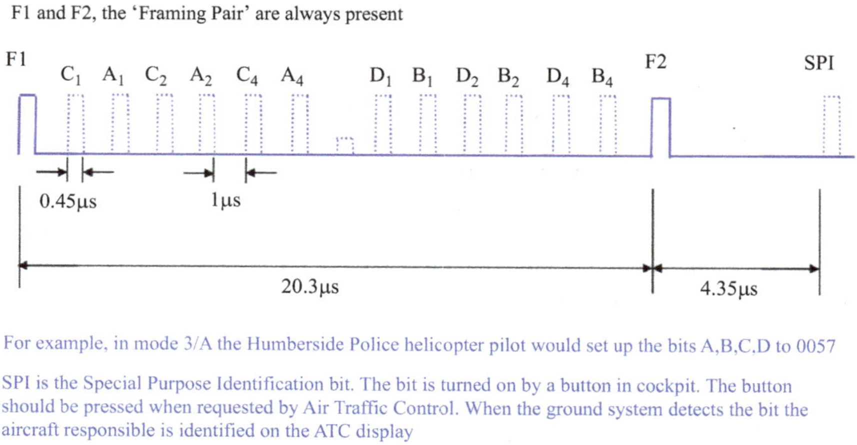

Figure 3.97 Aircraft SSR/IFF reply coding

Family. In other cases, the ATC operator will draw from a pool of codes allocated to the aerodrome; for example, codes between 6101 and 6107 are used by the military in the London FIR, 7030 to 7046 are used by the civilian controller for terminal control at Heathrow Airport and 7350 to 7365 are used for an approach to Manchester Airport. A number of squawk codes are allocated for emergencies, the pilot will select code 7500 to indicate that the flight has been hijacked, squawk code 7600 indicates that there has been a communications failure and 7700 indicates that there is an emergency.

Altitude messages are encoded using a Gillham Code: a type of binary coding where consecutive values only change a single bit. This type of coding is unlike conventional binary: for example, adding 1 to 3 changes the binary code from 011_B to 100_B.

3.10.4 Mode S

The legacy SSR/IFF codes have been in service for a long time and, while they provide important information about aircraft, it is limited and it has one major drawback: the interrogations are broadcast to all aircraft and it is not possible to address and solicit information from individual aircraft. Mode S is the selective mode: it allows controllers to solicit information from individual aircraft. The alternative name for Mode S is discrete address beacon system (DABS).

Aircraft are all allocated an ICAO code 24-bit[‡‡‡], this code provides the means of addressing aircraft. If the address code is all '1's, this is the equivalent of a broadcast to all aircraft. Other benefits of Mode S are that they allow data

[‡‡‡]The aircraft equivalent of a computer's IP address.

integrity to be checked by including error checking in the messages by use of a cyclic redundancy check (CRC) and it is possible to specify heights with greater precision.

There are two modes of operation of Mode S: all-call interrogations and roll-call interrogations. The normal method of operations is as follows:

- All-call period:
  - Aircraft that are newly entering the area will send an all–call reply with their 24 bit addresses to the ground station
  - The transponder will then '***lock out***' for a period of 18 s during which period, other all call interrogations are ignored.
- When the interrogator receives the all-call reply that aircraft is recognised as ***acquired*** and subsequently will be interrogated using a roll call message.
- Following the all-call period, the interrogator enters a roll-call period (addressing individual aircraft).
- As long as older SSR transponders are in use, a Mode S interrogator will also have to interleave Mode A/C interrogation: this behaviour is called a **Mode Interlace Pattern** (MIP).

The sequence for all-call interrogations is shown in Figure 3.98. The roll-call interrogation formats are shown in Figure 3.99, note that the data bursts in these formats are encoded in differential phase shift keying (DPSK). Aircraft equipped with older SSR/IFF equipment will be able to receive Mode S interrogations but by clever use of the P_2 bit, they will not respond.

Mode S reply formats are shown in Figure 3.100.

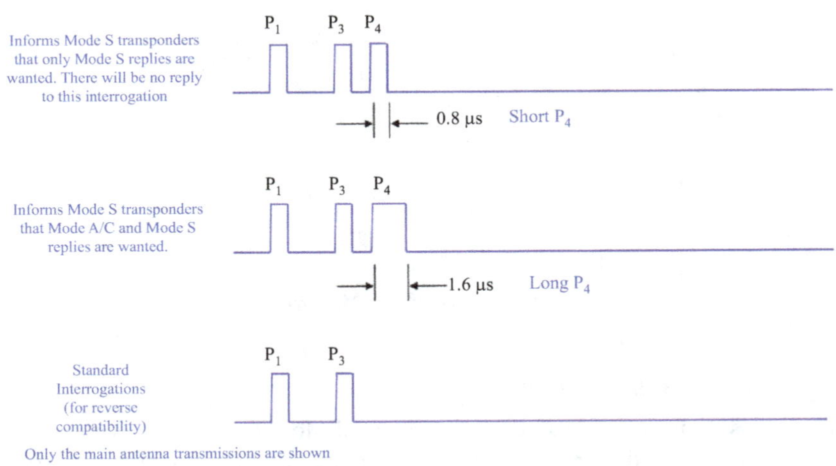

Figure 3.98 Mode S all call interrogation sequence

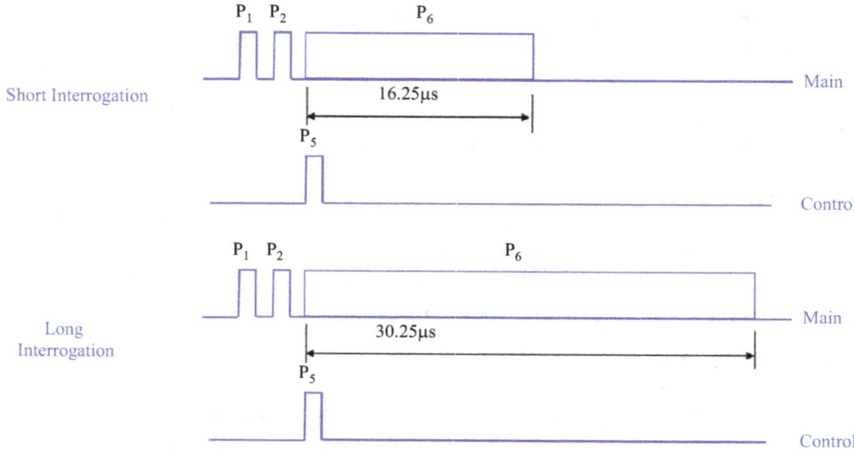

Figure 3.99 Mode S roll call interrogation sequence

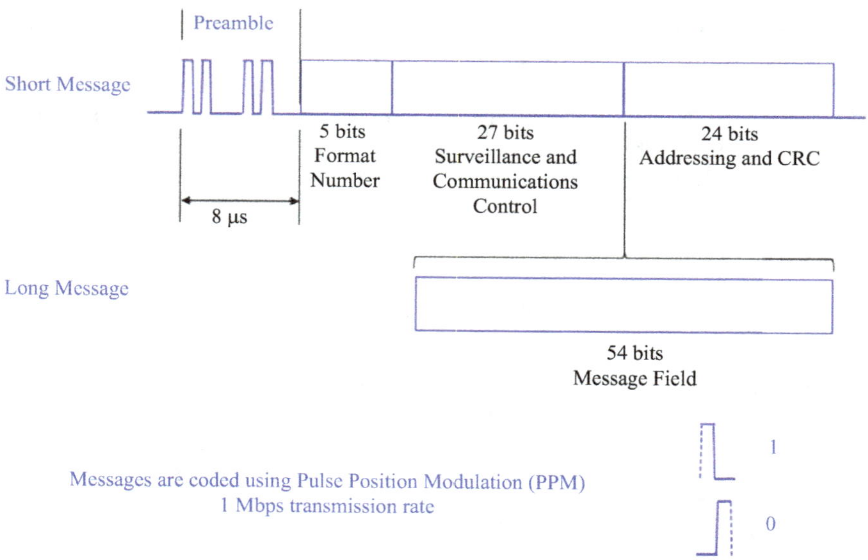

Figure 3.100 Mode S reply formats

3.10.5 Mode S message sets

Mode S message formats are used by conventional Mode S SSR, TCAS and ADSB: the latter two capabilities are described below. In addition to providing the Mode S equivalent of the legacy SSR/IFF messages, that is Ident and Altitude, Mode S has the ability to request a great deal more information from the aircraft. The interrogation messages can take on the role of a database inquiry language requesting different information stored by the aircraft. The protocol supports this capability with a standardised message set called Comm B Data Select (BDS) messages. An example will help explain how this works. In a conventional SSR system, interrogations alternate between a request for Ident and altitude. In Mode S, the protocol allows up to 255 different types of data to be requested. For example, one of the most commonly requested sets of data is the BDS message type 5, the track and turn request. When selected, the following data items are returned:

- Aircraft roll angle
- True track angle
- Ground speed
- Track angle rate
- True airspeed

In practice, not all 255 message types are currently used. Currently used message types are listed in Table 3.6. The most commonly used messages are the types 4, 5, 20 and 21 [101].

3.10.6 Advantages of SSR/IFF

The principal benefit of SSR/IFF is that the system is able to provide ground controllers with the identity of the aircraft and its height and it may be able to

Table 3.6 Mode S message set

Downlink format (DF)	Short/long format	Message type	Notes
0	Short	Short air-to-air surveillance	
4		Altitude reply	Mode S equivalent of Mode C
5		Ident reply	Mode S equivalent to Mode 3/A
11		All call reply	
16	Long	Long air-to-air surveillance	TCAS message
17		Extended squitter (ident, position, velocity)	ADSB message
18		Extended squitter – non-transponder	ADSB message
19		Military extended squitter	ADSB message
20		Comm B data selector (BDS) message with altitude reply	Supplementary Mode S information
21		BDS message with Ident reply	Supplementary mode information
24		Comm D long message	

provide supplementary information (discussed below). The height information is taken directly from the aircraft altimeter.

The fact that the information is provided to the ground by an active transmitting device on-board the aircraft means that the signal is much more powerful than the echo that would be returned to a primary surveillance radar. Therefore, the SSR is capable of operating at long ranges; typically, the maximum range of an SSR is 256 Nm (470 km). The codes transmitted by the ground are simple and the minimum ranges can be short, typically, 0.25 Nm (0.9 km); a figure that means an SSR service could be used on many aerodromes.

SSR uses different frequencies for the ground-to-air link (1,030 MHz) and the air-to-ground link (1,090 MHz) which provides greater resilience against false targets. However, these frequencies are sufficiently close that a single antenna can be used for both transmit and receive.

3.10.7 Disadvantages of SSR/IFF

While many aircraft are obliged to be equipped and to use SSR equipment, there are legitimate reasons why some aircraft do not. General aviation (GA) aircraft (light aircraft), gliders and hang gliders are not mandated to carry transponders; the main reason for this is that they do not fly above 10,000 ft.

Furthermore, transponders sometimes fail and they may not be turned on. The aircraft that attacked the Pentagon and the World Trade Center building during 9/11 had transponders disabled [102].

Therefore, the principal disadvantage of SSR/IFF is that not all aircraft can be relied upon to be transponding. Within the context of this book, this is a problem because SSR/IFF might otherwise provide a discriminant between aircraft and wind turbines. This argument is developed further in Chapter 4 and Chapter 6.

3.10.8 Why cannot wind turbines carry SSR like aircraft?

A question that frequently arises is, if SSR or the military equivalent, IFF, is used to identify aircraft, why cannot the same method be used to identify wind turbines? If this could be done, it would be possible to distinguish primary return echoes from wind turbines from aircraft.

Wind turbines are not equipped with SSR transponders and it is undesirable that they should. The message transmitted by an SSR or IFF transponder contains a number of bits that are determined by the request made in the interrogation request transmitted by the ground. It takes a finite time to transmit this information, 24.56 μs. For all practical purposes, the interrogation and reply messages travel at the speed of light. Consider then what would happen in the scenario set out in Figure 3.101.

Two wind turbines have been equipped with SSR transponders. An interrogation is transmitted from a ground station not shown in the figure. Let us suppose that there is a suitable reply message that can identify the wind turbine and that it starts to be transmitted as soon as the interrogation is received. It will take 24.65 μs for the reply to be transmitted. But the interrogation carries on travelling and a little while later reaches the second wind turbine. It too starts to reply giving

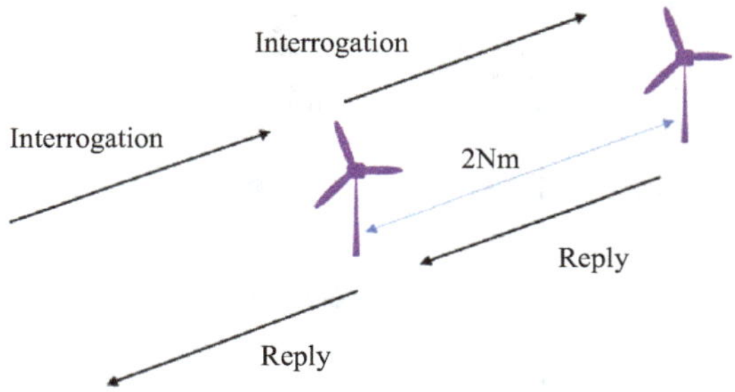

Figure 3.101 SSR transponder proximity

its identification as a wind turbine. The time taken for the interrogation to travel to the second wind turbine is determined by the separation distance which, in this case, has been deliberately chosen to be 2 Nm (3.7 km). The reply from the second wind turbine will reach the first wind turbine having travelled an additional 2 Nm. Travelling at the speed of light, the combined time taken for the interrogation and the reply to travel 4 Nm is almost exactly 24.65 μs. Therefore, the reply from the first wind turbine will just be completing as the reply from the second is starting. If the two turbines were any closer, the two messages would overlap and the messages from both turbines would be garbled.

Typically, the distance between the turbines in a wind farm is much less than 2 Nm. Therefore, if SSR was to be used in such a situation it would not be possible to have transponders on every turbine because of the resulting message garbling. And, if SSR transponders are not carried by every turbine, then it would not be possible to use their replies as a discriminant for primary echoes produced by turbines.

3.11 SSR derivatives

In 1956, there was a mid-air collision between a United Airlines Douglas DC-7 and a Trans World Airlines (TWA) Lockheed Super Constellation in uncontrolled airspace over the Grand Canyon. At the time, that was the worst-ever aircraft accident and all 128 passengers and crew of the two aircraft died. The focus of attention following the crash was on the state of ATC systems that might have avoided the collision. Conceptually, when an aircraft is capable of reporting its position to the ground, it is a simple step to being able to alert other aircraft and to develop methods for avoiding mid-air collisions. Ultimately, this led to the first generation of collision avoidance system, known as Beacon Collision Avoidance System (BCAS) which used replies to SSR interrogations. A further collision in 1978 led to the next generation of avoidance systems the Traffic Alert and Collision Avoidance System (TCAS), which is also sometimes referred to as Airborne Collision Avoidance System (ACAS).

The TCAS concept is illustrated in Figure 3.102.

There are three classes of TCAS:

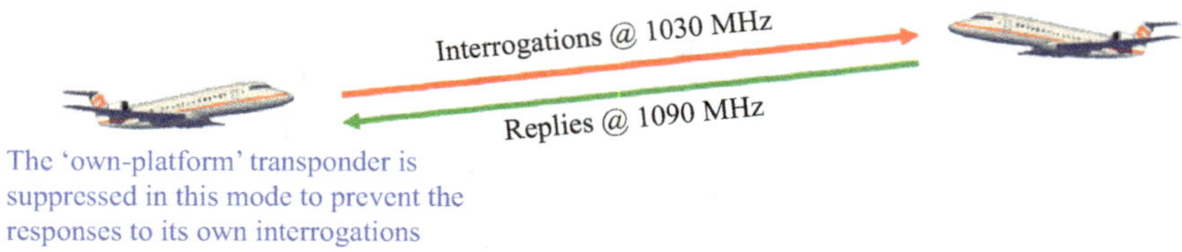

Figure 3.102 TCAS concept

- TCAS I
  - Intended for light a/c.
  - Can detect nearby a/c operating at the same flight level using SSR Mode C interrogations.
  - The pilot is warned and must decide on his own avoiding action.
- TCAS II
  - Intended for large air carriers.
  - Nearby aircraft are tracked in both the horizontal and vertical planes.
  - The pilot can be given advice (by the TCAS system) on any avoiding action required.
- TCAS III
  - Also known as enhanced TCAS
  - Intended for large commercial aircraft.
  - The system has a more directional antenna and can provide greater accuracy in avoiding manoeuvres.

It is axiomatic that in the highly unlikely chance of a collision occurring, then it will be in crowded airspace. This poses an engineering challenge. If there are numerous aircraft in relatively close proximity, they will receive interrogations close in time and their replies will overlap each other, **garbling** the replies. Garbling occurs when aircraft are less than approximately 2 Nm apart. A strategy is needed to manage the message traffic. The greatest risk of collision comes from those aircraft that are closest. Ideally, there needs to be a method of communication with close aircraft and having dealt with that risk, then consider aircraft that are farther away. This problem is solved by having two types of interrogations: a **whisper** and a **shout**. This concept is illustrated in Figure 3.103. Nearby aircraft are interrogated using a whisper and then more distant aircraft are interrogated using a shout. Nearby aircraft are prevented from replying to shout interrogations by transmitting a whisper interrogation pulse (P_1) and a shout pulse that appears to nearby aircraft to be a P_2 pulse, inhibiting their reply [98,103].

3.11.1 Automatic dependent surveillance – broadcast

Automatic dependent surveillance – broadcast (ADSB) is a form of electronic conspicuity. Aviation regulators see it as the long-term replacement for conventional primary and secondary surveillance methods. Automatic is in the ADSB title

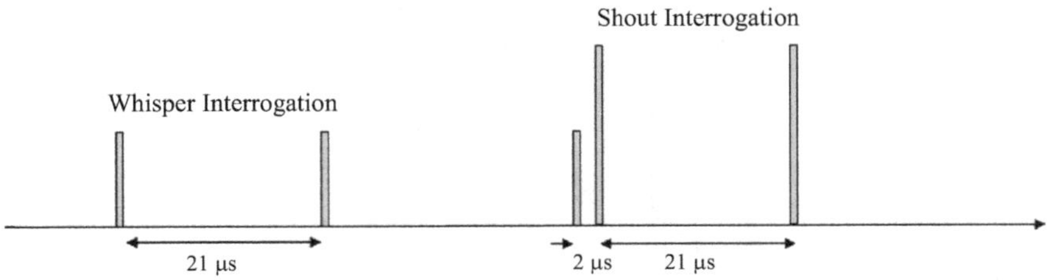

Figure 3.103 Whisper shout interrogations

because when enabled, it requires no interaction to continue reporting; dependent is included because ADSB requires satellite navigation to provide the position information of platforms using the service; surveillance data is included in the ADSB reporting and broadcast refers to the fact that it is not a point-to-point service, its broadcast and can be received by any suitable equipment.

ADSB operates as follows. Platforms determine their position using satellite navigation and then broadcast the position and other related information. This process is called ADSB-Out.

- ADSB-out: in this mode equipment on-board and aircraft
  - Periodically, broadcasts identification, position and velocity based on GPS information. The rate of transmission depends on the type of message. For example, Ident messages are transmitted every 5 s and position and velocity messages are transmitted twice a second.
- ADSB-in:
  - Which is the ability to receive ADSB signals from nearby aircraft

3.11.2 Multilateration (M-Lat) and wide area multilateration (WAM)

M-Lat is a technique that uses multiple ground receivers to measure the time difference of arrival (TDOA) of messages. In principle, M-Lat can be used by moving platforms to determine its location by measuring the TDOA of signals from the ground. Prior to the availability of satellite navigation, this scheme was widely used for marine and air navigation using systems long-range navigation (LORAN), Omega and Decca Navigator: all three systems were based on the Gee system developed during the Second World War.

M-Lat in this context refers to the use of multiple receivers on the ground being used to determine the location of an aircraft. Two receivers can localise the position of an aircraft to being somewhere on a hyperbolic arc. A third receiver provides two specific locations on the arc, often this is sufficient but a fourth receiver can unambiguously identify the location of the aircraft. The concept is illustrated in Figure 3.104.

M-Lat is dependent on the aircraft making transmissions and, hence, it is a cooperative system. A typical configuration combines M-Lat with an ADSB

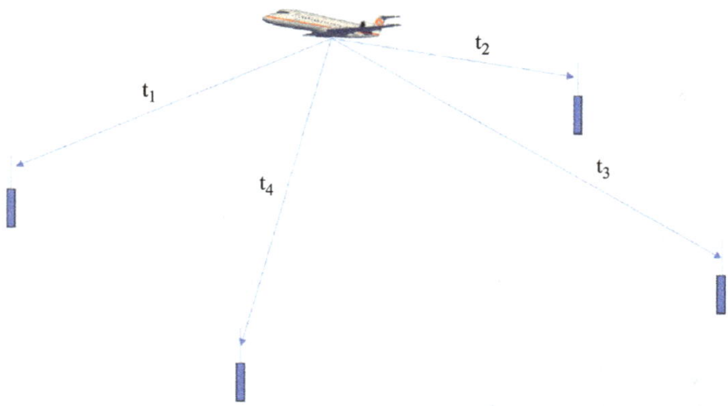

Figure 3.104 Multilateration concept

receiver. ADSB provides both the required messages to make M-Lat provide accurate tracking data as well as the benefits of ADSB, including, for example, the aircraft Ident.

WAM is a variation of M-Lat that is used, as the name suggests, over a wider area providing a service comparable with en-route sensor capabilities [104].

3.12 Air defence radar

The military user's requirements for an air defence (AD) radar are quite different from those in the civil and military ATC sectors. The following discussion illustrates the differences.

In ATC situations, air traffic is, normally, cooperative. There may be accidental, or even malicious, airspace incursions; however, these are not the norm and, usually, aircraft will: obey the rules of airspace; follow pre-planned designated air routes; in the overwhelming majority of cases, controllers on the ground can communicate with the pilot by radio and may be able to rely on secondary surveillance techniques to positively identify the aircraft. The radars employed for the detection of aircraft can operate from pre-established, fixed, locations, where they can generally assume the environment is benign; for example, radio interference may occur accidentally but this will be unusual.

In AD situations, cooperation of air traffic cannot be assumed and enemy air vehicles, aircraft, cruise missiles and un-manned air systems (UAS) may try to evade detection, rather than facilitate it, for example, by trying to exploit ground clutter and terrain masking. The enemy will not follow a disclosed path and may approach from any direction. Furthermore, as demonstrated by the events of 9/11, it cannot even be assumed that the enemy-controlled air vehicles will approach from outside home territory. Early detection and the ability to sustain contact are essential to allow maximum time to assess the threat, determine the enemy's intent and plan how to react to and engage the threat. There can be no reliance on the use of secondary surveillance sensors: the enemy may not only not cooperate but may try to spoof secondary surveillance. Sensors may be deployable and the enemy may intentionally attempt to degrade detection capabilities through the use of various types of jamming. Groups of sensors may be used in a network: the military equivalent of the civil en-route networks. Their information, when pooled, is used to create the recognised air picture (RAP).

There are some similarities between the military and civil technical requirements that flow down from the user's requirements; for example, military radars must still detect targets of a certain size and behaviour at a certain range. However, some of the differences are very basic; for example, while an ATC radar needs only provide 2-D information because it can be supported by secondary provision of information, an AD radar must be able to determine the air vehicle position and track information in 3-D.

The first matter dealt with here is the need to provide a 3-D picture, this is achieved through a technique called **phased array radar**. This technology is applicable to many other domains some of which have already been met here: including secondary surveillance antennas and ILS.

3.12.1 Phased array radar

To provide the ability to measure the height of an air vehicle, an AD radar has to be able to provide the ability to sense the environment at different heights. Once this function would have been carried out by height-finding radar: essentially a radar that nods up and down to find targets. A phased array carries out the same function by electronically steering the radar beam up and down. The most common approach used to build an AD radar is to combine electronic steering in elevation and mechanical rotation to provide azimuth coverage, this is illustrated in Figure 3.105.

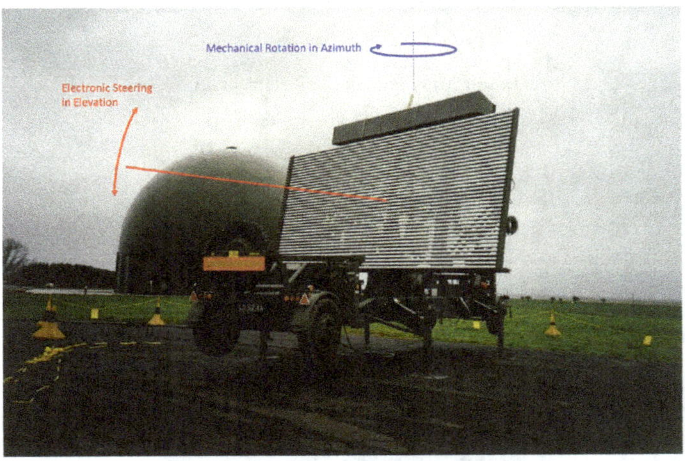

Figure 3.105 AD radar steering/rotation (image courtesy of RAF Boulmer)

A phased array is an antenna using more than one antenna element. Although widely regarded as a modern technology, it was invented by the German engineer Karl Braun (1850–1918) in 1905. Braun was awarded the Nobel Prize for its invention; sharing the prize in 1909 with Guglielmo Marconi for his work on radio. An alternative name for a phased array is an electronically scanned array (ESA).

Some useful rules about phased array antennas are that the greater the number of elements in the array:

- The higher the gain of the antenna.
- The narrower the beam that is possible.
- The greater the pointing accuracy of the beam.

To illustrate these factors, linear arrays of 8 elements and 32 elements were modelled and compared. The results are shown in Figure 3.106. The gain of the antennas modelled is shown in linear (not logarithmic) units that were used to illustrate better, the relationship between gain and the number of elements is linear. It is, however, more usual to show gain in dB and this is shown in Figure 3.107. Figures 3.106 and 3.107 also show that the energy is distributed between a dominant main lobe and 'sidelobes'. It is sometimes necessary to distinguish between close-in sidelobes and those farther away from the main lobe. In the literature, the first two, or sometimes three, sidelobes are called the **near sidelobes** and the remainder, **far sidelobes**.

Other advantages of phased array radar are:

- Because they use many discrete elements, each of which can include its own transmit and receive electronics, it is a useful means of exploiting solid-state electronics, with the associated high reliability.
- The elements may be organized in a linear array (which can steer a beam in 1 dimension), a 2-D planar array or even in a 3-D, volumetric, array. The latter two cases allow the beam to be steered in 2-D.

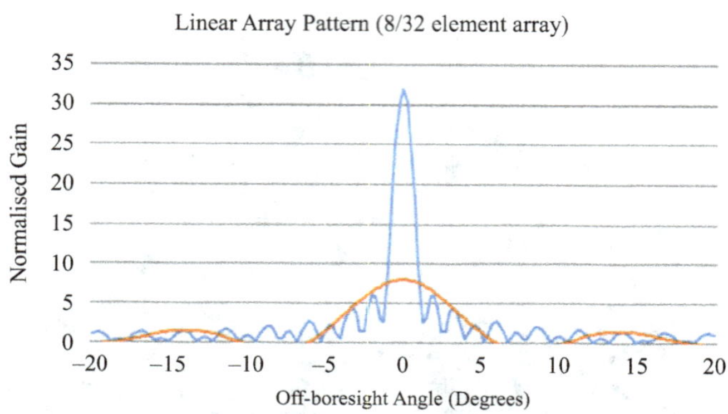

Figure 3.106 8- versus 32-element linear array gain

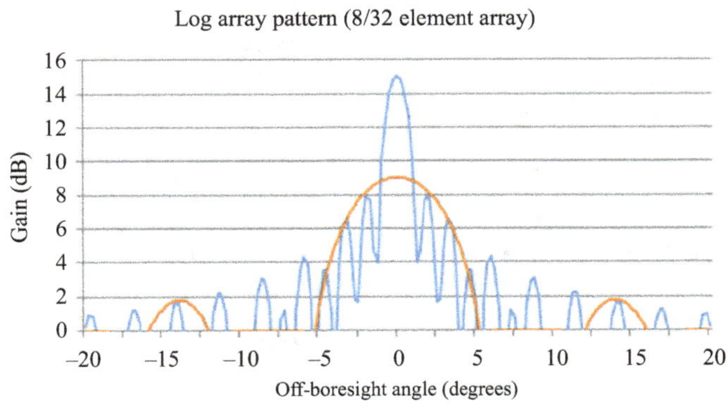

Figure 3.107 8- versus 32-element linear array gain in dB

- It is easy to organize the array so that it can provide monopulse: a process for providing a more accurate assessment of the location of a target by comparing the data from different parts of the array. This feature was described in the section dealing with SSR.
- It is easy to implement aperture control that provides the ability to manipulate the shape of the beam produced from an array. The aperture control may be fixed, for example, by introducing a taper, as described later. But some military radars are now being developed that allow dynamic control over the beam shape to reduce susceptibility to interference and jamming. This technique is called digital adaptive beamforming (DABF).
- The ability for rapid beam steering facilitates the implementation of multifunction technology, that is, the ability to interleave surveillance and tracking tasks [105].

3.12.1.1 Electronic phase steering

There are two principal ways to steer a beam electronically: by phase and by frequency. The following discussion explains how.

The simplest way to understand the operation is to consider a transmission being made from the phased array antenna. Suppose there are four elements in the array. All the elements are in a straight line. For the first part of this 'thought experiment', assume a pulse is transmitted from all the elements simultaneously. Shortly after transmission, the radio waves have propagated away from the antenna by a short distance. This is shown in Figure 3.108. The wavefront is moving away from the antenna, parallel to the plane of the antenna elements.

Now imagine the same array of antenna elements. Instead of simultaneously transmitting from each element, first a transmission is made from element 1; then after a brief interval, a transmission is made from element 2, and so on until all four elements have been used. Now consider the situation shortly after the final

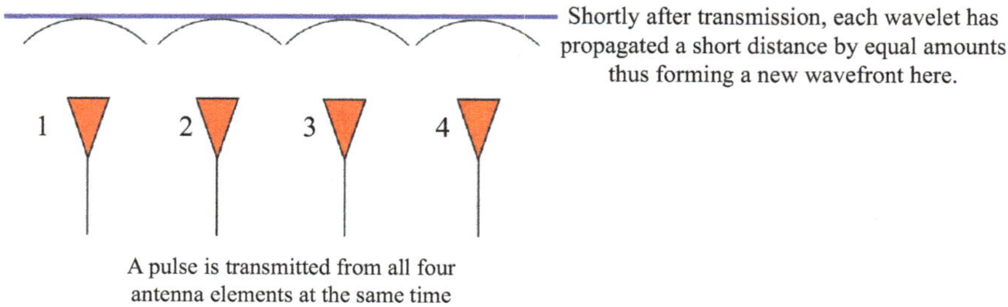

Figure 3.108 Phased array steering – 1

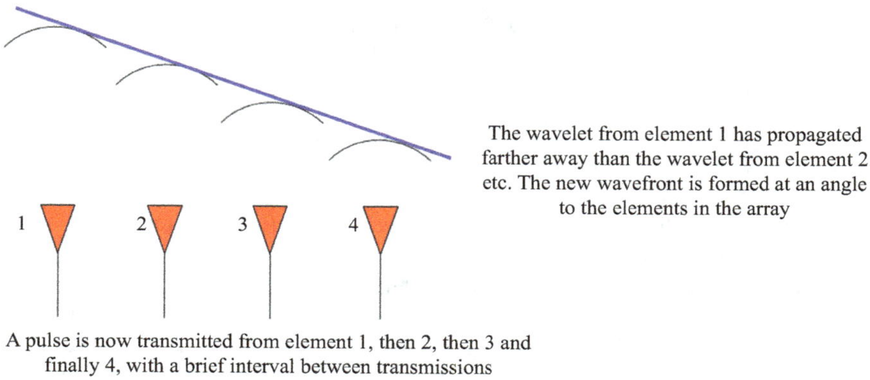

Figure 3.109 Phased array steering – 2

transmission from element 4. This situation is illustrated in Figure 3.109. The addition of incremental time differences between transmissions have caused the wavefront to be radiated at an angle from the array. The beam from the array has been steered at an angle off the normal to the array elements. This begs the question, what delay is required to produce a given steering angle?

3.12.1.2 Steering analysis

The most convenient way to analyse steering is to consider a signal being transmitted by an array. The array in this case consists of only two antenna elements labelled A and B. A coherent wavefront in which all the points along the wavefront are at the same phase is approaching the two elements. The wavefront is parallel to the plane of the two antennas and when it arrives at the antennas both receive signals that are in phase. The wavefront is said to be on **boresight**. When the two signals are summed, the sum is coherent and the output is maximized, see Figure 3.110.

Now consider the same arrangement, but this time, the wavefront approaches the two elements at an angle, that is off-boresight. The wavefront reaches element B before it reaches element A. The result is a phase shift, α, later than signal at

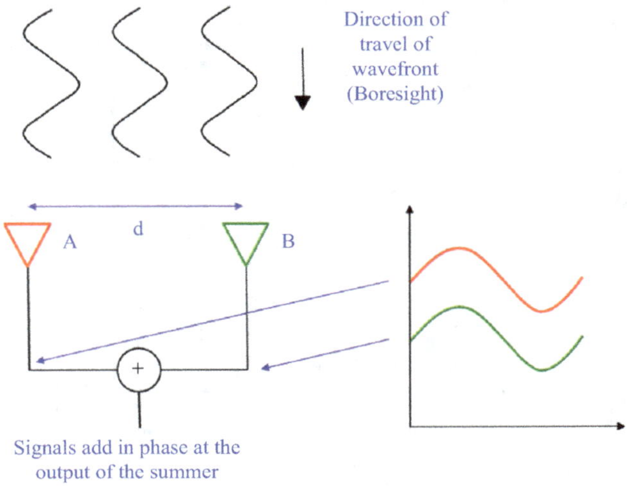

Figure 3.110 Phase steering on-boresight

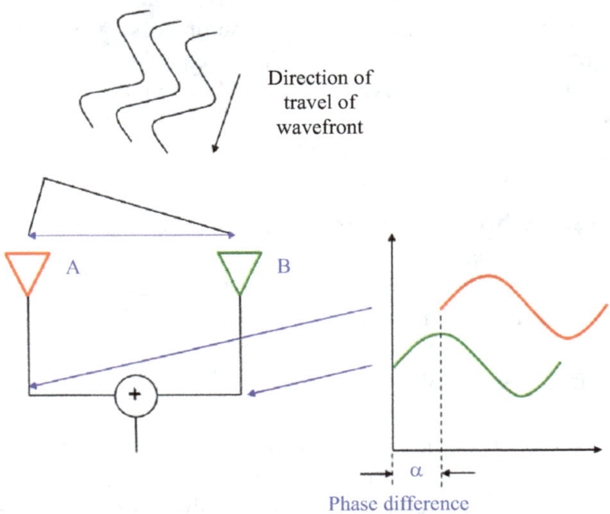

Figure 3.111 Phase steering off boresight

element A. Now the signals are out of phase, when summed, they destructively interfere and the output of the summer is not a maximum in the direction from which the wavefront arrived at the array, see Figure 3.111.

Suppose, now that a phase delay is added between element B and the summer device and the value of the phase delay is chosen to be α to match the delay caused by the wavefront approaching from off boresight. The results are illustrated in

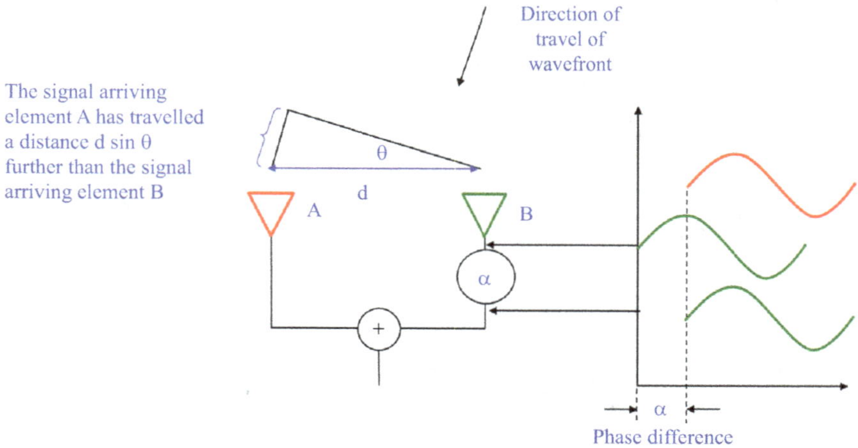

Figure 3.112 Phase correction

Figure 3.112. As might be expected, the phase correction at element B corrects for the delay experienced at A and the output of the summer is, once again, maximised.

The value of the phase shift α can readily be calculated as follows: the additional distance travelled by the wavefront to reach antenna element A, once it has reached antenna element B, is:

$$\text{Distance} = \text{d}\sin\theta \tag{3.15}$$

where d is the distance between the elements A and B and θ is the steering angle off boresight.

The value of α can now be calculated when the distance $d \sin\theta$ is converted to a proportion of the wavelength λ thus:

$$\text{Phase Shift to steern } \theta \text{ off boresight}, \alpha = \frac{2\pi \text{d}\sin\theta}{\lambda} \tag{3.16}$$

Therefore, it has been shown that an array of elements can be steered off boresight by inserting a readily calculable phase shift corresponding to an angle α. If more elements are added, then additional phase shifts are required as illustrated in Figure 3.113.

The analysis above assumed that the wavefront was being received, but the method works equally well in transmission. In practical terms, the usual way to use an array is to set up the desired steering angle, transmit a pulse of rf and then remain in that steering direction until echoes could have been received from the maximum desired range. Then reset the steering in preparation for the next transmission. Before discussing further how an array is used, two practical considerations must be addressed: how is phase steering implemented and how frequency steering provides an alternative?

3.12.1.3 Phase shifters

The first important point that must be addressed concerning phase shift is that the analysis presented implies that a steering angle is chosen and the precise phase shift angle is computed. In practice, the reality is that a phase shifter cannot assume any arbitrary phase shift, they are actually quantised and the desired phase shift must be approximated using a combination of quantised values.

The simplest way to delay a signal by a given fraction of a wavelength is to arrange a system where the signal must travel an additional distance. Such a scheme must be able to switch in and switch out those additional distances as required. A practical way of achieving this is by using a type of semiconductor called a diode. Depending on the voltage applied to a diode it can be made to open or close like a switch. This type of arrangement is shown in Figure 3.114. From this it can also be seen that four control signals would be required to control this network of phase shifts, using computer control this would be realised using four bits. The smallest phase shift that can be included by a network like this is 22.5°. This begs the question; does not this imply that the steering will be very granular?

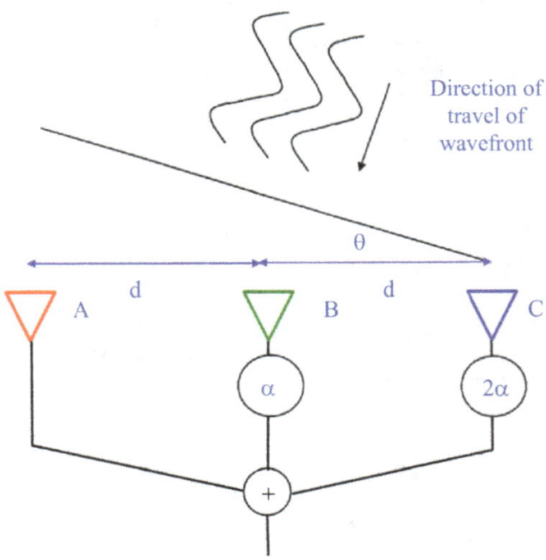

Figure 3.113 Additional elements

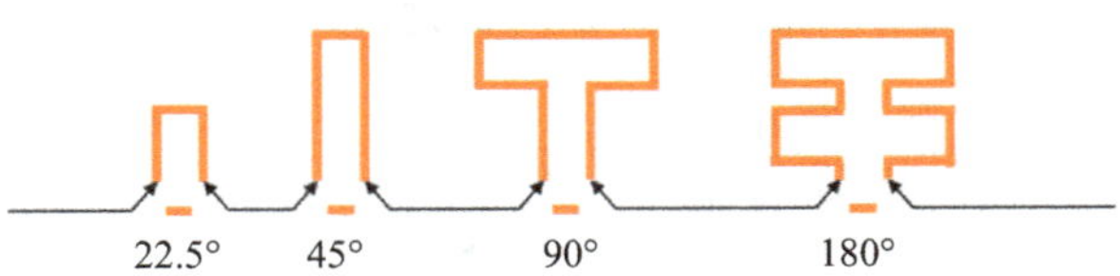

Figure 3.114 Digital steering

A practical example illustrates why this is not the case. Suppose there are eight elements, 0 to 7, in an array and the phase shift between elements to achieve the desired steering angle is 60°. Assume each element has four-bit phase control as illustrated in Figure 3.114. In this manufactured example, chosen to illustrate the principle, the average quantisation error was 0.94° over eight elements, as shown in Table 3.7. Note this was achieved with the granularity of the phase shifters being 22.5°. It can be shown that very small steering errors can be achieved with limited numbers of bit settings. Three-bit and four-bit control are the commonest forms of digital control [106].

3.12.1.4 Other types of phase shifter

Some materials have the capacity to slow radio waves and, therefore, have the ability to make it appear as though the wave has moved a greater distance. Ferrites have this ability and phase shifters can be fabricated from these materials. An illustration of this type of phase-shifting technology is shown in Figures 3.115 and 3.116.

Table 3.7 Four-bit phase shifting

Element number	Desired phase shift	Actual phase shift	Quantisation error
0	0°	0°	0°
1	60°	67.5° (45°+22.5°)	+7.5°
2	120°	112.5° (90°+ 22.5°)	−7.5°
3	180°	180°	0°
4	240°	247.5° (180°+ 45°+22.5°)	+7.5°
5	300°	292.5° (180°+90°+22.5°)	−7.5°
6	360° = 0°	0°	0°
7	60°	67.5° (45°+22.5°)	+7.5°
		Average quantisation error	0.94°

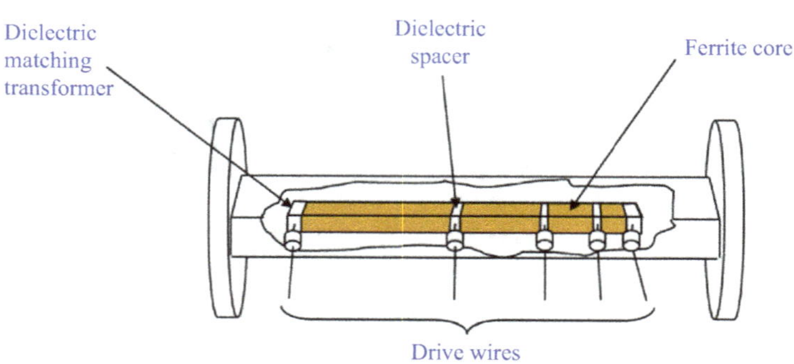

Figure 3.115 Four-bit ferrite phase shifter

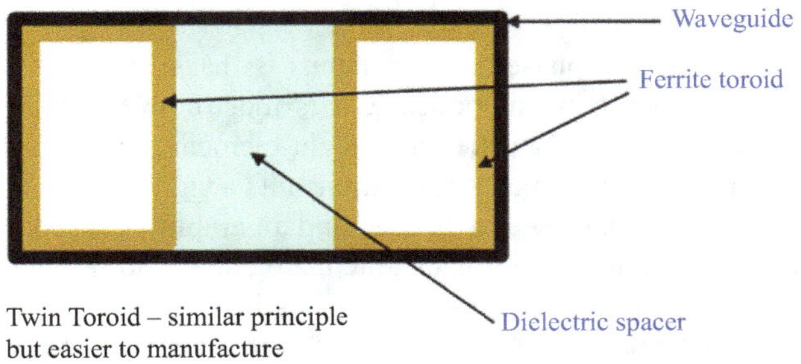

Figure 3.116 Twin toroid implementation of a ferrite phase shifter

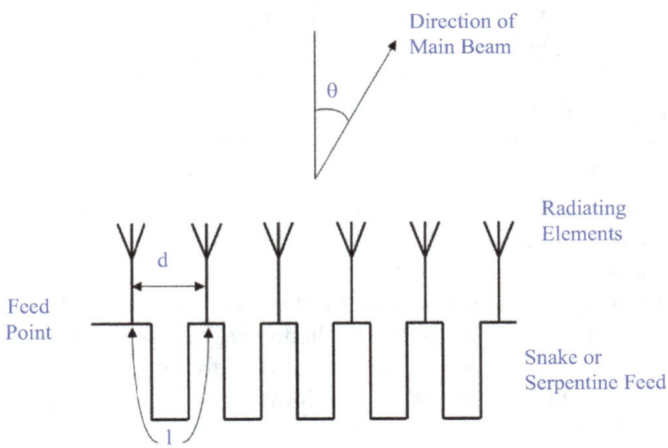

Figure 3.117 Serpentine feed

3.12.1.5 Frequency steering

Frequency steering of a phased array is also possible. Consider the network illustrated in Figure 3.117.

Between each antenna element, there is a feeder with a length, l. At any frequency, this length is a proportion of a wavelength, in other words, a phase shift. But as the frequency changes, this length is a different phase shift and would result in a different steering angle.

3.12.1.6 Phase shifter performance

Irrespective of the type of phase shifter used, they offer the benefit of being able to move the beam steering position very rapidly and without the effects of inertia associated with mechanical scanning. The diode-switched phase shifters referred to above can move the beam in microseconds.

3.12.1.7 Aperture control

One of the advantages of a phased array antenna is that it is easier to implement **aperture control**. To explain this concept, it is most convenient to consider an antenna with more elements than the basic eight-element array described earlier. Suppose, an antenna has 16 elements and it transmits a signal, say a radar pulse. In general, a high transmission power is needed and an amplifier is required. Assume that there is a separate amplifier for each antenna element. So far when discussing transmission from a phased array no thought has been given to the signal level provided to each antenna element for transmission: there has been an implicit assumption that the same signal has been provided to every element. This arrangement is called a **uniform illumination factor**. However, there are benefits to being able to vary the signal level provided to each element, this will now be described.

It was pointed out earlier that the antenna pattern (the distribution of radio energy in space) from an array is like any other antenna, it comprises a main lobe and sidelobes. The latter represents energy that provides no benefit either in transmit or receive; they are just an artefact of the transmission. Aperture control provides a means of controlling the distribution of energy. The reason the main lobe is 'main' is that it receives energy from all of the elements in the array. The sidelobes receive a contribution from all its elements but some elements contribute more energy than others. Figure 3.118 shows the pattern produced by a 16-element array with every antenna element receiving the same amount of power from its amplifier. Note that the main lobe and the first and second sidelobes have been identified in the diagram and the source of their energy is presented in Figure 3.119. This shows that the main lobe receives a contribution from every element. However, the sidelobes energy is not uniform, they receive a disproportionate amount of their energy from the edge of the array.

If the contribution to the antenna pattern from the outer elements of the array were reduced, this would have an impact on the main beam, removing some of the energy

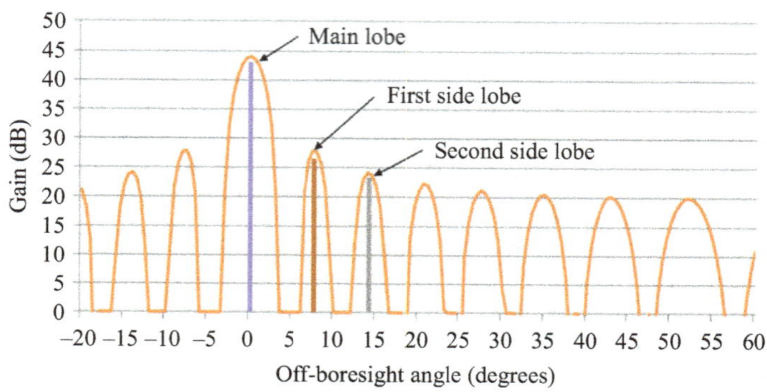

Figure 3.118 Pattern analysis

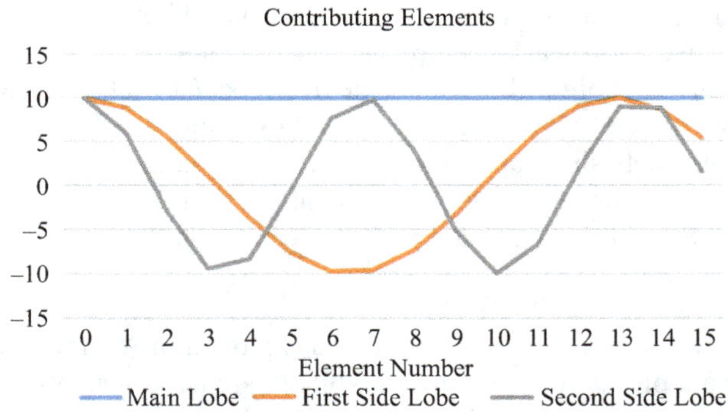

Figure 3.119 Energy distribution

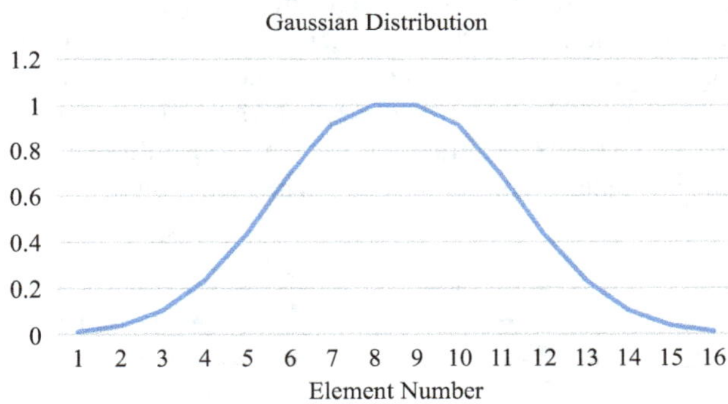

Figure 3.120 Gaussian taper

from it, but it would have a much higher impact relatively on the sidelobes. Using this strategy is called a **taper**[§§§]. There are many algorithms that can be employed to determine the value of the drive level to the elements, each conferring a different characteristic to the antenna pattern. The concept is illustrated using Gaussian Taper: a taper which obeys the following equation as illustrated in Figure 3.120:

$$f(x) = \frac{1}{\sigma\sqrt{(2\pi)}} e^{(-1/2(X-\mu)^2/\sigma)} \tag{3.17}$$

The antenna pattern created by a Gaussian taper is shown in Figure 3.121, comparing it with the uniform illumination pattern. The figure illustrates why a Gaussian taper might be used: there is a decrease in the main lobe but the side lobes

[§§§]In the sense of something that is narrow at one end.

are significantly reduced and the far sidelobes have effectively been eliminated. Note that the remaining (near) sidelobe has been broadened.

The different taper algorithms tailor the main lobe/sidelobe structure to satisfy the requirement for the system behaviour. For example, another commonly used taper, the Dolph Chebyshev taper, produces a wider main lobe and sidelobes that are all on the same level. This arrangement is useful for many applications including the ILS localiser antennas.

3.12.1.8 Using a phased array radar

The ability to switch beam positions rapidly is exploited in AD radar by being able to move the beam produced in elevations while the radar is scanning in azimuth. It is possible to create scan patterns like the one illustrated in Figure 3.122: what is, in effect, a synthesized cosecant squared antenna pattern.

The low-elevation beams are often segmented as shown in Figure 3.122. It is necessary to be able to detect objects (enemy aircraft) at long ranges. Long pulses are used to meet the desired probability of detection. However, they may be so long that they prevent detection at short ranges; they are said to **eclipse** targets. So, to detect short-range targets short pulse lengths are used. For low-level elevations, this would consist of setting the beam to the desired elevation, transmitting a short pulse to detect short-range targets, waiting for echoes, transmitting a long pulse and waiting for echoes to return from long range. The elevation setting would then be

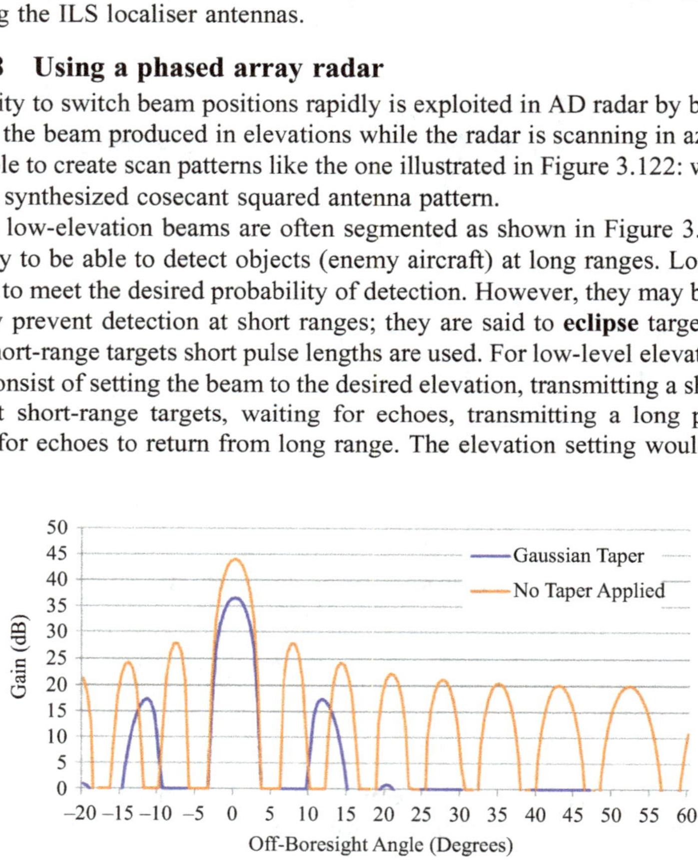

Figure 3.121 Antenna pattern without and with a Gaussian taper

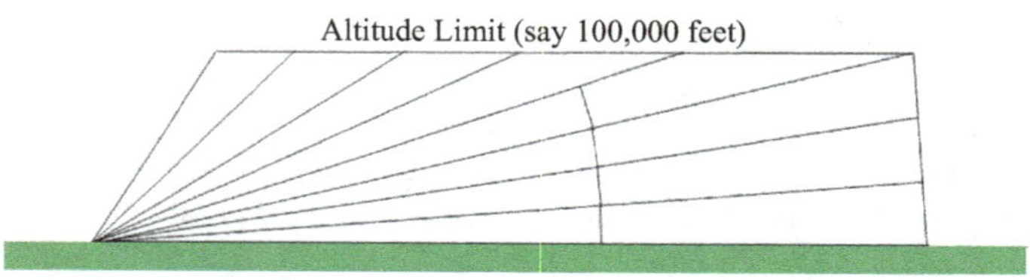

Figure 3.122 AD radar pattern

changed to a higher elevation and the process repeated. As the elevation increases, a point is reached where the time for long-distance echoes can be shortened because at this point, the distance takes the range beyond the height at which aircraft can fly. At the highest elevations, multiple transmissions may not be required.

The elevation ranges can also be negative for these radars to look down on targets. For example, if they are located on cliffs overlooking the sea, it is very desirable to look down on approaching aircraft or missiles.

Typical ranges for AD radar are similar to the radars used for enroute surveillance, typically, 250 Nm (450 km).

3.12.1.9 E-Tilt

The normal 'tactical' use of being able to shift the radar beam in elevation can be supplemented by being able to move the beam dynamically. For example, if interference is detected on a particular azimuth or if terrain intrudes into the radar coverage. This ability is referred to in the literature as **'e-tilt'**.

3.13 PAR

The same ability to switch rapidly from one beam position to another, both in azimuth and elevation, can be exploited for other applications. PARs, also referred to as GCA radars, are one such application. These radars are located on aerodromes to detect aircraft landing when the pilot cannot see the ground because of inclement weather. The radars provide a 3-D information on landing aircraft to ground controllers so they can provide lateral and vertical guidance.

3.13.1 PAR requirements

Requirements for PAR are set by ICAO [107]. Civil use of PAR is in steep decline and the International Air Transport Association (IATA) regard its use as redundant [108]. However, PARs remain in use by military pilots. Pilots must be guided from a distance of typically 20–30 Nm to the runway which is perhaps 30 m wide. To satisfy the need for accuracy, PAR/GCA operates in X-Band (9 GHz) which facilitates narrow beamwidths and lowers the uncertainty in aircraft position.

A key benefit of PAR is that it only requires a radio to be on-board the aircraft so the pilot can receive instructions from the ground controllers.

There is also a Doppler coverage requirement for PAR, $\pm$ 40–240 knots. PARs also often provide separate ways of detecting and tracking small and large aircraft to avoid large aircraft being seen as multiple targets.

3.13.2 PAR coverage

Precise coverage parameters are manufacturer specific. Typical tactical coverage is illustrated in Figure 3.123. The term tactical coverage is used here because PAR is usually designed to provide coverage of more than one runway, but only one runway at a time. Switching between runways only takes a short time, typically seconds, but if the wind changes direction and aircraft must use an alternative runway then a switch must be made, see Figure 3.124. To provide coverage for multiple

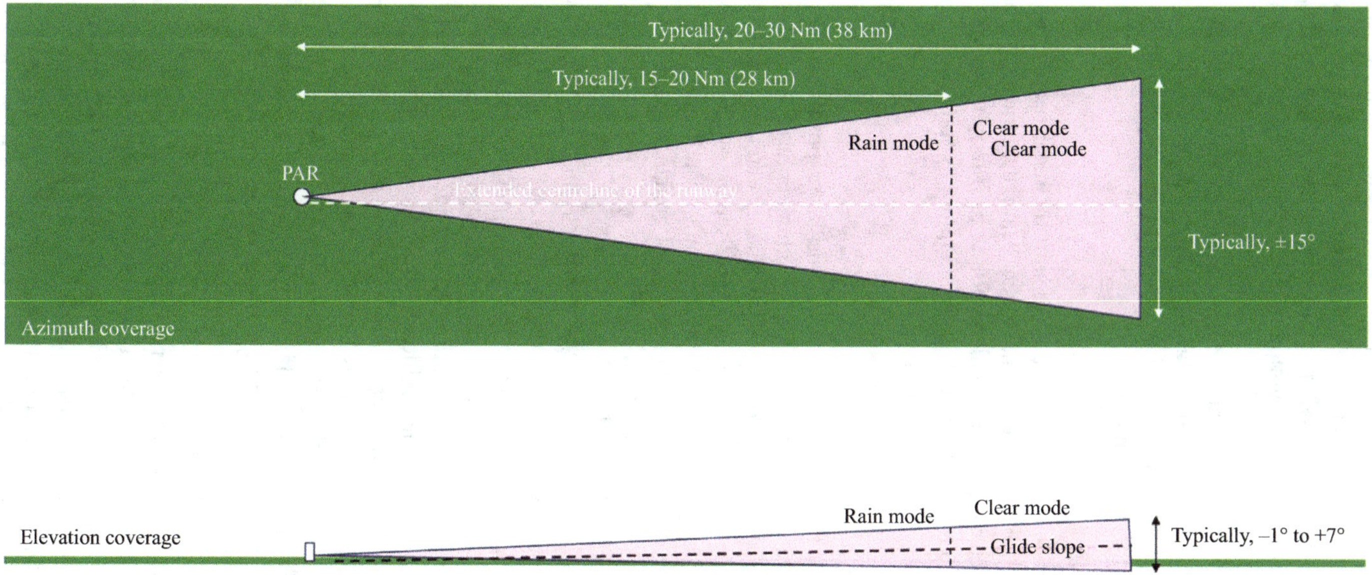

Figure 3.123 PAR tactical coverage

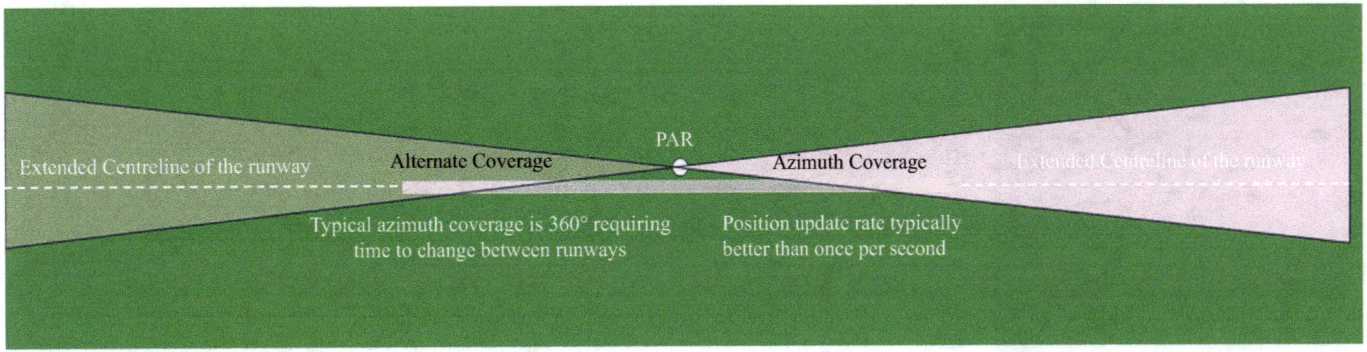

Figure 3.124 Support of multiple runways

Figure 3.125 A typical PAR (image courtesy John Hogan)

runways, this means that the equipment must be located close to the runways. However, being close to runways is a risk and ICAO regulations specify a minimum setback of 120 m.

A typical PAR antenna system is shown in Figure 3.125.

References

[1] ICAO (2006), *Aeronautical Telecommunications: Volume 1, Radionavigation Aids*, ICAO, Annex 10 to the International Convention on Civil Aviation, 2006.

[2] ICAO (2016), *European Guidance Material on Managing Building in Restricted Areas*, ICAO EUR Document 015, Third Edition, November 2015.

[3] EUROCONTROL (2014), *How to Assess the Potential Impact of Wind Turbines Surveillance Sensors*, Edition: 1.2.

[4] *FAA* (2023), *A Brief History of the FAA*, FAA. https://www.faa.gov/about/history/brief_history#:~:text=Birth%20of%20Federal%20Aviation%20Agency,-On%20May%2021&text=Two%20months%20later%2C%20on%20August,responsible%20for%20civil%20aviation%20safety. Retrieved 26 April 2023.

[5] CAA (2017), *UK Air Traffic Management Vocabulary*, CAP 1430, Civil Aviation Authority (CAA) Safety and Aerospace Regulation Group (SARG) January 2017.

[6] ICAO (2023). https://www.icao.int/wacaf/documents/apirg/sg/2012/afi_opmet_mtf4/docs/wp10_atta.pdf. Retrieved 21 April 2023.

[7] National Air Traffic Services (NATS).

[8] Airport Data (2004), *Paris Charles De Gaulle (Roissy) Airport*, Airport Data. Com. https://www.airport-data.com/world-airports/LFPG-CDG/. Retrieved January 2023.

[9] ICAO (1951), *Aerodromes: International Standards and Recommended Practices, Annex 14 to the Convention on International Civil Aviation*, International Civil Aviation Organisation, 1 November 1951.

[10] CAA (2022), *Licensing of Aerodromes*, CAP 168, Edition 12, Civil Aviation Authority (CAA) Safety and Aerospace Regulation Group (SARG), January 2022.

[11] FAA (2017), *Pilot/Controller Glossary*, Federal Aviation Agency (FAA), 12 October 2017.

[12] CAA (2007), *Visual Aids Handbook, CAP 637 Civil Aviation Authority (CAA) Safety and Aerospace Regulation Group (SARG)*, Issue 2.

[13] Fitzgerald, A. (ed.) (2010), *Air Crash Investigation: Tenerife Airport Disaster, The World's Deadliest Plane Crash Ever*, Mabuhay Publishing.

[14] ICAO (2023), *Vision and Mission*, ICAO. https://www.icao.int/about-icao/Council/Pages/vision-and-mission.aspx. Retrieved 16 May 2023

[15] NATS (2023), *Introduction to Airspace*, National Air Traffic Services. https://www.nats.aero/ae-home/introduction-to-airspace/. Retrieved January 2023.

[16] Danish Air Force (2019), *Flying in Denmark*, Royal Danish Air Force, Flight Information, Edition 35, 31 January 2019. https://dulfu.dk/wp-content/uploads/2020/04/flying-in-denmark-2019.pdf. Retrieved 16 May 2023.

[17] IAA (2023), *Guide to IAA Air Traffic Management Operations*, Irish Aviation Authority (IAA). https://www.iaa.ie/docs/default-source/misc/guide-to-air-traffic-operations.pdf?sfvrsn=6d570df3_0. Retrieved 16 May 2023.

[18] CAA (2022), *Guide to Visual Flight Rules (VFR) in the UK*, Civil Aviation Authority (CAA).

[19] NATS (2023), *Introduction to Airspace*, National Air Traffic Services. https://www.nats.aero/ae-home/introduction-to-airspace/. Retrieved 20 June 2023.

[20] Personal discussions between the author and Cdr John Taylor (RN Retired).

[21] ASI (2023), *Aerodrome Traffic Zones*, Aerospace Safety Initiative. https://airspacesafety.com/wp-content/uploads/2019/04/AerodromeTrafficZone.pdf Retrieved 20 June 2023

[22] CAA (2014), *Level Busts: Hazards and Defences*, CAA. https://publicapps.caa.co.uk/docs/33/SafetyNotice2014004.pdf. Retrieved 20 June 2023.

[23] EUROCONTROL (2019), *EUROCONTROL Specifications for harmonized Rules for Operational Air Traffic (OAT) under Instrument Flight Rules (IFR) inside controlled Airspace of the ECAC Area (EUROAT)*, EURO CONTROL. https://www.eurocontrol.int/sites/default/files/2019-11/change-5-eurocontrol-specificatitions-oat-ifr-rules-version_0.pdf. Retrieved 20 June 2023.

[24] FAA (2022), *Principles of Flight*, FAA. https://www.faa.gov/regulations_policies/handbooks_manuals/aviation/phak/media/06_phak_ch4.pdf. Retrieved 20 June 2023.

[25] CAA (2022), *Manual of Air Traffic Services – Part 1, CAP-493, Safety and Airspace Regulation Group (SARG)*, CAA, 10th ed.

[26] EASA (2023), *Easy Access Rules for Standardised European Rules of the Air*, *European Air Safety Agency*, European Union.

[27] ICAO (2005), *Rules of the Air*, Annex 2 to the Convention on International Civil Aviation, 10th ed., ICAO, July 2005.

[28] CAA (2023), *Guide to Visual Flight Rules (VFR) in the UK*, Civil Aviation Authority. https://www.bfgc.co.uk/VFR_Guide.pdf. Retrieved 20 June 2023.

[29] Hallion, R.P. (2017), *Solving the Problem of 'Fog Flying'*. https://www.historynet.com/solving-problem-fog-flying/. Retrieved 16 May 2023.

[30] Lindbergh, C.A. (1953), *The Spirit of St Louis*, Scribner.

[31] FAA (2023), *Flying Blind the Story of the First Takeoff, Flight and Landing Using Only Instruments as a Guide*, FAA Aerospace Pioneers. https://www.faa.gov/about/history/pioneers/media/First_Instrument_Flight_Doolittle.pdf. Retrieved 20 June 2023.

[32] Crane, C.J. and Ocker, W.C. (1930), *Bank and Turn Indicator*, US Patent US2053182A.

[33] Carter, L. (1938), *Artificial Horizon*, US Patent US2183133A.

[34] Nolan, M.S. (2003), *Fundamentals of Air Traffic Control*, Brooks/Cole.

[35] Binns, C. (2018), *Aircraft Systems – Instruments, Communications, Navigation, and Control*, John Wiley and Sons for IEEE Press.

[36] NATS (2020), *ATC 100: 100 Years of Keeping the Skies Safe*, NATS. https://www.nats.aero/news/features/atc100/. Retrieved February 2023.

[37] Stacey, D. (2008), *Aeronautical Radio Communications Systems and Networks*, Wiley and Sons.

[38] CAP 452 (2016). https://publicapps.caa.co.uk/docs/33/CAP452_Edition15_OCT2016.pdf.

[39] Ofcom (2017), *UK Frequency Allocation Table*, Ofcom, Issue 18v1.1 Updated 12 January 2017. https://www.ofcom.org.uk/__data/assets/pdf_file/0016/103309/uk-fat-2017.pdf.

[40] EUROCONTROL (2016), *8.33 kHz Voice Channel Spacing Communications*, EUROCONTROL. https://833radio.com/news/show/6. Retrieved February 2023.

[41] CAA (2023), *Aeronautical Radio Stations: Who Is Responsible and How Is It Managed?* https://www.caa.co.uk/commercial-industry/airspace/communication-navigation-and-surveillance/aeronautical-radio-stations/. Retrieved February 2023.

[42] Heathrow (2019), *ATC Radio Frequencies Control of Vehicles on the Manoeuvring Area*, Heathrow Airport Operational Safety Instruction. https://www.heathrow.com/content/dam/heathrow/web/common/

documents/company/team-heathrow/airside/operational-safety-insttructions/ASDRVE_OSI_010.pdf. Retrieved February 2023.

[43] EUROCONTROL (2009), *EUROCONTROL Guidelines on the Use of UHF for Air Traffic Control (ATC)*, EUROCONTROL Guidelines, Edition 1.0. Edition date: 02/06/2010 Reference nr: EUROCONTROL-GUID-138-2009. https://www.eurocontrol.int/initiative/ultra-high-frequency-air-traffic-control#:~:text=The%20ultra%2Dhigh%20frequency%20(UHF,cases%20to%20support%20OAT%20flights. Retrieved February 2023.

[44] ICAO (2016), *Procedures for Air Navigation Services: Air Traffic Management*, Document 4444, 16th ed.

[45] Stacey, D. (2008), *Aeronautical Radio Communications Systems and Networks*, John Wiley and Sons.

[46] Blake, D.K. (1988), *Modern Radar System Analysis*, Artech House.

[47] Toomay, J.C. (2012), *Radar Principles for the Non-Specialist*, SciTech Publishing.

[48] NATS (2023), *Introduction to Airspace*. https://www.nats.aero/ae-home/introduction-to-airspace/#:~:text=Upper%20air%20routes%20(UARs)%20sit,is%20Class%20C%20controlled%20airspace Retrieved February 2023.

[49] FAA (2019), *Communications Requirements in Oceanic Airspace Delegated to the FAA for Provision of Air Traffic Services*, Federal Aviation Administration, International Notices, 18 July 2019. https://www.faa.gov/air_traffic/publications/internationalnotices/intl_2_20000.html. Retrieved February 2023.

[50] Naqvi, A. and Ling, H. (2013), 'A study of radar features of wind turbines in the HF band', *Progress in Electromagnetics Research*, 143:605–621.

[51] See for example, CAA (2018), *Airspace Modernisation Strategy*, CAP 1711, [UK] Civil Aviation Authority, 17 December 2108.

[52] Swopes, B.R. (2018), *9 May 1932: Albert Francis Hegenberger*. https://www.thisdayinaviation.com/tag/albert-francis-hegenberger/ Retrieved 4 June 2023.

[53] USAF (1948), *Major General Albet F Hegenberger*, Biography. https://www.af.mil/About-Us/Biographies/Display/Article/106809/major-general-albert-f-hegenberger/. Retrieved 1 June 2923.

[54] ICAO (2006), *Aeronautical Telecommunications*, ICAO Annex 10 to the Convention on Civil Aviation, vol. 1. Chapter 2: General Provisions for Radio Navigation Aids.

[55] Airnav (2023), *NDB*, Airnav.eu. https://www.airnav.eu/index.php?stranka=NDBen#:~:text=L%20the%20NDB%20beacon%20in,of%2012%20times%20per%20minute. Retrieved 6 June 2023.

[56] ITU (2019), *Radio Noise*, ITU Publication, ITU-R P.372-14, August 2019.

[57] Nolan, M.S. (1999), *Fundamentals of Air Traffic Control*, 3rd ed., Brooks/Cole.

[58] Luck, D.G.C. and Norton, L. (1937), *Apparatus for Timing of Periodic Events*, US Patent Application 2121359, 31 March 1937.

[59] Curtis, D. (2016), *Is this the End for Stack Holding?* National Air Traffic Services, Airspace blog. https://nats.aero/blog/2016/05/is-this-the-end-of-stack-holding/. Retrieved 15 May 2023.

[60] King, J. (2015), *Has GPS killed off VOR*? UK National Air Traffic Services. https://nats.aero/blog/2015/05/has-gps-killed-off-the-vor/. Retrieved 11 May 2023.

[61] FAA (2023), *Very High Frequency Omnidirectional Range Minimum Operational Network (VOR MON)*, Federal Aviation Administration Navigation Programs. https://www.faa.gov/about/office_org/headquarters_offices/ato/service_units/techops/navservices/gbng/vormon. Retrieved 11 May 2023.

[62] ICAO (2006), *Aeronautical Telecommunications, ICAO Annex 10 to the Convention on Civil Aviation*, vol. 1 Chapter 2: General Provisions for Radio Navigation Aids.

[63] Wiki (2023), *VHF/UHF Plan for Aeronautical Radionavigation*. https://wiki.radioreference.com/index.php/Instrument_Landing_System_(ILS)_Frequencies. Retrieved 11 May 2023.

[64] Alford, A. (1942), *Antenna System*, US Patent Application 2,283,897. Submitted 26 April 1939 and granted 26 May 1942.

[65] Lee, P. (2016), *The Technical Wizardry of VORs: How They Work*, MIET. https://www.youtube.com/watch?v=tZ2gG1v9Xg8. Retrieved 22 May 2023.

[66] Buderi, R. (1996), *The Invention that Changed the World: How a Small Group of Radar Pioneers Won the Second World Ware and Launched a Technology that Changed the World*, Simon and Schuster.

[67] Lo. S., Chen, Y.H., Enge, P., and Peterson, B. (2013), *Distance Measuring Equipment Accuracy Performance Today and for Future Alternative Position Navigation and Timing (APNT)*, Federal Aviation Administration, Robert Lilley, Aviation Management Associates. https://web.stanford.edu/group/scpnt/gpslab/pubs/papers/Lo_IONGNSS_2013.pdf. Retrieved 7 June 2023.

[68] CAA (2022), *DVOR/DME/NDB Rationalisation: Guidance for the use of RNAV Substitution*, UK Civil Aviation Authority.

[69] FAA (2023), *Navigation Programs – NextGen DME*, US Federal Aviation Administration. https://www.faa.gov/about/office_org/headquarters_offices/ato/service_units/techops/navservices/gbng/nextgen_dme. Retrieved 7 June 2023.

[70] Airnav (2023), 'DME Infographics', Airnav.eu. https://www.airnav.eu/index.php?stranka=DMEen#:~:text=The%20DME%20operates%20on%20frequencies,are%20paired%20to%20response%20frequency. Retrieved 7 June 2023.

[71] DoD (1998), *Department of Defense: Interface Standard: Standard Tactical Navigation (TACAN) Signal)*, Mil Std 291-C. 10 February 1998.

[72] ICAO (2023), *History: The Beginning*, International Civil Aviation Organization (ICAO). https://www.icao.int/EURNAT/Pages/HISTORY/history_1910.aspx. Retrieved February 2023.

[73] A. E. B. (1928), 'Aids to air pilotage', *Science Progress in the Twentieth Century (1919–1933)*, 22(88):667–670. http://www.jstor.org/stable/43428603.

[74] Nature (1921), 'The leader cable system', *Nature*, 760–762.

[75] Aviation (1922), 'The Loth guide cable for flying in fog. French invention for guiding aircraft through fog described. System functions in preliminary trial', *Aviation (predecessor of 'Aviation Week')*, 12(15):422–423.

[76] Diamond, H. and Dunway, F.W. (1930), 'A radio beacon and receiving system for the blind-landing of aircraft', *Journals of Research*, 6:897–931.

[77] Conway, E.M. (2006), *Blind Landings: Low-Visibility Operations in American Aviation, 1918–1958*, John Hopkins University Press.

[78] IEEE (1964), *Citation for Ernst L Kramar*, IEEE, 1964 Pioneer Award from the IEEE. https://ieeexplore.ieee.org/stamp/stamp.jsp?arnumber=4502167. Retrieved 28 February 2023.

[79] Hollman, M. (2007), *History of Radio Flight Navigation Systems*. http://www.radarworld.org/flightnav.pdf. Retrieved 28 February 2023.

[80] Holm, R. (2017), *ILS Fundamentals*, CreateSpace Independent Publishing Platform; Edition 8.2.

[81] See for example, CAAM (2023), *Air Traffic Control*. https://www.caam.gov.my/resources/aviation-professional/air-traffic-control-atc/#1613615119676-482589e1-60fd. Retrieved 1 January 2023.

[82] See for example, Brookner, E. (ed.) (1988), *Aspects of Modern Radar*, Artech House.

[83] CAA (2023), *Aviation Safety*, UK CAA. https://www.caa.co.uk/consumers/guide-to-aviation/aviation-safety/#:~:text=Within%20controlled%20airspace%20there%20are,normally%20increases%20to%20five%20miles. Retrieved 3 January 2023.

[84] Skolnik, M. (2001), *Introduction to Radar Systems*, McGraw-Hill.

[85] *CAA Policy and Guidelines on Wind Turbines*, CAP 764, Civil Aviation Authority Directorate of Airspace Policy.

[86] Skolnik, M. (ed.) (1988), *Radar Applications*, IEEE Press.

[87] ITU (2018), *Mathematical Models for Radiodetermination Radar Systems Antenna Patterns for Use in Interference Analyses*, ITU-R M 1851, ITU.

[88] Barton, D.K. (1988), *Modern Radar System Analysis*, Section 1.5, Artech House.

[89] Skolnik, M. (2008), *Radar Handbook*, McGraw Hill

[90] Butler, M. and Johnson (2003), *Feasibility of Mitigating the Effects of Wind Turbines on Primary Radar*, Alenia Marconi Systems, ETSU W/14/00623/REP DTI PUB URN No. 03/976.

[91] Skolnik, M. (2001), *Introduction to Radar Systems*, McGraw-Hill.

[92] Neyman, J. and Pearson, E.S. (1933), 'On the problem of the most efficient tests of statistical hypotheses', *Philosophical Transactions of the Royal Society: London*, 231(Ser A):289–337.

[93] Chatfield, C. (1979), *Statistics for Technology: A Course in Applied Statistics*, Chapman and Hall.

[94] Finn, H.M. and Johnson, R.S. (1968), 'Adaptive detection mode with threshold control as a function of spatially sampled clutter level estimates', *RCA Review*, pp. 414–64.

[95] Hansen, V.G. (1973), 'Constant false alarm rate processing in search radars', in *IEE Radar Proceedings*, London, October 1973.

[96] Nitzberg, R. (1999), *Radar Signal Processing and Adaptive Systems*, Artech House.

[97] Gough, J. (1993), *Watching the Skies: The History of Ground Radar in the Air Defence of the United Kingdom*, HMSO.

[98] Stephens, M.C. (1998), *Secondary Surveillance Radar*, Artech House.

[99] CAA (2019), *Operation of IFF/SSR Interrogators in the UK: Planning Principles and Procedures*, UK Civil Aviation Authority and the UK MOD. CAP 761, January 2019.

[100] Meneses, G. (2014), *ORCAM CAM SSR Codes*, Annex B, 29 September 2014. https://www.icao.int/NACC/Documents/eDOCS/ATM/ORCAM%20CAR%20SAM%20-%20SSR%20Codes%20en.pdf. Retrieved 18 June 2023.

[101] Sun, J. (2021), *The 1090 Megahertz Riddle: A Guide to Decoding Mode S and ADS-B Signals*, TU Delft Open Publishing. https://mode-s.org/decode/content/introduction.html#:~:text=The%20Mode%20S%20communication%20protocol,the%20data%20block%20are%20presented. Retrieved 20 June 2006.

[102] Ritter, J. (2002), *Flight Path Study American Airlines Flight 77*, National Transportation Safety Board (NTSB), 19 February 2002. https://www.ntsb.gov/about/Documents/Flight_Path_Study_AA77.pdf. Retrieved 21 March 2023.

[103] FAA (2011), *Introduction to TCAS II: Version 7.1*, FAA, 28 February 2011. https://www.faa.gov/documentlibrary/media/advisory_circular/tcas%20ii%20v7.1%20intro%20booklet.pdf. Retrieved 19 June 2023.

[104] Kaune, R., Steffes, C., Konle, W., and Pagel, J. (2012), 'Wide area multilateration using ADS-B transponder signals', in *15th International Conference on Information Fusion*.

[105] Skolnik, M.I. (2001), *Introduction to Radar Systems*, 3rd ed., McGraw-Hill.

[106] Brookner, E. (1999), *Practical Aspects of Phased Array Antenna Systems*, Artech House.

[107] ICAO (2006), *Aeronautical Telecommunications: Volume 1 Radio Navigational Aids*, Annex 10 to the Convention on International Civil Aviation. 6th Edition July 2006. Chapter 3 and Attachment C.

[108] IATA (2017), *User Requirements for Air Traffic Services (URATS) Communications, Navigation, and Surveillance (CNS) Technologies*, International Air Transport Association, Edition 3, July 2017. https://www.osti.gov/servlets/purl/894743.

Chapter 4
The wind, wind turbines and wind farms/wind parks

4.1 Introduction

This chapter is concerned with the wind, wind turbines and wind farms or, as they are called in Continental Europe, wind parks. A definition of terms is provided. The components of an individual wind turbine are described; this is important because each component interacts with aviation radio and radar systems with differing effects. Planning of wind farms is described. It is during the planning process that possible interactions between a proposed wind farm and aviation interests are predicted and the need to predict interactions in advance is a critical consideration. The phases of construction of a wind farm are described; this is also important because when mitigation of the effects of wind farms is being implemented, the milestones of the mitigation project and the construction project must be coordinated. The radio signature of wind turbines and wind farms are described and the chapter concludes by examining how these signatures might affect aviation radio and radar systems.

The chapter begins with a discussion about wind: what causes it and why wind energy is considered renewable. Wind shear and turbulence are described. These properties of wind directly affect the size, tip height and hub height of turbines and give an insight into how these wind turbines will change in the future. A recognised method of classifying wind strength, and, therefore, one method of classifying wind turbines (and its limitations), is provided. Some features of the operation of a wind farm are discussed which, in some specific circumstances, may be a useful mitigation.

4.2 The wind

4.2.1 Causes of terrestrial wind

The wind is created by the following large-scale effects in the atmosphere:

- **Differential heating** of the Earth's surface by the Sun. The variations in the Sun's illumination are caused by diurnal changes (day/night) and variations throughout the year caused by the seasons. Another cause of differential heating is the difference in the rates at which the sea and the land warm and

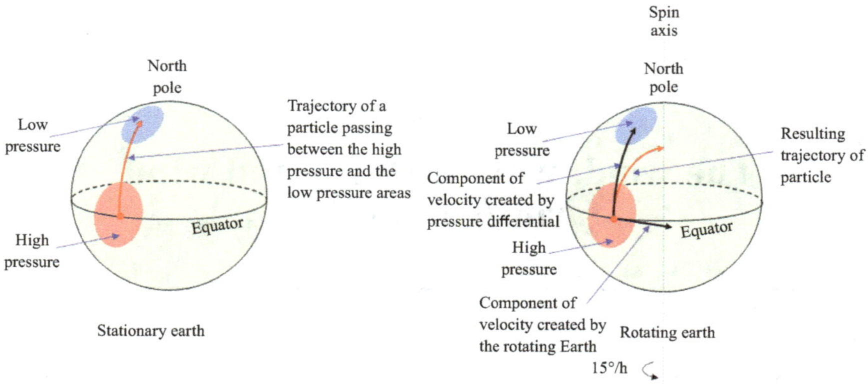

Figure 4.1 Coriolis effect

cool. The air above a warm surface tends to rise increasing its pressure and the opposite effects occur over cool surfaces, lowering the pressure,

- **Air pressure differences** cause the movement of air (wind) from areas of high pressure to areas of low pressure, the path taken by the wind is determined by the following:

The Coriolis Effect. Consider the path taken by a particle in the air flowing between a region of high pressure, at the equator, and a region of lower pressure further north. Assume, initially, that the Earth does not rotate. In this circumstance, a particle in the wind would move in a straight line between the two regions from the high pressure to the low pressure; this is shown in the left-hand case in Figure 4.1. The speed of the particle is determined by the pressure difference. In the second case, the Earth is spinning on its axis, completing one revolution every day; that is spinning 360° in 24 h which corresponds to 15° per hour. Note also that the spin is always to the East (hence the sun rising in the East and setting in the West). The angular rate of rotation is the same at every point on the Earth's surface. However, when translated to a linear velocity, a point on the surface at the Equator travels a distance equivalent to the circumference of the Earth (40,075 km) every 24 h. Improbable as it may sound, a point on the surface at the Equator, and the air immediately above it, is travelling eastward at a speed of 1,670 km/h. Moving north from the equator, even though the angular rate remains the same (15° per hour), the linear velocity of a point on the surface decreases with latitude until at the North Pole each rotation is completed with no change in the linear position at all, that is the linear velocity reduces to zero*. Consider then, the trajectory of a particle in the wind travelling from the high pressure on the equator to a region of lower pressure in the north. The particle's velocity has a northerly component caused by

*The Earth's spin axis is not completely stable; it has a periodic motion called precession. The period of this precession is 26,000 years. This motion has been ignored in the above discussion.

the pressure differential but it also has an easterly component caused by the linear velocity acquired from its origin at the Equator. The resulting trajectory is a curved path arching to the east. This path is illustrated on the right-hand side of Figure 4.1. Thus, wind does not flow from a region of high pressure, instead it swirls in a clockwise direction in the Northern Hemisphere. And, air flows around regions of cold pressure in the opposite direction. The Coriolis effect also applies to winds in the Southern Hemisphere but, in this case, the winds move in an anticlockwise direction around regions of high pressure and clockwise around areas of low pressure. The Coriolis effect gives rise to the prevailing weather patterns and the prevailing wind directions. A final comment on the Coriolis effect; it is often described in the literature (incorrectly) as a force; on the basis of Newton's first law which states that a force is required to change the motion of an object. No force is necessary for the Coriolis effect to occur because the particle already possesses the motion that causes the effect.

- **Centripetal force** acts on the wind moving around a centre of rotation, that is, either a high-pressure or low-pressure centre, compressing the size of the spiral as it rotates faster. A common analogy is the figure skater spinning and increasing in speed as their arms are drawn in [1,2].

All these phenomena will continue to create and determine the direction of winds as long as the Sun illuminates the Earth creating pressure differentials and as long as the Earth spins on its axis producing the Coriolis effect. In short, wind energy is renewable.

4.2.2 Friction and wind

The forces and effects listed above create wind and determine its direction of flow; however, they are not the only forces acting on the wind. Where wind touches a surface, it is acted upon by friction which reduces the speed of the wind and in some circumstances will result in the phenomena of turbulence. Both wind speed and turbulence play an important role in the design of a wind farm, affecting the type of wind turbine selected, their hub heights† and tip height. These factors, in turn, play a key role in determining the visibility of the wind turbines to aviation radio and radar systems.

To describe the relationships between friction, wind speed and turbulence, the term '**wind shear**' will be introduced. However, wind shear has a slightly different meaning to different communities. These differences are described first to avoid confusion.

4.2.2.1 Wind shear

The flow of a gentle wind on a calm day is **laminar**, a term derived from the Latin word 'lamina', a thin leaf. The wind blows as though it moved in discrete layers with no interaction between those layers (called the lamina); the idea is illustrated in Figure 4.2.

†There may be more than one hub height in a wind farm.

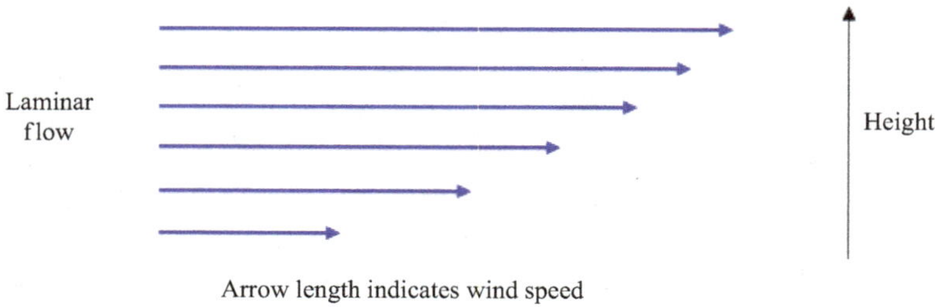

Figure 4.2 Laminar air flow

Friction causes the wind in the lamina touching the retarding surface to be slowed. It can be assumed that the retarding effect is in the opposite direction to the wind‡. The wind speed in laminae farther from the surface increases, as shown in Figure 4.2. A term for the rate of increase in speed is **wind gradient**. Another term used in meteorology for this phenomenon is **wind shear** [3].

In aviation, the term 'wind shear' has a different, albeit related, meaning. In laminar wind flow, the wind blows from a region of high to a region of low pressure and it is unidirectional. However, sometimes the wind direction can change abruptly. The following conditions can prompt these rapid changes:

- Weather fronts.
- Thunderstorms.
- Temperature effects, including inversions (normally, the temperature of the atmosphere decreases as height increases. In an inversion, regions of warm air are trapped under cooler air).
- Surface terrain and obstructions [4].

Wind shear in aviation refers to these abrupt changes in direction. The sudden changes in direction may be horizontal. However, they can also blow downwards. The reason why this condition is a concern to pilots will be obvious.

In the following discussion, the term wind shear is used in the less specific meteorological sense of wind gradient.

4.2.2.2 Wind speed and wind turbine sizes

As discussed above, the retarding effect of friction occurs when wind touches surfaces. The nature of the surface material determines the amount of friction and, hence, its retarding effect. Rougher surfaces create more friction than smooth surfaces; for example, the built environment is rougher than forest, forest is rougher than agricultural land and this, in turn, is much rougher than the sea. The measure of roughness is captured in a quantity called the **roughness length**, usually represented in formulae by the symbol z_0. It is an estimate of the height above the surface where the wind speed would be reduced to zero which is approximately one-tenth of the height of the objects present in the subject environment [5].

‡Strictly, wind is *y* a vector quantity having magnitude and direction.

Having measured the wind speed at a given height, using an anemometer, a common requirement is to estimate the wind speed at other heights above the measurement position. Assuming that temperature and pressure are held constant, this calculation can be carried out using the following formula [6]:

$$\text{Wind Speed, } v = v_{ref} \frac{\ln (z/z_0)}{\ln (z_{ref}/z_0)} \qquad (4.1)$$

where v is the wind speed at a height z above ground level, v_{ref} is the wind speed at the reference height, z is the height above ground level at which the wind velocity is required; z_0 is the roughness length in the direction of the wind; z_{ref} is the reference height (at which the wind speed was known).

Note: The speeds in the above equation are measured in units of metres per second and the distance values are measured in metres. However, v and z are referenced to ground level. It is commonplace for such values to be measured in metres above ground level (magl) or above sea level (masl). Another measure commonly referred to is metres above mean sea level (mamsl), although mamsl and masl are frequently assumed to be the same. The roughness length is a measure of distance and it is measured in metres.

This formula was applied for a notional wind speed measurement carried out at 100 m above the surface. An arbitrary value of 7.5 m/s was chosen. The rate of change of wind speed (the wind shear profile) was calculated for wind blowing over forested land, agricultural land and the sea, that is, z_0 of 1, 0.2 and 0.0002, respectively [7]. The results of this analysis are plotted in Figure 4.3, where the rate of change of wind shear is shown at increasing heights above ground level for different types of surfaces. The general trend is the same for all three surfaces (roughness values), that is that the rate of change of wind speed versus height is greatest closer to the surface where the surface is 'sticky' and retarding. The three plots show that, for the surfaces chosen, the most retarding surface is forested land and the least retarding surface is open sea.

The more general conclusions illustrated by this analysis are the following:

- Wind speeds increase with height, likewise the kinetic energy in the wind. Therefore, the taller a wind turbine is, the greater the potential energy yield is.
- Related to the above finding, even at a height of 100 m above ground level, the nature of the surface still influences wind speed and the more favourable conditions are over the sea (i.e. off-shore versus on-shore) and collection conditions over open land are more favourable than in regions of forestation.

However, the wind speed is not the only factor that must be taken into account by the designer of a wind farm. Recall that an implicit assumption in the above calculation is that the temperature and pressure were assumed to be the same for all three roughness values. In practice, this is unlikely to be the case. For example, local temperature variations near the surface will play a role. The margins of areas are also influential, that is the transition from, say, open land to forestation.

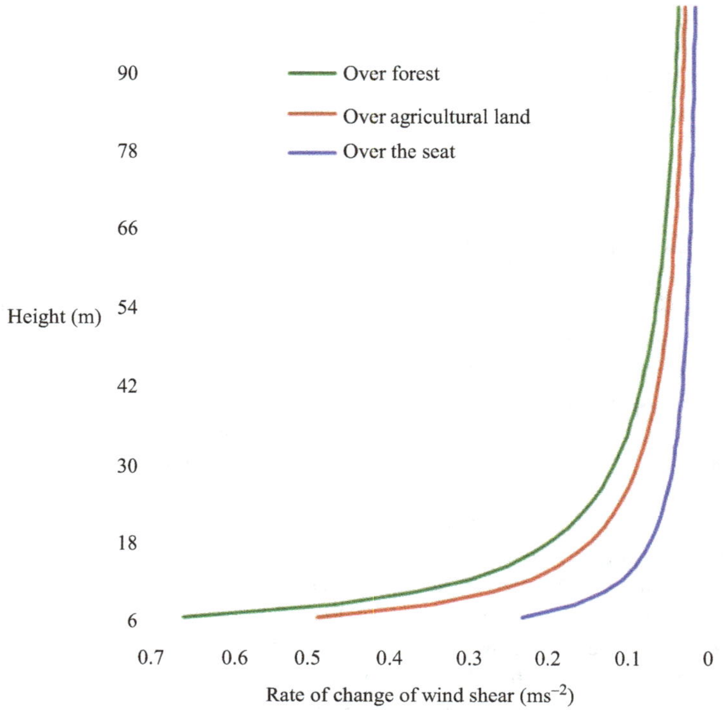

Figure 4.3 Rate of change of wind shear versus height for different surfaces

Moreover, the rate at which wind speed varies with height provides a greater potential for **turbulence**.

4.2.3 Turbulence

In practice, the speed of the wind varies continuously at all levels. If the changes in wind speed are rapid and short-term (gusty), then the laminar nature breaks down and this is referred to as turbulence. Figure 4.4 illustrates the difference between the laminar and the turbulent flow. The higher the wind shear, that is, the greater the rate of change of wind velocity with height, the greater the potential for turbulence. Temperature variations also play a role in creating turbulence by causing localised areas of higher and lower pressure. Therefore, turbulence can vary from location to location within a site. If turbulent air cannot be avoided by changing the location of a turbine when a wind farm is being designed, then there are two principal methods to deal with turbulence; tip height and the type of blades used.

The degree of turbulence that might be expected at a prospective wind farm site is a key factor in the design of the site and the selection of the turbines.

In summary, on-shore, wind flow is more likely to be turbulent because of greater friction and local temperature variations and this effect is exacerbated in

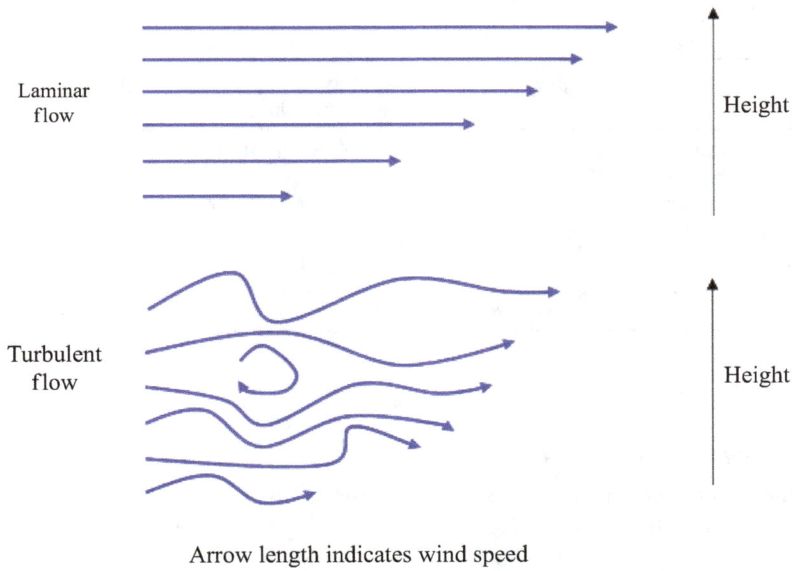

Figure 4.4 Laminar versus turbulent flow

forests. To reduce the effects of turbulence on turbine blades, the tip height of turbines can be increased, effectively lifting the rotor out of the way of the turbulent air. Turbines can operate in turbulent air but special-to-purpose (that is stronger) blades must be used and the life of the turbine blades will be less than those operating in a laminar flow. Off-shore, friction (roughness length) is orders of magnitude lower, the more uniform surface is less prone to localised temperature variations so hub heights and turbine tip heights can be lower.

4.2.4 Wind speed classes

The International Electromechanical Commission (IEC) has developed a large number of specifications for wind turbines and wind farms. In 2005, they published (inter alia) a classification system for wind speeds and turbulence conditions[§] [8]. The classification scheme is set out in Table 4.1; the class number indicates the wind speed ranging from class 1 which is high speed to class 4 which is low speed. The class also has a suffix which is either A or B which indicates the degree of turbulence, A is higher turbulence (up to 18% of the time) and B lower turbulence (up to 16% of the time). This specification is widely referenced in the literature but today it is more applicable to on-shore and smaller developments than off-shore, where much higher speeds are observed routinely; the cut-out speeds for many off-shore turbines is 25 m/s or greater, for example, the Vestas C-162-6-2. The turbines used in all new off-shore developments all turbines are IEC category 1A.

[§]This standard was revised in 2019.

Table 4.1 IEC wind classes

Class	Speed	Annual average speed at hub height	Once in 50-year extreme gust
1A higher turbulence 1B lower turbulence	High	From 8.5 m/s to 10 m/s and above	70 m/s
2A higher turbulence 2B lower turbulence	Medium	From 7.5 m/s to 8.5 m/s	59.5 m/s
3A higher turbulence 3B lower turbulence	Low	From 6 m/s to 7.5 m/s	52.5 m/s
4	Very low	Up to 6 m/s	42 m/s

4.3 Definitions

4.3.1 The wind turbine

In some early literature on the subject, machines for converting wind power to electricity were referred to as windmills, for example, the US Department of Defense (DoD) expressed the impact of State of the Art (SOA) windmill farms on military readiness in 2006 [9]. Today, it is generally accepted that the term for such a machine is a wind turbine or a wind turbine generator (WTG) and the term windmill is restricted to those machines that mill flour using wind power. The term has also been adopted for machines that lift water to enhance the drainage of land. The collective description for all these classes of a machine is that they are 'wind-powered'.

4.3.1.1 Turbine types

The conversion of the kinetic energy of wind to electricity is via the rotational kinetic energy of the turbine blades and the shaft to which they are connected. The shaft may be mounted horizontally, hence, the machine would be called a horizontal axis wind turbine or vertically creating a vertical axis wind turbine. More recently, designs are being developed based on vortex technology, the latter consisting of a cylindrical tube that is free to oscillate in the wind [10].

The rotating shaft may be connected to the generator via a gearbox or connected directly. The benefit of having a gearbox is that the relatively low rotational speed of the blades/wind shaft (typically 10–15 revolutions per minute) can be increased to a speed that is more suitable for generating electricity (typically 1,000 to 1,800 rpm). However, there are important benefits from the direct drive; gearboxes require regular maintenance and are a source of failure and they are heavy, adding to the complexity of construction. The disadvantages of direct drive arise from the increased complexity and weight of the generators. The drive shaft spins permanent magnets within a fixed stator that is wound with wire coils. As the lines of force from the magnets act on the coils, the electricity is generated. The arrangement would have been recognised by Faraday (see Chapter 2). The magnets need no excitation but they are made from rare earth materials such as neodymium and dysprosium which are expensive and the reduction in mass within the nacelle is, to some extent, offset by the increase in mass of heavier generators [11].

The different types of generators have led to one of the means of classifying wind turbines, that is by the technology used to generate electricity; the following classes represent increasingly sophisticated power generation [12]:

- Type 1: limited variable speed, squirrel cage induction generator (SCIG)
- Type 2: limited variable speed, wound rotor induction generator
- Type 3: variable power with partial power electronics, doubly fed induction generator (DFIG) or doubly fed asynchronous generator (DFAG)
- Type 4: variable power with full power electronic conversion
- Type 5: variable speed drive train with synchronous generator.

A characteristic of the direct drive generator is that the power produced is alternating current but the frequency and amplitude are variable. This power cannot be used by the electricity grid. To condition the power so that it can be exported to the grid, it is first rectified to produce direct current and then converted back to alternating current using an inverter. The inverter can be phase locked to the grid, greatly simplifying connecting and disconnecting turbines. The resulting power is alternating current at mains frequency. However, power conversion comes with disadvantages. Losses may be between five and ten percent. The associated semiconductor electronics must deal with high powers in confined spaces, may be unreliable and costly and difficult to repair.

4.3.1.2 Other classification systems

A simpler and common method of classifying wind turbines is by scale (the amount of electricity that the generators can produce):

- Utility scale (on-shore)
- Off-shore
- Small wind, definitions vary from country to country, the US Department of Energy [13] has endorsed a definition of this category as being wind turbines generating less than 100 kW, in the United Kingdom, the Microgeneration Certification Scheme (MCS) [14] considers systems generating up to 50 kW as small.

Notwithstanding the different classification methods listed above, probably the commonest classification is by the amount of power that the generator is capable of producing; for example, a wind turbine might be specified as an 8 MW, 11 MW, 14 MW or a 22 MW machine.

This is a convenient place to compare the power capacity of on-shore and off-shore turbines.

4.3.1.3 On-shore/off-shore power comparisons

Off-shore wind turbines will, typically, produce double the amount of electricity that an on-shore turbine will generate with a comparable rotor diameter. Each on-shore, utility-scale, wind turbine entering service in 2023 will typically generate 7 MW of electricity compared with each off-shore turbine which might produce 15 MW.

This relationship could apply for the foreseeable future. There are plans to operationalise 22 MW turbines offshore and wider developments in off-shore technology may facilitate greater rotor diameters and higher powers.

4.3.2 Wind farm/wind park

A wind farm is a defined area containing a group of wind turbines. In principle, two wind turbines might be considered a wind farm, in practice it is usually more. The area does not have to be on land, it may also be off-shore and the turbines need not be fixed to the sea bed; wind farms are being developed that use floating offshore wind turbines (FOWT).

The scale of a wind farm and its footprint varies considerably. Single turbines providing an alternative source of energy in a large domestic or small commercial setting such as the one shown in Figure 4.5 are common. Medium scale wind farms providing power for larger businesses are also commonplace such as the one illustrated in Figure 4.6.

However, large, utility-scale, development, such as the one shown in Figure 4.7, the 3400 turbine Tehachapi Pass wind farm at Mojave in California, which is the largest wind farm in the United States, is an extreme example. Today wind farms of that scale are becoming increasingly the prerogative of the off-shore environment.

Figure 4.5 Single turbine in an agricultural setting

Figure 4.6 Typical factory setting

Figure 4.7 Large-scale utility setting

Figure 4.8 Battery facility at Pen y Cymoedd Wind Farm (image courtesy Vattenfall PYC)

4.3.3 Combined energy farm

Wind turbines only produce power when the wind is blowing; they are not able to store the energy on their own and adjunct technologies are required to provide a persistent source of energy. Increasingly, modern wind farms will be built to include an integral battery storage facility; for example, Vattenfall's Pen y Cymoedd, 76 wind turbine, wind farm in South Wales includes a 22 MW lithium-ion battery system, see Figures 4.8 and 4.9. Such facilities are valuable tools for supporting grid operations. They may be used to stabilising the supply frequency and preventing over-reaction to short-term effects. Alternatively, in the event of a failure of supply, they can be used as a stopgap until other facilities can be brought on-line such as pumped storage generation.

This development is an example of a more comprehensive trend in which multiple energy capabilities share resources, such as grid access and road transport. Such sites might include wind turbine generation, solar cells, hydrolysers and battery storage [15].

4.3.3.1 Operation of a storage system

The frequency of the electricity on the grid must be maintained within a tight tolerance, for example, the UK National Grid operates at a frequency of 50 Hz and the maximum permitted tolerance is 1%, that is $\pm$0.5 Hz. In practice, the National Grid Company attempts to maintain the grid within an even tighter tolerance,

Figure 4.9 Battery banks at Pen y Cymoedd Wind Farm (courtesy Vattenfall PYC)

±0.2 Hz. Any variation in the grid frequency is indicative of an under or over supply. If too little electricity is being generated compared with the demand, the frequency goes down and vice versa [16]. Storage systems monitor the frequency of the grid and battery systems like the one at Pen y Cymoedd wind farm are organised so that they can either supply or store energy. The key to maintaining grid integrity is the speed of response. The associated energy conversions can start within a cycle of the mains frequency when requested.

4.4 Wind turbine construction

The principal components of a wind turbine are the tower, the nacelle and the blades; these are illustrated in Figure 4.10. These are discussed in turn below.

The foundations must also be considered; they are important for a number of reasons. Digging the foundations for on-shore wind turbines is one of the most

Figure 4.10 HAWT turbine components

time-consuming tasks in the construction of a wind farm; this affects the construction schedule and this, in turn, affects the time available for implementing mitigation. Furthermore, weather dependencies for construction and, what is called, balance of plant (discussed later in this chapter) are likely to be factors in determining key construction dates. Off-shore, foundation technology is evolving rapidly and this has longer term implications for the siting of wind farms; the size of the turbines and, therefore, the impact they may have on aviation radar. Foundations are introduced in this section but they are also discussed later in this chapter and in Chapter 6.

4.4.1 The tower

4.4.1.1 Tower material

The tower supports the whole structure; it is generally free-standing and may be of open lattice or a tapered, tubular steel construction. Some towers have been made from concrete but worldwide these are comparatively few in number; Miceli reports that concrete is favoured in Brazil where steel prices are exceptionally high

Figure 4.11 *Tower sections (courtesy Vattenfall PYC)*

[17]. Lattice towers offer some benefits: they use less material and they do not cast a shadow in the same way as solid alternatives. However, they are generally regarded as having a less acceptable visual impact than solid towers, they are not capable of supporting the heavy loads associated with higher power turbines. Tubular steel structures prevail in modern turbine construction [18].

4.4.1.2 Tower design

The taper in steel towers reduces the materials used but also makes the tower stronger. The taper angle (the amount off vertical) is normally between 1° and 1.5°. A door in the base of the tower allows access to the interior and to the other parts of the turbine.

4.4.1.3 Tower scale and transportability

As the tower sizes of on-shore turbines have increased, they have become harder to transport and many are now sectional and are assembled on-site (Figure 4.11).

However, with towers in excess of 90 m, the base section may exceed 5 m and many countries do not allow loads of this width to travel by road. Hence, the base section of a tower may be built from a number of components which must also be assembled on-site [17]. In this regard, off-shore turbines have an advantage as they are transported to the sites using special-to-purpose vessels designed to carry such large loads.

4.4.1.4 The tower in operations – deformation

Despite their length and their slender design, wind turbine towers are very stiff. In operations, even in gusty wind, modern towers, which may measure over 100 m, would not be expected to deflect by more than a few millimetres. Thus, from the perspective of the radio signature, the tower may be assumed to be stationary.

4.4.2 The nacelle

Mounted on the tower is the turbine nacelle¶. The nacelle covers the main drive shaft, if present the gearbox, and the electricity generator as well as the turbine control equipment. The absence of the gearbox means the nacelle can be shorter than those using a gearbox as illustrated in Figure 4.12.

Instrumentation on each turbine detects the wind speed and direction and the signal is provided via a controller to the yaw motor which rotates the nacelle on the tower to maintain the blades on the up-wind side of the tower (precisely the function of the fan tail on a windmill). The traditional mode of operation is to optimise each turbine's performance to extract the maximum energy from the wind for each turbine. However, this traditional method may lead to a situation where the wake created by a turbine on the edge of a wind farm may disadvantage turbines behind by placing them in more turbulent air. Adaptive wake control systems are available now that overcome this limitation; they adjust the yaw angle of the turbine and redirect the wake away from subsequent turbines. This process reduces the power available from individual turbines but optimises the overall performance of the wind farm and also increases the life of blades in inner turbines by reducing the time they spend in turbulence.

The yaw control motors must be capable of dealing with the combined masses of the nacelle and the blades. In a modern wind turbine, the nacelle might weigh up to 1,000 tonnes. Adding to the loading is the angular momentum (gyroscopic effect) of the rotating elements. Hence, yaw slew rates are low; typically, of the order of between 0.5° per second for the larger nacelles to perhaps 2° per second for smaller nacelles [19]. Anderson points out that in many modern turbine designs, the yaw is continuous, that is, there is no end stop and the nacelle might be able to rotate through more than 360°. This ability could result in problems for the cabling carrying the power between the nacelle and the base, so this must be monitored and additional controls applied if there is a risk of tangling [20].

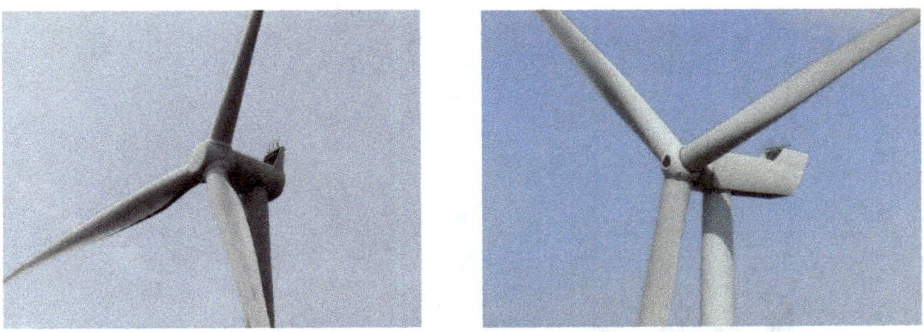

Figure 4.12 Nacelle types: direct drive on the left, with gearbox on the right

¶From the old French word for a small boat.

4.4.2.1 Yaw angle and a radar

If the turbine is in the coverage of a radar, the yaw angle selected by the nacelle will also determine the orientation of the blades with respect to the radar. This is an appropriate place to introduce some terminology that will be encountered in the literature.

It is usually the case that if the direction of the radar is perpendicular to the plane of the blades and the tower is behind the blades (in other words if the wind is coming from the direction of the radar) this is said to correspond to a yaw angle of 0°. This concept is illustrated in Figure 4.13.

To define any angle uniquely would require a convention to be adopted for nacelle rotation. For example, suppose it was deemed conventional for the nacelle to rotate anti-clockwise, then viewed from the direction of the radar if the nacelle had rotated through 90° then the tower would appear on the left the blades on the right (and the blades would be rotating in a fashion that pointed directly towards and away from the radar). This convention has been adopted later in this chapter:

Figure 4.13 Turbine axes of rotation

however. in practice, it is common to adopt the convention that 0° yaw angle and then consider angles to be ± an angle, for example, ±45° or ±90°. The value of being able to reference the yaw angle is discussed later in the chapter where the radar signature of wind turbines is discussed.

4.4.3 The blades

Mounted on the main drive shaft on the turbine hub are the turbine blades. Large turbines have been designed with single, two and three blades. Some smaller turbines have been produced with as many as five blades. Two-bladed designs can still be found in domestic and small business settings, see for example Figure 4.14, but three-bladed designs dominate, such as the turbine shown in Figure 4.10.

The blades are aerodynamically shaped and wind flow across the blades reduces the pressure on one side of the blade and increases the air pressure on the other side of the blade. The pressure differential produces the forces necessary to make the blade move, see the sketch in Figure 4.15 which includes some approximate proportions of the elements of blades.

The combination of both lift and drag forces allows the blade to move faster than the air driving the motion. For example, the Vestas V-162 turbine has a cut in wind speed (the speed at which the turbine starts to rotate) of 3 m/s and a cut-out speed of 25 m/s. Between those limits, the turbine will rotate at approximately 12–13 rpm, at which speed the blade tip motion can exceed 80 m/s. The different speeds are illustrated in Figure 4.16.

As wind turbine blades increase in size, they become heavier and in a relaxed state (not powered by the wind) they may be flexible; for example, see the off-shore wind turbine in Figure 4.17 and the on-shore turbine in Figure 4.18. However, because of their aerodynamic design, as soon as there is airflow over the blades, they become inflexible.

Blade length and tip speed are discussed in more detail below.

Figure 4.14 Tofts lane two-bladed turbine

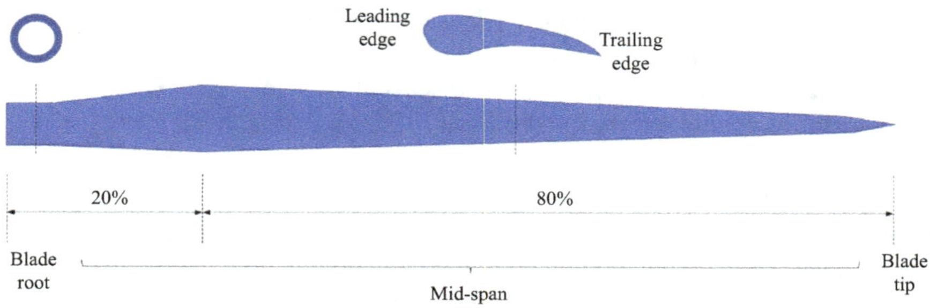

Figure 4.15 Blade plan and section sketch

Figure 4.16 Relationship between speeds

Figure 4.17 Turbine blade deformation (image Courtesy Ben Flett, Vattenfall)

Figure 4.18 On-shore blade shaping (courtesy Patrich Delaney Vattenfall PYC)

4.4.3.1 Blade speed

As rotor diameters increase in size, the blade rotational speed is reduced so that the turbine tip speed can be maintained. This reduction is modest, for example, a Vestas V80 turbine installed 10 years ago which had a rotor diameter of 80 m would have a nominal rotational speed of 16.7 rpm whereas the Vestas V162 has a nominal rotational speed of 12–13 rpm.

The tip speed for an on-shore wind turbine is usually close to 80 m/s. At higher tip speeds, the noise levels produced by turbines exceed those that are acceptable. Hence, avoiding higher tip speeds is a design goal. Higher speeds are possible off-shore where higher noise levels do not pose the same environmental challenges; typical speeds are 100 m/s. 80 m/s corresponds to 288 km/h and 156 knots, see Table 4.2.

The landing speeds of aircraft vary depending on many factors such as surface cross winds and whether or not there are any technical faults on-board the aircraft but the typical landing speed for an aircraft such as an Airbus A-320 or a Boeing 737 is 137 knots. The landing speed of larger aircraft and some military aircraft is typically a little higher; the normal landing speed of a Boeing 747 is 150 knots, similar to the normal landing speed of a Eurofighter/Typhoon. In summary, the tip speed of a wind turbine is comparable to the speed of a landing airliner. It may be concluded that velocity alone is not a good discriminant between an aircraft and a wind turbine.

The tip speed referred to above is in the plane of motion of the turbine blades. In general, the wind will strike the blades approximately perpendicular to the plane of their motion. However, the wind striking the blades at the top of their travel is moving more rapidly than at other positions in their cycle; this feature can be seen in Figure 4.18. The cause is wind shear and this places additional load on the main (low speed) shaft bearing; this was the same problem faced by the old millwrights discussed in Chapter 2. Giving the shaft a small inclination, as shown in Figure 4.19, mitigates the effect of this loading. The angle of tilt is usually approximately 4°. The tilt means that blades moving to the top of their travel move behind the hub and at the bottom of their travel they are in front of the hub. Consequently, turbine blades have a small forward and backward motion out of the plane of their rotational motion. Using a very simple model of this motion, that is

Table 4.2 Speed conversion table

m/s	km/h	Knots
60	216	117
70	252	136
80	288	156
90	324	175
100	360	194

Figure 4.19 Tilt angle

one which takes no account of the flexibility of the turbine blades[1], then the out-of-plane velocity would be calculated thus:

$$v_{oop} = \frac{2 \times \text{rotor diameter} \times \text{sine (tilt Angle)}}{2 \times \text{half the rotation period}}$$

[1]In practice, the deflection at the top of travel of the blade, where the wind speed is greatest, will be greater than that at the bottom of the travel.

Assuming a rotor diameter of 150 m, a rotational speed of 14 rpm and a tilt angle of 4°, then the average velocity for half a rotation of a blade is 4.9 m/s in the forward direction and then −4.9 m/s in the reverse direction. Note this value should be compared with the velocity thresholds for a typical Moving Target Detection system discussed in Chapter 3, there is further discussion on this topic later in this chapter where the signature of wind turbines is discussed.

4.4.3.2 Blade maintenance and leading-edge erosion

During the lifetime of a wind turbine, there is relatively little wear on the surface of the blade. However, prolonged exposure to hydrometeors (rain, hail, etc.), ultra violet (UV), dust and airborne particulates causes leading edge erosion (LEE); that is wear and tear on the part of the blade that cuts into the air as it rotates.

LEE increases the drag forces on the blade, reduces the blade efficiency and, therefore, the amount of electricity that the turbine can produce. A further undesirable effect of LEE is an increase in the noise produced by the blades.

To extend the life of the blade and mitigate noise, through-life maintenance is required: helicopter blades require similar maintenance. Specialist treatments are available for leading edge protection (LEP) and these can be applied to the blade either in or post production. LEP treatments include special coatings or tapes. Sometimes, serrated edges may be fitted to the leading edges. However, the LEP techniques do not completely solve the problem and they will themselves require periodic inspection and on-going maintenance.

4.4.4 On-shore foundations

The broader base of towers for larger turbines also requires larger foundations. These are discussed below because of the impact this has on the time taken to erect the wind turbine. The tower must be built on solid foundations and when sites are surveyed many core samples are taken to determine the suitability of the sub-structure. Despite taking samples, occasionally when a wind farm is built problems are encountered and small changes in the precise tower location have to be made. This situation is known as **micro-siting**.

Different nations' planning organisations take different lines on micro-siting. For example, in Sweden, turbines are required to fit within a 3-D volume but the turbine can fit anywhere within that volume; in the United Kingdom, an allowance is made for micro-siting and the final construction may take place a short distance from that planned; a typical allowance is 50 m. In the Netherlands, the rules are similar to the United Kingdom but in Germany, any change from the permitted location requires a new planning application. Rarely, but sometimes, micro-siting can make the difference between a turbine being visible or not to a radar. It is good practice to be alert to the possibility of micro-siting location changes when assessing turbine radar-visibility.

4.4.5 Off-shore foundations

There are two types of off-shore wind turbine: fixed foundation and floating; the depth of the sea bed determines which type can be used. In less deep waters that, with current technology, consist of shallow water which is 30 m or less in depth and transitional depths which are between 30 m and 60 m in depth, fixed foundation platforms are used. In water, deeper than 60 m, floating wind turbines can be used.

4.4.5.1 Fixed foundation platforms

The International Renewable Energy Agency (IRENA), an international, inter-governmental agency, compiled a summary of foundation types for off-shore wind turbines [21]:

- Those suitable for shallow water:
  - Gravity foundations: they consist of a heavy structure that sits on the sea bed.
  - Monopile foundations: which, as the name implies, consist of a pile driven into the sea bed.
- Those suitable for transitional depths:
  - Tripods: these are similar to monopile foundations but with three piles spaced out over the sea bed.
  - Jacket foundations: these structures are used widely for oil drilling rigs; they consist of what is, essentially, a steel tower anchored at the base with piles. The term 'jacket' is used because the structural components have a protective layer to reduce corrosion.

These foundation types represent a trend that has enabled operations in increasingly deeper water. In the literature, water depths are often associated with distances from the shore; however, this is not always the case. For example, the North Sea is comparatively shallow; the average depth is only 95 m and the waters 20–40 km offshore can still be exploited for electricity production by fixed turbine platforms. The Eastern seaboard of the United States is similar. However, there are less accessible parts of the North Sea and there are parts of California, for example, where the continental shelf is less than 1 km wide [22]. In these deeper areas, new technologies will be required; for example, floating platforms.

4.4.5.2 Floating wind turbines

FOWTs offer a number of advantages for the developer, they are able to operate in deeper water than fixed wind turbines and they can be built in port and towed to their site off-shore greatly simplifying construction and reducing costs.

IRENA also provides a classification for floating wind turbine platforms and also reports that the industry regards floating platforms as less disruptive of the sea bed ecology. Common features of all the types are that they float and have anchors that rest on the sea bed. It is worth noting that, although they have the advantage that they can be used in deeper water, it is also possible to use them in shallower water and their greater use could become a trend because of their ecological benefits. The types are:

- Barge: An ocean-going vessel with a large flat deck area for installing the wind turbine. The motion of the barge is damped by adding large baffles underneath the vessel, called heave plates.

- The semi-submersible: This type of vessel is commonly used for off-shore oil drilling because of its stability. Like barges, the semi-submersible can be fitted with heave plates to reduce motion. A key design feature is that the surface area exposed to the sea, and needing to be protected from the effects of the sea, is minimised.
- Spar: This is a large tube-like buoy. The buoy contains solid ballast to lower the centre of gravity and aid stability. Structures like this can be prone to oscillation under the action of wind and tide; this can be controlled by altering the wind turbine blade pitch and also by adding mass to the structure by taking on water. The increased mass of the buoy dampens oscillation and has the additional advantage of altering the draught of the vessel if required.
- Articulated multi-spar: As the name implies, this has multiple spar-buoys. The reference to articulation comes from the ability to mount a spar at each corner of the deck which is, typically, square. The mount is hinged allowing the spar to be folded along the length of the deck. This is beneficial because it allows the whole structure to be integrated into the port.
- Tension leg platform (TLP): This type of platform is the most recent type to be developed. The benefit of this approach is to minimise the size of the structure and, thereby, reduce cost. The TLP has heavier anchors which can be dropped to the sea bed and the anchor cable can be tensioned to restrict the platform motion.

4.4.5.3 FOWT motion

A concern that arises about FOWT is that they may move as a result of wind and wave action and that this may make mitigation more difficult or even impossible.

A floating wind turbine is capable of two forms of motion: long-term drifting motion causes the turbine to move its geographical location away from its nominal position; IBERDROLA reports that turbines can move between 20 m and 50 m from a fixed turbine centre point depending on the method used to anchor the turbine [23]. However, in addition, a wind turbine can move more dynamically as a result of wind and wave action. The six degrees of freedom of this motion are:

- Pitch: the angular motion associated with the rise and fall of the bow and stern of a vessel.
- Roll: the angular motion associated with the rise and fall of the port and starboard sides of a vessel.
- Yaw: the angular motion associated with the port to starboard motion of the hull of the vessel.
- Heave: the linear, up and down, motion of the vessel.
- Sway: the linear, left to right motion of the vessel.
- Surge: the linear, forward or backwards, motion of the vessel [24].

Wang *et al.* [25] simulated the six degrees of motion in different wind and wave conditions of a scale that might be encountered in the South China Sea; these correspond to Sea State 5/6 (rough/very rough), see Figure 4.20. Their

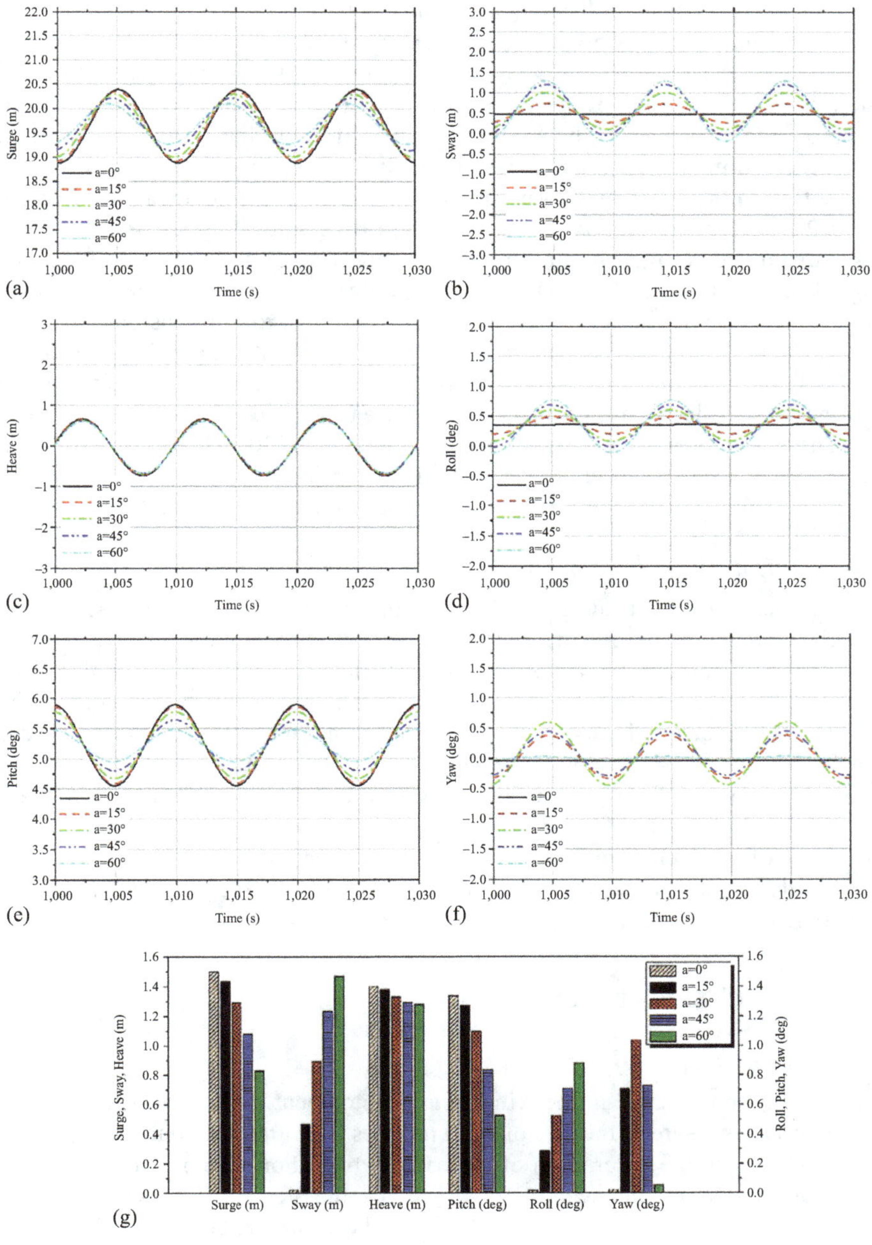

Figure 4.20 FOWT motion from © 2022 Wang et al. Creative Commons Attribution License (CC BY) [25]

findings indicate that both wind and wave action give rise to motion: wind determines the mean position of the motion, whereas wave action determines the amplitude of the motion. The relative proportions of sway and surge depended upon the angle of attack of the waves. For example, head-on waves gave rise to an amplitude of motion in the surge dimension of 1.5 m but virtually no motion in the sway dimension, as the angle of attack increased to 45° the amplitude of the motion in the surge and sway dimensions were very similar (approximately 1.1–1.2 m). Heave is relatively insensitive to the angle of attack and in all the cases examined resulted in amplitudes of motion of approximately 1.3 m. The angular degrees of freedom behaved in a similar fashion to the linear motions, that is, the angle of attack is important in the pitch and roll dimensions. Head on waves cause an amplitude of motion of 1.3° in pitch but little effect in roll. As the angle of attack increases, the amplitude of the motion equalises. If the angle of attack is 45° the pitch and roll amplitudes are 0.85° and 0.7°, respectively. The yaw angle was most affected with an angle of attack of 30° which gave rise to an amplitude of 1°. In all cases, the period of the motion was 10 s.

The implications of these results are discussed later in this chapter.

4.4.6 Lightning protection

On average, every wind turbine is hit by lightning once a year, much more frequently than might be experienced by an aircraft because the wind turbine is mounted on the surface and it can provide a path to electrical earth. To mitigate the effect of strikes, every turbine must have a lightning protection system (LPS). The LPS needs to protect all the components of a turbine; for example, the tower base and foundations must have an earth termination system and the blades have conductors running the full length of the blades. In the most modern turbine blades, there are not only longitudinal conductors running down the blade but there is also a network of transverse conductors running across the blade [31].

It follows that irrespective of the material used to manufacture them, the blades still contain a lot of metal, which has a bearing on their radio-reflecting properties.

4.5 Size of wind turbines

4.5.1 Metrics

It is custom and practice in the wind farm development industry to use the rotor diameter as a measure of the size of wind turbines and often manufacturers include this dimension in the description of the wind turbine. For example, the Vestas V-162 turbine has a rotor diameter of 162 m [26] and the Siemens Gamesa SG 7.0-170 turbine has a rotor diameter of 170 m [27]. In the United Kingdom, in Electricity Act consent and planning permission circles, it is more common to refer to tip height as the key metric. These different metrics are illustrated in Figure 4.21.

The different metrics are indicative of alternative concerns. The wind farm industry is concerned with maximising the electricity production from wind

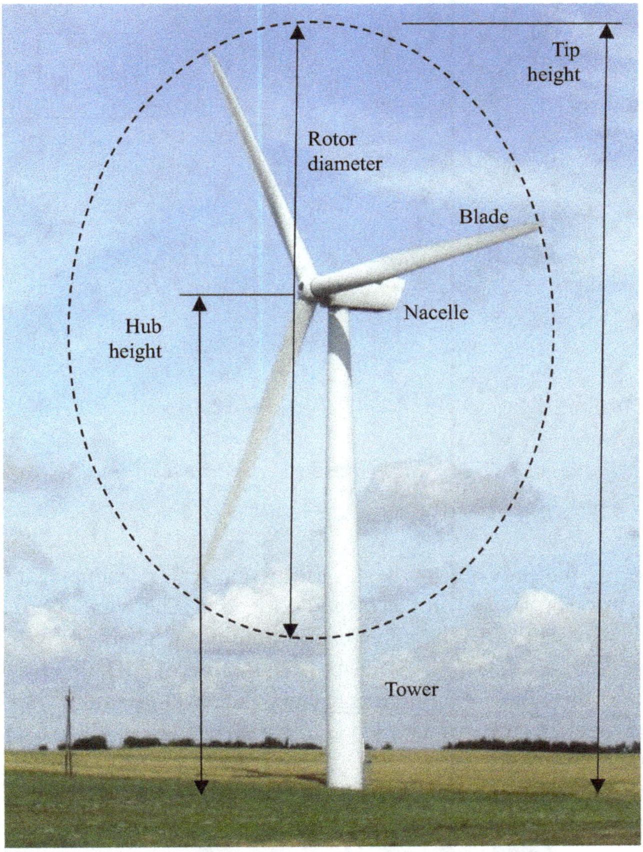

Figure 4.21 Wind turbine metrics

turbines and the scale of the rotors is material to achieving this. For the planning community, the visual and environmental impact, as well as the impact on aviation radio and radar systems, is determined by the tip height.

Part of the process of establishing the feasibility of a potential wind farm site is to measure the wind speed, direction and the amount of turbulence. Ideally, these measurements will be made over a complete year. The measurements are carried out using anemometers, wind vanes and, increasingly, light detection and ranging (LIDAR). In the case of anemometers and wind vanes, these are mounted at various heights on a slender, usually steel lattice, mast which is as high as the anticipated height of the turbine hub. The masts are often temporary and will be dismantled when measurements are complete but the duration of the measurement programme makes it one of the most time-consuming elements of construction. Usually one (or more on a large wind farm) is left in place after commissioning. One of the principal products created by the measurements is a wind rose, see, for example, Figure 4.22. This

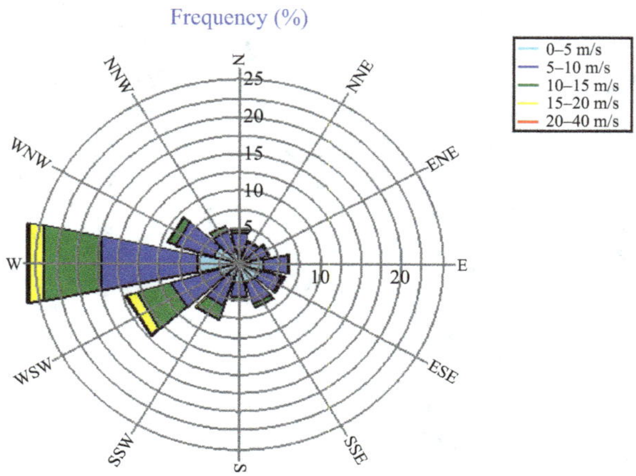

Figure 4.22 Wind rose Green Rigg Wind Farm (courtesy Wind Prospect Au and EDF Energy)

diagram captures the prevailing wind direction, as in this example, this is usually from the West or South West in the United Kingdom. The wind speed and relative frequency are also recorded. The construction time of the anemometer mast and the associated costs are leading to the widespread adoption of LIDAR measurements as an alternative because these measurements can be made at ground level. As the name implies, these devices are similar in operation to radars but they use optical frequencies to make measurements.

Another important piece of information provided by the wind measurement campaign is an evaluation of the turbulence at the site. As described above, this quantity is associated with the terrain. If the terrain at a potential site is not uniform, some parts of the site will experience different levels of turbulence from others and, in these cases, multiple anemometer masts will be used to plan the site.

In areas of higher turbulence, blades that are more tolerant of turbulence (stronger) may be indicated or an alternative is to use shorter blades on a tower with a greater hub height; again, the goal is to lift the turbine out of the turbulent air. Thus, while the tip height of all the turbines on any one site will be fixed, there may be different rotor diameter turbines operating in different locations on the farm with different hub heights [28]. For example, at the Pen y Cymoedd wind farm, two rotor diameters are used, 108 m and 113 m. The former used in the areas of greatest turbulence [29]. Off-shore, in the absence of terrain, air flow is usually more laminar and, as a consequence, tip heights are lower than the same rotor diameter turbines on-shore.

4.5.2 Other factors

Planning considerations may be a factor in determining the size of a turbine. For example, in the United Kingdom, a common turbine tip height is 147–149 m. This tip height is chosen because different planning considerations need to be taken into account above 150 m; turbines below 150 m tip height do not routinely need to be lit [30].

4.5.3 Trends

In 2023/2024, an on-shore wind farm coming into service might, typically, have rotor diameters of between 140 m and 160 m. However, the wind farm development industry plans a long way in advance and in 5 years' time are likely to have rotor diameters of between 170 m and 180 m. This is also likely to apply to older wind farms that are being repowered.

As the size of wind turbines increases, each component increases in size and weight. As rotor diameters increase in size, the individual blades increase in length; for example, when rotor diameters reach 175 m, then the individual blades will be 85 m long. The towers must increase in height to accommodate the longer blades but this also means that the tower diameters must increase to support the extra weight. The increased weight of the blades also places a greater load on the power train in the nacelle. All of these components have to be transported to the site and assembled on site. To mitigate the transportation problem, towers can be made sectional and now even blades are being made in the same way.

Off-shore, turbine rotor diameters are currently 236 m. In the next 5 years, manufacturers are developing machines capable of generating 18 MW of electricity and these will have rotor diameters of 260 m.

A corollary of the increase in height of wind turbines is their increased visibility to aviation radio and radar systems. In Chapter 1, an example was provided of the effect of increasing the tip height from 150 m to 200 m which increases the area of visibility by 33% (corresponding to 2,670 km^2). Because of the increased visibility, it cannot be assumed that because permission was granted for an original wind farm that permission will also be forthcoming if the wind farm is re-powered even if there are fewer wind turbines.

4.6 Wind farm layout and design factors

4.6.1 Turbine layout – on-shore

Turbines create their own turbulence as a result of the vortex formed behind the turbine blades. It is undesirable for adjacent turbines to operate in the vortex and space must be allowed between turbines to optimise performance. A traditional typical layout allows a spacing of three rotor diameters between turbines in the cross-wind direction and five rotor diameters in the prevailing wind direction as illustrated in Figure 4.23. As wind turbines get larger, this restricts the number of turbines that can be included in a given area and increases the inter-turbine distances.

4.6.2 Turbine spacing off-shore

Construction of turbines off-shore is a significant cost driver and the major factor in determining this cost is water depth; increasingly so as wind farms are built in

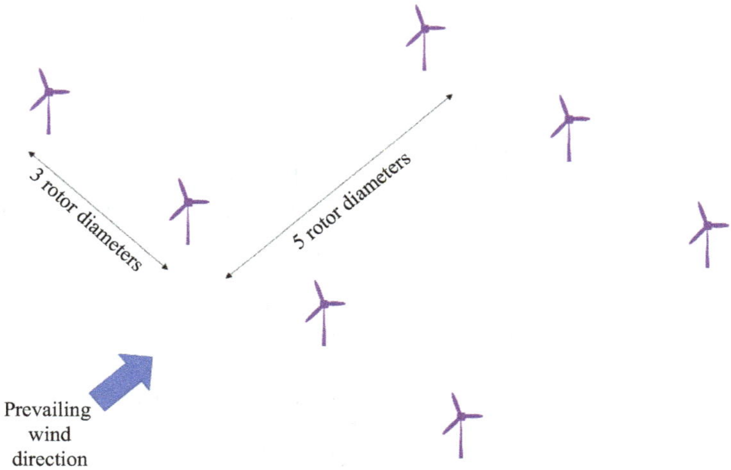

Figure 4.23 Inter-turbine spacing

deeper water; depths of up to 60 m are now possible with mono-pile construction [32]. Early off-shore windfarms favoured organised ranks of equally spaced turbines; in such circumstances, inter-turbine spacings are constants. However, in deeper water wind farms, sometimes even modest changes in the location from a fixed grid pattern to patterns where the sea bed is more auspicious can make a big difference in construction costs. Hence, in more recent off-shore wind farms, it cannot be assumed that inter-turbine spacing is a constant. Modifying patterns to reduce construction costs is facilitated by adaptive wake control technology.

4.7 Wind farm lifetime

Typical lifetimes of wind farms have increased and are still increasing. Ten years ago, the in-service lifetime of a wind turbine would have been 20 years. The turbines in a wind farm entering service in 2023/4 might be expected to last between 30 and 35 years.

4.8 Wind farm operations – curtailment

Curtailment is the process whereby blade rotation is slowed or halted for a period of time. From the perspective of the wind farm operator and the grid operator, curtailment is highly undesirable. For the operator, it reduces the amount of electricity generated, causes excessive wear and tear, and possibly even damage, to the turbine. All three effects have financial impact on operating costs. For the grid operator, if a wind farm ceases production of electricity because of curtailment the energy must be replaced from other sources; a problem greatly exacerbated if the curtailment occurs at short notice. Conversely, if the grid operator requests curtailment, the wind farm operators may receive some payment for lost revenue.

4.8.1 The emergency stop

The most extreme and most rapid form of curtailment is the emergency stop. A 7 MW turbine is a 10,000 horse power (HP) machine and suddenly stopping such a powerful machine creates extreme mechanical loading on the gearbox. A wind turbine is only capable of executing an emergency stop a limited number of times in its life before it fails. Therefore, as the name implies, this form of curtailment triggered by an automatic protection system or by manual intervention should only be undertaken in emergencies when there is a risk to life.

4.8.2 Slowing and stopping the turbine

The process of slowing the turbine in a more controlled fashion is part of a wider turbine management function of starting and stopping. The principal system for managing starting and stopping is the blade pitch control, that is, by altering the orientation of the blade with respect to the wind direction. An orderly shut-down of a turbine would be carried out by changing the blade pitch and allowing the blades to slow down. When the blades had reached the idle setting, braking would allow the turbine to be brought to a halt. The braking methods include brakes on the turbine hub, similar to those on a car and air brakes on the tips of the blades that are similar to the air brakes on aircraft. The whole process to stop the mechanical turbine typically takes one minute. The time taking to stop the turbine generating electricity might typically be 15 s.

To minimise the risks and reduce (but not eliminate) the costs of curtailment, the processes should be pre-planned and should be carried out as a staged process slowly reducing the speed of the turbine before, if required, halting it.

Curtailment is used to deal with a wide variety of conditions met in operations. These are listed below.

4.8.3 Acoustical noise

There are two sources of acoustical noise from wind turbines: there is the aerodynamic noise caused by sheer stresses in the airflow around the blades (usually described as a whooshing sound) and there is the mechanical noise created by the gearbox and generator. The former source of noise is related to the blade tip speed and if this exceeds a threshold, then curtailment is used to slow the turbines to restore the tip speed to below the threshold. Typically, this would be a blade tip speed not exceeding 80 m/s. The permissible noise limits vary between countries, although a common factor to all the countries is that a distinction is made between allowable maximums at night and during the day.

4.8.4 Shadow flicker

If the Sun is behind the blades of a wind turbine, it can cast a shadow which, when observed through an aperture (a window) can cause a flashing effect, which can be

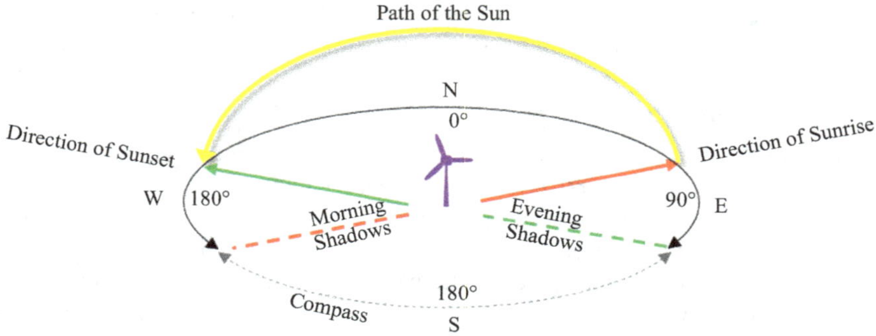

Figure 4.24 Shadow flicker mid-summer Northern Hemisphere

uncomfortable for the residents and may be dangerous to anyone who suffers from photo-sensitive epilepsy** [33]; this condition is known as **shadow flicker**. Anderson points out that shadow flicker is unlikely to be a problem if the turbine is more than 10 rotor diameters from a property and it is also dependent on the relative locations of the turbine and residences. This latter condition is illustrated in Figure 4.24 which depicts a wind turbine somewhere in the Northern Hemisphere in the Summer. During the middle of the day, the Sun will be high in the sky. But in the morning and evening, the Sun is low in the sky and can cast long shadows in the regions depicted in the figure [34].

All these conditions are predictable, the one condition that is not predictable is whether the Sun may be obscured by cloud but that can easily be determined by sensors. Taking all these factors into consideration, mitigating shadow flicker by curtailment can be accomplished in a timely fashion that prevents damage to the turbine.

4.8.5 Ecology

Curtailment can be used to mitigate some ecological concerns about birds and bats. The greatest risks to bats occur in late Summer, around dusk, when they hunt for food. The greatest risk to birds is during migrations. Both situations are predictable and curtailment can significantly mitigate risks to both types of animals at critical times.

4.8.6 Grid capacity

The amount of electricity required by users varies continuously. There are, for example, large diurnal changes; more electricity is used during the day than at night. But changes can occur in a less predictable manner and the distribution network operator (DNO) must make decisions about sources of supply of electricity. This can lead to a need for rapid curtailment of turbine supplies. Although

**Approximately 1 person in 3,000 suffers from photo-sensitive epilepsy.

battery storage is not capable of storing large amounts of electricity compared with the wind farm itself, batteries do provide a means of smoothing out rapid changes in demand.

4.8.7 Ice accretion

During freezing weather, it is possible for build-up of ice on the whole turbine structure. This build-up is problematic on the blades because as the ice melts the motion of the blades can cause the ice to sheer off with risks to people and property, and possibly damage to the blade/rotor caused by the uneven distribution of mass. Curtailment is employed to mitigate this risk.

4.8.8 Aviation objections

It is worth repeating that curtailment is undesirable because of wear and tear on the turbines as well as loss of revenue but it is, necessarily, part of routine wind farm operations. In extremis, and in a limited number of cases, curtailment agreements have been reached as a means of mitigating aviation objections. The sorts of situations where curtailment might be suitable are in a scenario in which in normal, day-to-day, circumstances mitigation is not required but in a very specific, narrow situations, the turbines may be slowed to prevent them from creating clutter. Often curtailment agreements are associated with wind farms situated close to an aerodrome and mitigation is required in a non-standard traffic situation. Agreements using this mitigation are reached on the basis that requests for curtailment should be infrequent and they should provide a period of notice, for example, 10 min to allow the turbines to stop in a slow, controlled manner.

4.8.9 Trends

4.8.9.1 On-shore

Many good on-shore sites have already been used up and developers are turning to less productive sites which would not have been considered in the past, for example, forests and other sites where airflow is more prone to turbulence. The best way to exploit these lower yield sites will be to increase the turbine tip heights. The corollary of this trend is that on-shore turbines will become more visible to aviation radio and radar systems.

As discussed in Chapter 2, there are two broad classes of turbines: HAWT and VAWT. VAWT are less efficient than HAWT and, despite the fact that they do not create turbulence in the same way as HAWT, they do not provide as much yield per unit area as HAWT. For the foreseeable future on-shore, large-scale electricity production is likely to come from HAWT. However, VAWT are viable in some small-scale settings and, in the future, there may be many more.

4.8.9.2 Off-shore

The US National Academy of Sciences commissioned a study on the impact of wind turbines on marine radar systems. The study was reported in 2022. One of the subjects addressed was a forecast of the types of wind turbines that would be in use in the future. They predicted that up to 2035, the three-bladed HAWT would be the 'marine standard'. However, beyond 2035, they predicted that in deep water, the VAWT would take over because it offers a lower centre of gravity [35].

There are a number of existing trends and some emerging in off-shore siting:

- The easiest trend to identify is the shift from on-shore to off-shore. There is less public opposition to off-shore and siting is easier to find.
- However, there is significant public opposition to applications for planning permission for landing sites. This leads to the need for combining outputs off-shore from multiple wind farms so that the electricity can come ashore at fewer locations.
- The earliest off-shore wind farms were built, by the present-day standards, close to shore. Technological improvements are facilitating wind farms farther off-shore.

As with on-shore wind farms, there is a trend to increasing the size of wind turbines off-shore. Normally, an increase in turbine size is associated with an increase in inter-turbine spacing. However, wake adaptation technology allows the wake to be steered, reducing wake losses, and improving the efficiency of electricity generation. Figure 4.25 is an image of the Horns Rev wind farm off Denmark. The weather conditions on the day this photograph was taken give a good illustration of the wake effects behind wind turbines and clearly indicate that being able to steer the wake by only a few degrees could yield benefits beyond the front row of wind turbines in a farm. It remains to be seen whether, in future wind farms, this could allow more dense turbine configurations.

Figure 4.25 Horns Rev (courtesy of Vattenfall)

4.9 Wind farm planning and construction considerations

4.9.1 Introduction

There are a number of stages to the construction of a wind farm: site identification, detailed site assessment, the planning process and construction itself. The following discussions provide an overview of these processes, with particular attention being paid to those processes that are time intensive.

4.9.2 Finding an on-shore site to develop

The first step in identifying a potential site is to find suitable land. The developer may be approached by either the owner of the land, or their agent, interested in whether his or her land might be suitable for a wind farm development, but it is more usual for the developer to have to identify land themselves. For utility-scale wind farms, with a capacity of, say, greater than 50 MW of electricity production, this can require a large amount of land; 250 hectares (over 600 acres) or more. Finding parcels of land this large on-shore is difficult.

Ideally, the land identified will be **unconstrained**. The following, non-exhaustive, list illustrates the factors that constrain land suitability for wind farms:

- Proximity to residential property; the definition and understanding of 'reasonable proximity' varies among developers, local planners, and those impacted. In general, noise can be audible in some circumstances as far as one to two kilometres from a windfarm.
- Plainly, any aggregation of turbines will tend to produce greater noise levels than a single machine. Turbine size and numbers are therefore important. Careful positioning and accurate noise level predictions against contemporary acoustic scientific standards and skilled professional guidance are essential.
- Published guidance allows up to a 5 dB 'penalty' for those 'financially involved' in the project.
- Topography; sites with a gradient exceeding 14° are generally regarded as unsuitable because of the increased complexity of construction.
- The ability to provide electricity to industry and private consumers.
- Vehicular access to the site should be sufficient to allow passage of heavy and long loads. If access roads are insufficient, then roads would have to be modified. Modifications would increase project costs and would need to be undertaken without causing too great an inconvenience to local people.
- Proximity to aviation interests, for example, civil or military aerodromes or airways may make a site more complex to develop or, in present circumstances, may preclude the site from being developed at all.
- Proximity to meteorological sensing sites; these sites include radars and other radio assets. However, in October 2022, the UK Met Office reported useful progress in the mitigation of wind turbine effects using a technique called cross-polarisation. This development should have the effect of freeing up a number of potential sites. With research, it may be possible to use this technique for other radar systems; this is discussed in Chapter 7 [36].

- In Scotland, proximity to the British Geological Survey (BGS) Eskdalemuir site is problematic. The site operates a synoptic meteorological station and a broadband three-component seismometer. The seismological array is sensitive and it requires safeguarding [37].
- Proximity to water courses; no turbine or access road should be built within 50 m of a watercourse because of the risk of pollution during construction.
- Protected areas including:
  - National Parks.
  - National Scenic Areas.
  - Areas of Outstanding Natural Beauty.
  - Special Landscape Areas.
  - Special Areas of Conservation (SAC); Special Protection areas (SPA); (together *Natura 2000 sites)* which include Sites of Special Scientific Interest (SSSI), the habitats of protected species, for example, some amphibians, bats and birds and Ramsar sites (protected Wetland areas).
  - Other protected or designated areas, for example, designated sites of archaeological or historical interest, such as battlefields [38].

4.9.2.1 General observations about land

It is possible to create broad categorisations from these constraints lists which may help prioritise site selection. Using the United Kingdom as an example, in England, electricity grid access and road access are relatively good, but sites are highly constrained because of the density of the population. In Scotland, the population is relatively low and, therefore, residential constraints are fewer but road access and grid access can be problematic. Naturally, these factors tend to be related; if few people live in an area, there may be little industry and so there is no call for grid supply or high-load bearing roads. A similar situation applies in Wales; there are parts of Wales where the population is lower but road and grid access are more difficult. And, in both Scotland and Wales, there are arguments about visual impact.

Some national bodies offer land for consideration, for example, Forestry and Land Scotland, but often these bodies lack the expertise to carry out any filtering of sites and some offerings may be unsuitable for extraction of the wind resource.

There is a strong trend for wind turbines to increase in size, that is, for rotor diameters, hub heights and tip heights to increase. This trend makes it possible to exploit less favourable wind resources which opens up more land opportunities for development, but the increases in sizes also make turbines more visible to human and radio systems.

Distinctions about the relative merits of land hide the simple fact that in some countries there is a shortage of land available for utility-scale wind farms, and this may incentivise off-shore wind farms.

4.9.2.2 Site location tools

The usual starting point for identifying land opportunities is the use of Geographical Information System (GIS) software tools. GIS tools were invented in Canada in the nineteen sixties and commercial packages became available in the nineteen eighties. They are now well developed and allow users to perform a large variety of analyses [39]. A GIS tool allows a user to research an area of

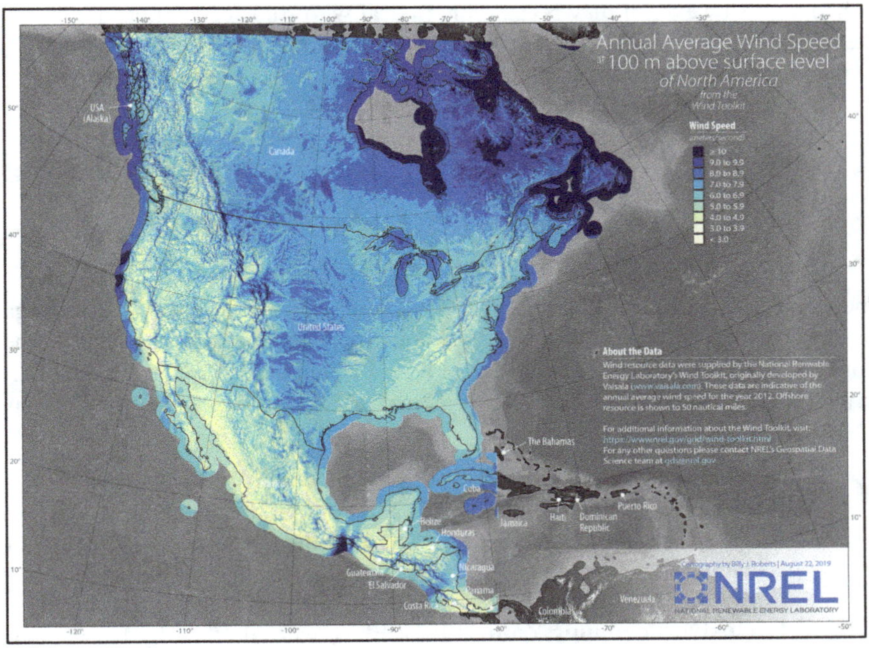

Figure 4.26 A wind Atlas overview (reproduced courtesy of the US Department of Energy National Renewable Energy Laboratory)

land starting with a map of the area and then overlaying the map with additional information. A limited number of examples might include the user being able to add the locations of landscape designations, settlements and housing, to create Zones of Theoretical Visibility (ZTV). Noise impact can be investigated. Recreation areas, water courses, soil types and airport consultation zones can also be plotted on maps.

GIS also provides a means of investigating wind resources. A number of respected wind databases are available including the Global Wind Atlas [40] and the European Wind Atlas [41]. In the United States, the Department of Energy [42] and the National Renewable Energy Laboratory [43] both publish wind data, see for example Figure 4.26. Information is also available for off-shore areas, for example, the UK Crown Estate publishes wind data for the Key Resource Areas (KRA) in the seas around the United Kingdom [44]. The US Energy Information Administration maps provide a good illustration of the scope of the information available [45].

Typically, these tools allow the user to identify regions where the winds fall into the different IEC categories and they also serve to identify land regions by roughness-length (described earlier in this chapter).

4.9.2.3 Candidate sites

Some sites will be eliminated from a list of candidates because constraints mean they are simply not viable or they are not economically viable with present technology. However, it may be possible to deal with some constraints and for the sites to remain viable. For example, there may be mitigations for aviation constraints and modified turbine layouts or reduced numbers of turbines may mitigate SSSI concerns. Therefore, on completion of such an initial assessment, a shortlist of sites will exist with their associated constraints. The next step is to secure rights to the land, often by the use of options, and when this is achieved a more thorough assessment of the site can proceed.

4.9.3 Finding an off-shore site to develop

The situation off-shore is different from on-shore. Usually, an individual or organisation has title to an area of land on-shore; whereas, off-shore, the legal framework for establishing ownership is set out by the UN in the 1982 Convention on the Law of the Sea (UNCLOS). The principal off-shore regions defined within UNCLOS are as follows:

- On the **continental shelf** (strictly, the maritime continental margin)[††]:
  - Within the 12 nautical miles (Nm) (22 km) limit, that is in the area 12 Nm from the coastline are the **territorial waters**; the nation owning the coastline has sovereignty.
  - Between 12 Nm and 200 Nm (370 km) from the coastline is an **Exclusive Economic Zone** (EEZ). EEZs were defined by the distance between nations separated by sea as less than 400 Nm (and two discrete 200 Nm EEZ cannot be accommodated) sometimes the median point is declared to be the boundary, sometimes rights are joint (that is shared between the nations concerned). However, coastlines are often irregular shapes and if there are international boundaries up to the coast, the boundary of the EEZ may be contested; for example, parts of the boundary between France and Spain in the Bay of Biscay are contested. There are also parts of the world where the whole EEZ is contested, for example around the Falkland Islands and Kuril Islands (off Japan) [46]. The nation owning the EEZ has sovereign rights for exploiting wind and wave energy within the zone.
  - The region between the edge of the continental shelf and the deeper ocean floor is referred to as the **continental slope** or the **extended continental shelf**. This region extends to a maximum of 350 Nm. The associated nation may have mineral rights in this area.
- Beyond the continental shelf/continental slope and beyond 350 Nm from the coastline the body with responsibility for protecting the seabed ecosystem and allocating mineral rights is the International Seabed Authority (ISA).

In practice, the regions of interest for FOWT are territorial waters and EEZ. In the future, this situation may change.

[††]This is a simplification because the size of the continental shelf varies from nation to nation.

Within the above legal framework, the process for identifying 'blocks' for off-shore wind farms and finding a path to consent varies among nations; however, they fall broadly into three categories:

- The developer identifies the site it wishes to develop and then is responsible for preparing the submissions to the Government for consent. For example, this is the process formerly used in the UK, or
- The Government identifies the site it wants to be developed, holds an auction round, and developers bid for the site and, if successful become responsible for preparing the submissions to the Government for consent in exchange for guaranteed prices. For example, this is the process followed today in the United Kingdom and in France [47], or
- The Government promulgates more general policy or strategic direction, identifies the site(s) it wishes to develop, carries out the assessments required for consent, and a developer then bids to construct and operate the site. For example, this is the process followed in the Netherlands [48].

4.9.3.1 Context

To put the distances of the legal entities, territorial waters and EEZ, in context, suppose an air defence (AD) radar is located on the coast, perhaps it is on the top of cliffs, and its antenna is 100 m above sea level. Further suppose, an off-shore wind farm has 200 m tip height blades and there is no other land or structures between the radar and the wind turbines. Then using the equation set out in Chapter 3, it is possible to calculate the distance the wind farm would have to be off-shore before the curvature of the Earth rendered the turbines invisible to the radar. In the scenario described the wind farm would need to be 54 Nm (100 km) away from the radar before the tips of the blades disappeared below the radar's horizon.

4.9.4 The planning process

4.9.4.1 Process complexity

Before a wind farm can be built it must receive consent or permission from local and/or national authorities. The terms are often used interchangeably but legally they are not the same thing. Obtaining consent or permission requires a complex set of processes to be followed and this can take a long time; from the identification of a suitable piece of land, to generating electricity can take as long as ten years, although it depends on the legal regime in place. Moreover, the process will be very costly; in no small measure because of the length of time the process takes and the need to bring in specialised expertise.

Some development companies specialise in different parts of the process and, it is commonplace that a site may change hands several times before construction begins. Moreover, turbine layouts and the number of turbines in a development may change as the development proceeds and the design is refined, even the name of a site may change. This situation can be confusing for the analyst supporting a developer. Taking the trouble to understand the history of a site can save an analyst a lot of time in the long run.

4.9.5 Preparation of a consent or planning application

4.9.5.1 Caveats

The following discussion illustrates the activities that might be required to secure permission to construct a wind farm; the discussion does not constitute advice on how best to proceed with the process, which is outside the scope of this book. Neither is this a legal textbook and no part of this discussion offers legal advice. However, some legal terms are introduced to help understanding the processes being described.

4.9.5.2 Preparing to submit an application

After the initial desk-based survey using the GIS tools, a more detailed assessment of a candidate site will be required leading up to a decision about whether the site is viable. Example issues to be refined include more detailed wind assessment, aviation, landscape and visual impacts on people in particular, birds and ecology, hydrology and hydrogeology, traffic and transport, heritage and history, and so on.

Aviation

Suppose, for example, that the initial survey identified potential aviation concerns. If there is a valid concern, then once the site is submitted for planning approval it would become the subject of a formal objection (discussed below). During pre-filtering, it might be premature to consult directly with any aviation stakeholder but, once it is decided to take the next step, it would be prudent to start discussions at the earliest opportunity. Early discussions are important because it is helpful to establish a cooperative and not a combative dialogue. Moreover, if mitigation is necessary, identifying, agreeing, developing and implementing it is likely to be a lengthy process; an early start will bear fruit later. The first imperative in any discussion is to confirm there is a problem and understand precisely its nature. It may seem like a statement of the obvious, but if the problem is not properly understood it is unlikely that a suitable solution will be found. It is also important to reach an agreement between the developer and the stakeholder about the nature and timescale of the solution. A useful expression that came up during research for this book was that it is important to 'agree to a scheme not simply to agree to agree' [49].

The following topics are discussed in more detail in Chapter 6. Once it is known that mitigation will be required, the developer and the aviation service provider will need not only to agree that mitigation is possible but also the technical elements of a scheme to ensure mitigation will be in place when it is required. In addition, it will be necessary to agree on how the mitigation will be funded. For example, a finding that may come from an early engagement is that the, say, airport already has some means of mitigation which was put in place for other development(s) in the area. This situation may not immediately offer a solution for a new development and a number of questions arise. Is the mitigation measure sufficiently extensible that an additional wind farm can be added without degradation of the service being provided; or is a new, more comprehensive, mitigation required? What sort of cost model is appropriate for sharing the mitigation? For example, if the existing mitigation measure was put in place to mitigate the effects of 50 MW wind farm and the new development is also 50 MW is a 50% share a fair distribution of costs?

Wind assessment

Consulting a wind atlas during the pre-filtering process will have provided the estimated wind speed for a site but a more detailed assessment would usually be considered necessary prior to deciding what types of turbines and blades would be appropriate. This step is another measure that needs to be considered early in the development process because of the length of time it takes; usually, the measurements are made for a year but this does not account for the time taken to submit planning permission for a mast (if required), for construction, de-construction and removal from the site as well as the time taken to analyse the data. But clearly, ultimate viability depends on the availability and consistency of the wind resource. Until recently, this process would also require the construction of infrastructure, namely an anemometer mast, as well as planning permission for the structure. Anemometers might still be used but, increasingly, this process is carried out by LIDAR. LIDAR does not preclude the need for an anemometer mast; sometimes the LIDAR instruments are mast-mounted and LIDAR is not always appropriate, for example, in forested sites.

Core sampling

Another time-consuming task that backs up initial assessments is taking core samples around the location of the proposed turbines to determine the exact nature of the sub-soil under the turbines.

Pre-planning inquiry

It is usual, in the United Kingdom, for an initial, pre-planning inquiry to be submitted to the relevant consent and planning authorities before submission of the full application. One of the key outcomes of pre-planning is establishing the scope of what will become the most complex pre-construction activity; the preparation of the project's Environmental Impact Assessment/Statement (EIA/EIS).

4.9.6 The EIA/EIS

The subjects that must be covered in EIA should be decided by a rigorous scoping process. The authorities setting out the scope will usually be able to draw on specialist advice available to them in-house. However, not every public authority has a panel of appropriate experts.

When the subjects have been decided, the developer must conduct the data gathering for the EIA. Like the authorities, the developer may have access to specialist expertise in-house but they will also need to access external expertise on the very specialised areas. Aviation expertise is typically outsourced because it is such a niche area of expertise.

The EIS must deal with all those elements of the environment that could be affected by the wind farm. In the United Kingdom and Europe, the European Union Regulations, in place since 1985 and amended from time to time, are regarded as the 'gold standard model'. In principle, the different investigations are carried out in parallel to save time but, in practice, there are many dependencies involved and preparation of the EIS is usually a long process resulting in a very compendious piece of work.

4.9.6.1 On-shore

An EIS for an onshore project must include a description of the development, including:

- For the proposed wind farm:
  - Its location.
  - Physical characteristics and land-use requirements.
  - Main characteristics of the operational phase, nature and quantity of the materials and natural resources (including water, land, soil and biodiversity) used.
  - An estimate of expected residues and emissions.
- A description of the reasonable alternatives studied which are relevant to the proposed project and its specific characteristics, and an indication of the main reasons for selection.
- A description of the relevant aspects of the current state of the environment and an outline of the likely evolution thereof without implementation of the development.
- A description of the factors likely to be significantly affected by the development:
  - Population.
  - Human health.
  - Biodiversity (e.g., fauna and flora).
  - Land (e.g., land take).
  - Soil (e.g., organic matter, erosion, compaction, sealing).
  - Water (e.g., hydro-morphological changes, quantity and quality).
  - Air.
  - Climate (e.g., greenhouse gas emissions, impacts relevant to adaptation).
  - Aviation and radar (the reference to radar as separate from aviation refers to the other radar systems such as meteorological radars and, if the wind farm is close to an estuary or harbour a Vessel Traffic Service (VTS) radar):
    - Infringement of safeguarded areas and air space
    - Collision risk
    - Primary and secondary radar impact
    - Radio communications and navigation systems
  - Radio and television services.
  - Microwave communications (fixed links) (these are a significant issue because there are a lot of fixed links in service. The links provide communications for, for example, utility companies which use the links to provide control and monitoring of their assets).
  - Material assets, cultural heritage, including architectural and archaeological aspects, and landscape.

- A description of the likely significant effects of the development on the environment resulting from, inter alia:
    - The construction and existence of the development.
    - The use of natural resources, considering the sustainable availability of resources.
    - The emission of pollutants, noise, vibration, light, heat and radiation, the creation of nuisances, and the disposal and recovery of waste.
    - The risks to human health, cultural heritage or the environment, including for example Shadow Flicker (discussed above).
    - The cumulation of effects with other existing and/or approved projects.
    - The impact of the project on climate and the vulnerability of the project to climate change.
    - The technologies and the substances used.

The description should cover the direct effects and any indirect, secondary, cumulative, transboundary, short-term, medium-term and long-term, permanent and temporary, positive and negative effects of the development.

- A description of the forecasting methods or evidence, used to identify and assess the significant effects on the environment.
- A description of the measures envisaged to avoid, prevent, reduce or, if possible, offset any identified significant adverse effects on the environment and, where appropriate, of any proposed monitoring arrangements.
- A description of the expected significant adverse effects of the development on the environment deriving from the vulnerability of the development to risks of major accidents and/or disasters.

4.9.6.2 Off-shore EIA

Off-shore EIS must, typically, address the following issues:

- Marine geology, oceanography and physical processes
- Marine water and sediment quality
- Off-shore air quality
- The following additional aspects of ecology
    - Benthic organisms (which live in sediments) and inter-tidal organisms
    - Fish and shellfish
    - Marine mammals ecology
    - Offshore ornithology

- Commercial fisheries
- Shipping and navigation
- Off-shore archaeology
- Aviation and radar
- Infrastructure and other users (e.g., harbours)
- Off-shore designated sites (e.g., Marine Conservation Areas)

4.9.6.3 Grid landing

Off-shore wind farms must also address all of the on-shore environmental issues because grid connections must come ashore somewhere. Developers have reported that this aspect of off-shore construction can attract a lot of public opposition, despite the fact that impacts on the local environment are short term because the grid landing sites are reinstated when cables have been laid. The cabling depends on the scale of the off-shore development and they can be either direct current (DC) or alternating current. Because of the opposition to grid landing sites, one approach is to share resources and bring the electricity from more than one wind farm at a single point. However, this approach also comes with some disadvantages. If large amounts of electricity come ashore in a single installation any failure could have serious supply and security implications [50,51].

On completion of the EIS, the resulting document will be a major part of the submission for planning permission.

4.9.7 Planning submission

4.9.7.1 Identifying the appropriate authority

Planning or zoning requirements differ from country to country and they may also vary depending on the scale of the proposed wind farm. The United Kingdom is a good example of how requirements may vary and of the complexity of arrangements. The following categorisations apply to on-shore developments.

In the United Kingdom, a domestic wind turbine does not require any planning permission. In England, wind farms with greater than 50 MW total power production are processed under National Energy Infrastructure rules and, in essence, determined by local planning authorities. Local 'buy-in' is therefore an essential component. By June 2023, only two have been commissioned in England.

In Scotland, wind farms exceeding 50 MW total power production are processed under the Electricity Act 1989, Section 36, and a planning authority is merely a consultee, although its objection will provoke a Public Inquiry. In Wales and Northern Ireland developments between 10 MW and 350 MW are subject to Developments of National Significance (DNS) rules.

Theoretically, developments below the thresholds listed above can be processed by a local planning authority (LPA) alone. Public support for the project is generally required. Furthermore, since 2015, LPAs in England were supposed to identify in their Local Plan areas which were suitable for wind farms. However, by 2022, only 11% of English LPAs had carried this out [52]. No such rules now apply in Scotland, where National Planning Framework 4 (2023) is now part of every Local Development Plan.

4.9.8 Planning conditions

4.9.8.1 Overview

Planning law in the United Kingdom and its practice throughout differs from place to place, and can be heavily nuanced by local preference. Both the consultee responses and local opinion can play an important role in mitigation, influencing both the nature of and the timescales for its implementation. Schedule 9 of the Electricity Act requires that an Applicant:

(a) *'shall have regard to the desirability of preserving natural beauty, of conserving flora, fauna and geological or physiographical features of special interest and of protecting sites, buildings and objects of architectural, historic or archaeological interest; and*
(b) *shall do what he reasonably can to mitigate any effect which the proposals would have on the natural beauty of the countryside or on any such flora, fauna, features, sites, buildings or objects'.*

A grant of consent under the Electricity Act may itself be conditional, and, in effect, creates what is deemed to be Planning Permission, to which Planning Conditions may be attached. The following discussion is a lay interpretation of the planning process to try to set out its influence on the technical implementation of mitigation strategies.

Consider a request for planning permission for a wind farm (or an individual wind turbine). In England and Wales, the application is sent to the relevant planning authority, as described above. In Scotland, it is sent to the Scottish Government's Energy Consents Unit (ECU). The authority or ECU will then inform statutory consultees about the proposal which (of relevance here) will include Air Navigation Service Providers (ANSP) and the Ministry of Defence[‡‡]. The statutory consultees are required to review the application and inform the Authority if they wish to object. They may express concerns without formally objecting.

4.9.8.2 Grampian conditions

If any statutory consultee objects, then this may lead to the application being refused, subject to discussions about enhanced mitigation, or to a public inquiry. However, an application may be granted subject to meeting a **planning condition**, which is a way of limiting and controlling how planning rules are implemented [53]. There are different types of conditions, some requiring pre-start assessment, for example, of ground conditions or Habitat Assessment, or the effects on aircraft or ground-based radar.

A relevant type of condition is sometimes called a **Grampian Condition or a** Suspensive Condition because of a legal precedent set in 1984. A farmer wished to change the use of some land from agricultural to industrial purposes. If the change had gone ahead, access to the site would have been via a road that was considered inappropriate and unsafe for the new industrial circumstances. There was no other reason why the development could not proceed, but the planning authority had no jurisdiction over roads and, therefore, the application was refused. On appeal. the House of Lords' opinion was that a condition could have been framed and applied

[‡‡]The whole purpose of the planning process is to take an holistic approach considering all the foreseeable consequences of development from the impact on ecology to the impact on road traffic. Consequently, there are many statutory consultees of whom ANSP and the MOD are just two.

in such a way that the application could not be allowed *unless* a change was made to the road. In short, such a condition means that it suspends approval, and implementation unless some action is taken.

4.9.8.3 Valid conditions

Such conditions must be fair, that is, they must not require the developer to do something that is impossible to achieve or have a purpose not relating to the problem identified, known in legal circles as a 'remote purpose' [54]. Before they can be imposed, conditions must pass a series of tests for them to be valid, they must be:

- Necessary
- Relevant to planning
- Relevant to permitted development
- Enforceable
- Precise
- Reasonable in all other respects

There are formal, legal, definitions of each of these tests. Planning circulars in England and Wales and in Scotland set out the tests for a lawful condition. For the purposes of a mitigation for an aviation radar or radio system, then, arguably, the key criterion is precision; that is the condition should ideally set out precisely what is required. But this can be a question, who has the expertise to judge?

Two further questions that arise from the imposition of a condition are: how is the condition to be discharged and how might the discharging of the condition affect the project to build a wind farm?

4.9.8.4 Agencies involved in discharging a condition

Implicit in the definition of a Grampian condition is the concept that the applicant seeking permission cannot itself carry out the work that allows the condition to be discharged and so third parties will be involved; however, the planning applicant will likely be responsible for paying for the work involved. Third parties include (but are not necessarily limited to) the agency that raised the objection, say, for example, an airport. Depending on the mitigation, equipment may be required that will be provided by another organisation, for example, a radar manufacturer. However, it must be demonstrated that it is safe to operate and validating and verifying safety is likely to be the responsibility of another organisation. Moreover, the safety must be established to the satisfaction of the airport regulator. Only when all these agencies have completed their work can the condition be discharged.

It will be seen that the impact of discharging a condition is likely to have a significant impact on the plan to build the wind farm.

4.9.8.5 The mitigation timetable

In the United Kingdom, to satisfy planning law, any mitigation must be identified in advance of the construction of a wind farm:

> If the permission is subject to conditions, for example, requiring you to submit for approval details of a specified aspect of the development which

was not fully described in the application, these must be dealt with before the development can begin [55].

In practice, this sets a timetable, and potentially a time limit, for the implementation of mitigation. In the United Kingdom, for example, if an application for planning permission has been approved, then development must be started no later than 3 years from the date of the approval. But this period can be extended on application.

In practical terms, once a suitable mitigation scheme has been agreed, it will be in all parties' interests to agree on a mutually acceptable, coordinated, project plan, although usually, the mitigation must be implemented before the completion of a wind farm. The developer may be able to negotiate with the ANSP or MOD and be allowed to commence some aspects of the construction before the completion of the mitigation scheme. However, it is likely that mitigations would have to be available before blades are fitted. This aspect is discussed in more detail in Chapter 6.

4.10 Construction of a wind farm

The project plan to build a wind farm will vary from project to project but some of the key milestones are common to all projects. Perhaps the most important aspects of defining the project are arranging those aspects of the project that are out of the hands of the developer; that is the project dependencies.

4.10.1 Ordering turbines

The work carried out in advance of planning will have identified the number and type of turbines that would be appropriate for a given site; at this stage the submission may only specify a tip height and a rotor diameter, not a manufacturer. To identify a 'candidate turbine' too early would be to lose a possible competitive economic advantage in the negotiation of the overall price. The planning process may have required changes to the submitted plan and, without planning consent, making a financial commitment by ordering turbines would be rash. However, once consent is granted, turbines must be selected, ordered and paid for. The delivery dates are a major milestone in the project plan.

4.10.2 Access works

If the development site is on bare earth, then the first construction activities are to create access and make routes that are capable of carrying large vehicles with heavy loads. Increasing shortages of suitable land are leading to the use of forested land for on-shore wind farm development. Ideally, on-shore wind farms would be built on open land because the wind flow is faster and more laminar when the site is clear; whereas the trees in a forest slow the wind, create greater wind shear and cause turbulence. Furthermore, the presence of trees on a site can create a need for on-going maintenance that is not required by a clear site. If a site is forested, then it will already have some access, forest tracks and windbreaks, and then first construction activities are to widen and extend the existing tracks. It will also, generally, be convenient to align cable trenches with the access road.

Construction does not have to lead to wholesale clearance of the land as can be seen in Figure 4.27. In addition to creating access, in a forest setting, a clearing must be created around the planned location of each wind turbine to create enough space to dig, then build the foundations and then, when the foundation is completed, to assemble the wind turbine. The space required is illustrated in Figure 4.28.

The time taken by these activities varies from site to site and even providing general guidelines is difficult but it can represent a significant proportion of the overall build time.

Although this section is not concerned with on-going maintenance activities, it is also worth pointing out that access roads will need to be used for the duration of

Figure 4.27 Road construction (image courtesy Vattenfall PYC)

Figure 4.28 Clearing around a wind turbine (image courtesy Vattenfall PYC)

the life of the wind farm. These roads will require maintenance. The task of maintenance will also be made more complex if the site is shared with commercial forestry extraction.

4.10.3 Turbine foundation works

A major time consideration in formulating the plan for a wind farm is the digging and preparation of the wind turbine foundations.

There are a number of different designs for the foundations of a wind turbine depending on the nature of the ground at the site. For example, if the soil is shallow, then the foundations may be anchored directly to the bedrock. This requires a specialised design. If the soil is soft or even marshy, the foundations might use

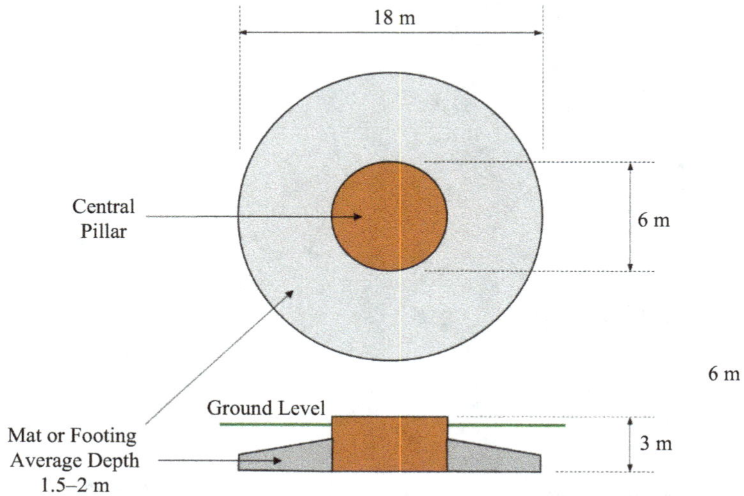

Figure 4.29 Shallow Mat foundation

piles. The most common type of on-shore foundation is the **gravity base** [56]. The design principle of gravity bases is simple, they must be heavy enough to counter the forces created by the wind loading on the turbine which, in effect, try to push over the turbine. A common gravity base design is called **'the shallow mat'**, see Figure 4.29.

To meet the requirement of having sufficient weight, the foundations typically need to be 3 m deep and would require the removal of over 600 tonnes of earth, not all of which will be replaced (Figure 4.30). Furthermore, even though core samples will have been taken in advance of construction, there may be unexpected findings and these can lead to delays. Once the earth has been removed, there are several hundred tonnes of concrete that need to be made, transported and laid. The concrete must then be allowed to set which can take a month. Taking all these factors into account, the digging of the foundations is one of the longest tasks in the construction of a wind turbine and it has a major bearing on the overall length of the project [57].

4.10.4 Cabling and the grid connection

The export of electricity from a site is via the electricity grid, operated by a transmission owner (TO) or a DNO. The grid is highly regulated and connection must be agreed and planned for in advance.

The DNO may be responsible for distributing electricity across a whole nation, for example in Finland, Italy and the Netherlands, or there may be several TO/DNOs each responsible for supply in part of a country. For example, in the United Kingdom, there are currently 11 TO/DNOs, but the structure of the industry is under review. In Belgium, there are eight [58]. In the lower 48 states of the United

Figure 4.30 Tower connection to the Central Pillar (image courtesy Vattenfall PYC)

States, there are 12 major supply operators with each supplying a number of sub-regions [59].

If there is no nearby access point to the grid, then planning the wind farm development must be agreed and coordinated with the DNO's grid development team. This can be difficult, time consuming and expensive. It may only yield a 'connect and manage contract'.

Moreover, electricity grids are segmented, with the segments being connected to nodes in the network. There is no point in increasing the power available to the grid in one segment if these nodes cannot pass more power to where it is needed. Therefore, future grid expansion is considerably more complicated than, simply, extending overhead lines (OHL). This is why it can take a long time (years) to plan for a new wind farm to be connected to the grid.

Preparing the necessary infrastructure to connect to the grid is a major part of the overall construction programme. It is possible to connect individual wind turbines to the grid but in the case of a wind farm, instead of connecting each turbine separately, it is usual to bring the power to a convenient collection point or location and then make a grid connection from that point. As the individual wind turbines in a wind farm are spaced out to take the best advantage of the wind available, cabling must be installed from the diverse turbine locations to a central point. The cables are usually buried underground and the cable runs marked to prevent accidental damage, see for example Figure 4.31.

The 'backbone' of the power grid of most nations is formed of OHL, usually supported on steel lattice towers, or pylons, delivering large amounts of power between different parts of a country or region. It is more efficient to deliver power

Figure 4.31 Cable runs (image courtesy Patrick Delaney Vattenfall PYC)

over long distances at very high voltage to reduce the current in the conductors for a given amount of power. This reduces losses caused by heating. These lines operate between 750 kV and 330 kV, for example, in the United Kingdom, the high-power backbone OHL operates at 400 kV. Connected to the backbone are lower power lines distributing power to consumers within a region. Distances are shorter and the power carried is less. These lines operate at lower voltages, usually between 275 kV and 110 kV, for example, in the United Kingdom, these types of lines operate at 275 kV [60].

To meet the demand for electricity, power must be provided to the grid. Wind farms are often found in the more remote locations where there is a lower demand for power and, hence, they are often connected to lower power, lower voltage, circuits, that is at typically 275–110 kV.

The electronics in each wind turbine can automatically match the frequency and phase of the grid. However, the power supplied by each turbine is, usually, only 600–700 V. This must be increased using transformers to increase the voltage to the figure required for export either to the grid or by the cabling that connects the wind farm to the grid.

Figure 4.32 shows the switchgear and transformers that would normally connect the wind turbines to the grid. Figure 4.33 shows the actual grid connection of the Pen y Cymoedd (PYC) wind farm which is made 2 km north of the wind farm at Rhygos in Rhondda Cyon Taf, Wales. At the time of writing, PYC is unique in the United Kingdom by being connected directly to the 400 kV backbone.

Prior to export, it is also necessary to control the **Power Factor** of the output from wind turbines. The power factor is a measure of the efficiency of power delivery. Ideally, the alternating current and voltage supplied by the grid are in phase; this corresponds to a power factor of 1 and the highest possible efficiency. If the voltage and current slip out of phase the power factor reduces and the capacity of the wind farm to deliver energy to the grid is reduced. This quantity can be controlled and part of the power conditioning plant at a wind farm controls Power Factor.

Figure 4.32 The transformers and switchgear at a wind farm (image courtesy Vattenfall PYC)

Figure 4.33 PYC grid connection (image courtesy Vattenfall PYC)

In addition to building the farm itself, the switch gear and safety equipment for turbines and other infrastructure must also be completed. Figures 4.34 and 4.35 show a local control centre in Wales. Increasingly, this level of monitoring and control is possible remotely from a site by using Internet access.

Figure 4.34 PYC wind farm Hwb (image courtesy Vattenfall, PYC)

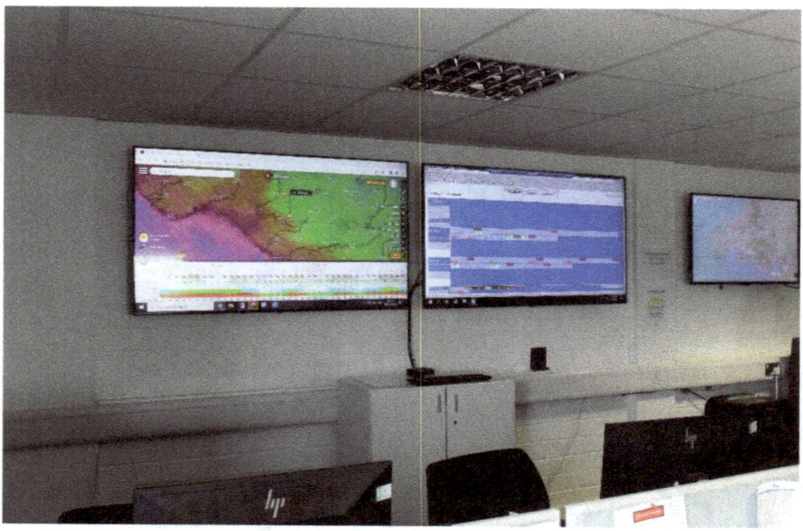

Figure 4.35 PYC control room displays (image courtesy Vattenfall PYC)

4.10.4.1 Delivery of turbines

It cannot be assumed that wind turbines are ready for delivery. Even if they were, it is usual to arrange a delivery schedule to the site to ensure turbines do not have to be stored for long periods and to optimise the use of cranage and other equipment used for movement and assembly. The whole process of delivering turbines to the site may be spread over weeks and months.

4.10.5 Coordination

4.10.5.1 Mitigation life-cycle

Before discussing the factors to be taken into account in coordinating the implementation of mitigation, an important point is that mitigation is not a single event. Mitigation needs to be considered and possibly reviewed over the life of the wind farm. The following considerations apply once the mitigation method has been agreed and selected:

- Initial commissioning of the mitigation (with all the associated planning and testing).
- The aviation system receiving the mitigation has a finite operational life. If that system needs to be replaced during the life of the wind farm, consideration must be given to whether the mitigation needs to be re-evaluated.
- The mitigation system itself has a finite operational life. For example, if the mitigation was an in-fill radar system (described in Chapter 6), then it may need to be replaced or upgraded during the life of the wind farm.
- If the wind farm is re-powered, then the mitigation will need to be re-evaluated.
- When the wind farm is decommissioned, the mitigation may need to be decommissioned as well.

In summary, the need for coordination is an activity that recurs for the whole life of a wind farm or until mitigation is no longer required because of technological improvements. It should also be added that at the time of writing this book, ideas about how mitigations are funded and, particularly the sustainment of mitigation may change.

4.10.5.2 Commissioning process

It is difficult to provide comprehensive guidance on the way to integrate the construction of a wind farm and the implementation of a mitigation scheme because there are many variables and no two wind farms are the same. Some general principles are described below.

Overall sequencing

The mitigation must be in place before the wind farm becomes operational and there is a potential for causing interference with air traffic control (ATC). However, several events will take place before the wind farm is fully operational that might cause interference. Example events are described below.

Commercial considerations

There will be commercial pressure to have a turbine generating revenue, as well as electricity, at the earliest opportunity. This situation might be dealt with by ensuring the mitigation is fully operational in advance of the wind farm. However, the situation may be more complex than it first appears. It is important to be able to demonstrate that the mitigation is effective, and it is likely to be decided that the only way this can be achieved is for it to be tested with all the turbines operational. To meet both ambitions, it may be necessary to carry out turbine commissioning and mitigation testing in stages.

Construction issues
For a radar, the effects that require mitigation will, ostensibly, coincide with the first time a turbine blade starts to rotate with the potential to create false returns that can appear aircraft-like. However, wind turbines are built using a crane and movement of the crane in the wind is likely to cause low-level interference during the construction phase. The effect is low level but it may be a distraction to ATC operators and this may be exacerbated if there is more than one crane on site.

Turbine testing
In advance of entering service, or being made ready to enter service, each turbine must be tested and this will involve blade rotation in advance of the turbine becoming operational. It may be necessary to agree that limited periods of turbine operation are permissible at specified times.

Flying
Testing mitigation was referred to above. This process will almost certainly involve flight trials. Flight trials are costly and, as with all flying, safety is paramount. If trials have to be postponed because of bad weather, this may lead to additional costs and programme delays. Moreover, if the turbine commissioning is carried out in stages. The whole process may be staged with some turbines being brought on stream earlier than others. More periods when flight trials are required multiplies the risks of weather delays.

4.11 The impact of a wind turbine on the electromagnetic spectrum

4.11.1 Scope

Four matters must be considered:

- The radar cross section (RCS) of wind turbines/wind farms and aircraft. The treatment here is a practical approach to the problem. Sources of alternative approaches are listed in the bibliography. It is important to be able to understand the RCS of wind turbines to be able to predict their impact on aviation radio and radar systems. It is useful to understand aircraft RCS because wind turbines must be discriminated from aircraft and also it is useful to understand how aircraft might be used in flight trials,
- The time intensity of RCS (how does the RCS vary with time). This is germane to why wind turbines pose a problem for aviation radar systems.
- The Doppler spectrum of the RCS of wind turbines is also germane to why wind turbines pose a problem for aviation.
- The shadow formed behind wind turbine blades.

4.11.2 General principles of RCS

RCS is a measure of the capacity of an object to reflect radio waves. The US Institution of Electrical and Electronic Engineering (IEEE) defines it as the 'reflective strength' of a target [61]. In principle, RCS is a simple concept but it is, arguably, one of the most complex subjects that a radar engineer must deal with.

4.11.2.1 RCS definition

A convenient model of the behaviour of the propagation of a radio wave is that, as the wave moves forward, it expands to illuminate an ever-greater surface area. The power in the wave, provided by the transmitter and the antenna, is shared over this area. As the distance from the transmitter and antenna increases, the area covered increases as the square of the distance. Thus, every time the distance travelled by the wave doubles, the area over which power is shared increases fourfold, as illustrated in Figure 4.36.

Power spread over an area like this is called a **power density**, it is measured in watts per square metre.

The amount of power available to be converted into a radio reflection by a target depends on how much power it can extract from the radio wave. Hence, the effective area of the target is critical. The emphasis is placed upon the word 'effective', it may have little to do with its physical area. Skolnik referred to it as a 'fictitious' area. The greater the effective area the more power there is available and vice versa. Strictly, the term ought to be called the Radar Cross-Sectional Area but it has become known, simply, as RCS [62].

4.11.2.2 A mathematical description of RCS

RCS is often described mathematically in the literature, and although it is not used here, the description can be a cause for confusion. It is important to avoid the confusion.

This leads to a definition of RCS as follows:

$$\text{Radar Cross Section, } \sigma = \lim(R \to \infty) 4\pi R^2 \frac{S_r}{S_i} \quad (4.2)$$

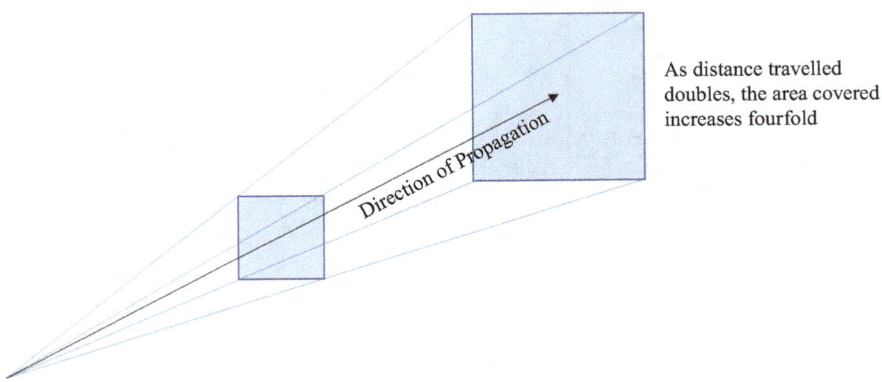

Figure 4.36 Radio wave propagation and effect on power density

where R is the distance between target and radar; S_r is the reflected power density (W/m^2); S_i is the incident power density (W/m^2) at the target; lim ($R \rightarrow \infty$) discussed below.

RCS is the factor that converts the incident power density into a reflected signal that is detected back at the radar. The equation is frequently quoted in electromagnetic theory a little differently, with the power density being replaced by the strength of the electric fields. However, note that this remains a power relationship and to convert the electric field intensity to its equivalent power the terms must be squared:

$$\text{Radar Cross Section, } \sigma = \lim(R \rightarrow \infty) 4\pi R^2 \frac{E_r{}^2}{E_i{}^2} \tag{4.3}$$

where R is the distance between target and radar; E_r is the reflected electric field strength (V/m); E_i is the incident electric field strength (V/m) at the target; lim ($R \rightarrow \infty$) discussed below.

4.11.2.3 Explanation

Rearrangement of (4.2) shows what the equation is describing:

$$\sigma S_i = \lim(R \rightarrow \infty) 4pR^2 S_r \tag{4.4}$$

The same principle can be applied to (4.3). The incident signal has a power density that impinges on a target with an effective area, σ. The resulting power is converted by the target into a reflected signal that is radiated equally in all directions. Figure 4.37, adapted from Shaeffer (2010), illustrates the process [63].

4.11.2.4 Confusion factors

A potential confusion arises from the distance from the radar to the target, R (strictly R^2), appearing in the equation implying that RCS may vary with range: *this is not the case*. The R squared is cancelled out by the expansion of $E_r{}^2$ or S_r which are inversely proportional to R^2. **RCS is independent of the range of the target**.

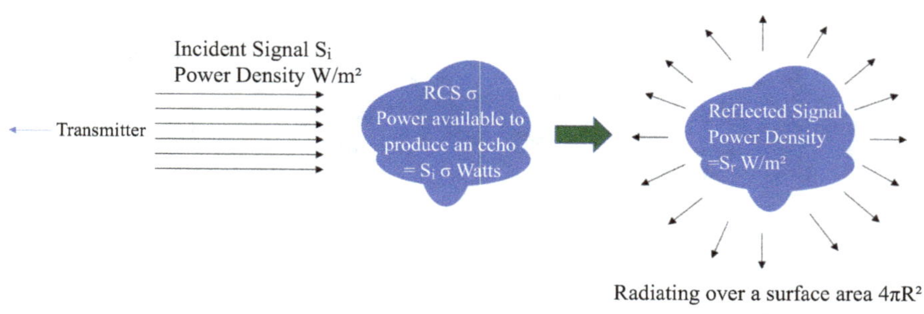

Figure 4.37 RCS definition adapted from Shaeffer (2010)

Confusion may also be caused by the reference to the 'limit as R tends to infinity'. The definition assumes the reflected signal is radiated equally in all directions by the target; so that, at a distance, the reflection illuminates the surface of a sphere and the illumination is uniform. This situation occurs when the target is a point source. A practical way of meeting this requirement is to make the target a long way from the radar so that its size is very small compared with the distance. It is worth repeating that RCS is the same at all ranges from the radar.

4.11.2.5 Factors affecting RCS

The complexity of RCS comes from the fact that it is modified by a number of factors, some of which are important to understand in the present context:

- Shape: Some shapes reflect radio waves in different ways than others, this will be discussed below in the context of the so-called canonical shapes (standard reference shapes) which can help predict the RCS of wind turbines.
- Size: As a general rule, larger objects have a larger RCS than smaller objects but this is a simplification and size must be considered further. Furthermore, when discussing the size of an object and its RCS, the size is with respect to the wavelength of the radar and, hence, frequency is also a determinant of RCS.
- Material/treatment: A common misunderstanding is that only metallic things reflect radio waves: this is not correct. The material is important in two ways, some materials, for example, can reduce RCS significantly (and can make things stealthy or can be used to make radomes which have very high transmissivity of radio waves).
- Surface: Rough surfaces and smooth surfaces reflect radio waves in different ways. Smooth surfaces are strong reflectors of radio waves creating **specular** reflection. Rough surfaces give rise to **scattering** reflections. It might be possible to apply such a surface to, say, wind turbine towers but it would not be suitable for turbine blades.
- Polarisation of the radio waves: This can have a large effect on the RCS of the target.
- Aspect geometry affects the direction of reflections. This can be very important if the target is complex, for example, an aircraft, and scattering from different parts of the target interact giving rise to constructive and destructive interference. The effects of geometry are discussed later.

4.11.2.6 Object shaping

A useful concept is that objects fall into one of two categories:

- Simple shapes: spheres, flat plates, cylinders, etc. These shapes are well understood and their RCS can be predicted with relatively simple formulae. These canonical forms can give great insights into, for example, the influence of size and wavelength, or,
- Complex shapes: such as aircraft and spacecraft.

4.11.2.7 Simple shapes

One of the greatest insights into how RCS varies with wavelength was provided by Mie [Gustav Adolph Feodor Wilhelm Ludwig Mie (1868–1957)] in 1908. Mie calculated the RCS of a sphere using Maxwell's equations (Chapter 2). The findings are illustrated in Figure 4.38.

It might be imagined that the results would be simple: as the sphere increased in size so might the RCS. However, Mie showed that there are distinct RCS regions:

- **Rayleigh scattering region**: If the object is small compared with the wavelength, the RCS is small and if the object is very small the RCS can be very small indeed and the object is difficult if not impossible to detect.
- **Mie scattering region**: As the size of the sphere increases, and the circumference of the sphere is approximately the same as the wavelength of the illuminating radar, the RCS enters a region of damped oscillation. The peak of the oscillation coincides with the wavelength of the radio wave and the circumference of the sphere being exactly the same. This phenomenon is described below.

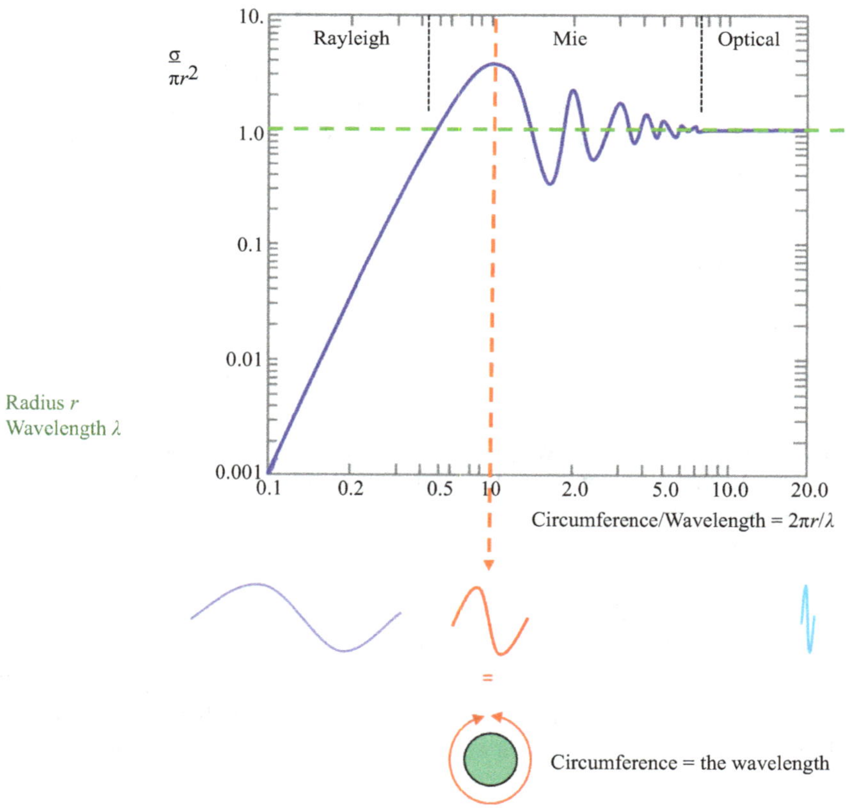

Figure 4.38 RCS of a sphere after Mie

- **Optical region**: When the oscillation has damped completely, coinciding with a region where the sphere is large compared with the wavelength the RCS levels off and no longer changes with increasing the size of the sphere.

4.11.2.8 Mie scattering

The Mie region illustrates an important feature of RCS: there is more than one mechanism at work producing the radio reflection. Consider a pulse of radar energy reaching a sphere. When the wavefront touches the sphere, it induces a current to flow and this creates another radiating signal. Part of this signal travels back to the returns to the radar; this return is called a **specular**. The current carries on flowing after the specular return has set off, radiating energy all the time. Eventually the current arrives back at the starting point. Part of the radiated signal will travel back towards the radar. If the circumference of the sphere is exactly the same as the wavelength of the incident wave, the **travelling wave** reflection and the specular will be in phase and will add constructively increasing the magnitude of the reflection. As the frequency increases and the wavelength gets shorter, eventually the wavelength will be half the circumference and a second peak in RCS appears. The process continues until the size of the sphere is large compared with the wavelength and the optical region is entered. The 'magnification' of the RCS has been exploited for the choice of the wavelength of radars.

4.11.2.9 Canonical shapes

Canonical shapes: that is, something from the canon of literature on the topic, provide useful standard shapes that help us to approximate real objects. Karabayr [64] proposes a model of wind turbines using canonical shapes based on flat plates and cylinders and produces results that are a good match with RCS predictions using more complex analytical solutions.

A particularly useful insight provided by this simple method is how RCS varies with frequency. For example, consider the RCS of a long cylinder; used by Karabayr to represent the turbine blades. The RCS of a long (more than 10 wavelengths long) cylinder is calculated thus:

For the normal case, that is the peak RCS when the source of the rf is normal to the alignment of the cylinder:

$$\sigma_N = \frac{2\pi r L^2}{\lambda} \tag{4.5}$$

and at observation angle θ,

$$\sigma(\theta) = \frac{\lambda r \operatorname{Sin}\theta}{8\pi \operatorname{Cos}(\theta)^2} \tag{4.6}$$

where r is the radius of the cylinder; L is the length of the cylinder; λ is the wavelength.

Figure 4.39 shows a plot of (4.5) and (4.6), assuming a cylinder of 30 m in length 1 m radius at S-Band. The shape described by the equation provides only the general shape of the RCS, in practice, the shape consists of a large number of closely spaced lobes, as illustrated in Figure 4.39.

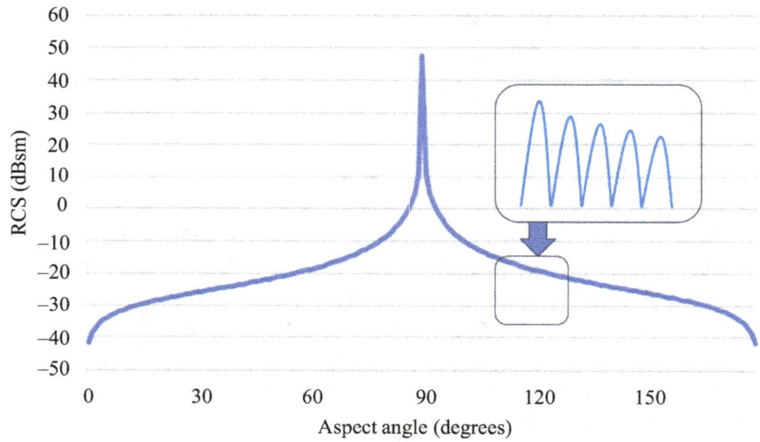

Figure 4.39 30 m cylinder RCS S-Band (assuming linear polarisation)

Table 4.3 RCS versus frequency of a 30-m cylinder

Frequency	Application	RCS (dBsm)
100 MHz	VOR	32.8
125 MHz	VHF communications	33.7
360 MHz	UHF communications	38.3
L-Band	AD and ARSR (and approximately DME)	43.5
S-Band	ASR	47.5
X-Band	Precision approach radar (PAR) (and some in-fill/wind farm tolerant radars)	52.8

The peak RCS is 47.5 dBsm. The calculation was repeated for a range of frequencies used for other purposes and the results are listed in Table 4.3. Indicated in these results is that RCS increases with frequency and this increase is approximately linear. That is, if the frequency doubles the RCS also doubles.

The same equations can be used to investigate the effect of size on RCS. Table 4.4 shows the RCS for different dimensioned cylinders calculated at S-Band. The trend, predictably, is for the RCS to increase as the size of the cylinder, surrogate blade, increases.

4.11.2.10 Material

There have been many initiatives to reduce the RCS of wind turbines using a variety of approaches. Fairley [65] reported a joint initiative by Qinetiq and Vestas which used a 5 mm layer applied to the tower and incorporation of two layers of radar-absorbing material in between the layers of glass fibre as the blade is built. The benefit of incorporating layers in this fashion is that it need not add to the weight or change the aerodynamics of the blades. The disadvantage of this method

Table 4.4 RCS at S-Band for cylinders of different dimensions

Cylinder	RCS (dBsm)
Length 30 m × radius 1 m	47.5
Length 50 m × radius 1.25 m	52.9
Length 75 m × radius 1.5 m	57.2

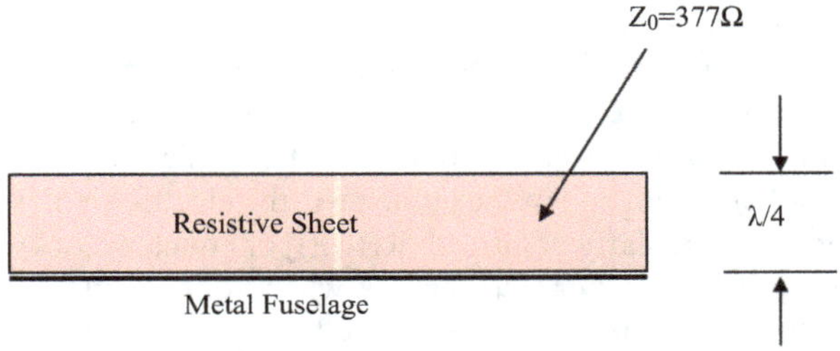

Figure 4.40 Salisbury screen

is that its performance is only optimized for a single frequency band. Norman [66] reported a reduction in the peak RCS of 15 dB. This reduction in RCS is useful and would reduce the potential for a condition called saturation (described below). However, with other mitigation measures it may not affect the outcome as the blade would still be detectable by radar. Moreover, the costs for implementation were estimated at 10% of the blade cost. Norman points out that in a large farm, this cost would be higher than the cost of an in-fill or a wind farm tolerant radar.

Where such a change in performance would be beneficial would be in reducing the impact on Communications and Navigation Aids. Unfortunately, the frequencies used for communications and navigation are relatively low, approximately 120 MHz and the mitigations are designed for radar frequencies.

Chambers and Tenant [67] developed an alternative technology based on a modified Salisbury Screen, illustrated in Figure 4.40 [68]. The Salisbury Screen applies a resistive sheet to the structure to have reduced RCS. The thickness of the layer is a quarter of a wavelength. The sheet should have the same resistance to radio waves (characteristic impedance) as free space (377 Ω). Provided this criterion is met there is very little reflection from the screen itself. An incident wave passes through the sheet, having energy absorbed as it passes through. It is reflected off the material underneath it and is reflected back where there is further absorption and very little is reradiated off the whole structure. The disadvantages of this method are the need for the application of a screening material and this is only effective at one frequency. A variation on this structure called the Jauman screen used multiple layers. The Chambers and Tenant method applied a phase-switched screen (PSS), to modulate a reflection changing its frequency to one outside the pass band of the receiver. They reported RCS reductions of approximately 50 dB. If it was possible to implement a screen like this on a wind turbine it might be possible to prevent the turbine from being detected by a radar. However, Chambers and Tenant only demonstrated this capability at a single frequency band.

4.11.2.11 Wave polarisation

An electromagnetic wave has two components, an electric field, shown with blue arrows in Figure 4.41, and a magnetic field, shown in green arrows. In free space, as the wave propagates. these two fields are orthogonal (at right angles to each other) and in phase (both grow in magnitude, peak and diminish in amplitude in sympathy as they move forward).

The transmitting antenna determines the orientation of the wave's electric and magnetic fields; the **polarisation** of the wave refers to the orientation of the electric field[§§]. Polarisation may be linear or circular; in the former case, the electric field may be vertical or horizontal and in the latter case, the electric field is rotated either clockwise [right-hand circularly polarised (RHCP)] or counter-clockwise [left-hand circularly polarised (LHCP)]. For clarification, when, for example, a RHCP wave is described as rotating clockwise that is as viewed by an observer looking in the same direction as the direction of propagation (away from the observer).

Linear polarized waves may be at any angle with respect to the horizon but, in general, they are either horizontal or vertical [some surface to air missile (SAM) radar systems use 45° linear polarization to mitigate the effects of clutter from the ground]. To extract the maximum amount of energy from a wave, the receiving antenna polarisation must be the same as the transmitting antenna, that is, horizontal to horizontal or vertical to vertical; in this case, the antennas are said to be **co-polarised**. If there is a mismatch in the polarisation, then the loss in signal strength is a function of the cosine of the mismatch of the angles. If the opposite polarization antenna is used, then the mismatch is 90° and theoretically, the loss

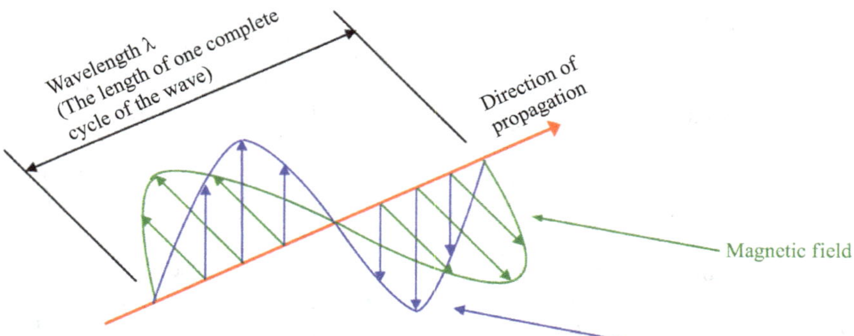

Figure 4.41 An electromagnetic wave

[§§]It is most common to measure the electric field and its orientation has been adopted as the standard.

should be infinite, in practice the polarisation mismatch is never perfect and the drop in signal strength might be expected to be between 20 and 30 dB. Similar results might be expected for a mismatch in a circularly polarised wave; if the receiving antenna is, say, LHCP and the wave RHCP rejection might be 20–30 dB.

The situation set out above applies in, say, a communications system where a signal is transmitted from one antenna directly to another. However, in the case of radar, reflection of a target needs to be considered. If a linearly polarised wave is incident on a smooth target, say, for example, a horizontally polarised wave, then the reflected wave will still be horizontally polarised but it will, in effect, be the mirror image of the transmitted wave; this makes no difference to the receiving antenna, the wave is still horizontally polarised. In practice, for practical targets, consider an aircraft for example consisting of many reflecting surfaces that are close together. Interactions between these surfaces mean that the reflected signal will have horizontal components but there will also be components in other orientations as well.

The concept of a reflection is more complicated in the case of a circularly polarized wave. To illustrate the point, consider the following example. An antenna produces a RHCP wave that propagates away from the antenna and is incident on drops of rain. For the purpose of this example, it will be assumed that the raindrops are spherical[¶¶]. The reflection from a smooth, spherical, object rotates in the same direction as the incident wave (to the right as viewed from the observer located at the radar) but it is now travelling in the opposite direction (back towards the radar) and, hence, is cross-polarised on arrival back at the radar antenna and is rejected by the radar[||].

Now consider what happens when a wave from the same antenna is incident on an aircraft. The complex shape and arrangement of multiple reflecting surfaces give rise to returns that are both co-polarised and cross-polarised when received by the radar.

Consider then the net effect on the radar. Circular polarisation provides a means of detecting targets in the presence of rain (and other forms of precipitation). This method is commonly used in ATC and PAR and occasionally used in AD radars for that purpose. Some radars use circular polarisation exclusively, some radars provide the operator with both linear and circular polarisation and the operator can select a mode consistent with the weather.

4.11.2.12 Polarisation effects on RCS

In September 1975, Turnbull reported an investigation carried out by the US FAA in response to concerns raised by the National Transportation Safety Board (NTSB) about the radar-detectability of some small aircraft and their prevalence in mid-air collisions when they should have been detected. The objective of the study was to investigate the practicality of enhancing their detectability using passive devices

[¶¶]In practice, this is a simplification. When first formed, raindrops are approximately spherical but as they fall and accrete more moisture, they start to flatten under their own weight. This distortion can be detected by radar and meteorologists use this to categorise hydrometeors (raindrops).

[||]A model to help understanding is the idea of a screw thread. A right-hand thread bolt only fits a right-hand nut.

such as trihedral corner reflectors; a method that is very effectively used in small marine vessels, see Figure 4.42. The conclusion of the study was that these measures were not effective and that SSR should be encouraged for light aircraft but the findings of the study are revealing.

The Turnbull study was carried out on three General Aviation (GA) (light) aircraft: a Piper Cherokee 140, a Cessna 150 and a Piper Super-Cub. The first two of these aircraft have metal surfaces and the latter aircraft has glass fibre surfaces. These aircraft were chosen because they were representative of a wide variety of light aircraft. Accurate RCS measurements were made at a US Air Force test facility (Holloman Air Force Base (AFB), New Mexico). The results of these measurements are shown in Figures 4.43 and 4.44.

The plot on the left is the fabric-skinned Piper Super Cub and the plot on the right is the metal-skinned Cessna. The RCS has been compared at two angles, 0° yaw angle (nose on) and 90° yaw angle (looking at the wing tip and the side of the fuselage). In all four cases, the linear polarisation has an advantage over the circular polarisation signal. The results are listed in Table 4.5.

These results are broadly consistent with Poupart's findings that RCS measured with circular polarisation were between 15 and 20 dB lower than the equivalent linear polarisation measurements.

Figure 4.42 Trihedral corner reflector (photo Dr Robert Rohlfs, reproduced with permission)

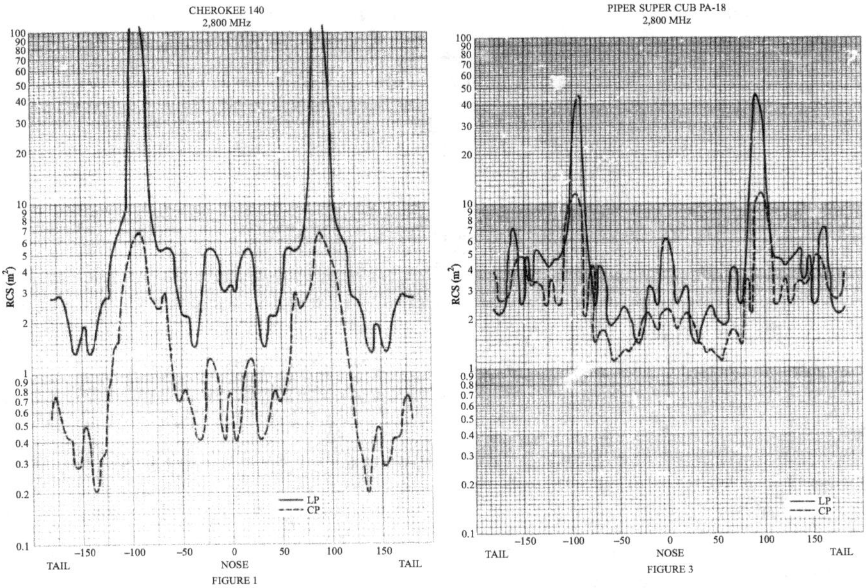

Figure 4.43 RCS v aspect angel showing the effects of polarisation, 2,800 MHz (information provided courtesy of the Federal Aviation Authority)

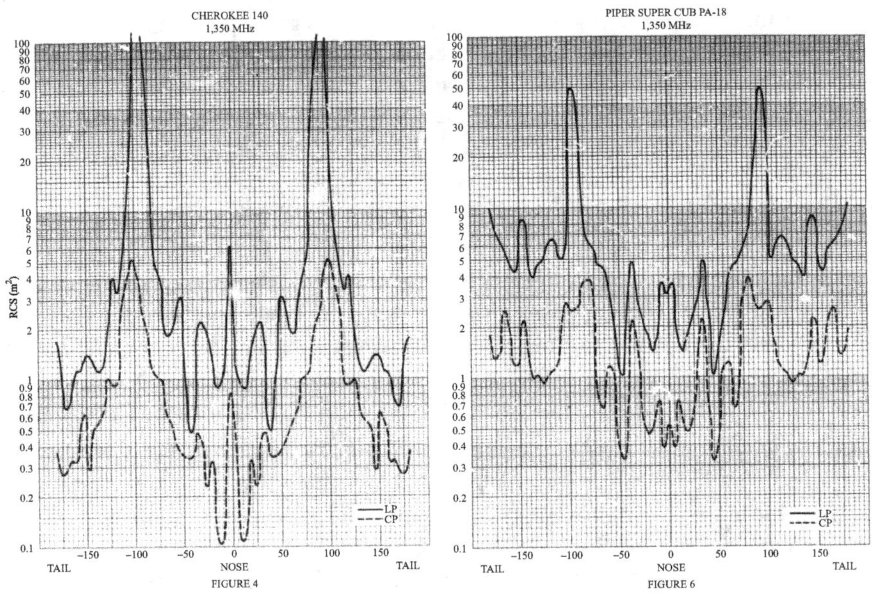

Figure 4.44 RCS v aspect angel showing the effects of polarisation, 1,350 MHz (information provided courtesy of the Federal Aviation Authority)

Table 4.5 Linear polarisation advantage (over circular)

Yaw angle	**Super cub**	**Cherokee**
Nose on	4.1 dB	6.5 dB
90°	5.9 dB	>11.7 dB

4.11.2.13 Effects of aspect geometry on RCS

Figures 4.43 and 4.44 illustrate not only the effects of polarisation on RCS but also the variability of RCS with aspect geometry. These plots are typical of RCS v aspect angle of aircraft. The variability is caused by different parts of the aircraft fuselage, engines, etc. each making their own contribution to the total RCS. These contributions add constructively and destructively depending on the geometry of each part with respect to the radar. However, consider the case where the aircraft is flying directly towards the radar. The geometry of each contributor and the radar change only slowly and, likewise, the RCS shows only very slow changes. Whereas, if the aircraft is flying across the radar's field of view, the geometry of each contributor and radar changes continuously, and the RCS becomes highly variable. The variabilities have been analysed by a number of researchers. Swerling, reported in [62] and [71], found that the RCS of aircraft generally conforms to one of two PDFs (the concept was introduced in Chapter 3). The slowly changing RCS, as an aircraft flies towards (or away from) the radar, is referred to as a *Swerling Case* 1 PDF. The low variability of RCS makes this a useful characteristic for conducting flight tests and for specifying the performance of radars. See, for example, the discussion in section 6.4.4 which describes performance metrics. The rapidly changing RCS when the aircraft flies across the radar field of view is referred to as a *Swerling Case* 2 PDF. This behaviour makes the aircraft harder to detect by radar and may be avoided in flight trials for that reason.

4.11.3 The RCS of wind turbines components, wind turbines and wind farms

One of the most comprehensive assessments of RCS was carried out by Poupart in 2003 in a study sponsored by the UK Department of Trade and Industry [69]. The work consisted of a mix of measurements and modelling and simulation. The tower, nacelle and blades of a turbine were addressed separately as well as the whole structure. The RCS of the whole structure was separated into total RCS, that is disregarding Doppler and with zero Doppler components removed. The principal findings were as follows:

- It is desirable to minimise the RCS of the static components and this is achievable through design. For example, the RCS of the tower is a major contributor to overall RCS, however, even small increases in the taper of the tower can reduce the RCS significantly.
- As a result of factors such as the tower taper, it does not follow that increasing the size of a wind turbine increases the total RCS on a pro-rata basis.
- The only way to control blade RCS is through construction by radar-absorbing materials (see discussion above).

- Yaw angle is a critical determinant of RCS:
  - A yaw angle of 90° and 270° reduces the probability of detection, but produces a large clutter return (maximises the Doppler footprint of the wind turbine).
  - A yaw angle of 0° and 180° increases the probability of detection but produces a smaller area of clutter (maximises the blade RCS which would tend to increase detectability but it also minimises Doppler which might lead to the signal being rejected by MTI/MTD).

The study also produced some interesting findings, some surprising:

- Consistent with Hunting's findings, circular polarisation leads to large reductions in RCS. Poupart reported a peak RCS for a single 26 m blade between 15 and 20 dB lower for a circular polarised wave compared with a linearly polarised wave.
- Peak RCS occurs when the illuminating signal is perpendicular to the length of the blade.
- RCS measurements and predictions are 'impulsive' (spiky).
- At least in some of the instances investigated, the highest measured RCS occurred when one of the blades was in the vertically downwards position.

Peak RCS measurements were reported 30 dBsm for both Enercon E-66 and Vestas V-47 wind turbines.

4.11.3.1 Wind farms

The radio signature of an individual wind turbine is of interest, if there are two wind turbines separated by, say, five rotor diameters and they present side by side to the radar, then at any distance greater than 23 km the two turbines will present to the radar as a single target***.

Randhawa and Rudd measured the RCS of a single wind turbine and a wind farm of 17 turbines as a part of a study for the UK Ofcom [70]. The study was interested in frequencies used for fixed link services and the measurement frequencies are not exactly in those frequencies used by radar but some are close. The results are set out in Table 4.6.

Table 4.6 Wind turbine and wind farm measurements after Randhawa and Rudd

Frequency (MHz)	Single turbine			17 Turbines		
	Forward scatter	**Intermediate scatter**	**Backscatter**	**Forward scatter**	**Intermediate scatter**	**Backscatter**
1,477	50	17–26	32	54	20–43	42
3,430				41	22–36	

***Assuming a rotor diameter of 120 m and an antenna beamwidth of 1.5°.

The principal finding from this measurement campaign is that the RCS of a wind farm does not scale up in a linear fashion. If it were to do so, then the wind farm measurements would be 17 dB higher than the individual measurements. Randhawa and Rudd attribute this to the problems posed by destructive interference of returns. The rotor diameters of the measured wind turbines/farms were either 80 m or 82 m.

4.11.3.2 Wind turbine clutter

The Doppler signature of a wind turbine is discussed later in this chapter. For the time being, it is asserted that an echo from the tips of the turbine blades can have the same Doppler frequency as, and, therefore, appear to be a return from a real aircraft. However, the RCS is highly impulsive and only on some sweeps of the radar will any given turbine appear. Any single turbine may only present a high RCS every 5 or 6 scans of the radar antenna. A common expression that is associated with the phenomenon is that the turbines appear to 'twinkle'.

In the case of a wind farm, on some radar scans, some turbines will appear and on others they would not. This can lead to a situation where, by fluke, over a number of scans, turbines will be observed that could appear to be moving like a genuine aircraft. In such a situation. a controller would have to assume that there is a risk, the clutter is a real aircraft and another close aircraft would have to be diverted to maintain separation.

4.12 The problem space

This section deals with the problems for aviation radio and radar systems because of their interaction with wind turbines and wind farms. In addition to the technical problems, there are some policy matters that are problematic. The topics addressed are:

- Radar technical interactions: the problems posed by the technical interactions between radar and wind turbines have changed little in the last 20 years. Primary and secondary radars must be considered as well as the secondary radar derivative systems. Moreover, the civil and military radars must be addressed,
- Radio technical interactions: in recent years, concerns have been expressed about how communications and navigation systems are affected by wind turbine and wind farms.
- Wider policy problems are posed. These problems centre on the fact that the behaviour of wind turbines must be predicted in advance of their construction.

4.12.1 Radar technical interactions

The principal problem posed by wind turbines, both individually and in wind farms, is because of their high RCS. This results in the following technical interactions:

- Primary radar effects
    - Saturation
    - Generation of clutter
    - Interactions with pulse compression
    - Processing overload
    - Obscuration, in some publications known as de-sensitisation
    - Track seduction
    - Shadow effects

- Secondary radar effects
    - Bearing errors
    - False targets

4.12.2 Saturation

The range of values of RCS across different types of aircraft varies greatly. At the frequency of an air surveillance radar (ASR) or AD radar [71]:

- A modern military 'stealthy' aircraft or a drone might have an RCS of −10 dBsm (a tenth of a square metre) or even lower.
- The RCS Of a General Aviation (light) aircraft may be only 0 dBsm to 3 dBsm (1–2 m^2).
- At the other extreme, the RCS of a wide-body jet like a Boeing 747 airliner is 19–20 dBsm (70–100 m^2).

Moreover, it is not just the target sizes that have a wide dynamic range, the distances at which these aircraft must be detected also vary a great deal. An aerodrome ASR, civil or military, must detect aircraft at only a few kilometres, and the maximum range is usually 110 km. An AD or an ARSR radar must detect aircraft at even longer ranges; perhaps up to 450 km. The amplification stages of a radar receiver must handle these extremes and still be able to process the target information accurately.

By comparison, an earlier section in this chapter pointed out that the RCS of wind turbine components can be very large. Depending on the blade sizes and their orientation, radar frequency and polarisation, the blades may have predicted peak RCS in excess of 50 dBsm (100,000 square metres), although measured peak values of RCS are often smaller.

If the same amplifiers in the radar receiver now have to process a signal which is very much greater than the normal target set of the radar and the receiver may enter 'saturation'. Conceptually, this is illustrated in Figure 4.45. In the left-hand part of the diagram, a 'normal' magnitude of the signal is being amplified within the linear portion of the amplifier gain. However, amplifiers cannot provide unlimited gain and a point is reached where it can no longer amplify the input signal: its gain levels off.

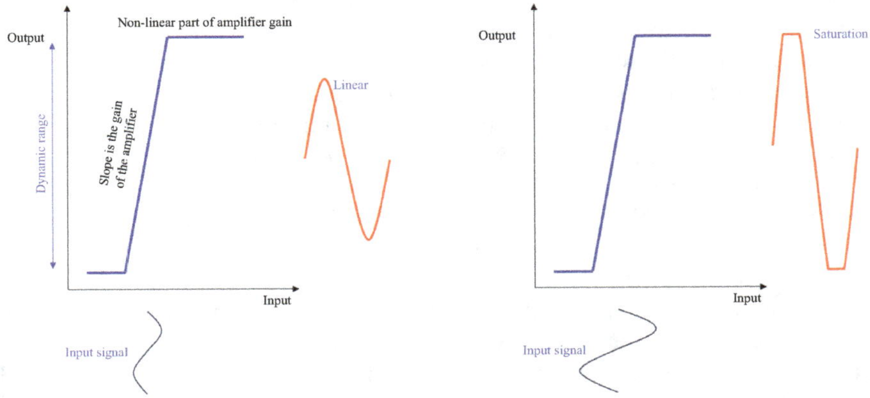

Figure 4.45 Receiver saturation

When this point is reached, the output signal from the amplifier is shown on the right-hand side of Figure 4.45. The peaks of the output signal are clipped, giving it a square wave appearance. A square wave occupies a much greater bandwidth than a sine wave. Therefore, when this type of distortion occurs, the output will consist, not only of the original frequency but it will be spread across greater bandwidth. This is problematic for the next stage of processing.

Once amplified, the usual next stage of processing in the receiver is to find out if the echo is from a moving source. The frequency of the echo signal frequency is compared with that of the transmitted signal to find out if a Doppler shift is present, caused by the motion of the target. However, if the output has now been corrupted through saturation, the Doppler information will now be mixed with all the frequency components associated with saturation. Therefore, if the input signal exceeds the linear limits of the receiver, called the **dynamic range**, then the signal amplitude will be distorted but the radar Doppler processing is also going to be compromised.

4.12.2.1 Dynamic range

The following equation provides an estimate of the dynamic range of a radar receiver [72]:

$$\text{Dynamic Range} = 10\text{Log}_{10}\frac{(\text{Max Target Size} \times \text{Max Range}^4)}{(\text{Min Target Size} \times \text{Min Range}^4)}\,\text{dB} \tag{4.7}$$

Consider the case of an ASR (a civil or military aerodrome ATC radar). Assuming that the maximum target size that might need to be detected is 100 m^2 and the minimum target size that needs to be detected is 0.1 m^2. The latter value is lower than most ASR would traditionally be required to detect but let us suppose that it is necessary to detect drone intrusions into airspace. The maximum range is 110 km, and the minimum range is 15 km. The minimum range value may seem high in value, given that an aerodrome radar must detect aircraft as they land and take off and the end of the runway is probably only going to be perhaps 2 km away. However, in a modern ASR, two different pulse lengths are used: a short pulse and a long pulse. A short pulse is used to detect targets close to the radar, such as aircraft arriving and departing from the aerodrome. A long pulse is used to detect targets out to the maximum range of the radar. The long pulse cannot detect objects close to the radar because the time taken to transmit a long pulse prevents near objects from being seen. The value of 15 km was chosen to represent the shortest range at which the long pulse would be used.

Given these assumptions and inserting them into (4.7), the dynamic range required for the example radar is 65 dB. This is an estimate, in practice, the radar will have other processes that prevent excessive signals from causing problems at short range. For example, sensitivity time control (STC) or swept gain can reduce the sensitivity of the radar at short ranges.

Now, suppose a wind turbine presents an RCS of 100,000 m^2 (50 dBsm) at a distance, chosen at random, 30 km away from the radar. Again, applying (4.7), the radar receiver would need to have a dynamic range of 72 dB to process the wind turbine in its linear region. But the radar was otherwise only required to have 65 dB and would have been designed with this requirement in mind. Faced with such a large target, the receiver would be driven into saturation by 7 dB and the amplified waveform would be clipped as illustrated in Figure 4.45.

Degrading of the Doppler assessment is not the only problem created by such high RCS targets being presented to the radar. The example used to illustrate the problem was premised on the idea of using a long pulse. The pulse is long because it uses pulse compression. This will be dealt with later. There is one other way in which saturation can occur and this will be discussed first.

4.12.2.2 Analogue to digital conversion

Saturation may not only occur in the analogue parts of the radar receiver. Modern radar receivers do a lot of signal processing using computer techniques. To make this possible, the received signal must be digitised using an analogue to digital converter (ADC). The design of ADC has improved a great deal in recent years. Now ADC is able to convert analogue signals faster and more accurately. Each conversion provides a binary number corresponding to the analogue value of a waveform at an instant in time. Given that radar is being considered, time corresponds to a range. Thus, another way of thinking about an ADC is that it is responsible for populating range cells with received measurements. Faster conversions allow the cells to represent smaller increments in range. Greater accuracy is provided by more bits in the digital words that are created. If the number of bits provided by the ADC is limited. The number of quantisation levels is 2^N where N is the number of bits. For example, an older 10-bit ADC has 1,024 quantisation levels. By comparison, a modern ADC might have 14 bits and the number of quantisation levels would increase to 16,384, a 16-fold improvement in dynamic range that is not available to older radars.

4.12.2.3 Dynamic range for different radars

A final observation is simply to make the point that the dynamic range of the radar is related to the maximum range at which the largest target must be detected (from (4.7)). The longest-range radars, ARSR and AD radars, must have commensurate dynamic ranges and, in general, this is true. However, these radars may have multiple pulse lengths and more complex pulse compression processing. This is discussed later in this chapter.

4.12.3 Clutter

An object with an RCS as high as a wind turbine, when illuminated by an aviation radar, has a very high probability of being detected.

There are many things in the environment that have a high RCS. For example, electricity pylons have some things in common with wind turbines, they have tall towers and they even have cross arms protruding from the tower. Pylons are visible to radars, see, Figure 4.46. This image shows the radar screen at an aerodrome close to a line of electricity pylons. At the time the photograph was taken, the radar was nearing the end of a period of routine maintenance and was in the process of being returned to service. Notice the S-shaped pattern snaking across the display. These returns are from electricity pylons. However, the pylons are permanent features in

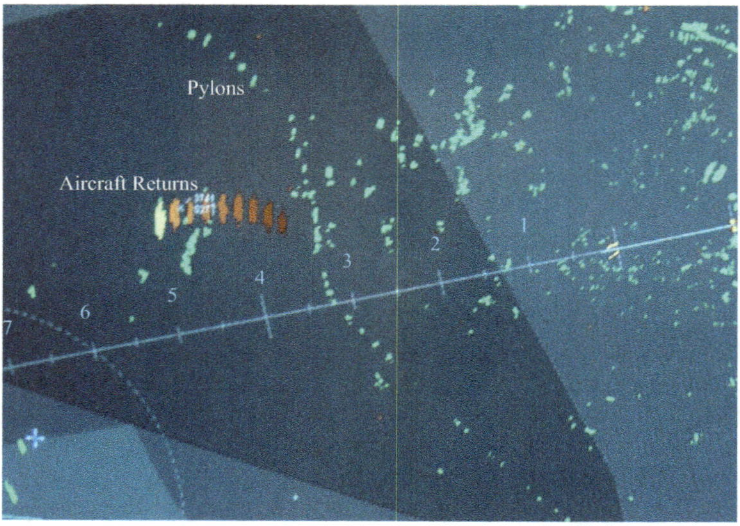

Figure 4.46 Pylons on a radar picture (courtesy RAF Brize Norton)

the landscape and the radar quickly adapts to these permanent features and the returns will be edited from the display.

The display shows some other interesting features.

- The first thing that would be observed by an air traffic controller is that this is an old type of display; the photograph was taken in 2009. These types of displays are sometimes referred to as 'analogue displays'. The radar data appearing on the display replicates the analogue signal received by the radar. A more modern display would appear quite different. Computer-generated symbols would replace the analogue/video. However, this photograph of an analogue display was chosen deliberately. It illustrates the analogue signal provided in these circumstances to digital processing and display.
- The background shading makes little sense on this scale of image: the darkest background is the aerodromes control zone, the general display background colour is a little paler and the palest shade indicates airways.
- The graduated line crossing the screen from left to right is the extended centreline of the runway, the numbers indicate the number of nautical miles from the end of the runway.
- There are a lot of other green-coloured returns: the majority are clutter returns from the landscape and some are due to weather.
- However, the principal feature is the presence of an aircraft. This is marked by a set of returns, the brightest on the left representing the return from the latest scan of the radar. Returns from earlier scans become progressively darker. The 'history trail' allows the controller to see the direction of the aircraft. These returns are the primary radar returns.
- In the middle of the aircraft ensemble of returns, the secondary return can be seen. The two rows of text indicating the aircraft is squawking with an Ident 3741, the altitude is Flight Level 027 and the aircraft is climbing.

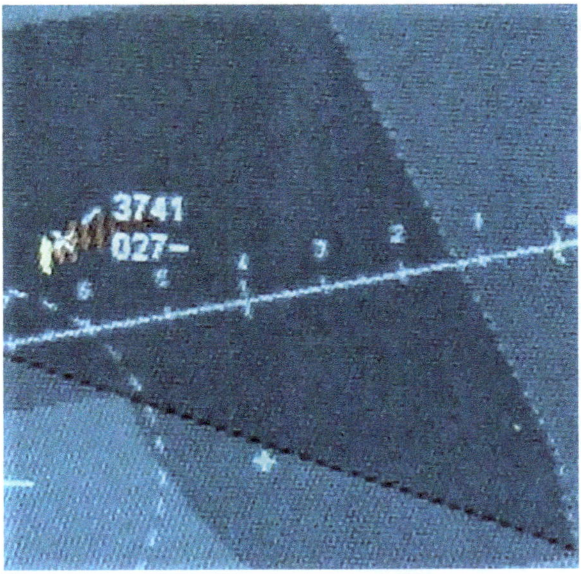

Figure 4.47 Moments later (image courtesy of RAF Brize Norton)

The image in Figure 4.47 was taken less than a minute after Figure 4.46. Sufficient time has now elapsed that the background averaging (clutter removal) system has eliminated the returns from the fixed structures including returns from the row of pylons. The font size of the secondary radar information has been adjusted in preparation for handing over to operations and it is now easier to see the 3741 Ident and the aircraft has now levelled off. Note also that there is a plus symbol in the bottom fifth of the screen. This symbol indicates that a return has been detected that does not correlate with any previous detection; something that is inconsistent with the background levels of detections in that area of coverage. Many things might cause an uncorrelated return. If there are many returns on, or adjacent to, a single bearing from the radar, this might be consistent with radio frequency interference. Even though a slow-moving clutter (low Doppler frequency) filter might be in use, sometimes vehicular traffic can cause this phenomenon. Even birds may be detected. The cause of the return in Figure 4.47 is unknown. During one flight trial observed by the author, a persistent source of clutter in a very precise location was eventually attributed to a crane being used to maintain electricity pylons on a windy day.

Notwithstanding these cases, Figures 4.46 and 4.47 illustrate the radar's filtering ability of static targets and many causes of clutter.

4.12.3.1 Has the clutter been removed?

Figures 4.46 and 4.47 show that high levels of clutter can be removed from a display, easing the job of the operator. But has the clutter been removed? Although the screen looks much 'cleaner' in Figure 4.47, the signals arriving at the antenna when it points at this area of airspace will be similar on each scan. In fact, part of the ability to remove the clutter so effectively is due to their persistence. What has been removed is the presentation of the clutter on the screen. There are a number of techniques to help remove the clutter, Doppler processing has already been discussed. Another technique is to create a map of persistent areas of clutter and edit them from the screen.

4.12.4 Pulse compression

Pulse compression was discussed in Chapter 3. As a brief recap, a pulse of coded rf energy is transmitted. Many early systems used pseudo-random binary sequences (PRBS) of phase shifts to code the transmitted pulse. Today the coding is almost invariably frequency modulation which may change to be linear with time (LFM) or non-linear (NLFM). These are called **chirps**. The chirps may increase or decrease in frequency with time, being referred to as up-chirps or down-chirps, respectively. The commonest form of chirp is the up-chirp. The pulse has a long, uncompressed, duration and in receive, the returns from targets illuminated with the long pulses are compressed to create a short pulse with superior resolution than the un-compressed. The compression process trades bandwidth for range resolution. The longer the pulse, the finer the range resolution.

A corollary of compressing the pulse is that **range sidelobes** are created (Figure 4.48). This was shown in Chapter 3 and is repeated here for convenience.

There are a number of ways for carrying out pulse compression and not all create the same level of sidelobe magnitude. It is possible to reduce the magnitude of the sidelobe at the cost of increasing the width of the main lobe. For example, one of the methods is to use NLFM, this reduces the sidelobes to approximately 20 dB below the level of the main lobe. Suppose the target being detected is a wind turbine and this presents an RCS of 50 dBsm. The signal present in the main lobe would, therefore, correspond to the signal coming from a target of 50 dBsm. It follows that the amplitude of the range sidelobes, assuming they are 20 dB lower than the main lobe, will correspond to a target with an amplitude of 50–20 dBsm, that is 30 dBsm. This is still a very large target. These 'ghost targets' are not real they are created by the process of compressing the pulse from a very large target. The ghosts appear on either side in range of the genuine target; to repeat the point they are *range* sidelobes.

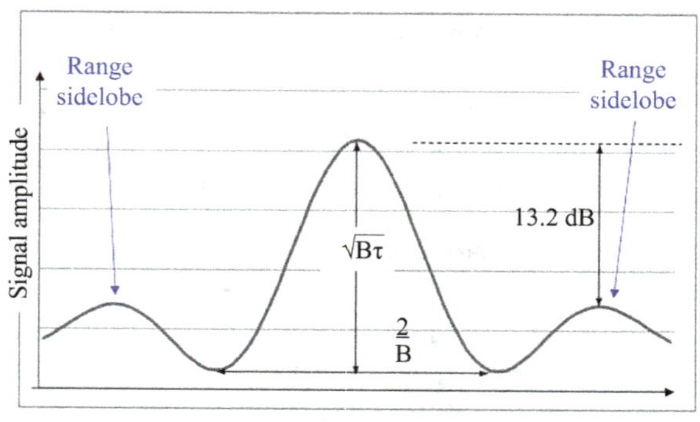

Figure 4.48 Range sidelobes created by pulse compression

A second feature associated with pulse compression is also related to the length of the uncompressed pulse. For the purpose of the discussion, consider the case of a Watchman ASR radar. This radar uses short and long pulse lengths used are 0.4 μs and 20 μs, respectively. Here the concern is the long pulse: 20 μs duration. This pulse corresponds to a distance of 6 km. A way of thinking about this is that as it is propagating away from the radar, wherever the leading edge of the pulse of energy, the trailing edge is 6 km behind it. When the leading edge of the pulse arrives at a large target like a wind turbine, as the magnitude of the return from the wind turbine influences measurements on either side of the target compounding the problem with range sidelobes.

Pulse compression allows radars to use lower transmitted powers but the transmissions are longer in duration. In the receiver, the pulses are compressed to compensate for the lower transmitted power and echoes from targets are "built" in strength. The ratio of the uncompressed pulse length to the compressed pulse length is given the name, **Pulse Compression Ratio (PCR)**. An alternative way of thinking about this ratio is that it is a measure of how the amplitude of the return from a target is spread across in time. The reduction in the signal strengths encountered in the radar receiver also affects the propensity for saturation and PCR must be taken into account by the analyst determining if a wind farm may cause saturation.

4.12.5 Processing overload

As a result of each radar transmission, there is a possibility a number of echoes will be detected. A modern radar incorporates a lot of computer processing. To process each target using a computer, the information content in each echo has to be stored in memory, much of the processing is intrinsically sequential and it takes a finite time to complete. This situation places demands on the amount of computer storage required and the time available to process echoes. The resources required are governed by a mathematical discipline called queuing theory [73]. It is not proposed to go into details about queueing theory but it yields some surprising results. If the average time of arrival of new echoes to process is the same as the time taken to carry out that process this leads to an infinitely long queue of echoes waiting to be processed. The rate of arrival of new echoes to process should not exceed 80% of the processing capacity. If the capacity to process data is marginal, even small increases in the number of clutter returns can overload the system.

4.12.6 Track data block obscuration

A problem that has been raised as a source of concern with a large wind farm adjacent to an airway is that if the track data block (TDB), that presents the information from SSR, tracks over the wind farm, then the information in the TDB may be obscured. It is possible to change the location of the TDB as it is presented on the operator's screen but that takes time and it may disturb the tempo of the controller. An obvious apparently useful solution is to use a RAG map to prevent the clutter from appearing. However, it is undesirable to do this because it is important to be aware of possible incursions into the airway.

4.13 Obscuration

Chapter 3 described the principles of CFAR, provided some illustrations of how it works and it also exposed some of its intrinsic problems. There are aspects of CFAR performance that are intrinsically problematic in the presence of wind turbines.

Chapter 3 also introduced the problem of masking when multiple objects were sufficiently close together that more than one could be present in either the leading or lagging cells in the computations to decide on the threshold. To find out what will happen when there is a wind turbine present, consider first the detection of an aircraft by a CFAR processor. See Figure 4.49, a target is present in the test cell, the threshold has been calculated based on the presence of noise alone. The value in the test cell exceeds the threshold and a detection is declared.

Now consider exactly the same situation with a wind turbine return present.

As illustrated in Figure 4.50, the signal level from a wind turbine echo is sufficiently large that it could cause the threshold to rise sufficiently, even though it

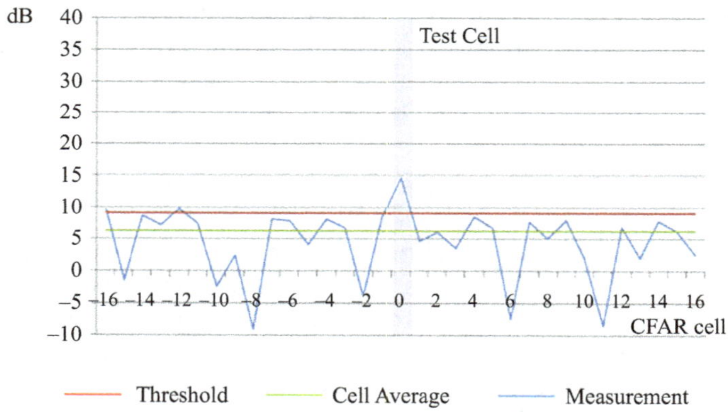

Figure 4.49 CFAR detection of a target

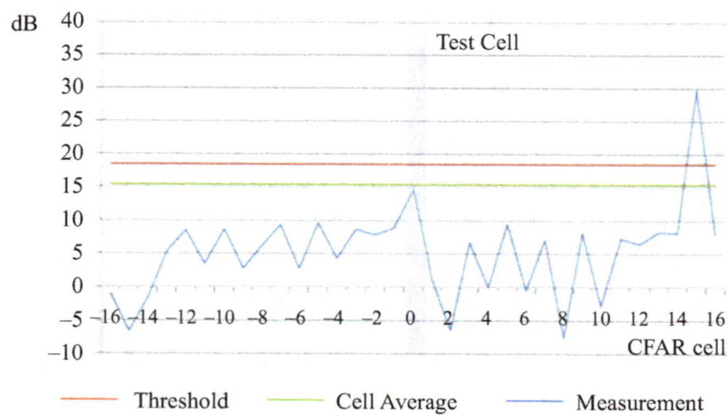

Figure 4.50 Target detection in the presence of a wind turbine return

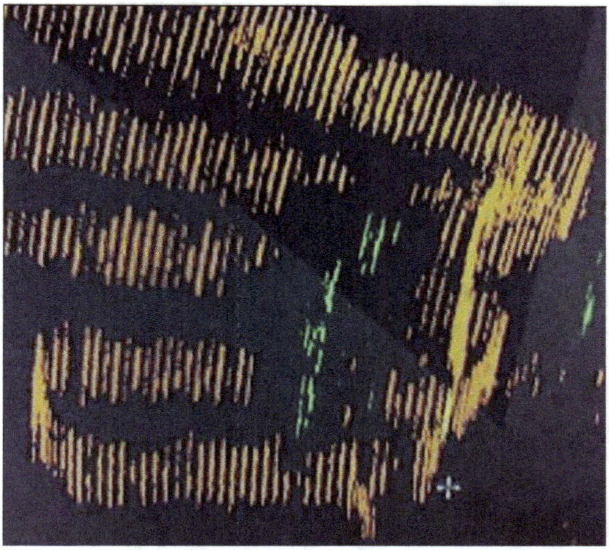

Figure 4.51 Practical observation of obscuration

may only appear in a single range cell that the genuine target cannot be detected. The same situation would occur with the wind turbine return appearing in any of the range cells that are taken into account in the calculation of the threshold. This effect is essentially the same as masking but it has become custom and practice to refer to this as **obscuration**, viz, the presence of the target (the aircraft) is obscured by the presence of the turbine return. It is also worth noting that the effect would be exacerbated if GOCFAR was used where only the leading or lagging cells, whichever produced the highest average noise estimate, were used.

This effect was reported in 2005 by the RAF in flight trials. Figure 4.51 shows a radar screen image of a trials-aircraft overflying a wind farm. The aircraft returns are shown in orange and they cross the vicinity of the wind farm five times. The radar has been configured so that the returns from the wind turbines themselves are shown in green. Note how as the aircraft approaches the wind turbines the aircraft returns disappear. This is the effect of obscuration [74].

Figure 4.51 shows an example of the complete obscuration of target returns from the presence of wind turbines. It is also possible that returns may just be diminished in the vicinity of turbines as shown in Figure 4.52.

4.13.1 Region and scale of obscuration

The preceding figures may suggest the effect only occurs when the aircraft is physically close to the turbines. It should be stressed that this does not necessarily imply that the aircraft has to be close to the turbine for this effect to be present. The ASR and ARSR radars in which this effect is most prevalent have no ability to

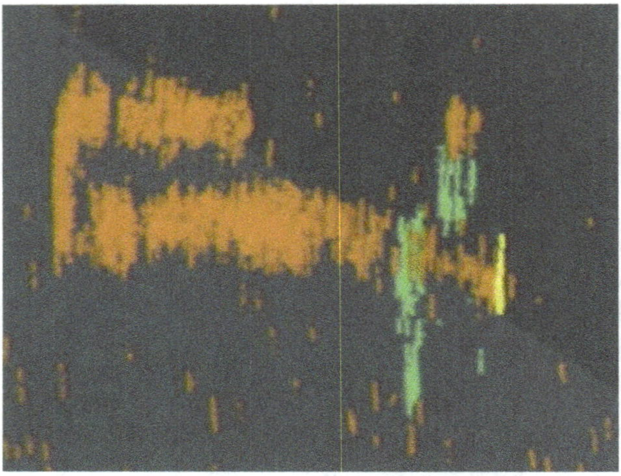

Figure 4.52 Partial obscuration

determine height. An aircraft might be thousands of feet above the turbine and obscuration could prevent its detection.

The lateral extent of obscuration is determined by the radar's range cell size. As analogue to digital converters have improved the size of range cells has decreased. For example, the Marconi S511 radar, a radar of the 1970s had a range cell size of 240 m. By the 1980s, the Watchman radar had a cell size of 60 m. In the case of the Watchman radar, the structure of the CFAR cells was to have 10 leading and 10 lagging cells and two gap cells on either side of the test cell. Thus, with a large wind turbine return detection could have been reduced to 720 m on either side of the test cell.

4.13.2 Tracking and track seduction

Once the prerogative of AD radars, a tracking function is common in most modern aviation radars. The principal function of the tracker is to associate returns from multiple transmissions and estimate the aircraft velocity. If only primary radar is available, then in a 2-D radar, the velocity can only be in 2-dimensions. In an AD radar, the tracker has height information available and the estimate can be a 3-D velocity. Although most commonly done in the radar data processor, some radars also fuse SSR data with primary returns to provide a composite track. A number of books have been dedicated to quantitative analyses of tracking such as Brookner's 'Tracking and Kalman Filtering Made Easy' [75]. The approach taken here is to take a qualitative approach to consider how the tracking functions are affected by wind turbines and wind farms. Tracking functions are performed in a computer program called the **tracker**.

Ostensibly, all the tracker must do is work out if a new radar return can be associated with one previously observed. In practice, their task is more complex and trackers can be very sophisticated. An illustration will illustrate the point. At a moment in time, there are no existing tracks being observed. A new radar return is received. This could be a false alarm caused by noise perhaps, or it is the first time an aircraft has been detected. The return has to be stored. In an ASR, usually 4 s later another return may be received. If it is a rotary-wing aircraft (a helicopter), then it may not have moved far or may even be hovering. If it is a GA (light) aircraft, it may not have moved far, but if it is a fast-jet it could have moved a long way. And, the new return may just be another false alarm. All options are valid and the tracker has to cater for all these eventualities.

To produce tracks the tracker has three basic functions:

- **Plot to track association or returns to track association (RTTA)**: This part of the tracking software decides if a new return is associated with an existing track or if it might be a candidate for starting a new track. There are several strategies for deciding whether **track initiation** can take place but, typically, at least three returns will be needed to start a track.
- **Track update**: RTTA has determined that a new return is part of an existing track and that track must be updated. Each track has an identification or an '**ident**'. The updating process will consist of updating the current position of the object and calculating its velocity. The form of the velocity varies from one system to another. If the radar is an AD radar or the height can be accessed by SSR or some other means, the velocity can be in three dimensions. If not, then a velocity in two dimensions is calculated. Typically, nowadays, in addition to the velocity, a measure will be created of the quality of the track; a statistical measure of whether the track is consistent or varying. This measure is called the **covariance matrix**. In some types of tracking software, the newly updated track can be very helpful to RTTA. If the track is known to be travelling in a particular direction, it is possible to predict where it will be at the time of the next anticipated update. The covariance matrix is also helpful because it can be used to work out how much uncertainty is attached to that prediction. The tracker will also perform housekeeping functions; for example, updating how many updates there have been.
- **Track termination**: The tracker cannot hold on to tracks indefinitely, there needs to be a strategy for working out when the track is no longer being updated. Typically, if there have been no updates for perhaps three or five revolutions of the antenna, the track will be declared to be inactive. This situation is problematic if the aircraft does not receive updates because, say, the aircraft is over a windfarm which de-sensitises the radar to aircraft updates. Coupled with the time the RTTA takes to declare a new track is present, this can be highly problematic for AD radars.

Different scenarios that might be met by the tracker are illustrated in Figure 4.53. In the first scenario depicted, Track 1 is travelling in a predictable manner and at scan $n + 3$, it is possible to predict that a future update is highly likely to be in the predicted position. Track 2 is receiving updates that have a high degree of uncertainty with them and the associated track uncertainty means the prediction of where the next update will be found has to be much larger. This does not stop the tracker from being able to smooth out the track. Track 3 stops receiving updates. The track is sustained by predictions of where the aircraft would be located but after (in this case) three occasions when there has been no update the track is dropped. In the final case, a new track is being started after three detections that can be associated.

Track seduction is illustrated in Figure 4.54. An aircraft has been tracked until it reaches a wind farm. RTTA is now faced with a problem. There are now multiple possible returns to choose from. In the absence of being able to distinguish aircraft and wind turbine returns, it depends on the return that is closest to the predicted location of the next plot. If RTTA makes the wrong choice, then the track is seduced away from the actual aircraft track.

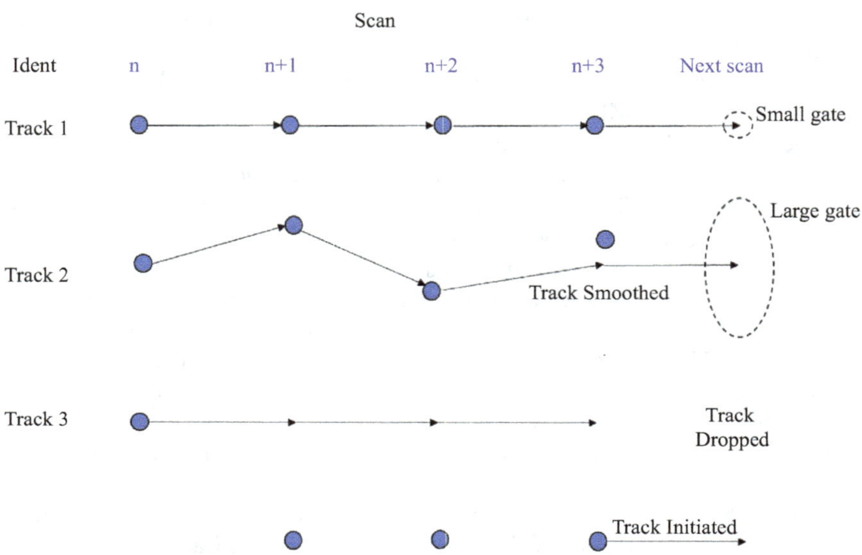

Figure 4.53 Track scenarios

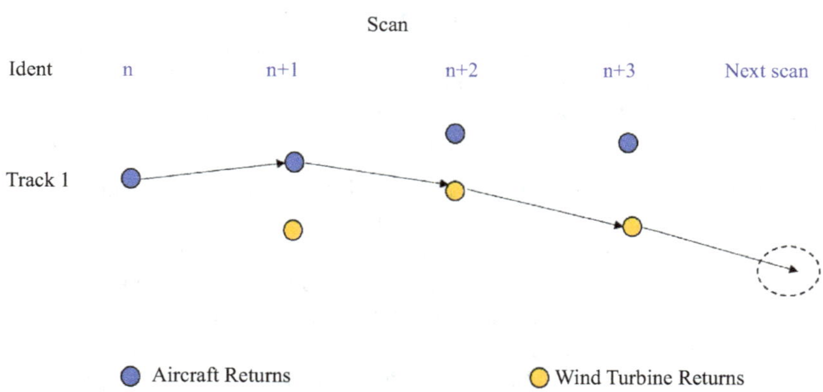

Figure 4.54 Track seduction

4.13.3 Processing overload

PAR was described in Chapter 3. Recall they provide a ground controller with the information necessary to instruct the pilot of an aircraft on the actions needed to stay on the glide slope and the extended centreline of the runway. There may be situations where a number of aircraft are to land in quick succession and these may appear in the coverage of the PAR concurrently. But, in general, a few aircraft will be present in the coverage of the PAR at any one time.

In such circumstances, there is no need for the radar designer to incorporate the ability to handle high volumes of aircraft returns to process concurrently. The design may include a modest safety margin for:

- Number of radar returns (echoes) that can be detected concurrently (which may also include natural phenomena such as a flock of birds).
- The communication of messages containing detection information to subsequent processing stages.
- The number of track files that can be maintained concurrently and the capacity for updating those files with new data.

Moreover, there is little (but not no) potential for mis-associating updates from clutter sources with genuine aircraft returns.

If a wind farm was built within the coverage of a PAR, even if not every wind turbine was detected every time the PAR scanned the sector containing a wind turbine, the additional detections might exceed the number that can be detected by the radar. If all the returns that can be handled by the radar are taken up with wind turbine returns it might even be possible either for genuine aircraft returns to be discarded because there is insufficient place to store the information.

If the wind turbine returns were associated with tracks, it might be possible for the radar's capacity for tracking to be exhausted.

Returns from wind turbines may present characteristics that cannot be distinguished from a genuine aircraft and a wind turbine plot may cause the aircraft track to be seduced.

There is very little that can be done about these problems without redesigning the radar algorithms and providing a higher throughput capacity computer to prevent the radar computer from overloading.

4.13.4 PSR shadow

Shadow is discussed later in this chapter.

4.13.5 PSR mitigations

4.13.5.1 ASR (civil and military) and ARSR

The options available for mitigation of these classes of PSR are discussed in Chapter 6.

4.13.5.2 PAR

The options for the mitigation of PAR are limited. However, the coverage of PAR is also limited and it may be possible to design a turbine layout that does not compromise the safeguarded area. In the United Kingdom, PARs are only safeguarded to a range of 20 Nm (37) and a total protected arc of 32° which is normally ±16°. However, this angle may be adjusted during commissioning and may be asymmetric.

Because of the nature of the PAR mission, a critical consideration is that the radar must operate close to the runways. Therefore, unlike an ASR which could be supplemented with an in-fill radar, assuming that there were no other factors precluding the option, a PAR cannot be relocated away from the aerodrome.

In theory, a PAR could be replaced with a wind farm-tolerant radar. However, there may be a number of complications and objections:

- The PAR requires a limited field of view; a high update rate, usually once per second and high track accuracy. The track information must be presented in both elevation and azimuth. In principle, this could be achievable for a wind farm-tolerant radar. However, the radar may need to be designed to operate in a different fashion from its normally intended purpose and this might complicate getting a safety case for this application.
- Crews are trained to operate with the existing fleet of PAR and special cases might be unwelcome.
- Special cases also need special maintenance arrangements which might also be unwelcome.

Furthermore, if the proposed wind farm was large enough to make buying a new radar financially viable, the presence of such a large wind farm close to an aerodrome may give rise to other objections.

4.13.6 SSR effects

To provide some context for the following comments, there are effectively three types of SSR systems (not counting the derivative technologies of ADSB and TCAS):

- Conventional (legacy) SSR using mode A/C
- Monopulse SSR (MSSR)
- Mode S SSR.

The CAA reports that SSR suffers fewer problems than PSR and these are normally only problematic if a wind farm is within 10 km of the SSR [76]. The CAA also identifies two broad classes of wind turbine effects associated with SSR:

- Diffraction effects, including but not limited to shadow.
- False targets.

EUROCONTROL [77] also raises the problem with MSSR bearing errors. Stevens [78] provides a relationship between the wanted to unwanted signal ratio and measurement errors.

4.13.6.1 Diffraction

The possibility of shadow was considered earlier in this chapter. There is an additional consideration in the context of SSR problems; this is illustrated in

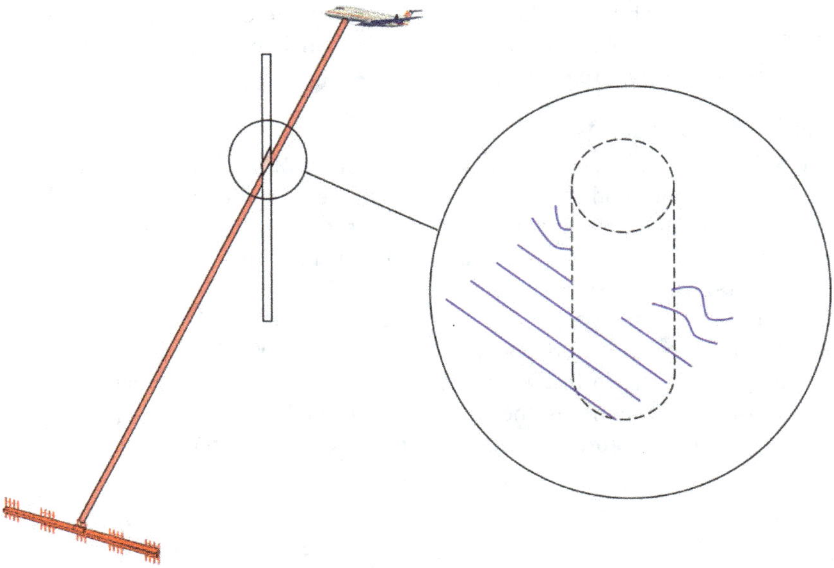

Figure 4.55 SSR diffraction effects

Figure 4.55. On either side of a cylinder, there is a region of airspace where the field (specifically the phase associated with the phase) is distorted.

Vinagre and Woodbridge measured the effect at Great Dun Fell of a mast 60 m from the SSR antenna and with an average diameter of 1 m. Within 2 degrees on either side of the tower, the greatest error measured was ¼ degree. The effect decays rapidly as the bearing from the tower increases [79].

Theil and Ewyjk [80] also investigated this phenomenon modelling the following scenario:

- A monopulse SSR; that is a system that can estimate the off-boresight angle (OBA) of a signal,
- An aircraft flying perpendicular to (across) the line formed between the antenna and the wind farm.
- The wind farm was 4 km from the antenna and the aircraft was 30 km down range.
- The aircraft 'flew' from the centre of the track to 1 km off the centre.

The maximum error occurred 500 m off centre and the error in the OBA peaked at 0.06°. However, this error could occur multiple times in a large wind farm. For comparison, CAP 670 sets a requirement of a standard deviation of azimuth measurement error of 0.15°.

These errors are most significant in the case of the legacy (Mode A/C) SSR that uses the reply from the transponder to determine the location of the aircraft. In this context, EUROCONTROL raises the issue of the strong reflections from wind farms affecting the ability of the monopulse to form an accurate azimuth bearing. EURCONTROL also observes that with Mode S, it is possible to derive location information from single replies.

4.13.6.2 False returns

Figure 4.56 illustrates a scenario where a reflection of a wind turbine structure is received by the transponder on an aircraft. Both the P_1 and P_2 pulses would have to be reflected in a similar manner to maintain the same amplitude relationship. Under these circumstances, a false target could be detected that is a response from a genuine aircraft on a false bearing.

A distinction should be made between the older Mode A/C SSR and Mode S. In the former case, SSR/IFF works out the location of the aircraft using the information from the SSR antenna. In the latter, the SSR antenna provides a means of systematically roll-calling aircraft but the aircraft provides their location in the response to the interrogation. Therefore, Mode S provides a means of mitigating this problem.

4.13.7 SSR mitigation

Options for mitigation of SSR are limited. Many SSR systems maintain what is effectively a clutter map. But EUROCONTROL points out that this technique has limitations with moving turbine blades. Observing safeguarding distances might in some circumstances provide a mitigation if turbine layout is permitted. However,

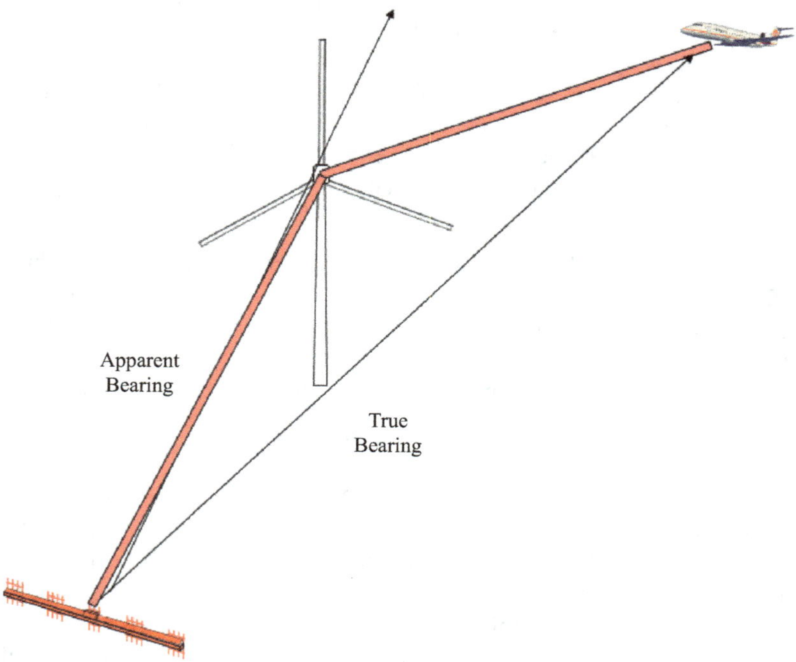

Figure 4.56 False SSR returns

unlike PSR, SSR is not obligated to be collocated with a PSR and many aerodromes operate without dedicated SSR systems. A feed from an alternative SSR may be a possible mitigation. Proliferation of SSR is undesirable but if a new system was considered necessary, detailed negotiations would be needed with the regulator. CAP 761 [81] provides guidance on SSR siting of SSR systems. The long-term mitigation is likely to come from electronic conspicuity techniques, discussed in Chapter 3.

4.14 Communications and navigation – fast fading and phase error

For the purposes of the following discussion, assume a transmission is being made from the ground (an air traffic controller is passing a message to the pilot of an aircraft) and the signal created is incident upon the antenna on the aircraft and also upon a nearby wind turbine. Wind turbines can have a high RCS, a factor that is equally important to radio communications as it is to radar, and the incident radio wave will cause an echo to be radiated (in all directions) from the wind turbine. A proportion of that echo will travel in the direction of the aircraft as well as all the airspace around it, and some of the echo signal will be incident on the antenna on the aircraft. This arrangement is illustrated in Figure 4.57.

Therefore, the signal produced by the antenna on the aircraft is the sum of the direct signal from the ground and the echo signal from the wind turbine. The latter signal causes interference to the desired signal and if sufficiently great will prevent or significantly delay the passing of information which would not be acceptable for the reasons set out above. The mechanism of creating the interference will now be described.

Figure 4.57 illustrates that the direct path to the aircraft antenna will always be shorter than the path taken by the interference signal. It follows that the interference signal will always arrive later than the direct signal by an amount proportional to the difference in the path length the two signals have travelled. This time difference corresponds to a number of cycles and a partial cycle of radio frequency of the

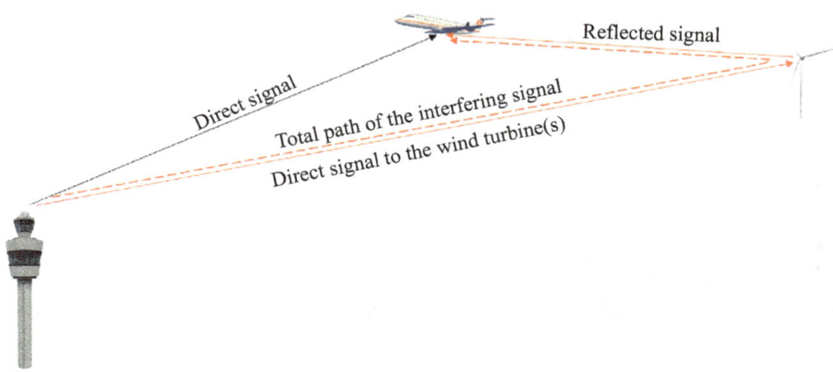

Figure 4.57 AGA communications and a wind turbine

interference signal. In other words, there will be a phase difference between the two signals. The aircraft antenna will be insensitive to the differences in complete cycles but will be sensitive to the phase difference between the two signals. Thus, when the signals arrive at the receiving antenna, this phase difference must be accounted for when the sum of the two signals is calculated; this is referred to as the vector sum.

A deliberately simple scenario has been modelled to illustrate the effect of this phase difference. The scenario is illustrated in Figure 4.58. A wind turbine is 30 km, a little over 16 Nm, from the control tower and the transmitting antenna. An aircraft is descending towards the airfield and at the start of the scenario is 1,540 m (approximately 5,000 ft) directly overhead the wind turbine. The aircraft is flying at 130 knots, which is 240 km/h. The first 15 s of flight have been simulated. In the first instance, in this simple scenario, no allowance has been made for a reduction in signal strength as the signal travels through the air (path losses) and it has been assumed that the wind turbine reflection is constant, that is there is no frequency difference caused by the Doppler effect as the rotating blades reflect signals and no variation in amplitude caused by the different strength of echoes from different parts of the cycle of rotation of the blades. These effects are discussed later. For the time being the intention is, simply, to understand the effects of the phase differences between the direct signal and the interfering signal reflected off the turbine.

The model being illustrated here does allow the relative proportions of the direct and interference signals to be varied. It is custom and practice in dealing with radio interference problems to refer to the reference signal as the 'carrier', in this case that is the direct signal between the tower and the aircraft. Thus, changing the proportion of the two signals in the simulation is to vary the carrier-to-interference ratio (C:I ratio). The frequency of the transmission has been assumed to be 127 MHz; this value is commonly used on VHF AGA communications problems because it is in the middle of the spectrum allocation[†††]. The transmission starts at the start of the scenario.

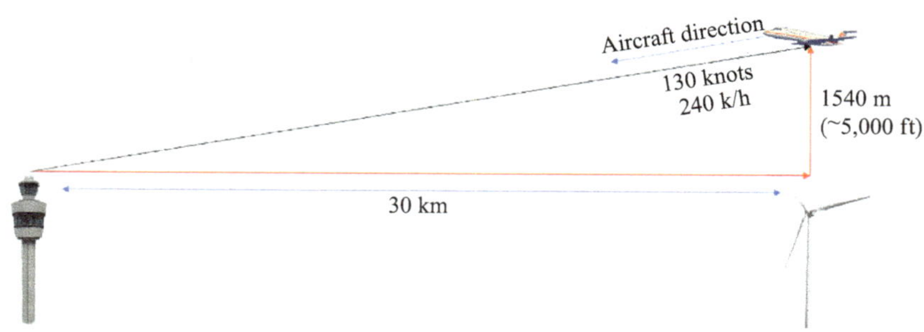

Figure 4.58 Illustrative scenario

[†††]The equivalent frequency for UHF calculations is 368 MHz.

The first results provided by the simulation are the phase relationships of the carrier signal and the interference signal at the start of the scenario, that is, at the specific location of the aircraft above the wind turbine. Assuming the phase of the signal as it first leaves the transmitting antenna was 0°, when the signals arrive at the aircraft antenna the phases are:

Carrier signal (direct path) phase: 259.9°

and

Interference signal (reflected path) phase: 336.0°

Recall that the phase difference arises because of the difference in distance travelled by the carrier and interference signals. At another point in space, the phase difference will be different. However, the phase differences at these different points in space, all other factors remaining the same, will be constant but as viewed from an aircraft moving between these points the phase difference will appear to change continuously. And, as will be shown, this is the case in the scenario being modelled. As the aircraft flies towards the tower, the distance between the tower and the aircraft decreases. The path of the interference signal is slightly more complicated. The distance between the tower and the wind turbine is constant but the path between the turbine and the aircraft will change with the motion of the aircraft. Because the aircraft is descending, perhaps counterintuitively, initially the path length between the aircraft and the turbine decreases. But once the aircraft reaches the point where the turbine direction is perpendicular to the trajectory of the aircraft, then the distance will increase continuously. This situation was chosen for the scenario because it is common for aircraft to descend as they get closer to an airfield and it is useful to understand the effects. This effect can be seen in Figure 4.59. As a direct result of the changes in path length, the relative phases of

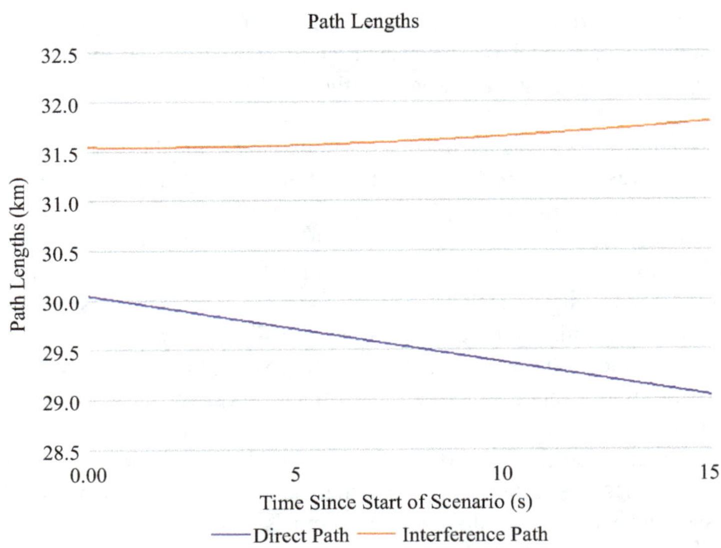

Figure 4.59 Path lengths as scenario evolves

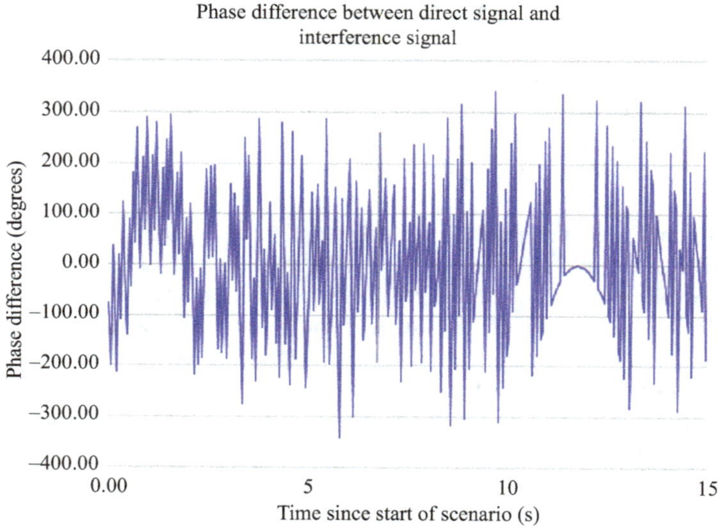

Figure 4.60 Phase difference varying with the aircraft position

the two signals vary. The model was then used to show the variation in the phase difference of the two signals as the aircraft moves have been plotted and are shown in Figure 4.60.

As far as the signal received by the aircraft is concerned, it is the sum of the carrier and interference signals that is important taking into account the effect of the phase difference illustrated in Figure 4.60. To establish the effects of interest, the next stage of the modelling was carried out in a number of steps. The first step was to normalise the level of the carrier signal (set it to 1) and eliminate the reflected signal and, as might be expected, the results plotted in Figure 4.61 show a continuous signal level.

Figure 4.61 provides a reference for comparison with the results that follow. The next step was to reintroduce the interference signal and its level has been chosen to match the carrier signal; that is the C:I ratio is set to 1 (usually expressed as 0 dB). In reality, this level would never be possible because the interference signal will always be smaller than the carrier signal but this step serves to illustrate how the interference signal affects the carrier in the worst (impossible) case scenario. The results of this simulation are shown in Figure 4.62.

Figure 4.62 shows the effect of the constructive and destructive effects of the presence of the interference signal, albeit with the level of the interference signal being set higher than would be realistic. When the two signals are in-phase (there is no phase difference between the two) the sum is 2 units. When the interference signal is out of phase the sum of the two signals is zero. It is also clear that there are lots of cases where an intermediate situation exists.

A further refinement of the model is now introduced. The C:I ratio has, thus far, been held constant (set to 1 or 0 dB). But as the aircraft approaches the tower

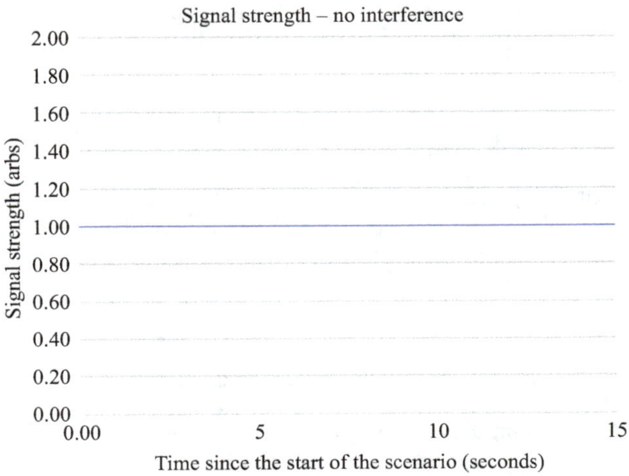

Figure 4.61 No interference signal

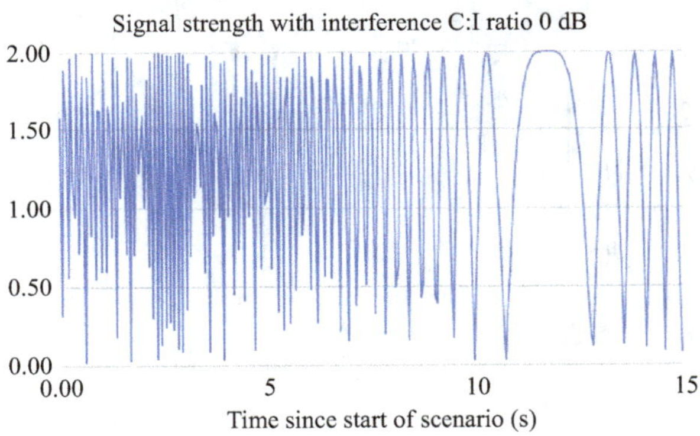

Figure 4.62 C:I ratio 0 dB

the carrier signal must get stronger and the interference signal correspondingly weaker because the former has a smaller distance to travel and the latter farther. In other words, the C:I ratio must vary with position. A simple change to the model accommodates this situation. For the time being, it has been assumed that both signals travel through unobstructed, free, space, discussed in Chapter 5. Free space losses, sometimes referred to as basic transmission losses, are incurred because the signal expands as it moves farther from the transmitter, sharing its energy over a larger area. This is accounted for by a loss in signal strength calculated as follows [82]:

Basic transmission loss, $L_b = 32.4 + 20 \times \mathrm{Log10}\ (f) + 20 \times \mathrm{Log10}\ (d)$

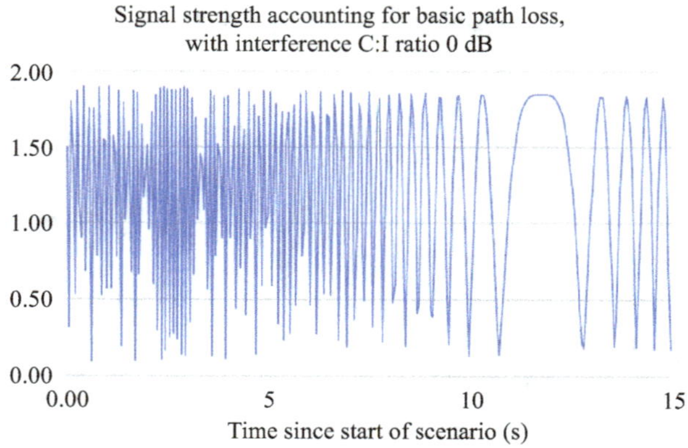

Figure 4.63 Signal sum, accounting for basic path losses and C:I ratio variation

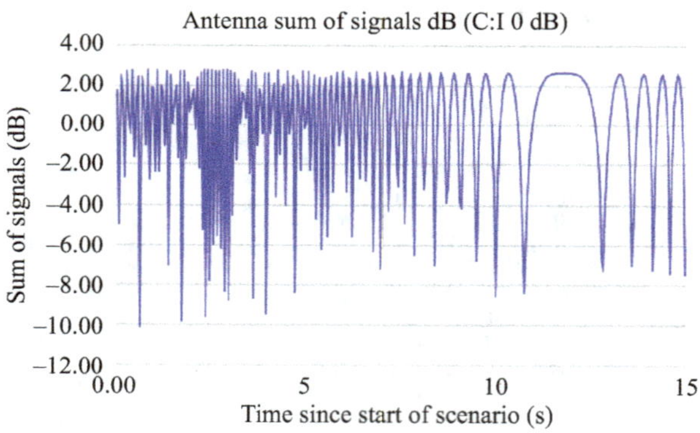

Figure 4.64 Signal sum in dB

where f is the frequency in MHz; d is the path length in km; 32.4 is a constant that is described later on but which is used to correct the units and account for other physical constants used to make the equation work[‡‡‡].

The results of accounting for basic path loss are shown in Figure 4.63; the points to take from this graph are that there is a general reduction in maximum signal levels and a slight downward trend as the scenario progresses.

It is more usual to present radio signals on logarithmic scales, that is, dB and the information shown in Figure 4.63 is reproduced in Figure 4.64.

[‡‡‡]The formula is presented by the ITU in a number of their recommendations. It is custom and practice for them to present such material using units that are most commonly encountered for the subject in question; in this case in kilometres and MHz. This can be quite confusing for readers more used to system international (SI) units for equations.

4.14.1 Fading

Figure 4.64 shows that the presence of high levels of interference signal can seriously degrade the performance of the communications link; in particular causing occasional large drops in signal strength; a condition known as fading. For anyone trying to find out more information on the subject, the terminology used to describe fading can be confusing. Sometimes the communications medium itself can cause fades (channel fading) but in this instance the cause is the arrival of components of the received signal from multiple directions; hence, this class of fading is called either **multipath fading** or **multipath-induced fading**. Another term used to describe this class of fading is **small-scale fading** to distinguish this phenomenon from the large-scale effects of signal reduction over long paths when features such as terrain screening may play a part.

There are two key features associated with this class of fading: the random nature of the occurrence of the deep fades and the frequency with which they occur. Thus, although a deep fade is caused by precise, deterministic states, in which the carrier is cancelled by the interference signal, it is the motion of the aircraft that creates the situation, and with the variations in aircraft height, distance and velocity and the radio frequency used, it is impossible to predict when a deep fade will occur; they are **random**. The frequency of fluctuation of the signal is an important characteristic of fading and leads to the terms fast and slow fading. Fast, and, therefore, slow, fading are most easily defined with digital communications where information is sent as bits (1 or 0) and it is possible to define degradation by comparison with the time taken to transmit each bit. The closest equivalent for voice communications is the phoneme which is the smallest element of speech. Unlike digital messages, where the bit rate and the duration of bits are constant, phonemes have different durations of the order of milliseconds to tens of milliseconds and sometimes even longer. But as Figures 4.63 and 4.64 illustrate from this simple scenario, fades can occur multiple times a second and lead to the classification of **fast fading**.

Another characteristic of this type of fading is that unlike other forms of interference which lead to a reduction in the signal-to-noise ratio (called additive interference), which can be remedied by using more powerful transmitters, fast fading is characterised as **multiplicative interference**. The term arises because it is as though the signal is being multiplied by a factor, which at times of the deepest fade effectively turns the signal off. Whereas, in the case of an additive interferer, it is possible to overcome the interference by increasing the power of the transmission. Unfortunately, this is not possible with multiplicative interference [83].

Although in the scenario investigated the cause of the reflection has been assumed to be a wind turbine, there are many other causes of such interference. Moreover, the communications receiver will include Automatic Gain Control (AGC) which will provide partial mitigation of the effect of fast fading.

4.15 AGA safeguarding

4.15.1 Principle

In the absence of a simple means of mitigating AGA interference from wind turbines, the alternative is to be better aware of whether or not a proposed wind farm (or an extension to an existing wind farm) might pose a problem. The favoured method is to predict where a proposed wind farm may give rise to low C:I ratios, that is, where the interference signal is large compared with the carrier and determine if these may pose an operational risk. The method is as follows.

Having confirmed the locations of the turbines in the proposed wind farm and established that it is somewhere within the AOR or where it may potentially affect part of the AOR, the aim is to create a set of maps, one for each of a number of altitudes. In the United Kingdom, the CAA recommends maps for 1,000 and 2,000 ft above ground level (AGL) and 5,000, 10,000 and 20,000 feet above sea level (ASL) [84]. The CAA recommends the limits of the maps should be when the signal level has dropped to 26 dBμV/m (which corresponds to a power of −89.8 dBm).

Each map should identify regions where the C:I ratio is less than 20 dB for a single wind turbine. For multiple wind turbines, the regions where the worst-case turbine (that is the one producing the lowest C:I ratio) C:I ratio is less than 23 dB should be identified and the aggregate C:I ratio for all the turbines should be identified if the C:I ratio is less than 14 dB. The CAA guidance stresses that the radio propagation modelling should be carried out using the Delta Bullington method and the results for the multiple turbine case should be presented with all the turbines presenting interference in phase. The Delta Bullington method is of wider interest than AGA communications and this is discussed in Chapter 5. In the following example, basic path loss has been used where appropriate. The method discussed produces coherent results by default. A further refinement that would be needed to make the simple example adequate for the purpose is to take into account the curvature of the Earth and the refraction of radio waves. As with the Delta Bullington method, this is of wider application and will be discussed in Chapter 5, in this case, a simple 'flat-Earth' approach has been adopted.

4.15.2 Power level calculations

Calculating the C:I ratio requires the calculation of the power at points in the map (at different altitudes) of the direct (carrier) signal and the indirect (interference signal) and then calculating their ratio [85]. There are two different methods that can be used to calculate these power levels: the **Power Equation** and the **Radar Equation**. The former is based on an equation from electromagnetic theory, the latter is slightly more complicated because it is necessary to account not just for the path between two points but also the signal reflected off the wind turbine. However, it will be seen that these two equations are functionally identical. The first equation to be discussed is the power equation.

4.15.3 Carrier power

As with the examples in Part 2 above, the assumption is made that a signal is transmitted from the control tower to a point in the sky (which in Part 2 had an aircraft present). The intent of this equation is simply to calculate the power available to a notional receiver located at the selected point. The method for applying the power equation is in the form of a **Power Budget**, it works thus:

Carrier power available to a hypothetical receiver at a point in space of interest,

$$P_{cr} = P_t + G_t - L_t + G_r - L_r - L_b \tag{4.8}$$

where P_t is the transmitter power (at the control tower); G_t is the gain of the transmitter antenna; L_t is any losses encountered in the transmitter and antenna assembly; G_r is the gain of the receiver antenna; L_r is any losses encountered in the notional receiver; L_b is the basic transmission loss described previously.

The first three terms in the equation (i.e., P_t, G_t and L_t) provide the power projected into the sky in the direction of the point of interest. The next two terms account for the contribution made by a notional receiver antenna and associated losses. As the wavefront created by the transmitter expands, it spreads its energy out over a wider area and this is accounted for by the remaining term in the equation, viz the basic propagation losses discussed earlier.

4.15.4 Interference power

The interference signal power at any point in space is caused by the carrier signal being reflected off the wind turbine(s). Therefore, the calculation of the power must take into account radio reflection; a process for which the radar equation is specifically intended. The radar equation is most often used for what is called the monostatic case, that is, where the same radar antenna is used for the transmission and reception of echoes. But this does not have to be the case and the equation described below is the **bistatic radar equation**. The method of applying the equation is first to calculate the power available at the location of the turbine, worked out as a power density, i.e., the power per unit area. The next step is to calculate the power radiating from the turbine and the final step is to calculate the power available at a point in the sky for which the C:I ratio can be calculated.

To understand how the equation works, it is convenient to derive it from first principles. The derivation starts with the idea that a transmitter of a given power is radiating energy equally in all directions; it is said to be radiating 'isotropically'. The power is being shared out over the surface of a notional sphere of which the transmitter is at the centre and the surface area of a sphere is $4\pi R^2$:

$$\text{Power Density at the turbine location} = \mathrm{P_t}/\ (4\pi \mathrm{R_{tt}}^2) \tag{4.9}$$

where P_t is the transmitter power (at the control tower); R_{ct} is the distance (range) from the control tower to the turbine.

No allowance has been made for the transmitting antenna having any gain in a particular direction or for the practical situation where there are losses. To allow for the gain and losses to be accounted for, the equation is modified to become:

$$\text{Power Density} = \mathrm{P_t G_t / L_t (4\pi R_{tt}}^2) \tag{4.10}$$

where G_t is the gain of the transmitting antenna; L_t is the losses between the transmitter to antenna.

The second step in the process is to calculate the power radiated by the turbine. The turbine is a complex shape with a number of constituent parts. A quantity called the RCS reduces this complexity into a single representative area. That is an area that has the same electrical effect as the complex shape of the turbine. RCS is commonly represented by the Greek letter σ. To calculate the power reflected by the turbine, the above equation is modified thus:

$$\text{Power reflected by turbine, } P_{ref} = P_t G_t \sigma / L_t (4\pi R_{tt}^2) \quad (4.11)$$

where σ is the wind turbine RCS.

The final step is to calculate the power available to a receiver at a point in the sky of interest (which will be somewhere on a surface at a given altitude). Note the above power is assumed to be radiated isotropically and the initial calculation is to calculate the power density at the desired point:

$$\text{Power Density at Point in Sky, } P_s = P_t G_t \sigma / L_t (4\pi R_{tt}^2)\,(4\pi R_{ts}^2)$$

where R_{ts} is the distance from the turbine to the point in the sky.

To allow the calculation of the C:I ratio, the power density must be converted into a power available to a receiver which is arrived at via an effective area for a given antenna. The relationship between the gain of an antenna and its effect area is as follows:

$$\text{Gain of the receiving antenna, } G_r = 4\pi A / \lambda^2 \quad (4.13)$$

where A is the effective area of the antenna; λ is the wavelength of transmission.

Rearranging this equation to make A the subject of the equation:

$$A = G_r \lambda^2 / (4\pi) \quad (4.14)$$

This expression can now be substituted into the power density equation above to yield the power available to a receiver at the desired point in the sky:

Interference power available to a hypothetical receiver,

$$P_r = P_t G_t G_r \lambda^2 \sigma / L_t L_r (4\pi R_{tt}^2)\,(4\pi R_{ts}^2)\,(4\pi) \quad (4.15)$$

4.15.5 *Significance of the power available*

The power results produced using the equations described above are all peak powers. If the ratio is needed for a multiple turbine wind farm, then the interference powers must be added. Two matters arise from this.

As the powers are peak, then it is thought that the contribution from each turbine is in phase with each other. This situation was the specified arrangement in the CAP 670 guidance. The approach taken is “worst-case”. In practice, it is highly unlikely that the contributions of multiple turbines would be coherent. As such, the results of summing the contribution from multiple turbines are likely to be pessimistic (indicating a higher interference power than is likely to be encountered in reality).

The second matter is a practical one. It is commonplace to calculate power values in decibels, that is, logarithmically. If multiple powers are to be added, then the decibel values must first be converted to linear values, the sum can be made and then the result converted back to decibels.

4.15.6 Illustration

The above method was used to calculate C:I ratios at different heights above a single large turbine which is located 30 km from an AGA transmitter on a control tower. It was assumed that the turbine RCS was 51 dB square metres (a value taken from CAP 670).

The results of this simulation are presented in Figure 4.65. Also shown in Figure 4.65 is the threshold value for an unacceptable/acceptable C:I ratio. The next graph, Figure 4.66, illustrates the resulting signal if the C:I ratio is 20 dB.

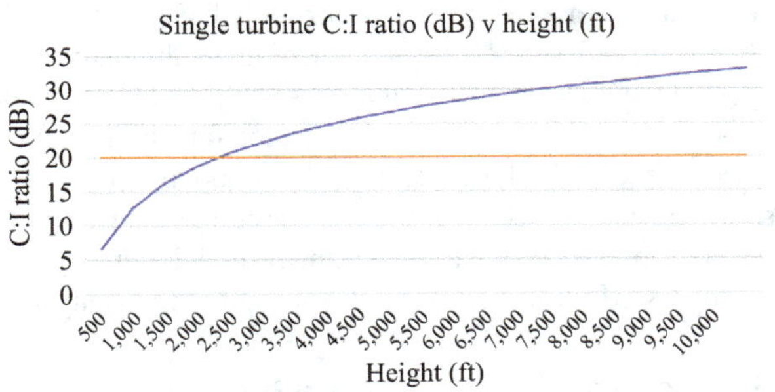

Figure 4.65 C:I ratio for a large industrial turbine versus height

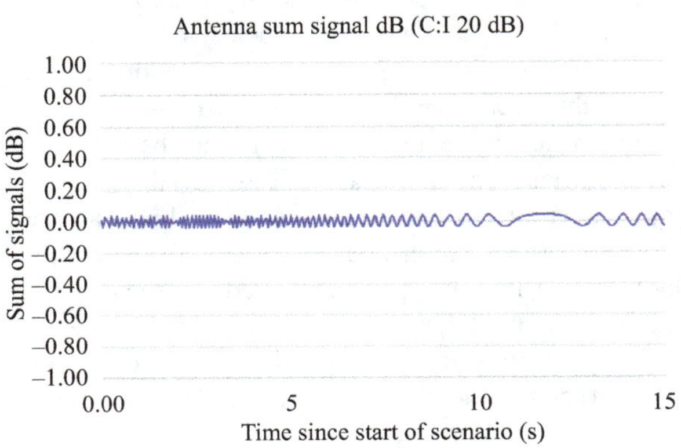

Figure 4.66 C:I ratio 20 dB

4.16 VOR and bearing error

Ben-Hassine [86] has analysed the effect of multipath interference caused by wind farms on bearing errors in VOR (both CVOR and DVOR). The principal findings of the work are:

- CVOR system errors are greater than those associated with DVOR.
- The worst effects are as a result of scattering off the wind turbine tower.
- The worst effects occur for aircraft flying at low elevations.
- For a wind turbine 3 km from a VOR beacon, the error in bearing assessment was 0.3° (at one standard deviation).

Wind turbines impact the accuracy of VOR navigation aids. The effects must be viewed in the light of the introduction of PBN.

4.17 ILS effects

Arguably the least well-understood effects occur with ILS. A standard approach that may be adopted is to calculate the effects of fading in the same fashion as AGA. Concerns expressed in the community are that these standard methods do not take into account modulation depth, which is the critical factor for an aircraft to determine its position with respect to the glide slope (left or right or above or below). Although there seems little reason to be concerned about the use of wanted to unwanted ratio, it would be prudent to conduct further research to determine if modulation depth can be affected. Ben-Hassine also suggested further work on ILS to mimic the analysis on VOR.

4.18 Doppler signature of a wind turbine

4.18.1 Doppler

A wave originating from a moving platform has a certain frequency and wavelength. The wave may be a transverse electromagnetic, for example, a radio or light, or a longitudinal sound wave. However, the frequency perceived by an observer, not onboard the platform, depends on the relative velocities of the platform and the observer. The traditional example of this phenomenon consists of two observers: one onboard a train and a second listening to the train as it moves past. To the observer onboard the train, the frequency of the engine noise does not change. However, the observer not on the train hears a shift from a higher pitched note to a lower pitched note as the train passes by.

This phenomenon was first described by the Austrian mathematician, Christian Andreas Doppler (1803–1853). Doppler postulated that the different colour of light from stars might be accounted for by their speed. This concept is only partially correct because the light from stars also depends on their constituents; however, the effect of speed is correct and the phenomenon is called the Doppler effect[§§§].

[§§§]Hence, in discussions of the effect, it is always spelled with a capital D because it was named after Doppler.

Aviation radar systems are able to exploit the Doppler effect to remove clutter, making it possible to distinguish between the echoes from moving targets (aircraft) and stationary targets (clutter such as the ground and the built environment). Moving target echoes are Doppler shifted and can go on to be processed further by the radar; stationary targets have no Doppler shift and can be discarded. An aircraft flying tangentially across the radar's field of view, despite travelling at high velocity, has no component of that velocity acting radially towards or away from the radar: a condition called the **tangential fade**. This is handled in other ways and need not be a concern to the current discussion. In the following discussion, it is assumed that the radar is **monostatic**; that is, it uses the same antenna to transmit and receive signals.

The Doppler shift may be calculated as follows. Consider an aircraft approaching the radar with a radial velocity v_r. The aircraft is a range, R, from the radar. The scenario is illustrated in Figure 4.67.

The total number of wavelengths and partial wavelengths in the round trip from the radar to the aircraft and back is:

$$= (2 \times \mathrm{R})/\lambda \tag{4.16}$$

where R is the range to (and from) the aircraft; λ is the wavelength of the radar transmissions.

A wavelength is one complete cycle of a signal which as an angular measure is 360° or 2π radians and partial cycles are a proportion of this. Eq. (4.8) can be expressed as an angular measure representing the total number of cycles and partial cycles in $2 \times R$. This is expressed in radians thus:

$$\text{the } \textbf{total angular excursion}, \ \varnothing = 2\pi(2 \times \mathrm{R})/\lambda = 4\pi\mathrm{R}/\lambda \tag{4.17}$$

If the aircraft is moving towards or away from the radar, then both ø and R will change. ø is a phase change and the rate of change of phase is frequency. This frequency is the Doppler frequency.

Differentiation equation (4.9) with respect to time:

$$\omega_{\mathrm{D}} = \frac{d\phi}{dt} = \frac{4\pi}{\lambda}\frac{dR}{dt} \tag{4.18}$$

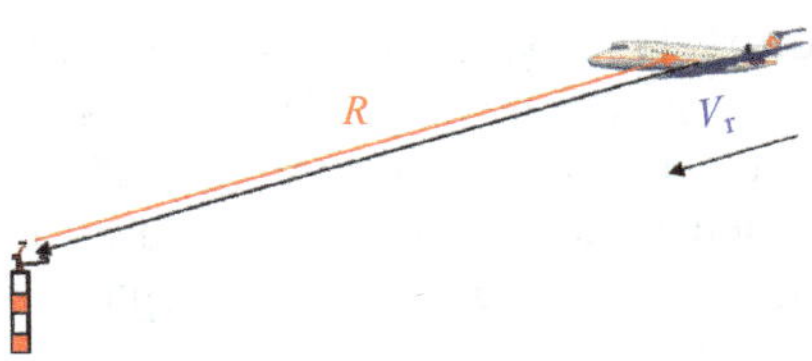

Figure 4.67 Doppler shift calculation

The rate of change of range is the radial velocity v_r:

$$\therefore \omega_D = \frac{4\pi v_r}{\lambda} \tag{4.19}$$

And the Doppler frequency, $f_D = \frac{\omega_D}{2\pi}$
Substituting into (4.11)

$$f_D = \frac{2v_r}{\lambda} \tag{4.20}$$

This may be expressed in terms of transmission frequency, f_t, instead of wavelength

$$f_D = \frac{2v_r f_t}{2} \tag{4.21}$$

For example, assume an aircraft is flying toward the radar at 400 km/h (216 knots) this corresponds to 111 m/s. The radar is S-Band with a frequency of 3 GHz which is a wavelength of 0.1 m. The Doppler frequency presented to the radar by echoes is:

$$\begin{aligned} &= \frac{2 \times 111}{0.1}\ \text{Hz} \\ &= 2,220\ \text{Hz} \\ &= 2.2\ \text{kHz} \end{aligned} \tag{4.22}$$

As the aircraft is flying towards the radar, the velocity is acting in the direction of the radar and the Doppler frequency is positive. In this example, the echo signal detected would have a frequency of 3 GHz plus 2.2 kHz.

4.18.2 Wind turbine Doppler signature

The equation for calculation of Doppler frequency, equation (4.13), derived above can be used to model the performance of a wind turbine Doppler signature.

4.18.2.1 Influence of yaw angle

Earlier in this chapter, the yaw angle was introduced; this is the direction in which the blades are pointing using the direction of the observer as the reference, illustrated in Figure 4.68.

If a radar observes the wind turbine along the yaw = 0° axis, when the blades rotate, there will be no motion towards, or away from, the radar (ignoring the slight backwards tilt of the blades). This situation would arise if the wind were blowing from the direction of the radar. In this case, all the blade motion is perpendicular to the radar.

If the wind direction changes and the turbine yaw angle moves to either 90° or 270°, the radar will see each blade in turn swing towards it and then away from it and will experience both positive and negative Doppler shifts. The motion with respect to the radar will vary along the length of each blade. At the hub, there is no radial motion and, at the blade tip, the motion is greatest.

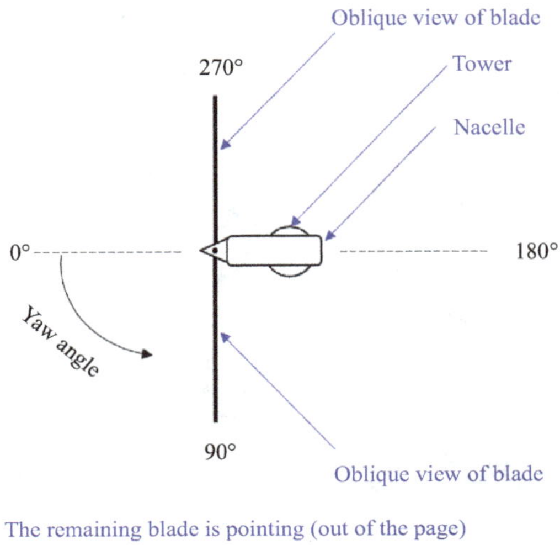

Figure 4.68 Yaw angle geometry

A Doppler frequency model was created for a single blade from a turbine with a rotor diameter of 80 m. The following assumptions have been made:

- The wind turbine is a horizontal axis, three-blade machine.
- From the centre of the hub to the tip of the blade is 40 m.
- The blade rotation rate has been assumed to be 17.1 rpm.
- The radar frequency is 2.8 GHz.
- The radar antenna turns at 15 rpm (4 s update rate).

4.18.2.2 Doppler spread across a single blade

For the first step of the analysis, the blade is assumed to be vertical at the top of its travel and rotating towards the radar. At this point, the velocity towards the radar will be at a maximum. Because the motion is towards the radar, the Doppler will be positive and the echo frequency will be higher than the frequency of the signal transmitted. The analysis was carried out by dividing the distance, from the centre of the hub to the tip of the blade, into ten 4 m long segments. The Doppler signal produced at the middle of each segment was calculated using equation 6 derived above and the result was plotted on a graph shown in Figure 4.69.

4.18.2.3 Doppler spread as the blade rotates

As expected, the Doppler shift is a minimum for the segment closest to the hub and a maximum for the segment closest to the blade tip. This graph shows the Doppler signal as a snapshot when the blade was in the vertical position. The Doppler frequencies are all positive because the blade is moving towards the radar. As the blade continues its rotation, it moves away from the vertical, its velocity with

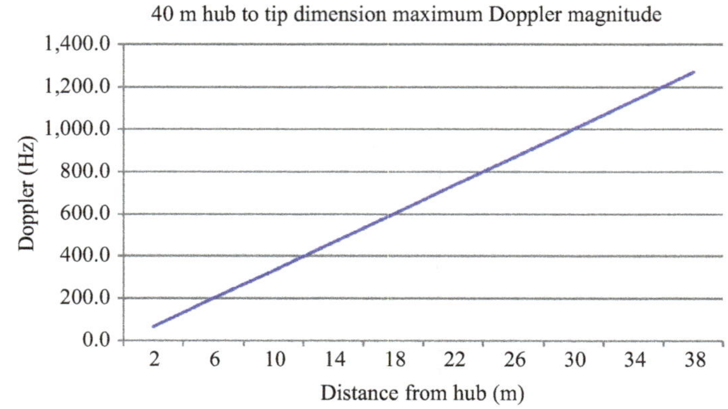

Figure 4.69 Doppler shift versus distance from the hub

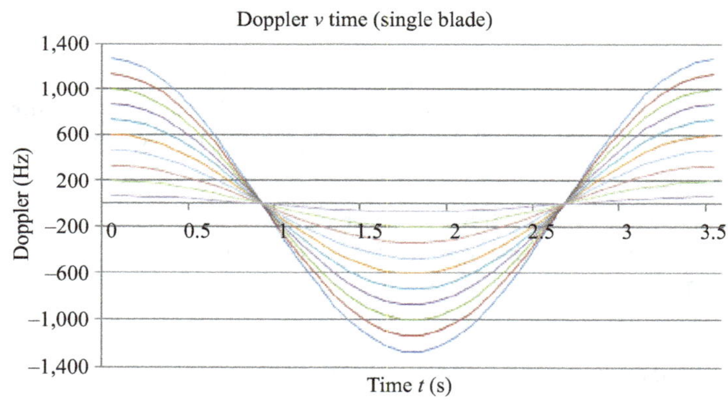

Figure 4.70 Doppler spectrum for one cycle of a single blade

respect to the radar decreases, reaching zero when it arrives at the horizontal position. When the blade passes through the horizontal, its direction of travel with respect to the radar reverses. The maximum velocity, away from the radar, occurs when the blade is pointing vertically downwards. Then, as the blade climbs, the velocity decreases again, reaching zero in the distant horizontal position. The cycle completes when the blade reaches the top of its travel again. At a rotation rate of 17.1 rpm, the whole cycle takes a little over 3.5 s. Taking into account the blade's motion, the Doppler was calculated again, for each segment, throughout a complete cycle of rotation of the blade; the results are plotted in Figure 4.70.

Although the segments are shown as separate, discrete, curves, the radar will see a continuum of frequencies between zero and the maximum Doppler. Note also that both positive and negative Doppler are present.

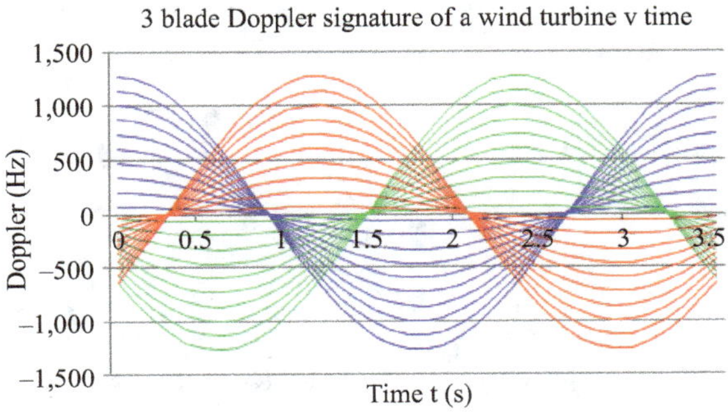

Figure 4.71 Doppler spectrum for one cycle of three blades

4.18.2.4 Doppler spread from three blades

The analysis has, so far, only considered one turbine blade. However, there would be three blades, not one. The process used to create in Figure 4.70 was repeated for all three blades. The results are shown in Figure 4.71.

Figure 4.71 illustrates the complexity of the Doppler spectrum that would be received by the radar and how this varies with time over a single rotation of the turbine blades. Some general 'rules' are apparent in the figure, during each complete rotation:

- There are six maxima during each cycle caused by there being three blades each of which can reach a maximum Doppler at two positions in its rotation.
- There are three periods when the highest positive Doppler is observed in returns.
- In between these three periods are three other periods when the highest negative Doppler is observed.
- Within each peak, there is a spread of Doppler signals from each segment of the blade.

4.18.2.5 Doppler presentation to the radar

However, a radar transmitting pulses and then receiving echoes from the wind turbine 'sees' this time history in a series of snapshots, not a continuous moving image. Figure 4.71 does little to help understand what the contents of the snapshot would look like. Further analysis was carried out to gain an understanding of how a radar would perceive the Doppler signature of a wind turbine.

A radar would make its detections with a periodicity that depended on the type of radar; for example, an aerodrome ATC radar would 'see' the turbine as a series of snapshots taken every 4 s, or an AD radar every 10 s. The cycle of the blade rotation used in this example is approximately 3.5 s. This cycle would repeat as long as the wind was blowing to cause the rotations. However, the radar observations and the blade rotations are not synchronised. Hence, the radar would make its observation at what are, for practical purposes, random times within these blade rotation cycles. The following illustrates what those observations might look like.

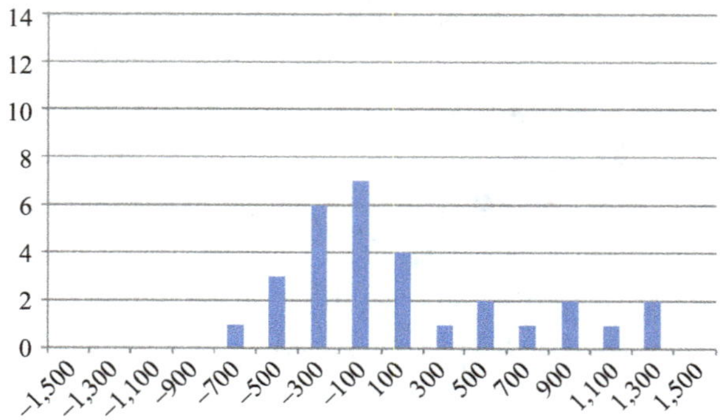

Figure 4.72 0.8 s into the cycle

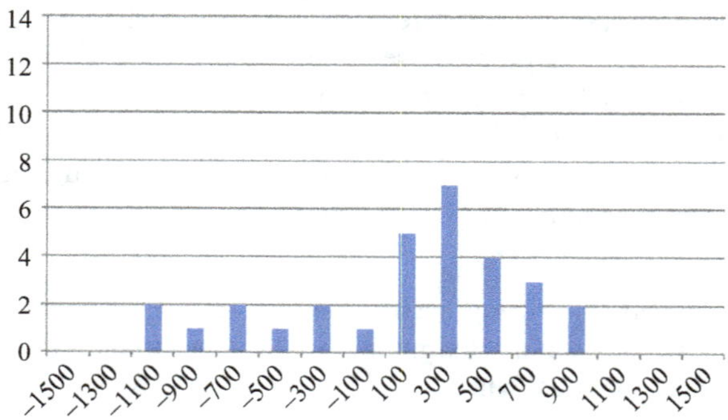

Figure 4.73 2.5 s into the cycle

The blade rotation cycle was broken up into 0.1 s snapshots. At each snapshot, the Doppler spectrum (that, in reality, is continuous) was analysed in samples with a bin size of 200 Hz (chosen randomly but considering computational complexity). If a blade segment created a Doppler within that 200 Hz wide bin that segment was added to that bin. Over the complete cycle a histogram is built showing the frequency with which that Doppler is encountered. Only a few of these samples are shown but the following figures show the relative proportions of Doppler frequencies occurring.

Figure 4.72 shows a sharpshot when the greatest number of samples is relatively low negative Dopplers between –100 Hz and –500 Hz. In the same snapshot, there is a only limited number of relatively high positive Doppler samples up to 1,300 Hz. This can be interpreted thus: one blade is close to the maximum positive Doppler (approaching the radar) position while the other two blades are moving away from the radar.

Figure 4.73 shows a snapshot where the opposite effect is occurring which would arise when one blade was close to the maximum negative Doppler position and the other two blades were swinging towards the radar but still some way off the maximum position for either blade.

Figure 4.74 shows a time when one blade is close to the zero Doppler position (in this case, pointing directly away from the radar) and the other blades are in the middle of their travel one producing a low level of positive Doppler and the other negative Doppler.

4.18.2.6 Comparison with measured data

Poupart [69] reported in 'Wind Farms Impact on Radar Aviation Interests – Final Report', Gavin Poupart, Qinetiq, FES W/14/00614/00/REP, DTI PUB URN 03/1294, Crown Copyright, 2003, the results of measurements of a wind turbine at Swaffham (Figure 4.75). The results are presented in Figure 4.76. The ordinate axis (y-axis) is

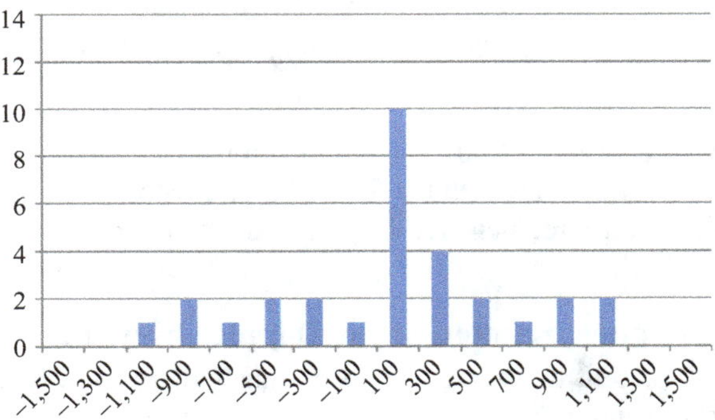

Figure 4.74 1.7 s into the sequence

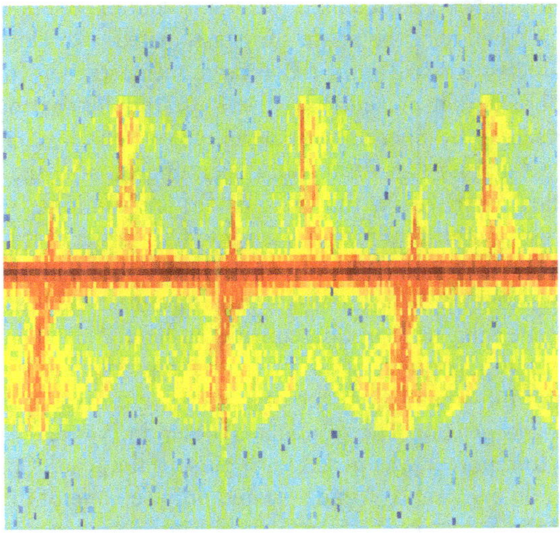

Figure 4.75 Doppler time intensity (adapted from Doppler spectrum from Poupart (2003) Crown Copyright)

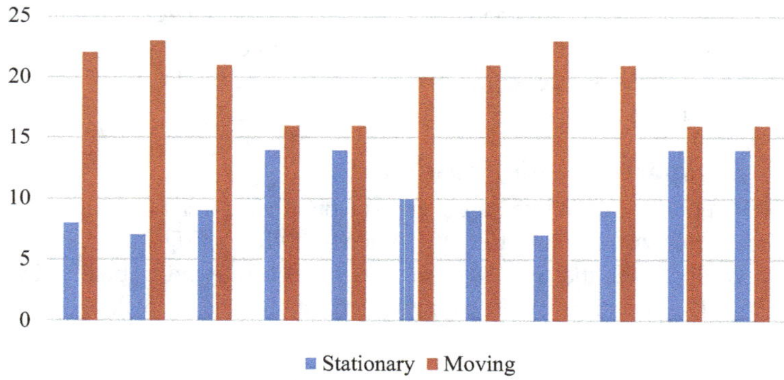

Figure 4.76 Stationary/moving target filtering

Doppler Frequency between ±1,500 Hz and the ordinate axis (x-axis) is time over a 5-s period. Note the similarities with the foregoing discussion; there are three positive Doppler excursions and three negative Doppler excursions in the complete cycle of rotation of the blades.

4.18.2.7 Radar interpretation of the Doppler presentation

Figures 4.72–4.74 present the mix of Doppler signals that would be received at the radar antenna from wind turbine, blade and echoes. Legacy (i.e., non-wind farm tolerant) radars process the signals received at the antenna and, at least, attempt to discard any stationary targets. Discarding ground clutter was a significant technological advance, before being able to do this, the pictures presented to the operator were very cluttered and identifying aircraft required great skill and experience. The decision is based on Doppler frequency or the phase shifts associated with Doppler shift.

It is undesirable to discard only those targets that present no Doppler shift. There are many objects that present low levels of Doppler that do not need to be seen on the radar screen by operators, such as motor vehicles and even leaves on trees will present a Doppler signature. But depending on how the design is implemented, for practical purposes, a threshold is set and stationary or slow-moving targets are discarded and all that remains is a moving target that can be displayed.

More modern legacy radars are able to perform more complex filtering in an attempt to discriminate and discard other forms of clutter such as weather returns but that type of processing need not be considered here.

The data from the previous examples was reprocessed so that instead of dividing the returns into many Doppler bins they were divided into either a stationary or a moving target bin. The results are shown in Figure 4.76. Because there are only two bins to report it was possible to present the results from a number of snapshots (eleven) on one graph. A radar would only see one of these time samples on any one scan.

The influence of the rotating blades can be seen in Figure 4.76 but information has been lost. It is no longer possible to determine from the filter outputs what the blades are doing. This point will be taken up later in this chapter when the implications of the wind turbine signatures are considered.

4.19 Wind turbines and radio shadow

The term 'shadow' has been applied to a number of phenomena that occur when wind turbines affect the operation of aviation radio and radar systems. For example, the term shadow was used extensively in AWC (2005) linked to the 'large [Radar Cross Section] RCS' of wind turbines and reduced probability of detection, and CAP 764 (2016) refers to shadows being created 'above and beyond the wind farm' by turbine blades [74,87]. Phenomena in which reduced probability of detection associated with wind turbine RCS and effects which occur other than behind a wind turbine are discussed later in this chapter. Butler and Johnson [88] use the term 'shadowing' in a sense of terrain creating a shadow that partially obscures wind turbines. This use of the term is more closely related to the discussion here because it is dependent on diffraction over terrain affecting the wavefront that illuminates a wind turbine and this affects how or whether it might be detected by radar. There is a discussion about the use of the term in this context in Chapter 5.

In this section, the term shadow is used to refer to signal losses that occur in the volume of airspace behind a wind turbine, when its structure obstructs the direct passage of radio waves from a radar. This section only considers the shadow effects at radio frequencies. Optical shadows were discussed earlier in this chapter as the cause of 'shadow flicker'. Poupart [69] investigated shadow in this context and the findings are discussed below.

4.19.1 Misconception

A possible misconception concerning the shadow region behind a wind turbine is that the whole sweep of the blades is shaded; this is not the case as examples show.

Assume a wind turbine is 10 km from an airfield ATC radar. An aircraft is 100 km beyond the wind farm. The path between the radar, the wind turbine and the aircraft forms a straight line. The aircraft is, therefore, 110 km from the radar. This distance was chosen because it is very close to the maximum range of typical ATC radars: 60 Nm (111 km). A pulse is transmitted by the radar on the bearing of the wind turbine and the aircraft. The distance from the wind turbine to the aircraft and back is 200 km as illustrated in Figure 4.77. Radio waves, travelling at the speed of light, take 667 μs to cover this distance.

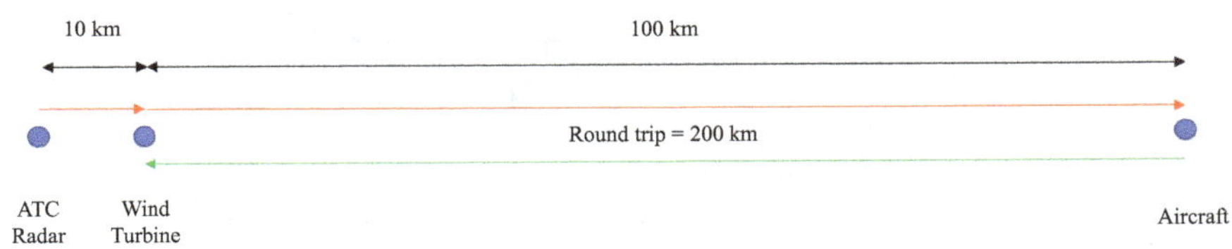

Figure 4.77 Simple scenario

Using the example provided earlier in this chapter, let us further assume that the wind turbine blades are rotating at 16.7 rpm and the distance from the centre of the hub to the turbine blade tip is 40 m. In the time taken for radio waves to travel from the wind turbine to the aircraft and return to the wind turbine, the turbine blade will have rotated through an angle of 0.07°, corresponding to the blade tip having moved 5 cm. For the purposes of investigating the shadow created by a wind turbine, a useful approximation is that the blade is stationary every time it is illuminated by the radar.

However, in successive radar scans, the orientation of the blades change, as will the shadow patterns they create as described below. In this example, the radar is an airfield ATC radar. The usual scan of rate of this type of radar is 15 revolutions per minute, giving an update rate (the frequency of the picture being refreshed) of once every 4 s. The orientation of the blades in successive updates, assuming a 4-s update rate is shown in Figure 4.78 (the blades are rotating clockwise).

The update rate varies for other types of radar: for example, air route surveillance radars (ARSR) and AD radars are slower to update because they are longer range, there is a greater elapsed time between updates and the blades move further. The following image shows the blade positions if the update rate was 10 s; typical of an AD radar and some ARSR. Note, in each case, one blade has been identified differently (red colour) as a reference point. As in the previous case, the blades are rotating clockwise (Figure 4.79).

4.19.2 Diffraction

Waves, in general, such as sound or on the surface of water, as well as radio waves, can be blocked by obstructions. However, the shadow edges will not be immediate

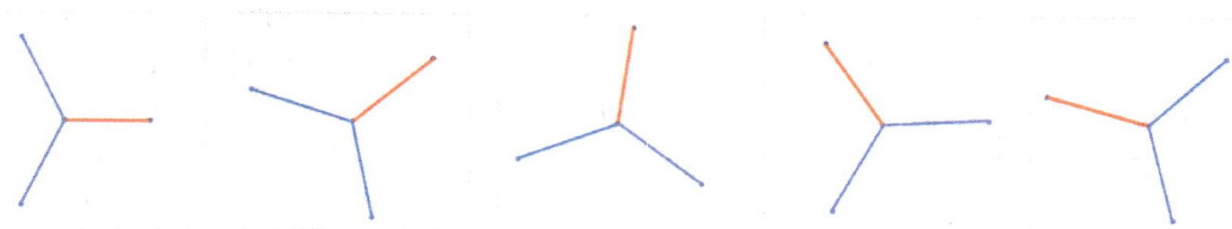

Scan 1 Scan 2 Scan 3 Scan 4 Scan 5

Figure 4.78 Blade positions – 4 s update rate

Scan 1 Scan 2 Scan 3 Scan 4 Scan 5

Figure 4.79 Blade positions – 10 s update rate

transitions to a complete absence of the wave. Diffraction is the mechanism that causes waves to spread into the region behind the obstruction. Introductions to diffraction can be found in Bacon (2003) [89], Haslett (2008) [90] and ITU Recommendation P-526-15 [91]. The following analysis illustrates the extent to which diffraction fills in the shadow behind a wind turbine when illuminated by a radar.

4.19.3 Aim

The aim of the analysis described was to characterise the radio-shadow behind wind turbine blades at different distances behind the wind turbine blades and at different radar frequencies.

4.19.4 Analysis

The following, previously unpublished, analysis was carried out by Bacon¶¶¶ between 2011 and 2012 and is reproduced with his kind permission.

The Fresnel–Kirchhoff Integral (described in the references and in Chapter 5) was used to calculate the amplitude and phase of signals behind a representative wind turbine blade structure.

4.19.5 Approximations/assumptions

The following approximations and assumptions were made:

- The blades were assumed to be 'thin'; they are represented as 2-D structures. More complex representations would be required if the scattering (reflecting characteristics) by blades was being considered. It is worth emphasising that an object's radio shadow is not an alternative representation of its RCS. The two are quite different quantities.
- The blades were assumed to absorb all the energy incident upon them. In practise, scattering will take place but the effects are considered small enough to ignore and the assumption of total absorption is appropriate.
- The blade size modelled was 20 m × 2 m. This size is modest by comparison in modern wind turbine blades. The size was chosen partly because of the computational complexity and also to simplify the presentation of the results.
- The modelling is based upon propagation using Huygens' construction described in Chapter 5, and it is assumed that the wind turbine blades are illuminated by a coherent wavefront (all the wavelets comprising the wavefront are parallel and in phase). At any finite distance from the source of the electromagnetic radiation this will be an approximation. Provided that the distance of the wind turbine blades is in the far field of the radar antenna, the distance is arbitrary. In each of the following analyses, it is assumed the wind turbine is 10 km from the radar system.
- The frequencies of interest are:
  - L-Band (30 cm wavelength) is typical of some AD radar, many en-route ATC radar and some in-fill radar types.

¶¶¶Bacon was the former head of Technical Computing and Propagation Modelling at the UK's Ofcom.

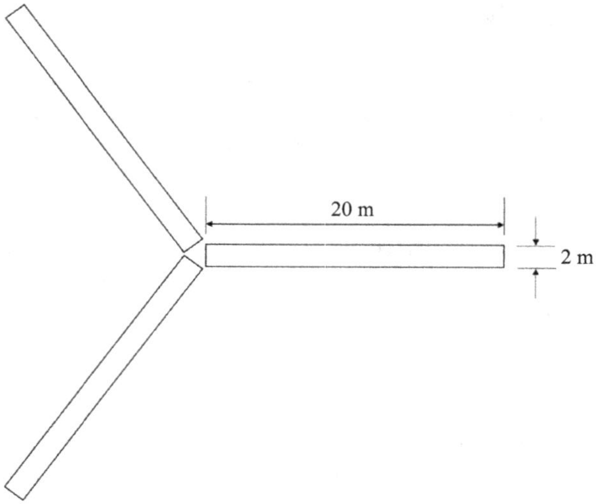

Figure 4.80 Turbine dimensions

- ○ S-Band (10 cm wavelength) is consistent with the majority of terminal ATC radars and some AD radars.
- ○ X-Band (3 cm wavelength), these frequencies are used by PARs and some in-fill radars.

The wind turbine blade dimensions modelled are shown in Figure 4.80.

The analysis geometry is shown in Figure 4.81. The plots created are for one-way propagation.

4.19.6 Signal amplitude results

The first set of results is the shadows created at S-Band.

Figure 4.82 shows a colour-coded plot of signal amplitudes. All the values are with respect to free space which is the signal level that would be present in the absence of the obstruction. Each colour represents a signal level in a bin between two values. For example, in Figure 4.82, the blue background colour represents signal levels between 0 dB and +1 dB with respect to free space losses and grey shows values between −8 and −5 dB with respect to free space. The bin sizes are not uniform to reduce the palette of colours required and to avoid over-complicating the figures. The plot represents 60 m × 60 m (±30 m) centred on the turbine hub. The distance of the plot behind the wind turbine blades is 100 m. The blade positions are outlined by thin black lines. Attention is drawn to the following information illustrated by the plot:

- The signal strengths in the majority of the airspace around the wind turbine blades remain between 0 and 1 dB (i.e., shown in the blue background colour).
- The figure shows that small enhancements in signal strengths are present, as well as losses. This effect is expected of diffraction.

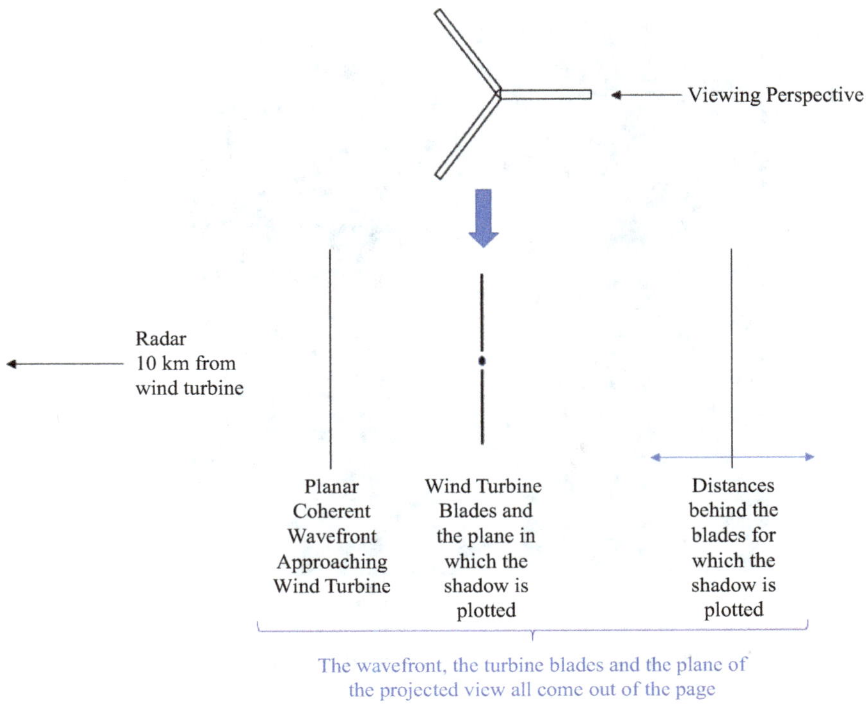

Figure 4.81 Shadow calculation geometry

- The signal level features in the shadow can be small in scale, some features smaller than a metre across.
- Even though the plot is only 100 m behind the turbine blades, the shadow is already filling in. In very small regions at the edges of the blades, and at the location where the hub would be, there are signal losses between 8 and 11 dB below free space losses. These regions are shown in grey.

The following figures illustrate how the shadow evolves as the distance behind the wind turbine increases from 200 m to 20 km (Figures 4.83–4.89). Note the shadow is filling in, it gets more shallow as the distance behind the turbine blades increases. To clarify the point further, the greater losses in signal strength close to the blades do not add to the losses further away: the losses are not cumulative.

Figures 4.83–4.90 illustrate the following features:

- At 200 m behind the wind turbine blades:
  - The deepest part of the shadow 100 m behind the turbine (−8 to −11 dB) has disappeared.
  - The regions of deepest shadow have grown in size but, as they have grown, they have filled in and the losses in the regions of deepest shadow are now between −5 and −2 dB with respect to free space.

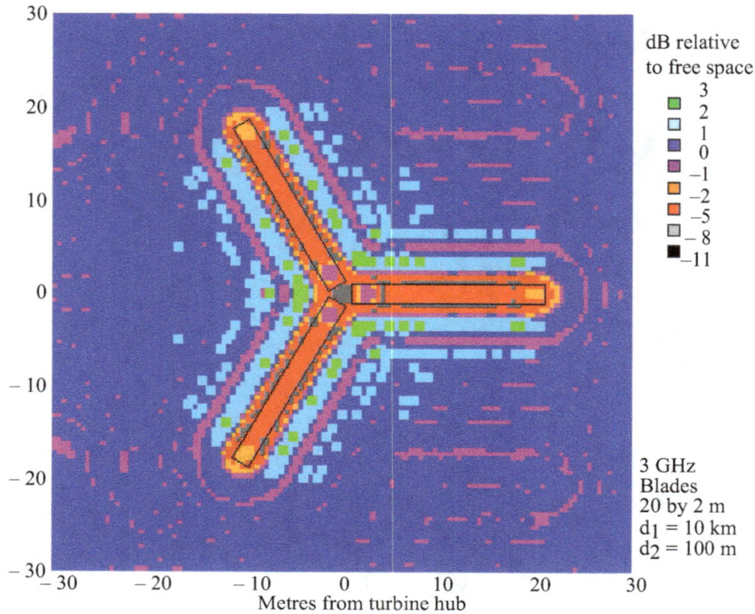

Figure 4.82 S-Band shadow 100 m behind the blades

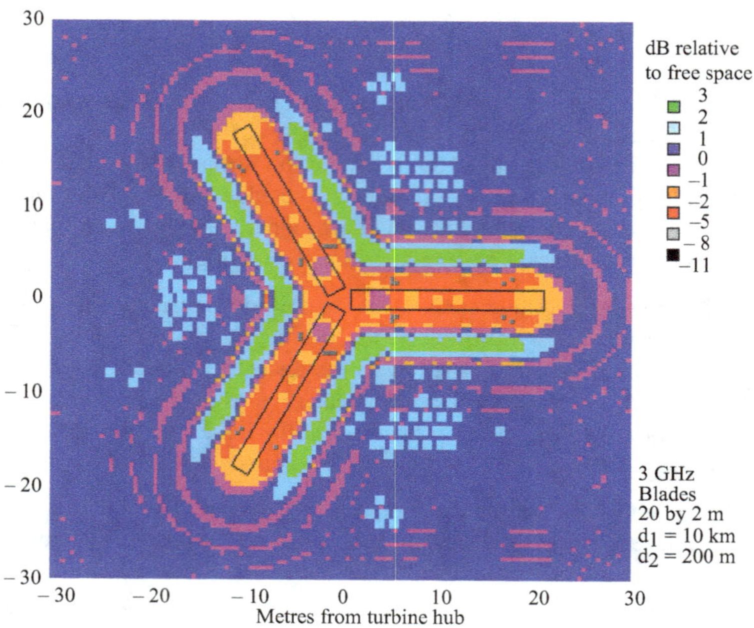

Figure 4.83 S-Band shadow 200 m behind the blades

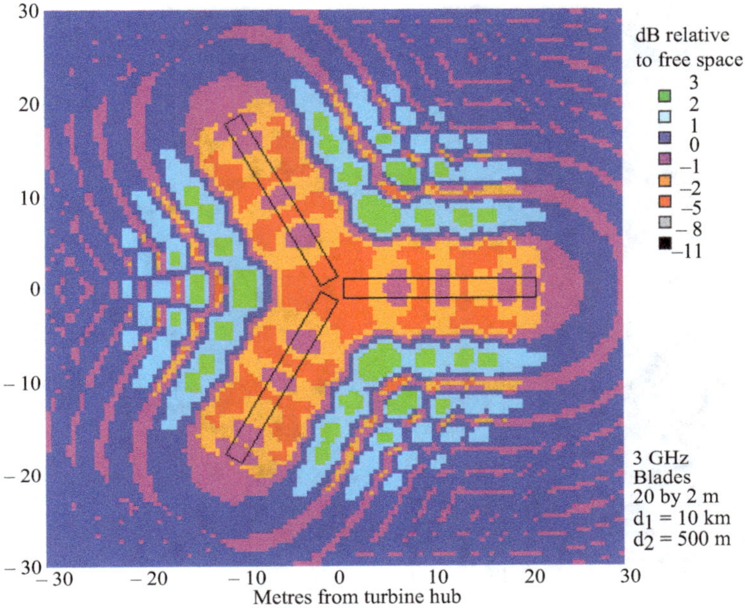

Figure 4.84 S-Band shadow 500 m behind the blades

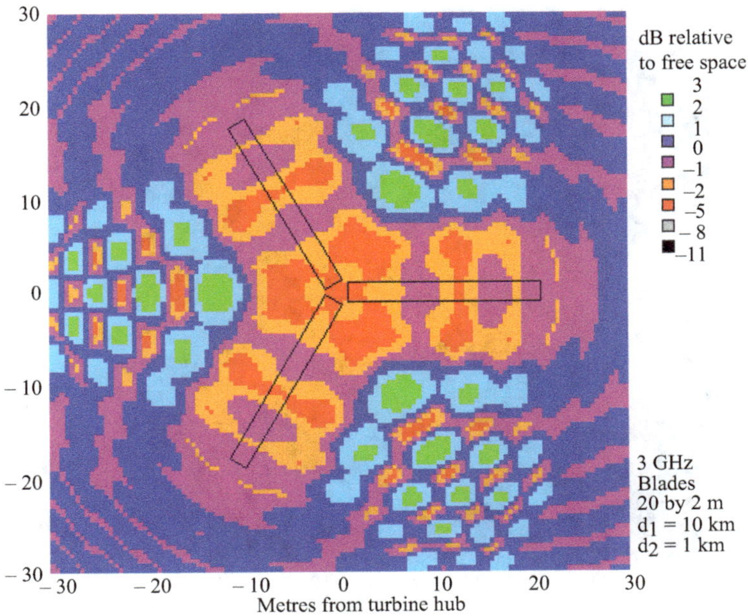

Figure 4.85 S-Band shadow 1 km behind the blades

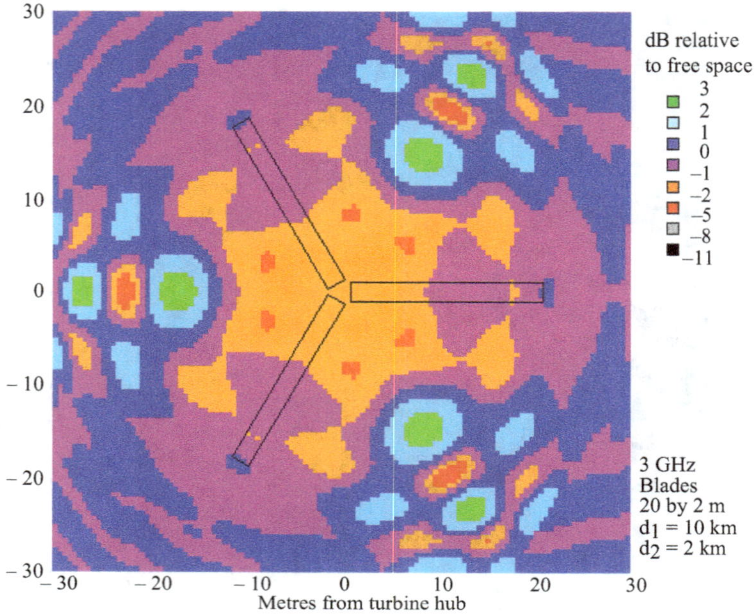

Figure 4.86 S-Band shadow 2 km behind the blades

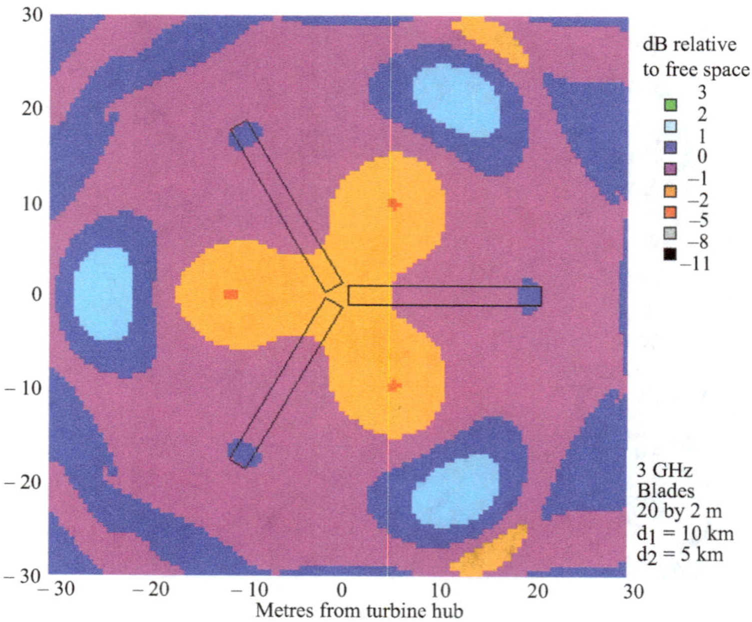

Figure 4.87 S-Band shadow 5 km behind the blades

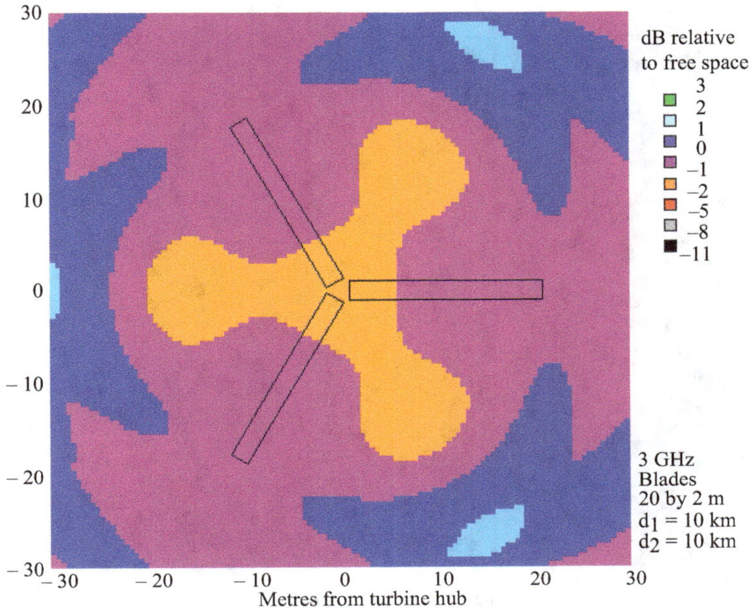

Figure 4.88 *S-Band shadow 10 km behind the blades*

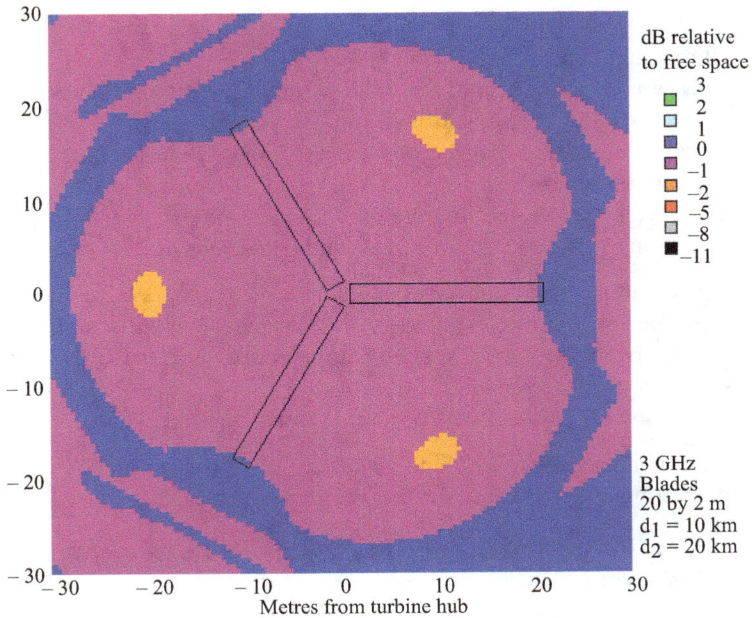

Figure 4.89 *S-Band shadow 20 km behind the blades*

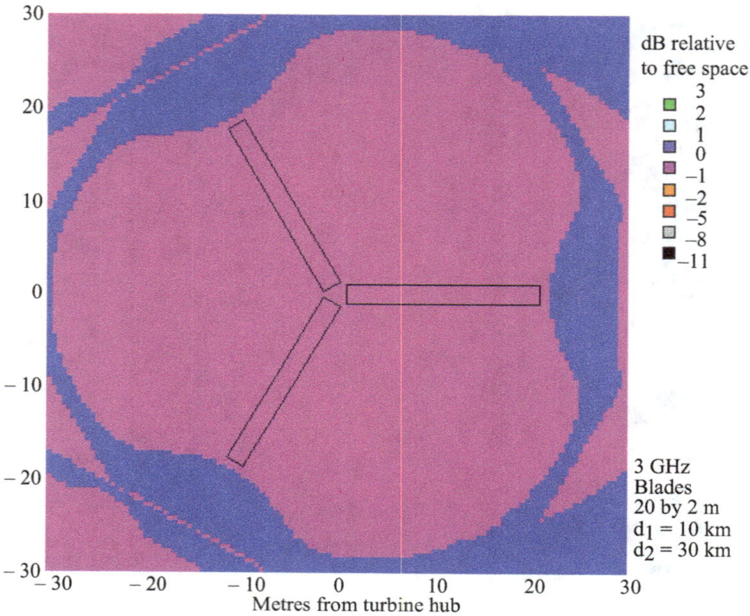

Figure 4.90 S-Band shadow 30 km behind the turbine blades

   - Surrounding the regions of deepest shadow are clearly defined regions where the signal level is enhanced.
- At 500 m behind the wind turbine blades:
   - Even immediately behind the blades (shown by the thin black lines) the deepest shadow region is filling in.
   - Regions where the shadow is between −5 and −2 dB are still present but they are becoming smaller.
   - The image creates the impression of waves moving out from the blade alternating between regions of loss and enhancement.
- 2–5 km behind the wind turbine blades:
   - The deepest shadows shrink as the distance increases and the small regions of greatest loss of signal strength are surrounded by regions of enhanced signal strength.
   - The shape of the shadow is becoming less blade-like.
- 10–20 km behind the wind turbine blades:
   - As the distance increases further, the losses decrease.

4.19.7 The effects of frequency on diffraction and shadow – signal amplitude

The foregoing plots were all created for S-Band wavelength (10 cm). The following collection of images illustrates diffraction and shadow as wavelengths vary.

Figures 4.91, 4.92 and 4.93 show the shadow region 2 km behind the wind turbine blades at wavelengths of L-Band, S-Band and X-Band, respectively. The effect illustrated is that predicted by diffraction theory: as the frequency of the electromagnetic wave decreases, the shadow region fills in more rapidly (or conversely as the frequency increases, the shadow fills in more slowly). This finding is also reported by Oswald and Baker, noting the benefits of L-Band over higher frequencies [92].

Figures 4.94 and 4.95 show the same effect in a different fashion. Figure 4.94 shows the shadow formed by an L-Band wavelength 200 m behind the wind turbine blades, and Figure 4.95 shows the shadow formed by an X-Band wavelength (one-fifth that of L-Band) at a distance of 1 km behind the wind turbine blades (5 times that of the L-Band plot). The similarity between the two images is obvious and they illustrate the effect of wavelength on diffraction.

4.19.8 Summary of amplitude results

Minimum and maximum signal strengths at each frequency were calculated for each shadow distance. The results are plotted in Figure 4.96 against the log of the distance behind the wind turbine blades. Bacon notes that the results at short distances are likely to be under-sampled. But these results show the obvious impact of shadow at short distances, at 1 km and less, lower signal strengths dominate. However, between 1 and 10 km, the effects of diffraction mean that losses and enhancements even out. Beyond approximately 5 km (2.7 Nm), the maxima and minima are almost equal in magnitude.

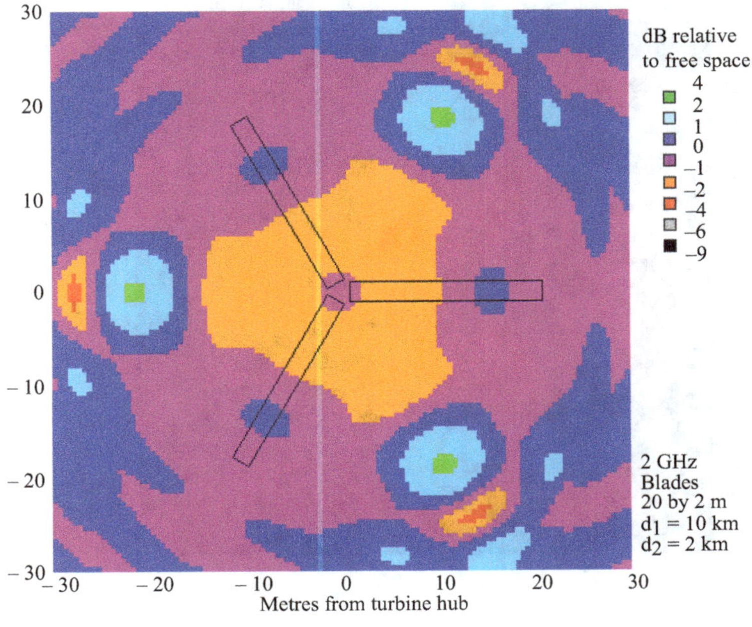

Figure 4.91 L-Band shadow 2 km behind the blades

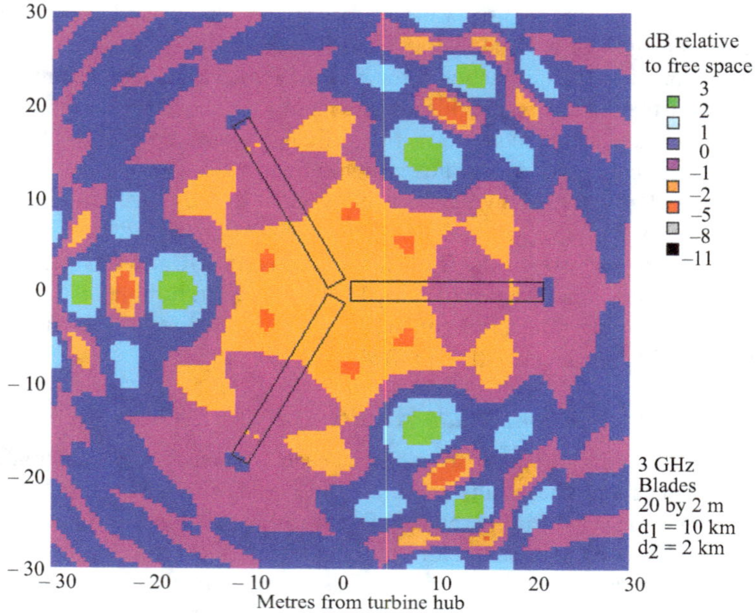

Figure 4.92 S-Band shadow 2 km behind the blades

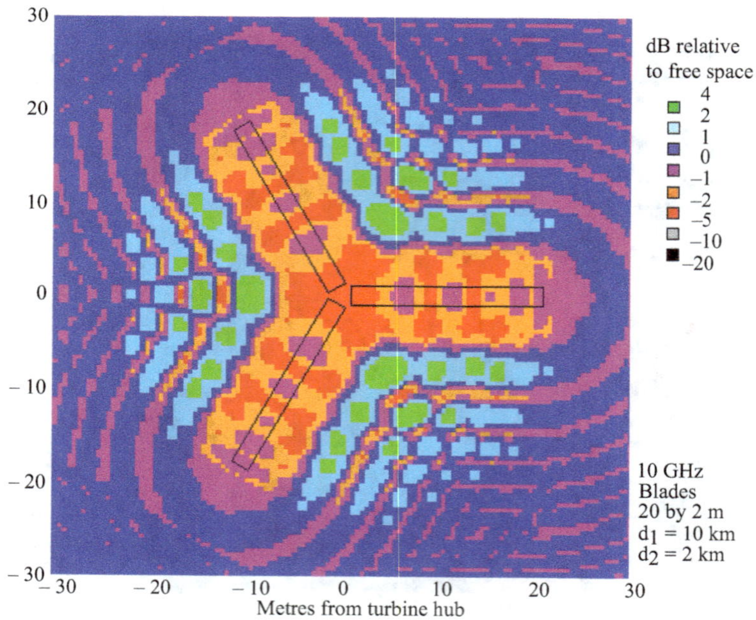

Figure 4.93 X-Band shadow 2 km behind the blades

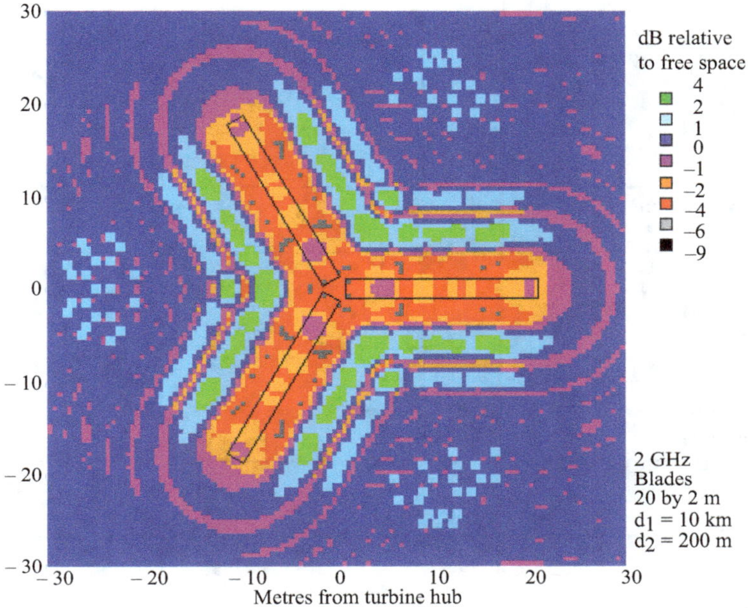

Figure 4.94 L-Band shadow 200 m behind the blade

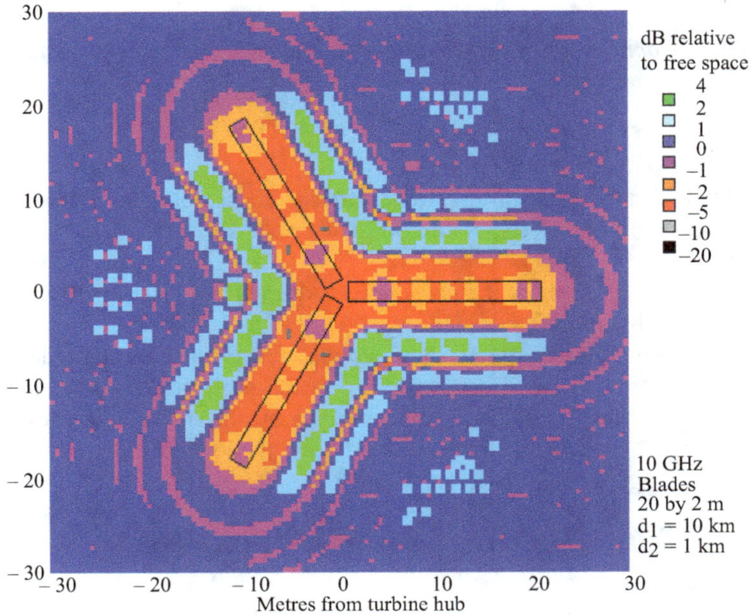

Figure 4.95 X-Band shadow 1 km behind the blades

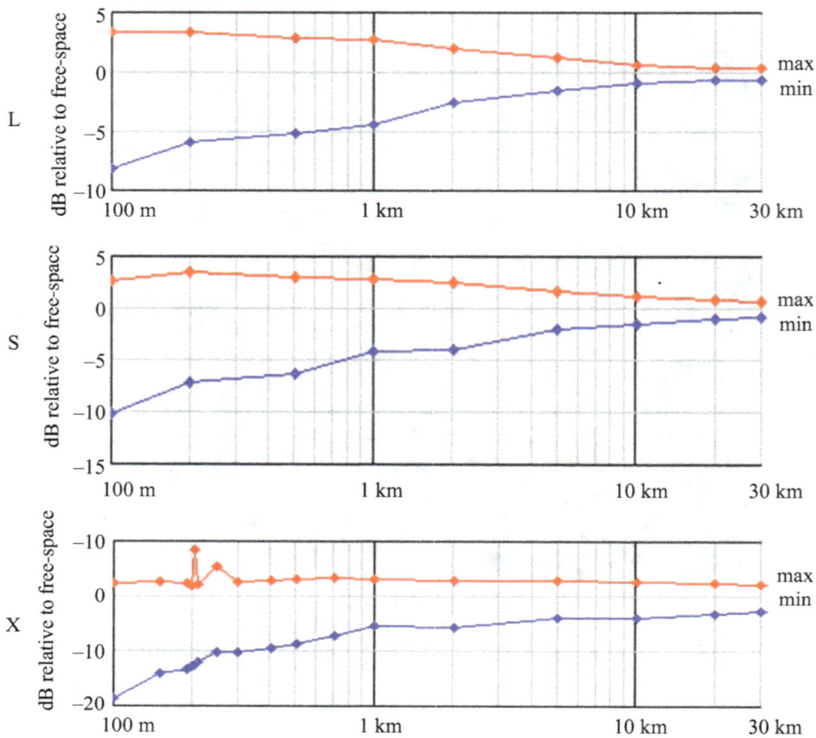

Figure 4.96 Comparison of maximum and minimum signal strengths with frequency

4.19.9 Shadow phase effects

Figures 4.91–96 all deal with amplitude effects. The amplitude effects are also accompanied by phase effects. As before, the results for S-Band wavelength are presented first.

Figures 4.97–4.105 show the phase effects at ranges of 100 m–30 km behind the wind turbine blades.

The figures for phase effects show similar results to the amplitude effects. Initially, the greatest phase effects are close to the projected wind turbine blade positions. As the distance behind the blades increases, the shadow phase evolves in a wave-like structure radiating from the blades. At the longest range analysed, the effects are no longer blade-like in appearance.

4.19.10 Effects of wavelength on shadow phase effects

Using the same approach as for the amplitude results, Figures 4.106, 4.107 and 4.108 show the shadow phase results at L-Band, S-Band and X-Band wavelengths, respectively, at 2 km distance behind the blade. Figures 4.109 and 4.110 show the

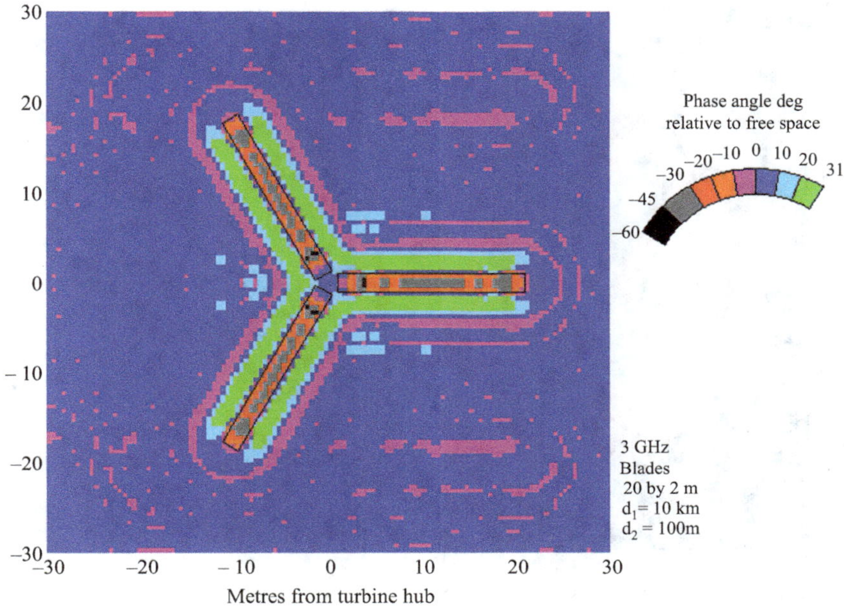

Figure 4.97 S-Band shadow phase 100 m behind the blades

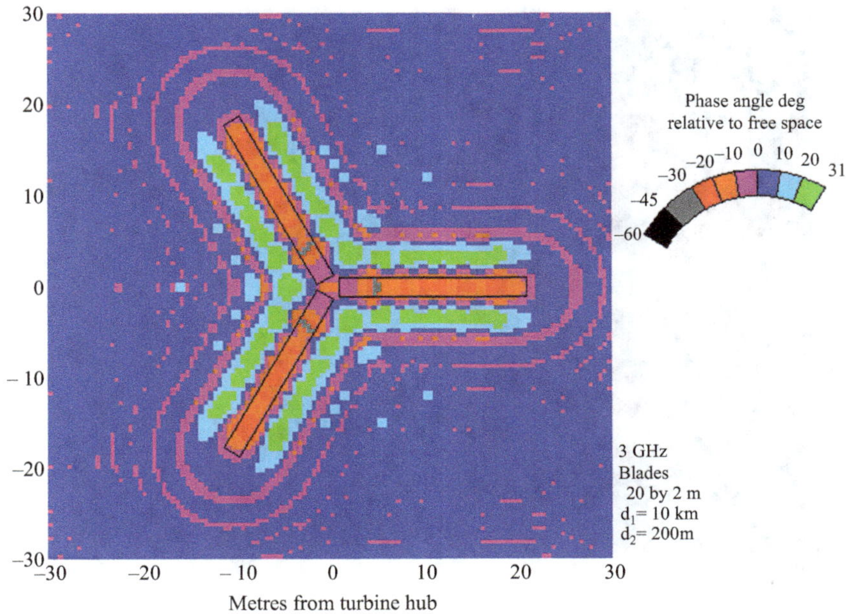

Figure 4.98 S-Band shadow phase 200 m behind the blades

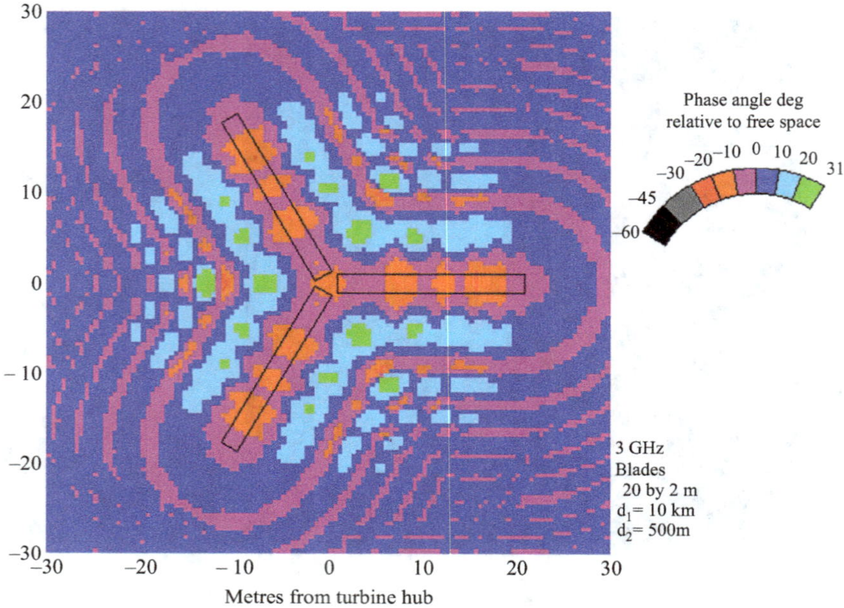

Figure 4.99 S-Band shadow phase 500 m behind the blades

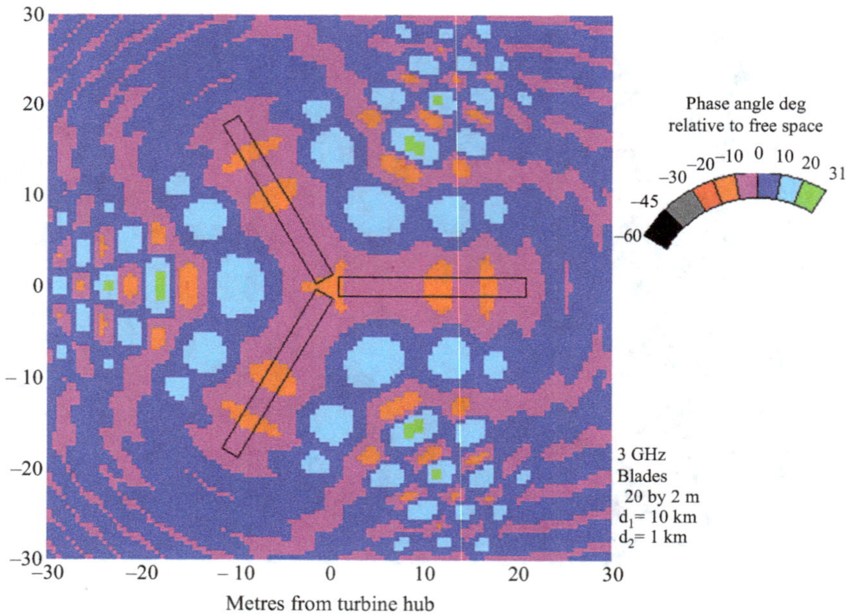

Figure 4.100 S-Band shadow phase 1 km behind the blades

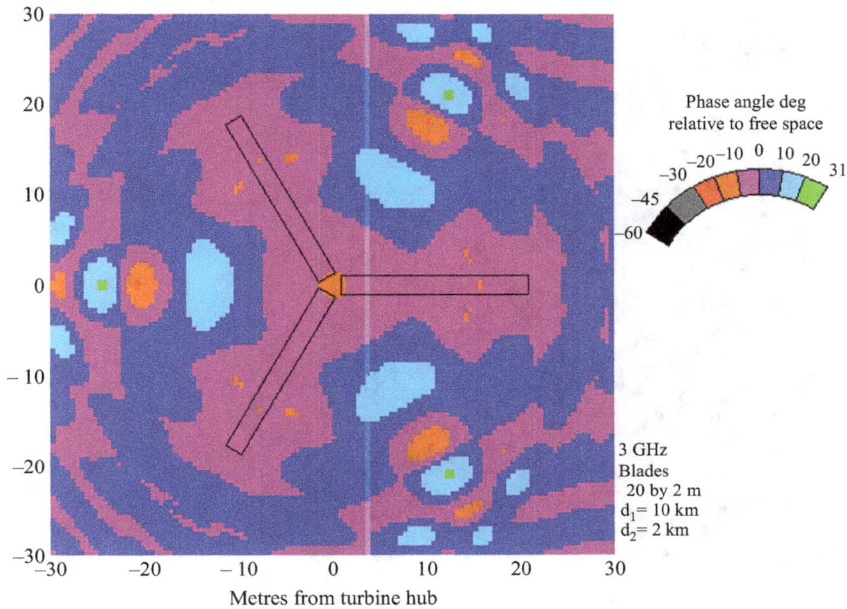

Figure 4.101 S-Band shadow phase 2 km behind the blades

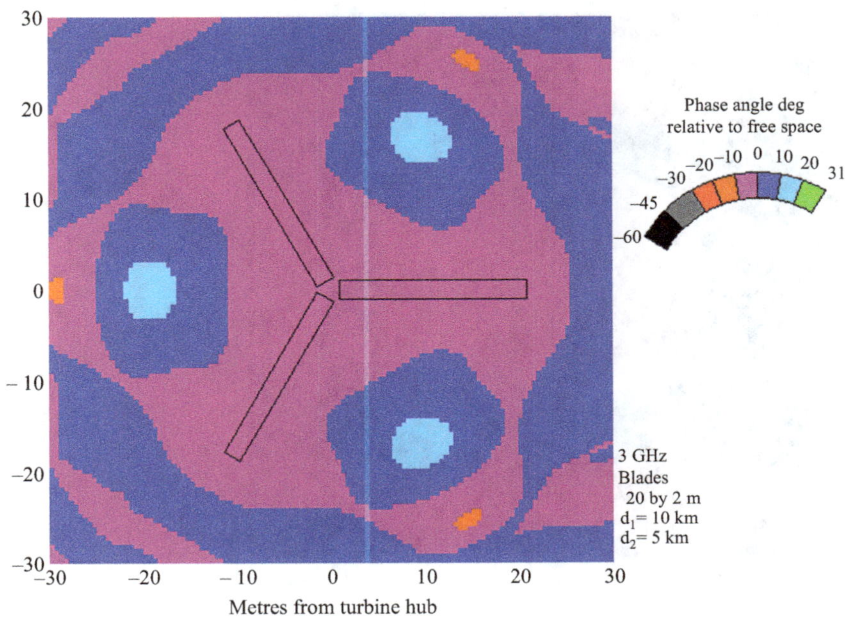

Figure 4.102 S-Band shadow phase 5 km behind the blades

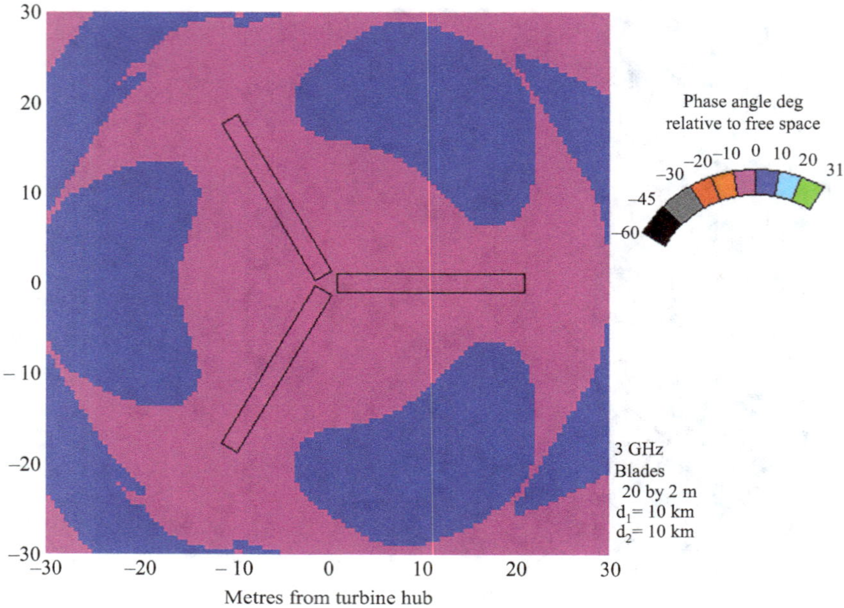

Figure 4.103 S-Band shadow phase 10 km behind the blades

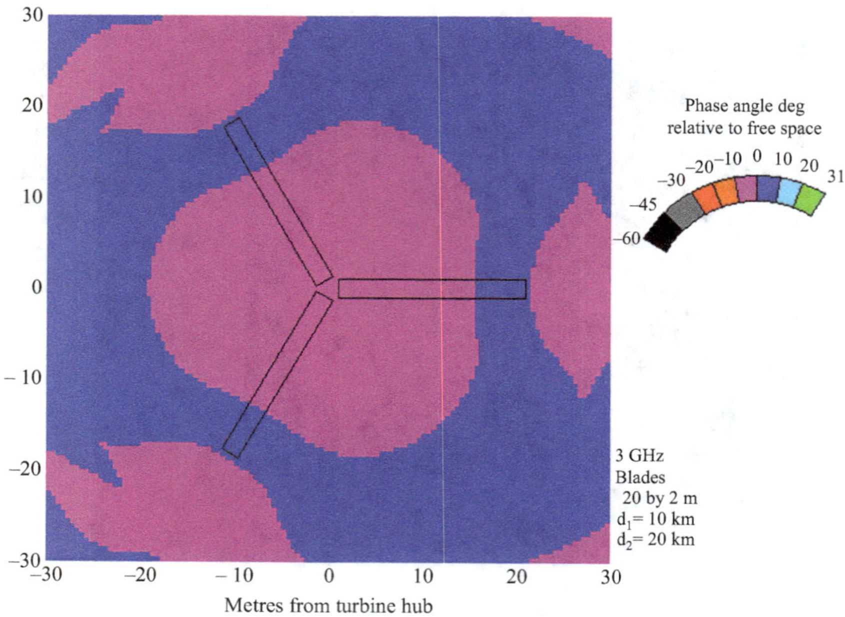

Figure 4.104 S-Band shadow phase 20 km behind the blades

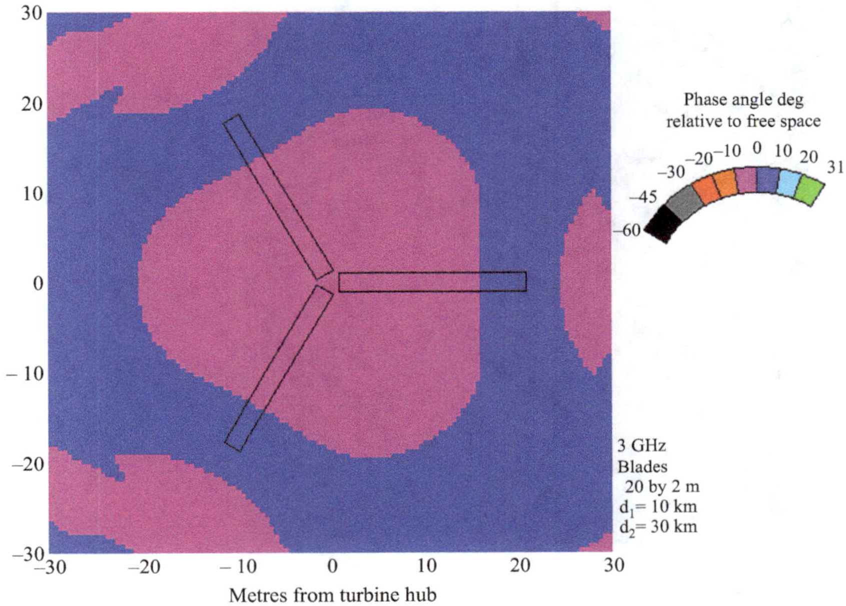

Figure 4.105 S-Band shadow phase 30 km behind the blades

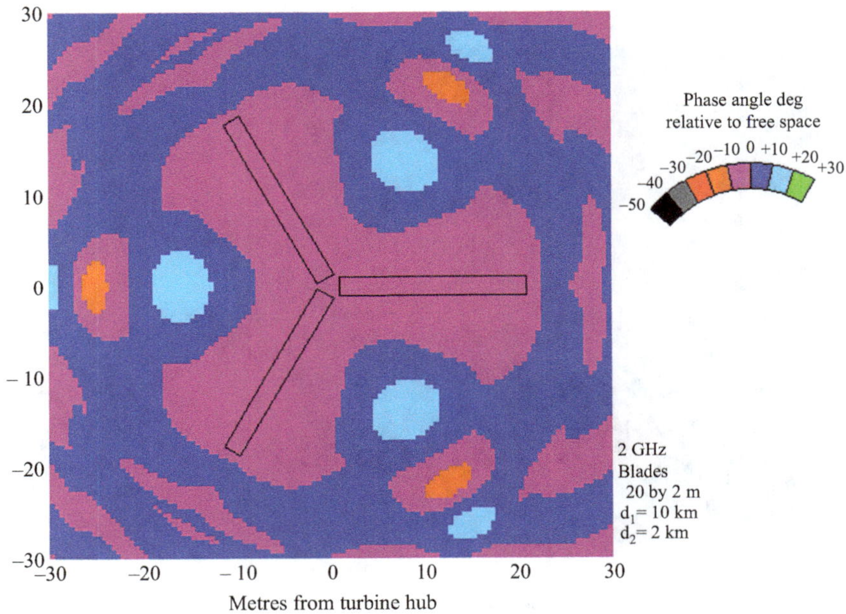

Figure 4.106 L-Band shadow phase 2 km behind the blades

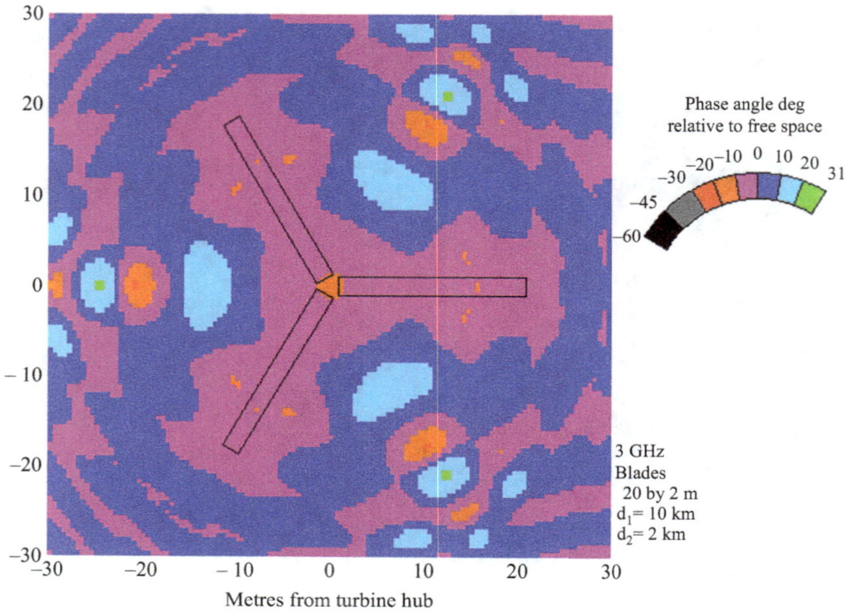

Figure 4.107 S-Band shadow phase 2 km behind the blades

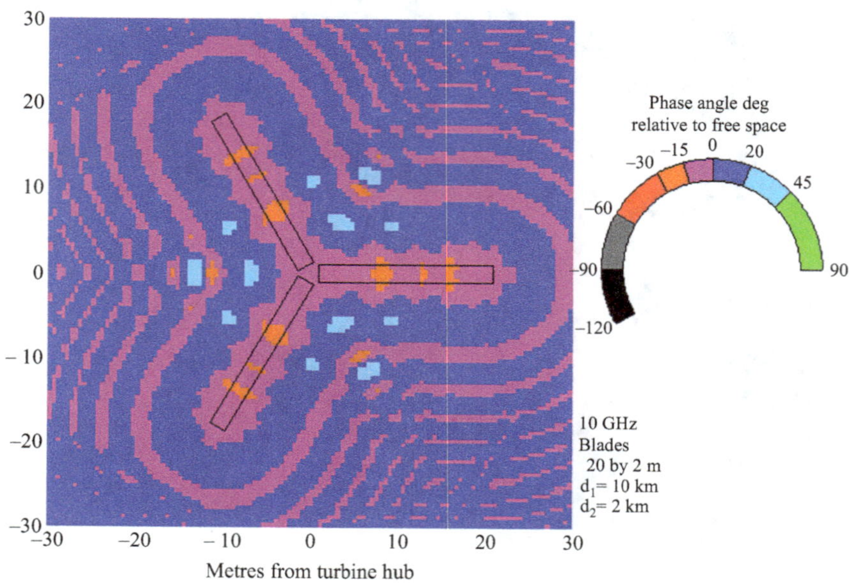

Figure 4.108 X-Band shadow phase 2 km behind the blades

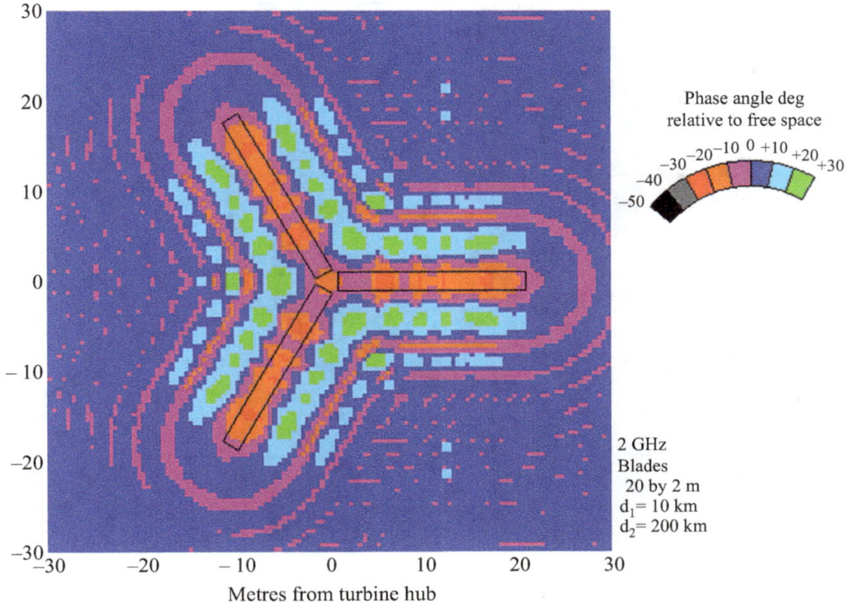

Figure 4.109 L-Band shadow phase 200 m behind the blades

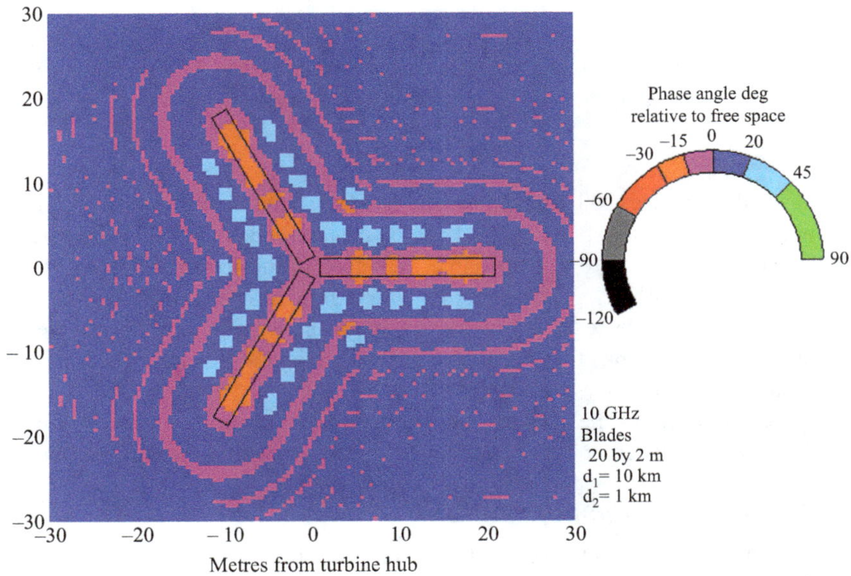

Figure 4.110 X-Band shadow phase 1 km behind the blades

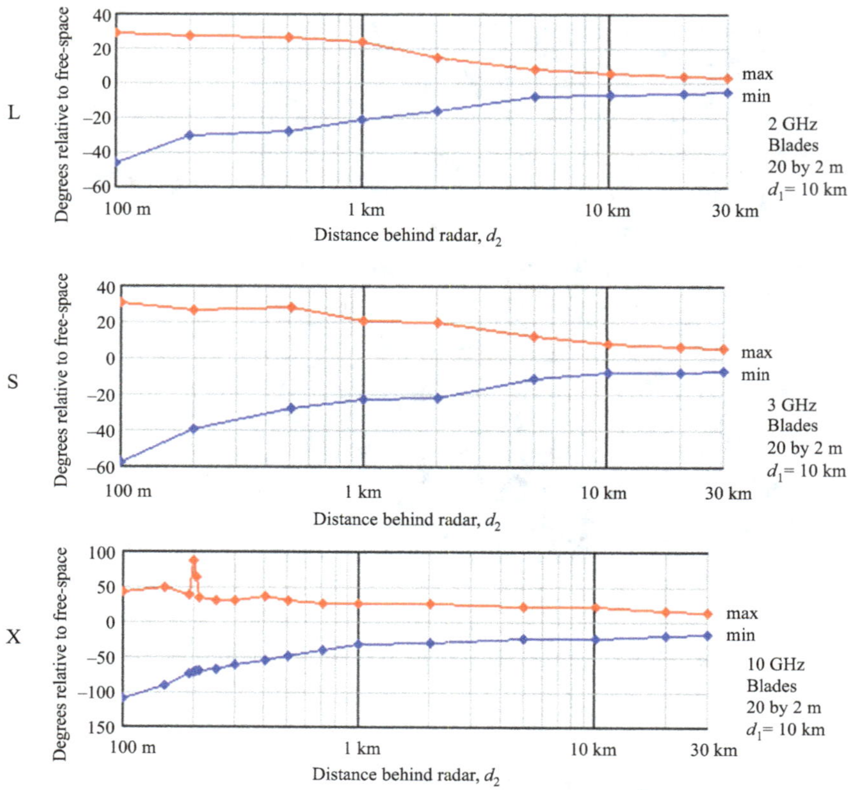

Figure 4.111 Comparison of maximum and minimum phase with distance and wavelength

L-Band shadow phase 200 m behind the wind turbine blades and the X-Band shadow result at 1 km distance behind the blades, respectively. Note that the bin sizes in these two plots are not the same thus accounting for the apparent differences in the phase effects. Taking this consideration into account, the results are similar to the amplitude results. Illustrating again the dependence on wavelength expected from diffraction, that is, the longer wavelengths penetrate into shadows more rapidly than shorter wavelengths.

Figure 4.111 shows the comparison of maxima and minima phase differences from free space at different wavelengths and distances behind the turbine blades.

4.19.11 Interpretation of results

Amplitude and phase effects created by diffraction of radio waves were analysed. Wave bands common in aviation radar systems were analysed at different distances behind wind turbine blades using the Fresnel–Kirchhoff Integral. The analysis considered only diffraction and the obstruction causes by the blades. No scattering effects were considered. Although three wave bands were analysed, only a single frequency within these bands was analysed. The results were amplitude and phase plotted with reference to free space. In practice, other factors affect signal strengths and phases available for detection of aircraft and these would include multipath effects and the mitigation for multipath provided by frequency diversity.

In the shadow region behind wind turbines, there are volumes of airspace where the signals available to detect aircraft are decreased in amplitude. Very close to the wind turbine blades, the losses can be significant, for example, at S-Band 100 m behind the turbine, there are very small volumes of airspace where the one-way loss is between −11 and −8 dB. This effect is offset by the larger volumes of airspace adjacent where the signal levels are enhanced by between 1 and 2 dB. As the distance from the wind turbine blades increases, the shadow fills in and, at distances beyond 5 km behind the wind turbine blades, only very small volumes exist where the losses are greater than 2 dB and maxima and minima are approximately the same at both L-Band and S-Band. Only at X-Band do the minima slightly outnumber the maxima.

Recall, the distance of the wind turbine from the radar used for the analysis was 10 km. Thus, the results at 10 km indicate the losses that would be experienced on the reciprocal path. For example, at S-Band, close to the blade root and the hub, the one-way additional loss compared with free space is between 1 and 2dB, a two-way path loss of 2–4 dB.

In practice, the volume of airspace where a consistent loss would be experienced is only a few metres wide. Moreover, if an aircraft return was subject to that loss on a single pulse, by the time of the next scan, the blade would be in a different position, as shown earlier in this section. Theil and Ewijk considered the possibility of small aircraft being able to deliberately exploit shadow [80]. Their work is interesting for considering the effects of the tower and the nacelle. Their conclusion is that at short distances between the radar and the wind turbine (less than 11 km) the target is not screened but at longer distances (>16 km) the tower may be able to screen the target provided the target is able to maintain a position behind the tower the tower with respect to the radar. Although Poupart [69] only reported 3 GHz results, the findings were similar. Poupart concluded, ‘dark shadows’ only exist for a few hundred metres behind a turbine and are only a few metres wide. At greater distances from the turbine blades, Poupart reported the reduction in power and time variation of the signal should not prevent detection of targets unless they were very small. An alternative way of expressing this is to say that the detection of targets should not be affected unless there is some other contributory factor, for example,

- If the aircraft detections were being made in the presence of strong ground or sea clutter in addition to shadow effects.
- If the ambient conditions were supplemented by other effects such as multipath propagation (recall the above signal strengths were referenced to free space losses).

These caveats are not intended to demur from the point that shadow is unlikely to result in loss of detection, unless detection is already being compromised by another factor.

Chapter 7 discusses some possible research work that would contribute to the wider understanding of the problem.

In many radar systems, aircraft are discriminated from clutter by processing the Doppler shift in their returns. Doppler shift can be treated as a rate of change in phase shift. The phase shifts introduced in the diffraction patterns will appear as phase noise on the Doppler shift. Close to the blades, this could present as a bias. Beyond approximately 2 km, the phase shifts are of equal maxima and minima eliminating any potential long-term bias.

4.20 Wider concerns

While this book was being researched, a wide variety of members of the civil and military aviation/wind farm community were asked for their opinions on key issues concerning the interactions between aviation radio and radar systems and wind turbines and wind farms. The following discussion is based on these consultations. The subjects raised were the views of the individuals based on many years of experience, they were not necessarily reflecting their companies and organisations.

4.20.1 Scope

There was a consensus that the problem had to be treated in an holistic fashion, it is not a primary radar problem or a secondary radar problem: all the systems have to work in the presence of wind turbines and wind farms:

- Radio locating systems: primary, secondary, AD, ASR, ARSR and PAR radars
- Communications, navigations and surveillance (CNS)
- What are the effects on sensor networks and networking, in particular the impact on the recognized air picture (RAP)

The scope of this book was deliberately limited to aviation radio and radar systems but there was a feeling that the problem was wider and further consideration was needed of point-to-point links, coastal and marine radars and spacewatch radars.

4.20.2 The greatest challenge

There was a consensus among all consulted that the greatest problem was to understand the interactions of a specific radar or radars and radio systems and a specific planned windfarm in advance of the wind farm construction. The understanding had to include the effects of terrain and the widest possible range of conditions that would be met.

Currently every aspect of this desire is challenging. Some more than others, for reasons set out earlier in this chapter, choosing a specific type of wind turbine is usually postponed.

4.20.3 Acceptable levels of confidence

If the radar is to provide mitigation of the effects of the proposed wind farm, then there has to be confidence that the mitigation would work to an acceptable degree. There was a great deal of discussion about what constituted acceptable. There was a consensus that this had to be scientifically based but also backed up by practical evidence. An interesting observation concerned the performance of radars 50 years ago, in the presence of ground clutter.

4.20.4 Precedent

These early radars were not capable of providing a guaranteed level of service in the presence of echoes from the ground and the built environment adjacent to the radar. As their technical maturity improved, radars became increasingly capable of operating in the most diverse of environments. Today, new radars can be delivered with an intrinsic capability to prevent clutter from the ground, or from local infrastructure, having an adverse effect on operations and there is confidence that that will be the case. It is accepted that all radars need to be commissioned into service but once that work is completed, there is confidence that they will be able to perform their role. The was a feeling that this was the sort of confidence needed for operations in the presence of wind turbines.

The reference to an historical precedent is interesting. There was no single technological advance that provided the ability to reject ground clutter: dual horn antennas, sensitivity time control and MTI/MTD all play a role.

4.20.5 Digital twins

The use of digital twinning technologies was thought to offer opportunities. However, it was acknowledged that this was complex. Simplifying the task is the extensive use of software in modern radars. The greatest challenges were seen to be modelling the environment and gaining an understanding of those factors that were critical at driving performance. An example might help explain the point, tracking an aircraft flying through (from the radar's perspective) wind turbine clutter is difficult. What are the factors that make it hard? Is it the inter-turbine spacing? Or, perhaps, it is the variability in the signature of the wind turbine in different levels of turbulence? Most likely it is a combination of both these factors and others besides. Knowing which are critical will be important for, designing trackers and developing test regimes that will provide confidence in advance of construction.

4.20.6 Common concerns

A concern expressed by a number of consultees was that it was generally accepted that radars had performance limitations but wind farm mitigation might be being held to a higher standard. The concern is better understood by considering an example.

Typically, a primary surveillance radar (PSR), the term is described in Chapter 3, must be capable of being able to detect an aircraft of a particular size, at a desired range, assuming that the aircraft is behaving in a particular way. It is not assumed that the radar must be perfect, a probability of detection is allowed. The aircraft size is measured as an area which is perceived by the radar and in typical specifications this is either 1 or 3 square metres (for this purpose, the value is not important, but they are consistent with a small propeller driven aircraft such as a Cessna or a small business jet like a Learjet) [71]. The aircraft should be flying towards the radar. The range of detection depends on the precise application but might typically be 100 km. The probability of being able to detect P_d the aircraft under these circumstances is typically either 80% or 90%. Allied to the probability of detection is the probability of a false alarm (P_{fa}). These figures always sound improbably small; one false alarm in a million possibilities. The reason for this small number is explained in Chapter 3 but, in practice, what this means is that on every scan of the radar there is likely to be one or two false alarms. This is an level of performance that has been acceptable for many years.

But what about detection of wind turbines? As the description of the aircraft detection criteria suggests, there are two obvious aspects to the problem; namely, will the presence of turbines reduce the probability of detecting genuine aircraft or will they increase the probability of false alarms. Is any degradation of these numbers acceptable? What if the false alarms are actually detections of wind turbines, might this be considered grounds for rejecting the radar? The more the problem is considered the more complex it becomes. If under some adverse weather conditions, there are more wind farms detections than during 'normal' conditions is such a degradation acceptable? The obvious answer is that the radar should be able to meet its detection criteria all the time. But in reality, in countries in mid-latitudes, such as much of the continental United States, most continental European nations and the United Kingdom, the atmosphere will cause anomalous propagation of radio waves for a number of days a year in normal years. Is there any reason why wind turbines should be treated specially?

References

[1] PMF IAS (2016), *Factors Affecting Wind Movement: Coriolis Force*, PMF IAS, https://www.pmfias.com/wind-movement-factors-affecting-winds-coriolis-force/#:~:text=Centripetal%20acceleration%20creates%20a%20force,of%20high%20and%20low%20pressure. Retrieved 3 April 2023.

[2] Wilson J.L. (2019), *How Does Wind Affect Weather?*, Sciencing, Leaf Group. https://sciencing.com/wind-affect-weather-6394667.html. Retrieved 3 April 2023.

[3] Met Office (1998), *Turbulence and Gusts*, UK Met Office, Forecasters Reference Book, Chapter 6. https://digital.nmla.metoffice.gov.uk/IO_05fb15c7-9b99-49c4-8dba-597bba31b561/. Retrieved 21 May 2023.

[4] FAA (2008), *Wind Shear*, Federal Aviation Agency Fact Sheets: FAA -P-8740-40, AFS-8. https://www.faasafety.gov/files/gslac/library/documents/

2011/aug/56407/faa%20p-8740-40%20windshear%5Bhi-res%5D%20branded.pdf. Retrieved 21 May 2023.

[5] AMS (2019), *Aerodynamic Roughness Length*, American Meteorological Society, 25 September 2019. https://glossary.ametsoc.org/wiki/Aero dynamic_roughness_length. Retrieved 3 April 2023.

[6] DWIA (2003), *Roughness and Shear*, Danish Wind Industry Association, 1 June 2003. http://xn–drmstrre-64ad.dk/wp-content/wind/miller/windpower%20web/en/tour/wres/shear.htm#:~:text=In%20gener.al%2C%2.0the%20more%20pronounced,the%20wind%20down%20a%20little. Retrieved 3 April 2023.

[7] WMO (2021), *Guide to Instruments and Methods of Observation: Volume I – Measurement of Meteorological Variables*, World Meteorological Organization, No. 8, 2021.

[8] INTERNATIONAL STANDARD IEC 61400-1, 3rd ed., August 2005.

[9] US Congress (2006), "The effect of windmill farms on military readiness 2006", A report to the Congressional Defense Committees by the US Department of Defense, 2006.

[10] Yáñez D.J. (2018), *VIV Resonant Wind Generators*, Villarreal, June 2018. https://vortexbladeless.com/. Retrieved 3 April 2023.

[11] EERE (2019), *Advanced Wind Turbine Drivetrain Trends and Opportunities*, US Office of Energy Efficiency and Renewable Energy, 3 July 2019.

[12] UOttawa (2023), *Wind Turbine Generators for Wind Power Plants*, IEEE PES Wind Plant Collector System Design Working Group. https://www.site.uottawa.ca/~rhabash/ELG4126WindGenerators.pdf. Retrieved 4 April 2023.

[13] See, for example, US NREL (2023), *Utility-Scale Wind Energy*. https://windexchange.energy.gov/markets/utility-scale#:~:text=Utility%2Dscale%20turbines%20are%20usually,Department's%20wind%20energy%20research%20portfolio. Retrieved 5 April 2023.

[14] MCS (2019), *Small Wind Turbines*, MCS Product Certification Scheme Requirements, Issue 2.2, Microgeneration Installation Standard (mcscertified.com). Retrieved 6 April 2023.

[15] Vattenfall (2021), *Hybrid Parks – New Opportunities with Multi-technology Facilities*, Vattenfall. https://group.vattenfall.com/press-and-media/newsroom/2021/hybrid-parks–new-opportunities-with-multi-technology-facilities. Retrieved 6 April 2023.

[16] ESO (2023), *What Is Frequency?* UK: National Grid Electricity System Operator, 2023. https://www.nationalgrideso.com/electricity-explained/how-do-we-balance-grid/what-frequency#:~:text=If%20the%20frequency%20is%20not,window%20either%20side%20of%2050Hz. Retrieved 3 April 2023.

[17] Miceli F. (2012), *Wind Farms Construction*. https://www.windfarmbop.com/wind-turbine-tower/#:~:text=They%20normally%20have%20a%20conical,from%2020%20to%2030%20meters. Retrieved 26 March 2023.

[18] DWIA (2003), *Wind Turbine Towers*, Danish Wind Industry Association, Factsheets 2003. http://xn–drmstrre-64ad.dk/wp-content/wind/miller/windpower%20web/en/tour/wtrb/tower.htm#:~:text=The%20towers%20are%20

conical%20(i.e.,materials%20at%20the%20same%20time. Retrieved 26 March 2023.

[19] Aho J., Buckspan A., Laks J., *et al.* (2012), *A Tutorial of Wind Turbine Control for Supporting Grid Frequency through Active Power Control*, US National Renewable Energy Laboratory.

[20] Anderson C.G. (2020), *Wind Turbines: Theory and Practice*, Cambridge University Press.

[21] IRENA (2021), *Offshore Renewables: An Action Agenda for Deployment*, Abu Dhabi: International Renewable Energy Agency.

[22] National Geographic (1996), *Continental Shelf*, Education Resource. https://education.nationalgeographic.org/resource/continental-shelf/. Retrieved 24 March 2023.

[23] IBERDROLA (2023), *Floating Offshore Wind Power: A Milestone to Boost Renewables Through Innovation*. https://www.iberdrola.com/innovation/floating-offshore-wind#:~:text=The%20wind%20turbine%20is%20not,metres%20around%20a%20central%20point. Retrieved 89 May 2023.

[24] OET (2023), *Beyond the Wow: The Six Types of Ship Motion*, Ocean Exploration Trust, https://nautiluslive.org/video/2020/12/09/beyond-wow-six-types-ship-motion. Retrieved 9 May 2023.

[25] Wang B., Li Y., Gao S., Shen K., Hu Z., and Zheng X. (2022), "Motion characteristics and aero-elastic responses of floating offshore wind turbine under coupling action of waves and winds", *Frontiers of Environmental Science*, 10:965334. doi: 10.3389/fenvs.2022.965334.

[26] VESTAS (2023), *Discover the Enventus Platform*, VESTAS Datasheet. https://www.vestas.com/en/products/enventus-platform/v162-6-2-mw. Retrieved 4 April 2023.

[27] Siemens (2023), *SG 7.0-170 On-shore Wind Turbine*, Siemens Gamesa Product data sheet. https://www.siemensgamesa.com/en-int/products-and-services/onshore. Retrieved 4 April 2023.

[28] AVDC (2013), *Guidance Note on Planning Applications for Wind Energy Developments*, Aylesbury Vale District Council, March 2013, https://buckinghamshire-gov-uk.s3.amazonaws.com/documents/Guidance-Note-on-Wind-Energy-Developments-March_2013_0.pdf. Retrieved 30 March 2023.

[29] Information courtesy of Mr Patrick Delaney, Vattenfall, Pen Y. Cymoedd, personal discussion with the author, May 2023.

[30] CAA (2017), "Lighting of Onshore Wind Turbine Generators in the United Kingdom with a maximum blade tip height at or in excess of 150m Above Ground Level", *Safety and Airspace Regulation Group Policy Statement*, 1 June 2017.

[31] Dehn and Sohn (2018), "Lightning and Surge Protection for Wind Turbines", White Paper. https://www.dehn-international.com/sites/default/files/media/files/lightning-surge-protection-wind-turbines-wpx023-e.pdf. Retrieved 24 March 2023.

[32] BOSLAN (2022), *The Next Generation Monopile Foundations for Off-Shore Wind Turbines: Manufacturing, Design and Handling Challenges*, Boslan

Engineering and Consulting. https://www.boslan.com/wp-content/uploads/2022/06/BOSLAN_monopile_foundations.pdf. Retrieved 14 April 2023.

[33] ES (2019), *Photosensitive Epilepsy*, Epilepsy Society Factsheet, September 2019. https://epilepsysociety.org.uk/about-epilepsy/epileptic-seizures/seizure-triggers/photosensitive-epilepsy?gclid=Cj0KCQjw27mhBhC9ARIsAIFsETFPben0IhZ1_8GN8bHQGHjqJINTu00J77eZ2MPd9LzJNakirCTpEiUaAmTNEALw_wcB. Retrieved 6 April 2023.

[34] Anderson C.G. (2020), *Wind Turbines: Theory and Practice*, Cambridge University Press.

[35] National Academies of Sciences, Engineering, and Medicine (2022), *Wind Turbine Generator Impacts to Marine Vessel Radar*, *Washington*, DC: The National Academies Press.

[36] IET (2022), IET Radar Conference 2022, Question and Answer session following a presentation by the Met Office.

[37] BGS (2023), *Eskdalemuir Magnetic Observatory*, British Geological Survey.

[38] JNCC (2023), *Special Areas of Conservation*. https://jncc.gov.uk/our-work/special-areas-of-conservation/. Retrieved 18 April 2023.

[39] ESRI (2023), *What Is GIS?* ESRI. https://www.esri.com/en-us/what-is-gis/overview. Retrieved 15 April 2023.

[40] DTU (2023), *Global Wind Atlas*, Technical University of Denmark. https://globalwindatlas.info/en. Retrieved 15 April 2023.

[41] NEWA (2023), *New European Wind Atlas*, NEWA. https://map.neweuropeanwindatlas.eu/. Retrieved 15 April 2023.

[42] Roberts B, (2019), *U.S. Wind Power Resource at 100-Meter Hub Height*, US National Renewable Energy Laboratory (NREL). https://windexchange.energy.gov/maps-data/324. Retrieved 15 April 2023.

[43] Drexl C. Hodge B.M., Clifton A., and McCaa J. (2015), *Overview and Meteorological Validation of the Wind Integration National Dataset Toolkit* (Technical Report, NREL/TP-5000-61740). Golden, CO: National Renewable Energy Laboratory. https://www.nrel.gov/gis/assets/images/wtk-100-north-america-50-nm-01.eps. Retrieved 15 April 2023.

[44] Crown (2023), *Fixed Wind KRA*, Crown Estate, 5 January 2023. https://opendata-thecrownestate.opendata.arcgis.com/. Retrieved 15 April 2023.

[45] UEIA (2023), *US Energy Information Administration Website*.

[46] Flanders (2019), *Marine Boundaries Geodatabase*, Flanders Marine Institute.

[47] CMS (2023), *Offshore Wind Law and Regulation in France*, CMS Law Expert Guide. https://cms.law/en/int/expert-guides/cms-expert-guide-to-offshore-wind-in-northern-europe/france. Retrieved 9 May 2023.

[48] Netherlands (2023), *Offshore Wind Energy*, Government of the Netherlands, 2023. https://www.government.nl/topics/renewable-energy/offshore-wind-energy. Retrieved 9 May 2023.

[49] Personal conversations between the author and a multinational wind farm developer, April 2023.

[50] OWIC (2020), *Grid-Access Technologies for GB Offshore Wind Industry*, Offshore Wind Industry Council (OWIC). https://www.hvdccentre.com/wp-content/uploads/2020/01/Grid-Access-Technologies_V3.pdf. Retrieved 22 May 2023.

[51] Sunak R. (2017), *Undersea Cables: Indispensable, Insecure*, Policy Exchange.

[52] UWE (2022), *National Planning Policy Limiting Creation of New Onshore Wind Farms in England*, Bristol: University of the West of England. https://info.uwe.ac.uk/news/uwenews/news.aspx?id=4220. Retrieved 16 April 2023.

[53] Planning (2023), *Approval (Discharge) of Conditions*, UK Department for Levelling Up, Housing and Communities. https://www.planningportal.co.uk/planning/planning-applications/consent-types/approval-discharge-of-conditions. Retrieved 17 April 2023.

[54] Planning (2019), *Use of Planning Conditions*, UK Department for Levelling Up, Housing and Communities. https://www.gov.uk/guidance/use-of-planning-conditions. Retrieved 31 January 2023.

[55] Planning (2023), *The Decision Making Process*, UK Department for Levelling Up, Housing and Communities with TerraQuest Solutions Ltd, 2023. https://www.planningportal.co.uk/planning/planning-applications/the-decision-making-process/if-permission-is-granted. Retrieved 31 January 2023.

[56] Anderson C.G. (2020), *Wind Turbines: Theory and Practice*, Cambridge University Press.

[57] CNBM (2018), *Wind Turbine Foundation: 5 Foundation Types Explained For Onshore Wind Turbine*, China National Building Materials, 9 May 2018. http://www.steelwindtower.com/wind-turbine-foundation-5-foundation-types-explained-for-onshore-wind-turbine/. Retrieved 18 April 2023.

[58] Chanel P. (2008), *Overview of Electricity Distribution in Europe*, Summary from Capgemini's 2008 European Benchmarking Survey, 2008. https://www.capgemini.com/wp-content/uploads/2017/07/tl_Overview_of_Electricity_Distribution_in_Europe.pdf. Retrieved 25 March 2023.

[59] Roberts D. (2020), *A National US Power Grid Would Make Electricity Cheaper and Cleaner*. https://www.vox.com/energy-and-environment/2020/6/20/21293952/renewable-energy-power-national-grid-transmission-microgrids. Retrieved 25 March 2023.

[60] National Grid (2017), *The Grid Code*, Issue 5, Revision 21. https://www.nationalgrid.com/sites/default/files/documents/8589935310-Complete%20Grid%20Code.pdf. Retrieved 5 May 2023.

[61] Jay F. (ed.) (1984), *IEEE Standard Dictionary of Electrical and Electronic Terms, ANSI/IEEE Std 100-1984*, 3rd ed., New York, NY: IEEE Press, 1984 and Wiley Interscience.

[62] Skolnik M.I. (2001), *Introduction to Radar Systems*, 3rd ed., McGraw Hill.

[63] Shaeffer J.F. (2010), "Target reflectivity", in Richards M.A., Scheer J.A., and Holm W.A. (eds.), *Principles of Modern Radar: Basic Principles*, Georgia Institute of Technology, Scitech Publishing, an imprint of the IET.

[64] Karabayr O. (2017), *Wind Turbine Effects on Radar Performance*, Lambert Academic Publishing.
[65] Fairley P. (2009), *Stealth-Mode Wind Turbines*, MIT Technology Review.
[66] Norman E.D. (2010), *Assessment of the Wind Farm Impact on the Radar*, Final Year Project Report, École Nationale Supérieure des Ingénieurs des Etudes et Techniques d'Armement (ENSIETA), Brest, France.
[67] Chambers B. and Tenant A. (2005), "A smart radar absorber based on the phase-switched screen", *IEEE Transactions on Antennas and Propagation*, 53(1):394–403.
[68] Knott E. (1988), "Radar cross section", in Brookner E. (ed.), *Aspects of Modern Radar*, Artech House (Chapter 8).
[69] Poupart G. (2003), *Wind Farms Impact on Radar Aviation Interests – Final Report*, FES W/14/00614/00/REP DTI PUB URN 03/1294. Crown Copyright.
[70] Randhawa B.S. and Rudd R. (2009), *RF Measurement Assessment of Potential Wind Farm Interference to Fixed Links and Scanning Telemetry Devices*, 2008-0568 (Issue 3).
[71] Nathanson F. (1999), *Radar Design Principles: Signal Processing and the Environment*, SciTech.
[72] Wolff C. (2023), *Dynamic Range of a Receiver*, https://www.radartutorial.eu/html/author.en.html. Retrieved June 2023.
[73] Kleinrock L. and Gail R. (1996), *Queueing Systems: Problems and Solutions*, Wiley Interscience.
[74] AWC (2005), *The Effects of Wind Turbine Farms on ATC Radar*, Report AWC/WAD/72/665/TRIALS.
[75] Brookner E. (1998), *Tracking and Kalman Filtering Made Easy*, John Wiley.
[76] CAP 764 (2016), *CAP Policy and Guidelines on Wind Turbines*, 6th ed., UK: Civil Aviation Authority, Safety and Airspace Regulation Group, CAP 764.
[77] EUROCONTROL (2014), *How to Assess the Potential Impact of Wind Turbines Surveillance Sensors*. Edition: 1.2 Edition.
[78] Stevens M. (1988), *Secondary Surveillance Radar*, Artech House.
[79] Vinagre L. and Woodbridge K. (1998), "Modelling and prediction of obstacle shadowing on the secondary surveillance radar target", in *Radar Systems Modelling*, IEE Colloquium.
[80] Theil A. and van Ewijk L.J. (2007), *Radar Performance Degradation due to the Presence of Wind Turbines*, TNO Netherlands, IEEE Radar Conference 2007.
[81] CAA (2019) *Operation of IFF/SSR Interrogators in the UK: Planning Principles and Procedures*, *MOD/CAA*, January 2019.
[82] ITU (2019), *Calculation of Free Space Attenuation*, *ITU-R PN 525-2, Annex* 2.
[83] Sakian A. (2011), *Radio Wave Propagation Fundamentals*, Artech House.
[84] CAA (2019), *Air Traffic Services Safety Requirements*, CAA, CAP 670.
[85] Nahvi M. and Edminster J.A. (2019), *Electromagnetics*, McGraw Hill.

[86] Ben-Hassine (2020), "Multipath and receiver models for assessing the VOR bearing error: application to wind farms". Signal and Image Processing. Université Paul Sabatier – Toulouse III, 2020.

[87] CAP 764 (2016), *CAP Policy and Guidelines on Wind Turbines*, 6th ed., Civil Aviation Authority, Safety and Airspace Regulation Group, CAP 764.

[88] Butler M. and Johnson J. (2003), *Feasibility of Mitigating the Effects of Windfarms on Primary Radar*, Alenia Marconi Systems, ETSU W/14/00623/REP, DTI PUB URN No. 03/976, Crown Copyright.

[89] Bacon (2003), in Barclay L. (ed.), *Propagation of Radiowaves*, 2nd ed., Institution of Engineering and Technology.

[90] Haslett C.J. (2008), *Essentials of Radiowave Propagation*, Cambridge University Press.

[91] ITU (2019), *Propagation by Diffraction*, ITU-R-P.526-15.

[92] Oswald G. and Baker C. (2021), *Holographic Staring Radar*, IET Scitech.

Chapter 5
Analysis

5.1 Introduction

This chapter introduces some techniques that are of value in analysing the effects of wind turbines on aviation radio and radar systems. The following topics are considered:

- Conversion of useful units
- Radar frequencies
- Radar performance
- Near-field/far-field boundary calculations
- Propagation
- Mapping

5.2 Conversion of useful units

5.2.1 Nautical miles and kilometres

It is custom and practice to use the Nautical mile (Nm) as a distance unit for air navigation. A Nm is one minute of latitude. The exact conversion between Nm and km is:

1 Nm = 1.852 km
1 km = 0.54 Nm

For anyone who favours using mental arithmetic to estimate answers, converting Nautical miles to kilometres is a simple process. If the value in Nautical miles is doubled and then 10% subtracted, the result is within 2.6% of the correct value. For example, 12 Nm

The exact conversion = 22.224 km
Double 12 Nm = 24 km
Ten percent of 12 Nm = 1.2 km
Approximate answer = 22.8 km

5.2.2 Decibels

5.2.2.1 Logarithmic values

The concept of logarithms was invented by the Scottish landowner John Napier (1550–1617)* and was first published in 1617. The logarithm (log) of a number is the power to which a base number must be raised to obtain that number. For example, two raised to the power three is eight ($2 \times 2 \times 2 = 8$). Expressing this as a log value, the log of 8 to the base 2 is 3 and this would be written:

$$3 = \log_2(8)$$

Different types of logarithms have different base numbers. Radar performance analysis using decibels uses base 10. For example, the log of eight to the base ten is 0.9:

$$0.9 = \log_{10}(8)$$

The base of 10 is by far the commonest in use and it is usual to omit the 10 when the log is written down. If the term log is seen, it may be assumed that it is the log to the base 10.

To multiply numbers together, the logs of the numbers are added. To divide numbers, the logs of the numbers are subtracted. The real value of using logarithms is when the numbers being multiplied and divided are either very big or very small or a combination of the two. An example is provided at the end of this section after decibels have been explained.

5.2.2.2 The decibel

As well as having different bases, there are different types of logarithms. The type that is of interest here is the decibel† (dB). A decibel is a measure of how much bigger (or smaller) a value is compared with a reference value. The conversion of a number to its value in dB is as follows:

$$\text{Value in dB} = 10 \times \log_{10}(\text{Value}/\text{Reference value})$$

The use of the factor of 10 in the calculation is because the measure is in tenths of a Bel.

An example helps; the radar cross-sections of objects are expressed with respect to a one square metre object. Thus, a 2 square metre target:

$$\begin{aligned}\text{Value in dB} &= 10 \times \log(2/1)\text{dB} \\ &= 10 \times \log(2)\text{dB} \\ &= 10 \times 0.3\text{dB} \\ &= 3\text{dB}\end{aligned}$$

*Napier is often described as a mathematician. Napier was the 8th Laird of Merchiston. Mathematics was Napier's hobby but despite this he made significant advances in spherical trigonometry as well as logarithms.

†The first use of decibels was promoted by Bell Laboratories for measuring losses in telephone and telegraph circuits. The name decibel was adopted in 1928 for Alexander Graham Bell for his work on telephony.

5.2.2.3 Decibel types

It is common practice to add a qualifier to dB to indicate the nature of the value. The previous example happened to be a radar cross section which was referenced to one square metre. To identify this fact the unit is written **dBsm**. Thus, a 3-dBsm object has a radar cross-section corresponding to an area of two square-metres.

Other quantities expressed logarithmically also use qualifiers. Log units of signal power are measured with respect to one Watt and the unit is written as **dBW**. But for most received signal powers 1 W is a very large unit and a more practical unit is to refer the power to 1 mW and the log value unit is written as **dBm**.

The focussing power of an antenna, its gain/directivity, is most often expressed with respect to an antenna that does not focus signals, that is the idealised, non-directive, isotropic antenna. Gains using this reference would be written as **dBi**. However, care is required when referring to antenna gain since the idealised isotropic gain is a theoretical concept that cannot be implemented in practice. An alternative reference that can be realised in practice is the very commonly used half-wave dipole antenna, see Figure 5.1. The qualifier for antenna gains expressed with respect to a dipole is **dBd**. The dipole has gain; the radiation pattern is a toroid shape with the dipole threading the toroid. That means the maximum transmit and receive energy is perpendicular to the alignment of the two components and the minimum energy is in line with the antenna components. The dipole's gain can be expressed in dBi as well as dB_d, it is equal to 2.15 dBi. That is 0 dB_d = 2.15 dB_i. Therefore, it will be clear that it is important to be aware of whether a gain is referenced to a dipole or an isotropic antenna.

5.2.2.4 Voltage or power

A source for confusion is how to convert values to decibels if the quantity being converted is a voltage or a fraction of a volt (V) such as a millivolt (mV) or microvolt (μV). For consistency with power measurements in decibels, voltage (and current) measures must be squared because the power flowing in a circuit is

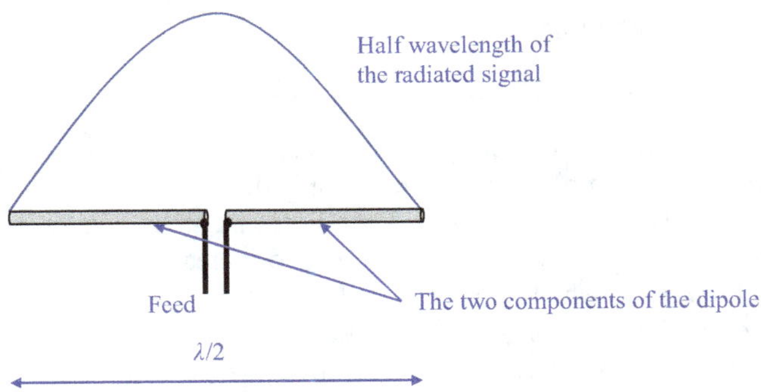

Figure 5.1 Half-wave dipole

proportional to the square of either current or voltage:

$$10 \times \log_{10}(P/P_{\text{ref}}) = 10 \times \log_{10}(V/V_{\text{ref}})^2$$
$$= 20 \times \log_{10}(V/V_{\text{ref}})$$

5.2.2.5 The benefit of using logarithmic values – an example

An example will explain the benefit of using logarithmic-based units (decibels). Later in this chapter, the radar equation is explained. To determine the distance at which an aircraft can be detected part of the calculation requires the radar power, which might have a numerical value of 50,000, to be multiplied by the focussing power of the radar antenna squared. When squared this numerical value which might typically be in excess of 6,000,000 is divided by a constant which has a numerical value of 0.0000000000000000000000138. Using logarithmic values this reduces to, 47+68−(−229). This calculation yields an answer in dB and this must be converted back to a linear number:

$$\text{Linear number} = 10^{(\text{dB value}/10)}$$

The solution to the foregoing example, which is only part of the calculation of the distance at which an object is visible, is 23 followed by 33 zeroes (a large number!).

5.3 Radar frequencies

5.3.1 Radio spectrum

Radio frequencies are the lowest frequency form of electromagnetic waves. Their span is from any positive non-zero frequency to the generally accepted maximum frequency of 3,000 GHz. An example definition is provided by the ITU, viz, a wave of arbitrary frequency below 3,000 GHz [1]. Higher frequency electromagnetic waves are considered to be infra-red, that is, heat.

There is no standard agreement on the lowest *usable* radio frequency but it is in low kilohertz. The UK Communications regulator (Ofcom) and the US Federal Communications Commission (FCC) have a lowest allocated band of 8.3–9 kHz (for meteorological aids) [2,3].

5.3.2 Radar frequency selection factors

The choice of radio frequencies for radar applications is driven by two competing considerations with a number of constraints applied.

Specific attenuation: The first consideration is that as frequency increases; the attenuation of radio waves by the atmosphere also increases. This phenomenon is illustrated in Figure 5.2 which is reproduced from ITU-R P.676-11. The ordinate (y-axis) of the graph is the [one-way] specific attenuation in dB/km. The specific attenuation is the attenuation caused by hydrometeors (rain, snow sleet, etc.). The abscissa (x-axis) is the frequency in GHz. The values plotted are all for a zenith ray,

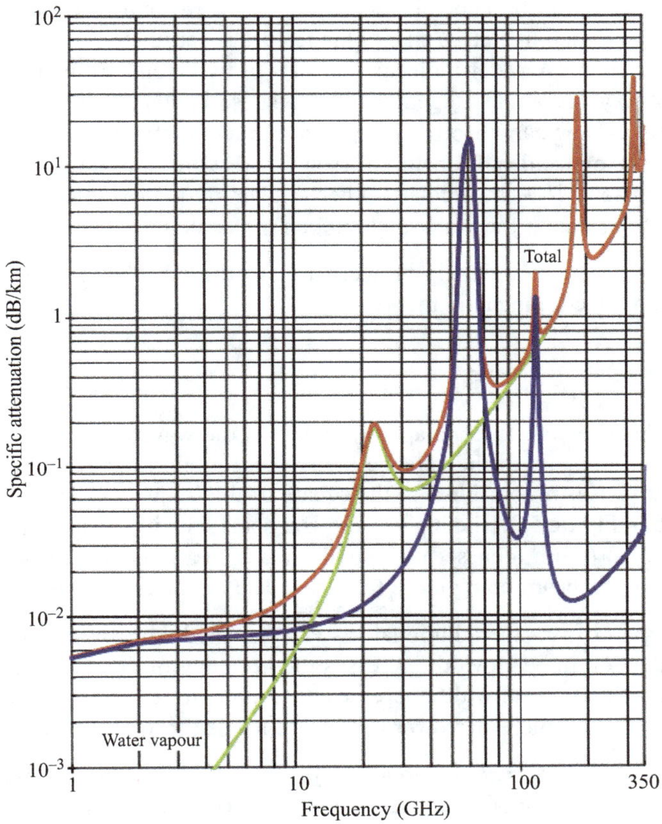

Figure 5.2 Specific attenuation from ITU-R P.676-11

that is a ray directed vertically up through the atmosphere. The blue plot is the attenuation for a ray in a dry atmosphere. The green plot is the attenuation due to water vapour and the red plot is the sum of these two effects, that is the total [one-way] attenuation.

A number of features can be seen in Figure 5.2:

- Specific attenuation increases with frequency.
  - The gradient is modest between 1 GHz and 10 GHz.
  - Above 10 GHz, the gradient is much greater (by approximately an order of magnitude).
- The above features are driven by the effects of water vapour which can be seen starting around 3 GHz.

- Superimposed on the general trends, are strong **resonance absorption lines**. There is one such line at 22 GHz. This feature is caused by the water vapour in the atmosphere. A water molecule consists of two atoms of hydrogen and one atom of oxygen bonded together (H_2O). These bonds behave as though they are elastic. They can absorb energy at a frequency of approximately 22 GHz, making the molecules resonate. Lines at 60 GHz are caused by a similar effect occurring to the O_2 molecules. Other, higher frequency lines are caused by other resonances in the water and oxygen molecules.

Directivity: The relationship between the aperture and gain of an antenna is available from electromagnetic theory:

$$\text{Antenna gain, } G = \frac{4\pi A}{\lambda^2}$$

where A is the effective antenna aperture and λ is the wavelength.

Therefore, the gain of the antenna is inversely proportional to the wavelength squared, that is, it is proportional to the square of the frequency. This equation has been plotted in Figure 5.3. The higher gain is associated with the antenna becoming more directional (more focussed) which also leads to a reduction in beamwidth; a direct result of the beam being more focussed. The corollaries of this situation are:

- Higher gain offsets the additional losses associated with specific attenuation.
- The reduction in beamwidth increases the accuracy of the measurements.
- The reduction in beamwidth increases the time required to surveil a volume of airspace, that is, more narrower beams are needed to fill a given volume.

Other general observations about frequency selection: The lowest radar frequency considered in the above discussion is 1 GHz. In the more general case, in

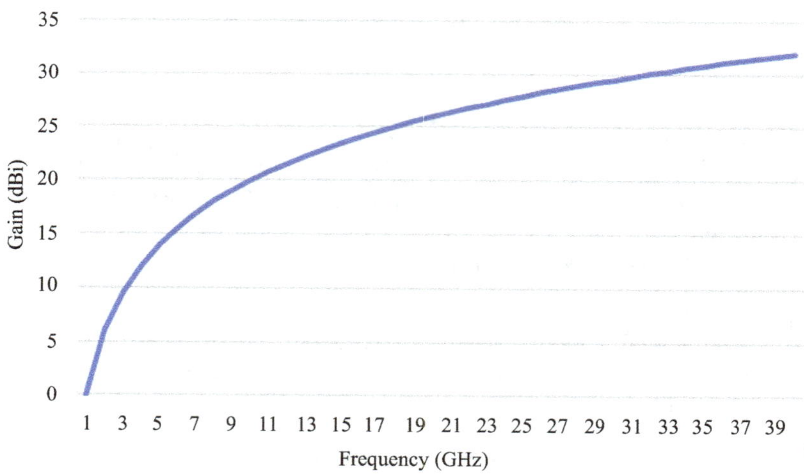

Figure 5.3 Antenna gain versus frequency (normalised to 1 GHz)

which, lower frequencies are considered, other factors should be taken into account. Some radar frequencies are more propitious than others for detecting particular sizes of the object, that is when the objects to be detected are comparable in size to the wavelength of the radar. It is important for aviation radars to be able to detect Doppler signatures coming from targets (to isolate them from a static background). At very low frequencies, extracting the Doppler is problematic. Neither of these constraints is important in this case because aircraft are much larger than the wavelengths of interest and the frequencies are high enough so Doppler extraction is not problematic.

Summary: If lower propagation loss, long maximum range, and wider beam-widths to favour surveillance are useful characteristics, a lower frequency radar is indicated. If greater accuracy is a requirement and a lower maximum range is less important, then a higher frequency radar is indicated.

Applying these factors to the radars of interest here, long-range surveillance radars used for either civil or defence applications tend to operate at L-Band. However, precision approach radars (PAR) or ground-controlled approach radars and airfield surface detection equipment radars, both of which require accurate tracking but where long range is less of a concern can operate at X-Band (8–10 GHz) and Ka Band, respectively. ASR falls between these extremes and generally operates at the S-Band. This terminology is explained in Table 5.1.

Table 5.1 Radar frequency bands

Radar band				**Application**
IEEE		**NATO**	**WRC/ITU**	
Terminology	**Frequency limits (GHz)**	**Terminology**	**Frequency allocation (source UK FAT)**	
L-Band (long wave)	1–2	D-Band	1.215–1.4 1.35–1.4 shared in region 1	Long-range surveillance ARSR/AD
S-Band (short wave)	2–4	E/F Band	2.7–2.9 Civil/military 2.9–3.1 Marine 2.9–3.4 Military	Surveillance ASR
X-Band (fire control)	8–12	I Band	8.5–10.5 10.5–10.68 Secondary use 10.5–10.55 Primary in Region 2/3	Precision tracking PAR/GCA
Ka-Band (Kurz-Above)	27–40	K-Band	31.8–36	Short range surface movement ASDE

5.4 Radar performance

5.4.1 A model of radar performance

Understanding how far away objects of a particular RCS can be seen is a key performance parameter (KPP) for the designer. It is also critical to the safeguarding of radars from structures such as wind turbines. Two methods of calculating radar performance will be described: the standard method, usually referred to as the Radar Equation, and the Blake method. The rationale for having two methods will also be described.

5.4.2 The radar equation

The radar equation is derived as follows [4].

Starts with the concept that the transmitter radiates power equally in all directions, that is, assuming the antenna is an isotropic‡ radiator.

The power source illuminates the sphere of space around it and at a distance R from the source, the power of the transmitter is shared equally over the whole of the surface area of a sphere:

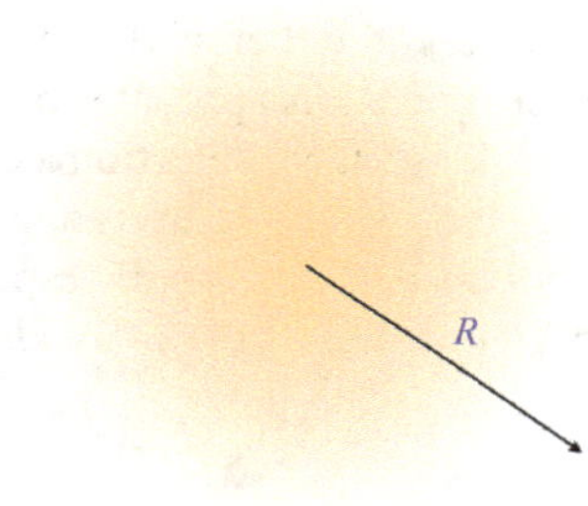

$$\text{Area of a sphere} = 4\pi R^2 \tag{5.1}$$

If the transmitter has a power P_t, then the power density, that is, the power per unit area, at any point on the surface of the sphere is:

$$\text{Power density} = \frac{P_t}{4\pi R^2} \tag{5.2}$$

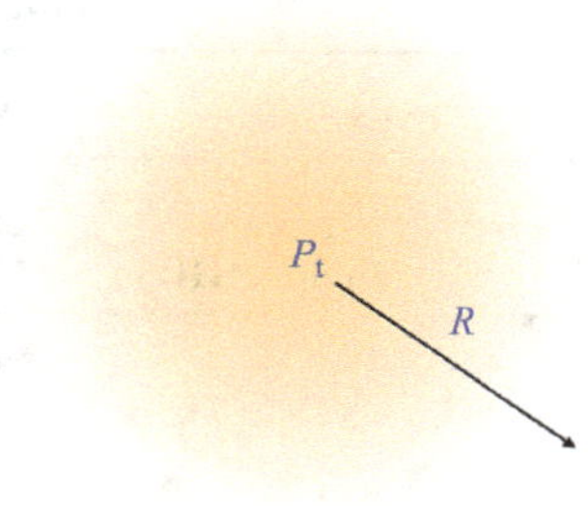

In electromagnetic theory, this equation is called the **power equation** [5].

In general, the radar will have an antenna that focuses its power on a beam§. The focusing ability comes from the antenna having a gain, G_t.

A good approximation to the antenna gain is the ratio of the surface area of the sphere over the area into which the power is

‡From the Greek, equal in all places.

§This may not be the case in the future. Radars are being developed that do not create a beam.

being focused:

$$\text{Gain} \approx \frac{4\pi}{(\delta\theta\delta\phi)} \tag{5.3}$$

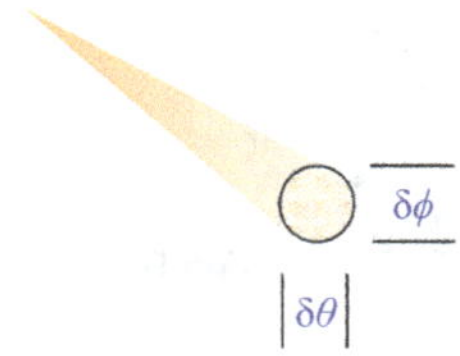

Taking the gain into account, at the point at which the antenna is pointing, the power density becomes:

$$\text{Power density} = \frac{G_t P_t}{4\pi R} \tag{5.4}$$

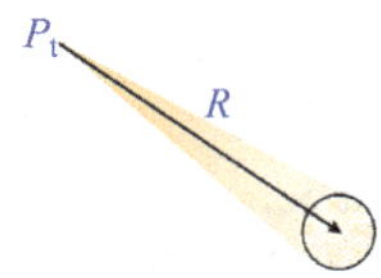

The power at the point at which the antenna is pointing is illuminating a target (an aircraft) which has a radar cross-section σ. Thus, the power available to be reradiated is:

$$\text{Power to be radiated} = \frac{G_t P_t \sigma}{4\pi R^2} \tag{5.5}$$

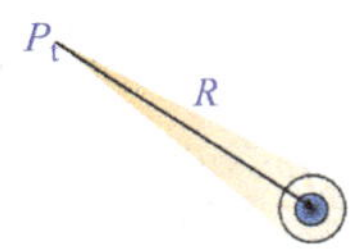

This power is radiated out over the sphere of space surrounding the target and, at a distance R from the target (i.e., back at the radar antenna), the surface area of the sphere is = 4 π R^2 and, therefore, the power density at the radar antenna is:

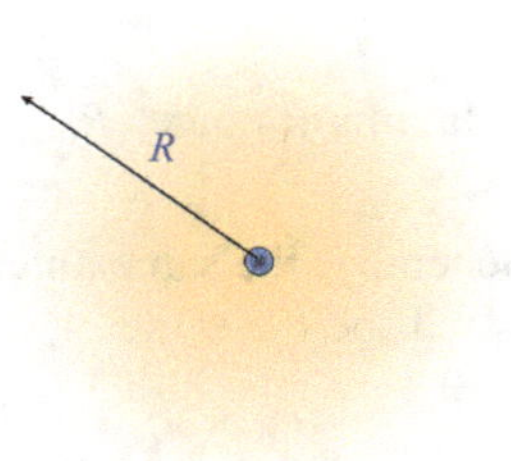

$$\text{Power density} = \frac{G_t P_t \sigma}{(4\pi R^2)(4\pi R^2)} \tag{5.6}$$

The power available from the antenna at the input to the radar receiver is a function of the power density created by the target radar cross section (RCS) and the area (called the aperture) of the antenna, A. Therefore:

$$\text{Signal power available} = \frac{G_t P_t \sigma A}{(4\pi R^2)(4\pi R^2)} \tag{5.7}$$

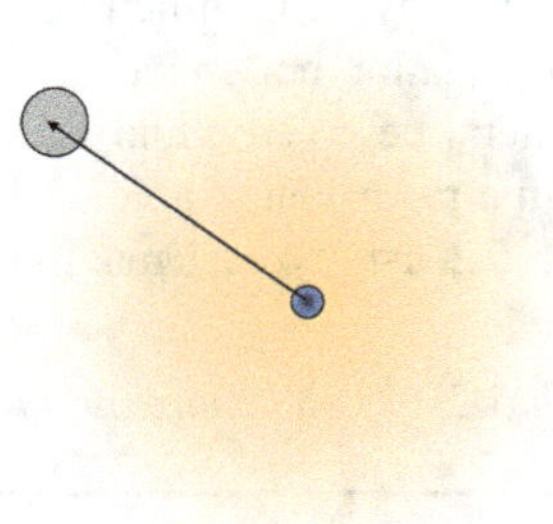

Referring to earlier in this chapter, the relationship between the aperture and the gain of an antenna is available from electromagnetic theory:

$$\text{Antenna gain } G_r = \frac{4\pi A}{\lambda^2} \tag{5.8}$$

where A is the effective aperture of the antenna; λ is the wavelength of the frequency of interest.

Rearranging this equation:

$$\text{Effective aperture, } A = \frac{G_r \lambda^2}{4\pi} \tag{5.9}$$

Substituting this relationship into the equation for the signal power available to the receiver:

$$\text{Power available, } S = \frac{G_r G_t P_t \sigma \lambda^2}{(4\pi)^3 R^4} \tag{5.10}$$

This equation can be rearranged again to make any of the variables the subject of the equation. A common use is to determine the range of the target the subject, thus:

$$\text{Range, } R = \frac{(G_r G_t P_t \sigma \lambda^2)^{1/4}}{(4\pi)^3 S} \tag{5.11}$$

There is one value of R that is of particular importance to the designer. If the signal available to the receiver is the lowest possible to detect an object (the minimum detectable signal (MDS), S_{min}) then that also corresponds to the maximum range at which an object with RCS, σ, can be detected by the radar. This leads to the following equation:

$$\text{Maximum range, } R_{max} = \frac{(G_r G_t P_t \sigma \lambda^2)^{1/4}}{(4\pi)^3 S_{min}} \tag{5.12}$$

Generally, the same antenna will be used for both transmit and receive and G_r and G_t will be the same:

$$\text{Maximum Range, } R_{max} = \frac{(G^2 P_t \sigma \lambda^2)^{1/4}}{(4\pi)^3 S_{min}} \tag{5.13}$$

This equation is **the radar equation**. Because it can be rearranged in many different ways it is sometimes referred to as a family of equations. The only variable in the equation not under the control of the designer is the RCS, σ. The designer will in general be given σ and R_{max} as design goals. That is, he or she will be required to design a radar that is able to detect an object of RCS, σ, at a maximum range, R_{max}. Table 5.2 lists typical maximum range values for different types of radar. Another

Table 5.2 Typical maximum radar ranges

Radar type	Typical maximum range
Surface movement radar (ASDE)	2 Nm (4 km)
Precision approach radar (PAR)	20 Nm (40 km)
Airfield surveillance radar (ASR)	60 Nm (110 km)
Air route surveillance radar (ARSR)	250 Nm (460 km)
Air Defence Radar (AD)	250 Nm (470 km)

term that is often used is **maximum instrumented range**, indicating that the radar is set up to detect objects out to the specified range. However, this may be beyond the range at which it will be able to detect the specified target RCS.

The radar designer will then have to decide on appropriate levels of power, antenna gain and frequency to meet the design goal.

5.4.2.1 Discussion of the basic radar equation

The radar equation illustrates some interesting properties of radars. If it is necessary to double the detection range of a radar, all the other variables being held constant, then the power would have to increase by a factor of 16. The range capability of the radar could be increased by making the receiver more sensitive (i.e., to decrease S_{min}) but this makes the radar more susceptible to interference or deliberate jamming by an adversary. Increasing the antenna size would increase its gain and, therefore, the range at which a particular RCS target could be detected. However, increasing the antenna size would also decrease the antenna beamwidth which might mean the antenna would have to slow down its rotation speed to maintain the same consistency of coverage. Furthermore, the antenna size may be constrained by other factors such as the tolerance to withstand high winds.

However, the equation derived above is optimistic; it overpredicts the range at which a target of a given RCS can be detected by the radar. Its two principal shortcomings are that it takes no account of either noise or losses. The first topic is noise and the method described allows noise to be accounted for in the radar equation.

5.4.2.2 Accounting for noise

Noise is present in all electronic systems, even the best-designed and manufactured radar receivers add a measure of noise. An important figure of merit, called the **Noise Factor**, can be defined for a real receiver design. Noise factor compares the noise performance of a real receiver with a conceptual, idealised receiver, that adds no noise of its own.

Hence, the noise factor is defined as follows:

$$\text{Noise factor, } F_n = \frac{\text{The noise out of a real receiver}}{\text{The noise out of an ideal receiver}} = \frac{N_O}{N_i} \tag{5.14}$$

If an ideal receiver could be created, then it would have a noise factor of 1 and any real receiver would be greater than 1. The noise factor is measured as a linear quantity; a related term the **noise figure** (NF), is the noise factor presented in decibels. That is:

$$\text{Noise Figure, NF} = 10\log10(F_N) \tag{5.15}$$

To develop the expression of the noise factor, the first step is to calculate the noise out of an ideal receiver and initially it will be assumed that this is accounted for by **Thermal Noise** first described by Bert Johnson (1887–1970) and Harry Nyquist (1889–1976) in 1928 [6].

5.4.2.3 Thermal noise

Thermal noise, which some authorities refer to as Johnson Nyquist noise, is caused by the thermal motion of electrons; the higher the temperature, the greater the agitation of the electrons and the higher the noise power. The relationship between energy and temperature, thermodynamics, was proposed by Ludwig Boltzmann (1844–1906) in 1877 but the fact that it is a linear relationship was discovered by Max Planck (1858–1914) and Planck named the gradient of that linearity after Boltzmann, that is Boltzmann's Constant = $1.3806485279 \times 10^{-23}$ Joules per degree Kelvin and given the symbol k. Thus, the noise power is calculated:

Thermal noise power per Hertz

$$= k\,T\,W(\mathrm{W/Hz}) \tag{5.16}$$

where k is Boltzmann's constant, 1.38×10^{-23} J/deg and T is the temperature in degrees Kelvin.

The ambient temperature, T_0, is usually taken to be 290°K. This figure leads to one of the most important relationships in radar:

Thermal noise power per Hertz

$$= -204\ \mathrm{dB_W Hz^{-1}} \tag{5.17}$$

Watts are usually too large a unit for noise (or received signal) power and it is commonplace to express power in milliwatts. Hence:

Thermal noise power per Hertz

$$= -174\ \mathrm{dBm\ Hz^{-1}} \tag{5.18}$$

Thermal noise power is linear with bandwidth; if the bandwidth of a circuit is doubled, the noise power is also doubled. But thermal noise power is independent of frequency. i.e., for any given bandwidth device, the noise power would be the same irrespective of the frequency at which the device was operating. An example may help to understand this concept; if a receiver circuit has a bandwidth of 50 kHz (typical of some of the older ASR radars still in service at the time of writing) then this corresponds to a thermal noise power of −127 dBm, it does not matter whether the radar operates in L-Band, S-Band or X-Band, the power is still −127 dBm. This property of the noise power being constant at all frequencies is referred to as **white noise**. Furthermore, as the noise power is determined solely by temperature, provided the temperature of the system remains constant, the noise power and its statistical properties, mean, variance, etc. also remain constant. In the statistical terminology, it is said to be **stationary**.

5.4.2.4 Bounding where noise must be accounted for

It was pointed out earlier that noise affects all electronic circuits and that signals are measured with respect to noise. However, in all modern radar systems, after the received signal has been conditioned (amplified and filtered), there is a point when the analogue signal is sampled and digitised. At that point, the signal can no longer be degraded by noise. Therefore, this provides the bounds of the description of a

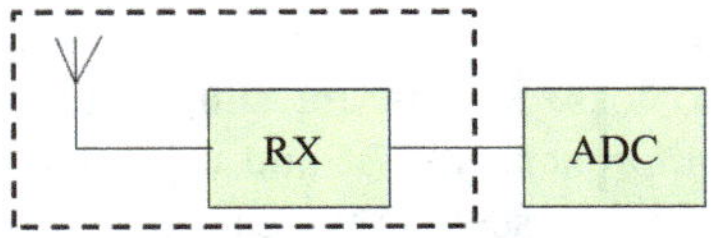

Figure 5.4 Receiver boundary

receiver; that is, it is the circuitry within the radar which processes the received signals up to the point where the signal is digitised as illustrated in Figure 5.4.

5.4.3 Incorporating noise factor

The noise out of an ideal receiver is the gain of the receiver multiplied by the noise going into the receiver:

From (5.16) above, the noise out of an ideal receiver

$$= G_{RX} k\, B\, T \tag{5.19}$$

where G_{RX} is the gain of the receiver; K is Boltzmann's constant 1.38×10^{-23} J/deg; B is the bandwidth in Hertz; T is the temperature in degrees Kelvin.

Including this relationship in (5.14) above,

$$N_F = \frac{N_O}{G_{RX} k\, B\, T} \tag{5.20}$$

The receiver gain is the ratio of the signal out (S_O) and the signal in (S_I), substituting this into the expression for Noise Factor, (5.18):

$$N_F = \frac{N_O}{k\, B\, T\, S_O/S_I} \tag{5.21}$$

Equation (5.19) can now be rearranged to make the Signal going into the receiver the subject:

$$S_I = F_n k\, B\, t\, S_o/N_o \tag{5.22}$$

If the input signal level is the minimum that the radar can detect, then (5.20) becomes:

$$S_{mIn} = F_n k\, B\, t\, S_o/N_o \tag{5.23}$$

This expression is intuitively correct because its signal is expressed in terms of a signal-to-noise ratio. The expression can now be substituted into the radar equation (5.13):

$$\text{Maximum range, } R_{max} = \frac{(G^2 P_t \sigma \lambda^2)^{1/4}}{(4\pi)^3 F_n k\, B\, t\, S_o/N_o} \tag{5.24}$$

Equation (5.22) accounts for thermal (Johnson Nyquist) noise. The next step is to add the effects of losses.

5.4.3.1 Losses

All systems are subject to loss. Radars are subject to a wide variety of losses; Skolnik [7] provides a comprehensive list and typical values. The following discussion identifies the principal losses of a typical radar. A simplified block diagram of a typical radar is shown in Figure 5.5.

5.4.3.2 The transmit path

The transmitter provides the high-power waveforms to be transmitted. In the radars of interest here, the antenna is usually shared by both the transmitter and the receiver. A device, called a **circulator** or a **duplexer**, directs the high-powered signals from the transmitter to the antenna for transmission and the low-powered echo signals received by the antenna are directed to the receiver. In addition, the circulator performs another important role of isolating the sensitive receiver from the high-power signals being transmitted. Circulators typically insert losses of between 0.5 dB and 1 dB.

Transmit signals, leaving the circulator, are passed to the antenna via a rotary joint which will use either a rotary waveguide or slip rings. Rotary joints typically add a further 1 dB of loss.

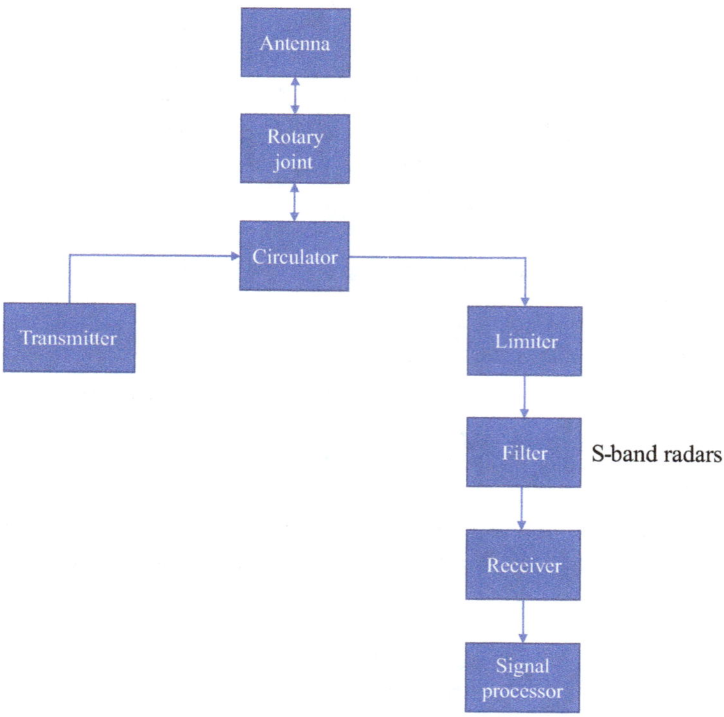

Figure 5.5 Simplified radar block diagram

In addition, the paths from the transmitter to the antenna will use waveguides or a mix of waveguides and coaxial cable. The losses of feeders are usually expressed in terms of dB per metre. Each connection adds another small contact loss.

Taking all these factors into account, losses in the transmit path might be at least 2 dB. To illustrate the effect of this loss: if a transmitter generated a pulse of 1 MW peak power (consistent with an older radar with a vacuum device transmitter such as a magnetron), only 630 kW would be available at the antenna. If the transmitter was a more modern, solid-state system that generated a 20-kW pulse then less than 13 kW would be available at the antenna to transmit.

5.4.3.3 Antenna pattern loss

In the derivation of the radar equation, it was assumed that the antenna was pointing directly at the aircraft target but it is more likely that the target is not on the beam centre. Recall also, that the beam is three dimensional and a target may be off beam-centre in both the azimuth and elevation planes. These losses are referred to as **beam shaping loss** or, in older literature, **pattern loss**. The loss accounts for the reduction in the signal returned to the radar from a target located off the beam centre compared with a target on the beam centre. Therefore, the loss does not need to be accounted for twice in both the transmit and receive paths.

Blake reports values for beam shaping loss from a number of researchers; ranging from 1.5 dB to 3.2 dB. Blake also points out that, in many radars, a number of pulses are integrated to form a target report and this tends to reduce beam shaping loss. Taking these factors into account, 1.6 dB is a good representative value for beam shaping loss [8].

Whether this loss should be included in the radar equation depends on the purpose of carrying out the calculation. If the intention is to determine whether a radar might be able to detect an aircraft then it would be appropriate to include this loss. However, if the intention is to safeguard the radar, the view might be taken that the worst-case situation would be if, say, a wind turbine was detected at the peak of the radar beam. In that case, it would be more appropriate to leave the loss out of the calculation.

5.4.3.4 Propagation path loss considerations

It was shown above that part of the radar equation is the power equation which calculates power density at a range of interest. This is effectively assuming that the propagation between the radar and the target is subject only to free space path loss. For an aircraft being detected by a radar, the assumption of a free space path loss is valid in many situations. However, if terrain starts to intrude into the path between the radar and the aircraft, then additional losses must be accounted for. Terrain is also likely to intrude into the path between a radar and a wind turbine. In either of these situations, diffraction losses will need to be calculated and added to the normal free space path losses.

Another type of loss that might be experienced when the pulse is ‘in-flight’ is **absorption loss**. This loss is caused by the atmosphere absorbing energy either by specific attenuation or resonance absorption described above. As with some other classes of loss, whether or not this might be accounted for in the radar equation depends on purpose. If the intention is to determine whether an aircraft might be detected, then including as much information as possible about possible losses is prudent. Conversely, if the intent is to determine if a wind turbine might be visible, then the worst-case assumption would be that the loss was not present. More information on accounting for absorption loss is included later in this chapter.

5.4.3.5 The receive path

The first part of the receive path from the antenna to the receiver is identical to parts of the transmit path and a 2 dB loss might be expected as a result.

To provide further isolation from the high power of the transmitted waveforms, the circulator may then be followed by a limiter circuit. However, there is an additional consideration for S-Band ASR. The spectrum used by these radars (discussed earlier in this chapter) is close to that used by 4G cellular telephones. To prevent interference to radars from these services, filters are built into the receive path and may form part of a combined limiter/filter circuit [9]. The loss incurred by these elements might typically be 0.4 dB.

Unlike noise, which ceases to be a problem after the signal has been digitised, losses can still occur after this stage, for example, straddling loss. The concept of straddling was introduced in Chapter 3 when CFAR was introduced. For convenience, Figure 5.6 is a repeat of the diagram from Chapter 3; it shows an aircraft straddling two range bins. Under these circumstances, neither of the straddled bins may contain a full measure of the target signal and this may have a direct impact on the probability of detection. Straddling loss may be as high as 3 dB.

An aircraft velocity may also be such that its Doppler signature may fall within two Doppler filters. This loss is referred to as Doppler straddling loss which might be up to 2 dB.

Straddling losses should be accounted for in the radar equation if the purpose is to determine performance in predicting aircraft detection. However, if the purpose is safeguarding the radar against wind turbines these may be disregarded, because

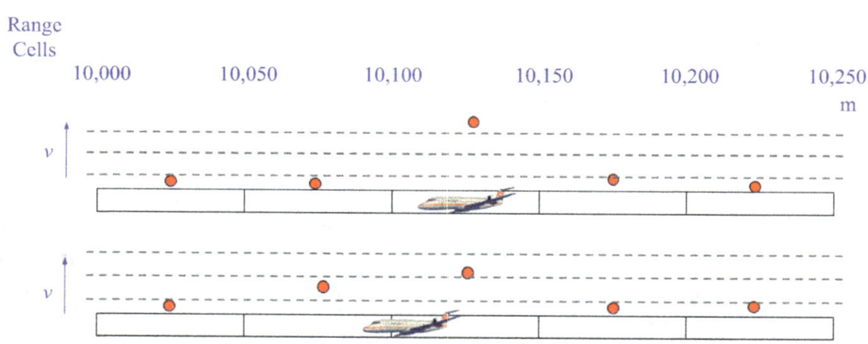

Figure 5.6 Straddling loss

the worst case would occur if the target (now a wind turbine) appeared in the middle of a bin.

5.4.3.6 Summary of losses

Typically, a radar will experience a 2 dB loss on transmit and 2.4 dB loss on receive. Other losses may occur which depending on the purpose of the calculation may need to be accounted for in the radar equation.

5.4.3.7 Accounting for loss in the radar equation

Accounting for losses in the radar equation is simple. The losses are totalled (L_{Tot}) and that added to the denominator of (5.22) as follows:

$$\text{Maximum range, } R_{max} = \frac{\left(G^2 P_t \sigma l^2\right)^{1/4}}{(4\pi)^3 L_{Tot} F_n k\, B\, t\, S_o/N_o} \tag{5.25}$$

5.4.3.8 Integration

If the return from a distant aircraft is similar in amplitude to the noise level, it will be difficult to distinguish the aircraft return from noise. There is a high probability of missing a detection. However, the fact that noise is random can be exploited here. In an ensemble of noise returns, the number of high noise values will be matched by an equal number of lower noise values. Therefore, the average of an ensemble of noise returns will tend to the average noise level. Recall from Chapter 3 that this principle is exploited in CFAR, in which a number of samples on either side of the sample under test are averaged as part of the process of computing a threshold. This principle can also be exploited if the values in range cells are saved from a number of transmissions. This will create an ensemble of cells for each particular range, each member of the ensemble coming from a different transmission. The values of each ensemble can be summed. If the values at any range are just noise, the sum will tend to the average noise level times the number of pulses being integrated. If, on the other hand, the values in the cells are from a range where there is an aircraft, then the signal component will be integrated, enhancing the SNR. An example will illustrate how this works.

Before looking at an example to illustrate the benefits of integration, it has been assumed reasonable to expect that an aircraft may be present in the same range cell over a number of pulses. A typical ASR scans the horizon every four seconds and may have a pulse repetition frequency (PRF) of 1 kHz. Even a fast jet does not move far in the range between pulses that are 1 ms apart and they will tend to present in the same range cell. However, the question arises, how many times might an aircraft be illuminated by pulses in each four-second period. There is a simple equation for calculating the number of pulses that will illuminate an aircraft on each sweep. An implicit assumption of this method is that the target will only be detected within the beamwidth (that is the half-power beamwidth):

$$\text{Number of pulses} = \frac{\text{PRF} \times \text{Half power beamwidth}}{360^\circ \times \text{Rotation rate}} \tag{5.26}$$

Taking a typical ASR as an example. The PRF is 1 kHz, the beamwidth is 1.5° and the rotation rate is 15 revolutions per minute (rpm) which corresponds to 4 s per revolution and 0.25 revolutions per second. Inserting these values into the equation gives 16 pulses (rounded down to the nearest whole pulse). This result shows there would be no purpose served by integrating more than 16 pulses; actual figures are likely to be less than that.

5.4.3.9 Multiple pulse integration strategies

There are two broad strategies for performing multiple pulse integration; coherent integration, in which signal amplitude and phase are processed; and non-coherent integration in which only the signal amplitude is considered. The **integration improvement factor** provides a measure of the enhancement in the SNR of each method. In both cases, the improvement depends on the number of pulses being integrated. If using coherent integration, the theoretical limit in the improvement is linear with the number of pulses being integrated. If four pulses are integrated, there will be a four-fold improvement and if eight pulses are integrated, the improvement factor will be eight. However, non-coherent integration is dependent on the SNR of the signals being integrated. The higher the SNR the closer the integration improvement factor is to the theoretical maximum. If the SNR is low, then the coherent integration improvement factor tends towards the non-coherent integration improvement factor which is the square root of the number of pulses being integrated. Thus, if four pulses were integrated, the integration improvement factor would be 2 and if eight pulses were integrated, then the improvement factor would increase to 2.8 (square root of eight).

The following figures illustrate the benefit of integration. A simple mathematical model of radar returns processing was created. The returns in four hundred range cells were simulated for eight successive pulses. Three simulated targets were inserted into the same range cells for all eight pulses; the target amplitude was chosen to be the same as the average noise level; without integration, detection of such a target would be very unreliable. Figure 5.7 shows two randomly chosen transmissions, pulse 2 and pulse 7. Note that there is little visible evidence of the presence of targets. Figure 5.8 shows the effect of integrating eight pulses.

However, the analyst must take care that the factors included in the radar equation include the factors that are important to the modelling required.

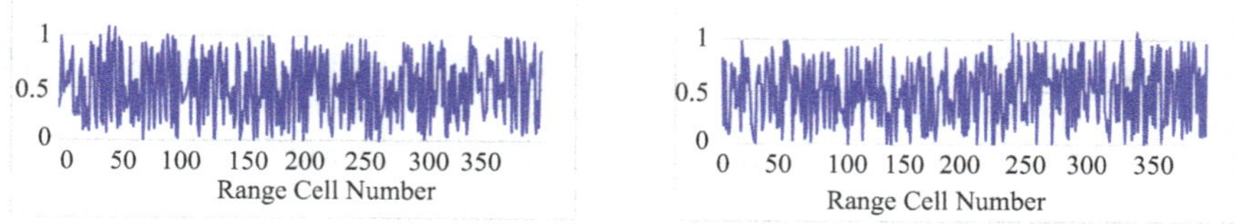

Figure 5.7 Individual pulses prior to integration

Integration is a good example of why caution is needed. Radar performance figures discussed in the literature are often based on a single pulse, in which case multiple pulse integration would not be accounted for in the radar equation.

5.4.3.10 A variation on the radar equation

A common variation on the radar equation for surveillance radars is to replace the peak power with the product of the average power (P_{ave}) and the pulse length (τ).

5.4.3.11 Evaluating the radar equation

Expressing units in dB offers a simple way of evaluating the radar equation. This method is based on Blake's method and demonstrates the benefits of using decibel values; greatly simplifying the calculation. See Table 5.3, the numerator and the

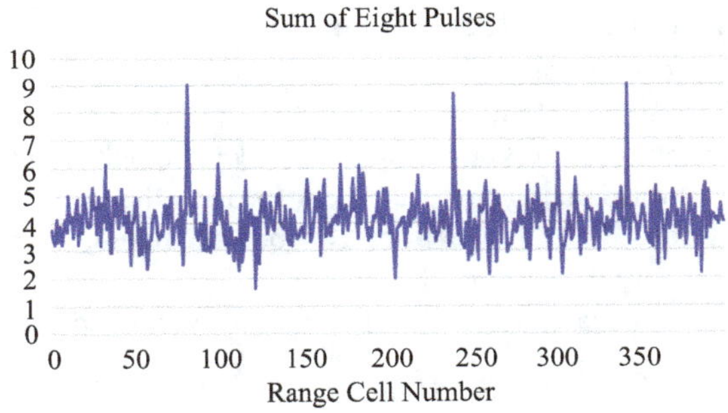

Figure 5.8 Multiple pulse integration

Table 5.3 Evaluation of the radar equation

Quantity	Symbol	Value	Unit	Note	Numerator	Denominator
					dB	dB
Peak power	Pt	730	kW		58.6	
Antenna gain	G	35	dB	Squared	70.0	
Wavelength	λ	0.1	m	Squared	−20.0	
RCS	σ	2	m^2		*3.0*	
Range	R	100	km	^4		200.0
Boltzmann's constant	k	1.38E−23				−228.6
Temperature	T_0	290	kelvin			24.6
Bandwidth		1,000	kHz			60.0
Noise figure	NF	5	dB			5.0
Losses		5	dB			5.0
$4\pi^3$		1984.4				*33.0*
				Sum	111.6	99.0
				SNR	*12.6*	

denominator of the equation are summed separately and then the denominator is subtracted from the numerator. In this case, the result is the SNR for all the values listed. It will be clear that this method lends itself to being evaluated by spreadsheet.

The equation used above values consistent with an ASR using a vacuum transmitter device, that is, a high-powered, short pulse with a commensurately high receive bandwidth. Moreover, the noise figure and losses are a little higher than might be found today. A more modern transmitter might have a lower peak power, say 15 kW, longer pulse lengths and, therefore, a lower bandwidth receiver, say 50 kHz. One might expect a lower noise figure, say 3 dB and lower losses, say 3 dB. Using these figures, the SNR remains almost the same, reflecting the fact that the technology has changed but the requirement remains the same. Having arrived at an SNR, this begs the question, is this reliably detectable by the radar?

5.4.3.12 The minimum detectable signal

From the foregoing discussion, it is clear that there is some signal level that corresponds to the minimum that the radar can detect: that is the **Minimum Detectable Signal**. But how do you know what is detectable? Boltzmann's constant provides a means of formalising that statement through the definition of a quantity called the **noise floor**. If the noise floor can be calculated, there is a means of estimating when the target becomes detectable.

The noise floor is usually stated in dBm and is calculated thus:

$$\text{Noise floor} = 10 \log (kT_0B) \text{ dBm} \tag{5.27}$$

where k is Boltzmann's constant; T_0 is the reference temperature (usually taken to be 290°K); B is the bandwidth.

The expression in (5.24) is the noise component of the SNR arrived at by application of the radar equation. However, when a target SNR just reaches the noise floor, this corresponds to being only just detectable, that is, detectable with a very low probability of detection. A useful rule of thumb is that the SNR must exceed 10 dB over the noise floor to be detectable. An alternative and practical method of knowing the detectability of a target is to consult textbooks such as Skolnik's [10] and Blake's [11]: these contain graphs relating SNR to the probability of detection. If these are consulted an SNR of 12.6 dB corresponds to a probability of detection of approximately 80%.

Some ANSP are prepared to release the Minimum Detectable Signal (MDS) of their sensors. If this is available, then a short-hand method for determining whether a target is on the edge of being detectable would be to use the following equation. In this example, let us assume it is necessary to know if a wind turbine is detectable:

$$\text{Signal presented by a wind turbine} = P_{\mathrm{T}} \times K \times G_{\mathrm{T}} \times G_{\mathrm{R}} \times L_{\mathrm{P}} \times L \times \sigma \tag{5.28}$$

where P_{T} is the transmit power; K is the conversion from W to mW (MDS are usually quoted in dBm); G_{T} is the transmit antenna gain (for the antenna elevation of this piece); G_R is the receive antenna gain (for the antenna elevation of this piece); L_P is the two-way path losses; L is the losses (a combined transmit and receive loss); σ is the wind turbine RCS.

If the signal level calculated by using (5.25) exceeds the MDS, then the wind turbine is on the threshold of detectability. If it exceeds the MDS threshold by 5 or 6 dB, then the probability of detection will be high.

An alternative form of the equation is to use a piecewise linear model (PLM) of the wind farm to compute the signal returned from pieces of a wind turbine and then perform a power sum of all the components to arrive at a signal power value for the whole structure. Both methods could be used to calculate the signal levels for static and moving components.

This method and the radar equation are based on the same core principles of physics. The radar equation has built into it the electromagnetic power equation. This assumes that the propagation mechanism is by simple expansion of the electromagnetic field assuming free space propagation and no atmospheric absorption. If diffraction losses are to be accounted for, they must be added to the implicit free space path losses. The equations would need to be augmented if a better propagation model was required. A further question arises; are there any other ways that the problem could be approached that might give a more accurate result?

5.4.4 Blake's method

The first systematic treatment of radar sensitivity is credited to Norton and Omberg in 1943. Their paper, like many wartime papers associated with radar, was republished after the war in 1947 [12]. This work was extended by Blake and was reported in a number of publications of which the most comprehensive is Blake (1980) [13]. Blake's contribution is twofold. The method uses a more comprehensive model of the radar system. In addition, the Blake method calculates the range using a chart that is in effect an early, handwritten, spreadsheet. The method has become an industry standard for calculating the performance of a radar.

Barton has identified a number of factors in the superior performance of the Blake method versus the radar equation methods [14]:

- The radar equation derived above assumed that the only source of noise came from Johnson/Nyquist or thermal noise. Blake's method uses a more comprehensive method of dealing with noise using noise temperature.
- There is a more comprehensive treatment of the radar design parameters and environmental interactions.

The methods used by Blake are illustrated below: noise temperature first.

5.4.4.1 Environmental noise

In older textbooks, Johnson Nyquist noise is often assumed to be the only source of noise. However, as receiver design improves, it is also necessary to consider noise from the environment.

The main difference in characteristics between thermal noise and environmental noise is that environmental noise is frequency dependent. It follows that a more comprehensive treatment must account for radar frequency and, therefore, the figures must take into account the radar's purpose. In other words, an ASR radar may use a different treatment of noise compared with, say, a space surveillance radar operating at a different frequency.

There are two principal sources of environmental noise: anthropogenic and ambient; the latter is sometimes called **sky noise**.

5.4.4.2 Anthropogenic noise

Anthropogenic noise covers sources including:

- Unintended radiation from electrical machinery, electrical and electronic equipment.
- Power transmission lines.
- Internal combustion engine ignition.
- Electric fences.

The ITU provides a useful model of anthropogenic [human-made] noise which is only valid for frequencies between 300 kHz and 250 MHz. An anthropogenic source of noise is essentially a low-frequency phenomenon. Hence, it might be problematic for AGA communications and navigation systems but it is unlikely to be a cause for concern for aviation radar systems [15].

5.4.4.3 Ambient noise

There are three principal sources of ambient noise: Galactic, or Cosmic, noise; Solar noise and atmospheric noise. Galactic noise and solar noise can be ignored above UHF. Therefore, like anthropogenic noise, it might be problematic for AGA communications and navigations systems but it is unlikely to be a cause for concern for aviation radar systems.

5.4.4.4 Noise temperature calculation

Blake's method assumes the antenna to have a main lobe and sidelobes pointing in different directions, in simple terms, the former pointing at the target in the sky and the latter pointing at the ground. He calculates the system noise temperature as follows:

$$\text{System input noise temperature, } T_{\text{S}} = T_{\text{a}} + T_{\text{r}} + L_{\text{r}} T_{\text{e}} \tag{5.29}$$

where T_{a} is the antenna temperature; T_{r} is the plumbing temperature = T_0 ($L_{\text{r}}-1$) L_{r} is the plumbing losses:

$$\text{Antenna noise temperature } T_{\text{A}} = \frac{\alpha_{\text{A}} T_{\text{A}}}{L_{\text{Ant}}} + \frac{\alpha_{\text{G}} T_{\text{G}}}{L_{\text{Ant}}} + \frac{T_0(1-1)}{L_{\text{Ant}}} \tag{5.30}$$

$$\text{Receiver effective temperature } T_e = T_0(F_N - 1) \tag{5.31}$$

where F_{N} is the system noise factor.

Inserting the values from Table 5.4 into equation (5.27) produces an antenna noise temperature of 83°K and a system noise temperature of 803°K.

Table 5.4 System noise temperature after Blake

Term	Value	Notes
Main lobe antenna factor α_a	0.876	Typical value proposed by Blake
Side lobe antenna factor α_g	0.124	1-mainlobe factor
Sky noise T_a	48 K	From Blake (1972) [16] for a 2.8 GHz radar with a 2.5° antenna pointing angle
Antenna loss	0.1 dB (1.023)	Typical value
Plumbing loss	2.5 dB (1.786)	Typical value
T_G and T_0	290 K	By convention
F_N	2.9 dB (1.949)	Typical value

The range equation used by Blake after Kerr is:

$$\text{Maximum range, } R_{\max} = \frac{(P_t \tau G_t G_r \sigma l^2 F_f{}^2 F_r{}^2)^{1/4}}{(4\pi)^3 k T_S D(0) C_B L} \tag{5.32}$$

Before explaining the terms in the equation, Blake recommends a method of using 'useful' units. These are accounted for by introducing a portmanteau correction factor. With modern computing facilities, this may no longer be strictly necessary but the method is commonly reported and this is the form described below. A final observation before explaining the method. Moving away from a simple vacuum physics, power equation, method of calculating the power density requires the atmospheric loss to be known. Atmospheric loss is reported in dB/km. Therefore, it is necessary to know the range of the target to calculate the atmospheric loss. But the equations are designed to provide the range of the target. The contradiction is obvious. The approach taken by Blake is to calculate a range to the target based on vacuum physics, use that range to estimate the path loss and then recompute the actual range to target.

The modified version of (5.29) using the 'useful units' approach is:

$$\text{Maximum range, } R_{\max} = \frac{129.2\,(P_t \tau G_t G_r \sigma F_f{}^2 F_r{}^2)^{1/4}}{f^2 T_S D(0)\, C_B L} \tag{5.33}$$

where P_t is the radar peak power in kW, τ is the pulse length in μs, σ is the radar cross-section in square metres, $F_t{}^2$ is the correction factor for the target not being at the peak of the transmit antenna gain (G_t), $F_r{}^2$ is the correction factor for the target not being at the peak of the receive antenna gain (G_r).

If the same antenna is used for transmit and receive, then the terms G_t, G_r, F_f^2, F_r^2 reduce to G^2, F^4; $D(n)$ is a detectability factor which is $D(0)$ for a single pulse and $D(n)$ for the non-coherent integration of multiple pulse integration of n pulses. The factor $D(n)$ also includes a consideration for the variability of the target over n pulse. For historical reasons this is sometimes referred to as the visibility factor, v, (in the sense of an operator's ability to see a return), f is the

Table 5.5 Blake chart example

Parameter	**Value**	**Value in dB as it appears in equation 29**
Power (KW), P_T	10	10.000
Pulse width (us), τ	75	18.751
Gain, transmit, (dB), G_T	34	34.000
Gain, receive, (dB), G_R	34	34.000
RCS, m^2, σ	2	3.010
Frequency (MHz), f	2,800	−68.943
System temperature (deg K)	803	−29.052
Visibility factor (dB)	9.01	−9.010
Bandwidth correction loss, dB, C_B	0.8	−0.8
Transmit loss (dB)	1.75	−1.750
Beamshape loss (dB)	1.6	−1.600
Other losses (dB)	0	0.000
Blake's correction factor	40 log1.292	4.451
Sum		−8.103

frequency in MHz, T_S in degrees Kelvin is system noise temperature (calculated above), $D(0)$ is a detectability factor, see comment on C_B, C_B is a bandwidth correction factor in case the receive bandwidth is not optimal. This value is often assumed to be 1.2 (0.8 dB). Taken together with $D(0)$ this makes the distinction that the matched filter which is supposed to optimally detect the received waveform based on the bandwidth may, in practice, be sub-optimal. $D(0)$ is a single hit version of this correction factor. If multiple pulse integration was being accounted for this would be replaced by $D(n)$ where n is the number of pulses being (non-coherently) integrated, and L are the total remaining losses not accounted for in the previous list.

An example of its application is helpful.

The result of the calculation illustrated in Table 5.5 must then be converted to the radar range in a vacuum which in accordance with Blake's method is carried out by multiplying the antilog of the number divided by 40 by 100. This gives a value of 62.7 Nm.

To arrive at the final answer, this must be corrected for the losses in the atmosphere. It has been calculated assuming a loss per kilometre of 0.009 dB/km. The total two-way loss is, therefore, 1.05 dB. The correction factor is 0.94. The correct value of the range is 59 Nm (109 km).

5.5 Near-field/far-field calculation

The terms near-field and far-field have been used a number of times in this book. For example, it was assumed the wind turbine was in the far-field, when the potential for shadows was discussed. This section considers the definition and importance of this concept.

A transmitter and an antenna create a wavefront. The wavefront is developed as the energy from the antenna passes through three regions of space in which different physical processes take place. These regions are:

- Very close to the antenna, a region called the reactive near-field.
- The near-field.
- The far-field.

5.5.1 The reactive near-field

The reactive near field is illustrated in Figure 5.9. Current from the transmitter passes into the antenna, in this case, a simple dipole. An electric field, *E*-field, is set up across the dipole caused by the potential difference (voltage difference) across the dipole. Two conductors separated by an air gap in this fashion behaves like a capacitor and E-field is determined by the flow of electrical charge into this capacitance. A current also flows into the dipole creating a magnetic field as it does so. This 'H-field' circles the dipole¶. The magnetic field set-up is determined by the inductance of the dipole elements. The inductive and capacitive fields have the electrical property of 'reactance.' Hence, why this region is called the 'Reactive' near-field.

Very close to the antenna, within a distance of $0.62 \times \sqrt{(D^3/\lambda)}$ from the antenna, there is no radiation. At the antenna, the *E* and *H* fields are out of phase and not aligned. As they pass through the region, they become progressively more aligned and when they reach the boundary of the region, they have formed a transverse electromagnetic (TEM) wave. That is, both fields are in phase but orthogonal to each other. When the TEM is formed, it passes the boundary of the reactive near field entering the near field region. To make the distinction clear, this region is sometimes referred to as the radiative near-field.

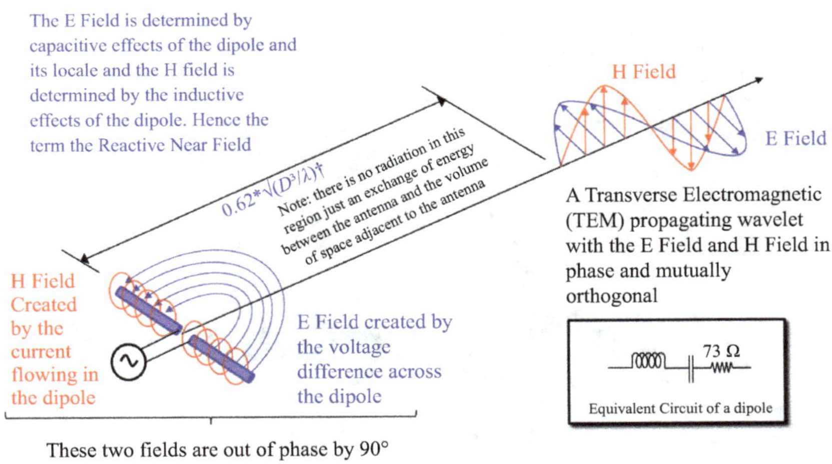

Figure 5.9 Reactive near-field

¶The letter H was chosen by Maxwell, it does not stand for anything.

5.5.2 The near-field

The near-field is the region between the reactive near-field and the far-field. In this region, there is a destructive interference between wavelets from different parts of the antenna and the total field strength is reduced. Furthermore, if a measurement of the antenna factor (or array factor in the case of a phased array) is taken, it may be distorted.

5.5.3 The far-field

The far-field is the region sufficiently distant from the antenna that all the wavelets from the radio source are assumed to be travelling parallel to, and in phase with, each other. When this condition is met, the wave is said to be 'coherent'.

Finding more information on these three regions is made more complicated by the use of different terminology in the literature. Figure 5.10 illustrates some of the alternative nomenclatures that will be encountered.

5.5.4 The location of the near-field/far-field boundary

Knowing the location of the near-field/far-field boundary is useful. If an antenna pattern (array factor) is to be measured, it should be carried out in the far field. It is also a basic assumption for other calculations such as computation of shadow and knife edge diffraction (essentially the same subject) and it is an assumption of Huygens construction discussed later. It is widely regarded as being useful too if an obstruction such as a wind turbine is in the far-field not the near-field. Then the presence of the structure does not unduly influence the signal field created by the radar, or other transmitting device.

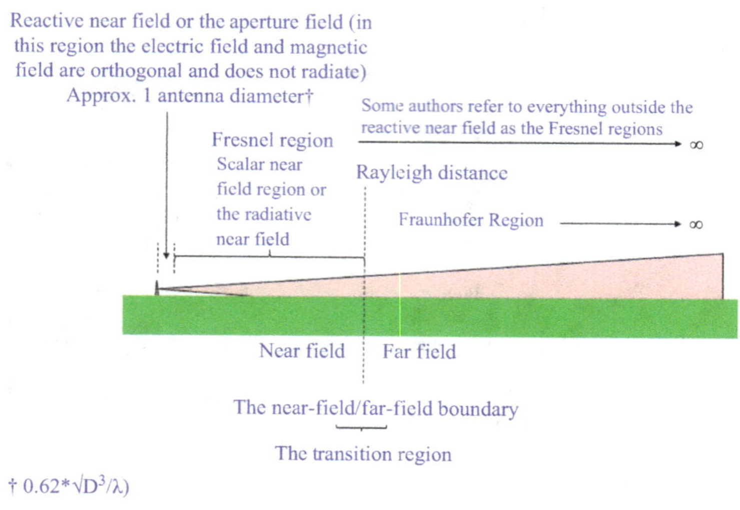

Figure 5.10 Near-field/far-field nomenclature

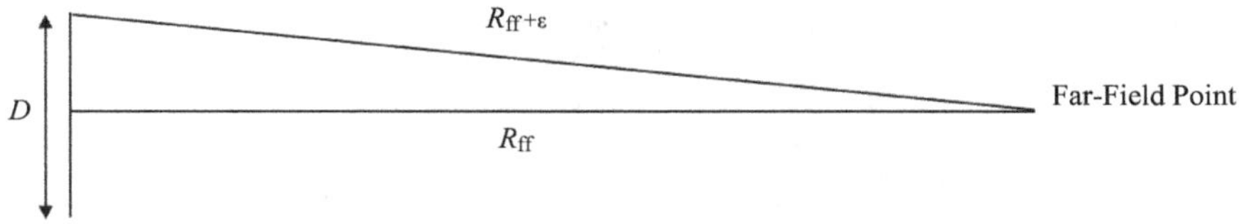

Figure 5.11 Far-field definition construction

The mathematical evaluation of the boundary is based on the construction shown in Figure 5.11.

An antenna has a maximum dimension, D. It must be stressed that this is not the width or height of the antenna, it is the maximum dimension. For example, if the antenna is a planar phased array, it is the diagonal distance across the antenna.

A measurement location in the far-field is a distance, R_{FF}, from the antenna. A wavelet from the edge of the antenna travels a distance of $R_{FF} + \varepsilon$ to the measurement location. Where ε is a small phase difference the wavelet must travel to the measurement point compared with the direct path length R_{FF}. Thus, a right-angled triangle is formed, with sides, $D/2$, $R_{FF} + \varepsilon$ and $R_{FF} + \varepsilon$. By Pythagoras, the triangle must satisfy the relationship:

$$(R_{FF}+\varepsilon)^2 = (D/2)^2 + R_{FF}^2 \tag{5.34}$$

$$\therefore R_{FF}^2 + 2\varepsilon R_{FF} + \varepsilon^2 = \frac{D^2}{4} + R_{FF}^2 \tag{5.35}$$

ε is small and ε^2 is very small and is ignored,

$$\therefore 2\varepsilon R_{FF} \approx \frac{D^2}{4} \tag{5.36}$$

$$\therefore R_{FF} \approx \frac{D^2}{8\varepsilon} \tag{5.37}$$

Therefore, an approximation to the far-field boundary distance could be formed if a value for ε was known. In other words, when is a small phase difference actually small? By custom and practice, ε is usually taken to be either an 8th or a 16th of a wavelength. Substituting these values into (5.34) gives [17]

$$R_{FF} \approx \frac{D^2}{\lambda} \tag{5.38}$$

$$R_{FF} \approx \frac{2D^2}{\lambda} \tag{5.39}$$

5.5.4.1 Interpreting the results

The starting point for this discussion was the idea of finding a distance from an antenna where the field it creates can be assumed to be coherent. The first thing that

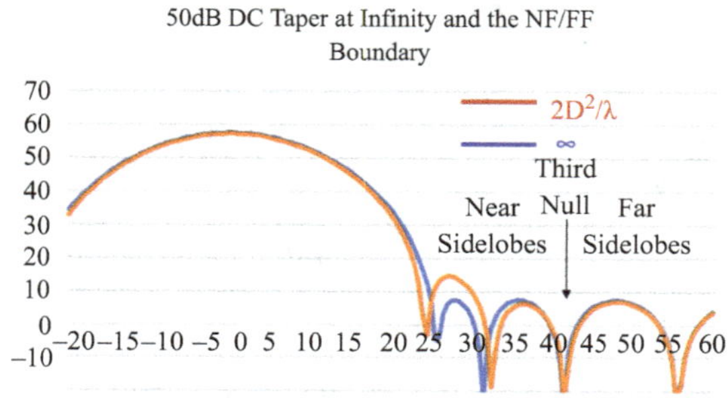

Figure 5.12 Gerlock's far-field findings

can be concluded is that, for the field to be coherent, then ε must be zero. This is the definition of a coherent wavefront that all phases along the wavefront are the same. Therefore, the construct set out above shows that this can only be true at infinity. Any other value must be an approximation; an engineering judgement of what is good enough.

However, it is important not to lose sight of the fact that it is an approximation. Gerlock investigated the near-field/far-field boundary under different circumstances:

- Is the value valid for all types of antennas?
- Is the value valid if the antenna is viewed off-boresight?

Gerlock's results are important [18]. Gerlock used as a vehicle for the investigation a phased array which used a Dolph Chebyshev (DC) taper. This is a taper that gives rise to equal amplitude sidelobes (performance used, for example by ILS Localisers). Gerlock showed that at a distance $2D^2/\lambda$ from the antenna, the sidelobe structure is still not fully formed, see Figure 5.12. This finding demonstrates a basic rule, although not stated, that the derivation of the near-field/far-field boundary distance assumed that the antenna was uniformly illuminated. The example in this case used a DC taper and by definition a taper is not uniform illumination.

A second of Gerlock's findings can also be observed in Figure 5.12. At a distance $2D^2/\lambda$ from the antenna, the near sidelobes are not fully formed but the far sidelobe structure is. Thus, the near-field/far-field boundary can be relaxed when the antenna pattern is viewed from off-boresight.

5.6 Propagation

5.6.1 The troposphere

The propagation of radio waves is affected by the medium through which they pass. For the cases of interest here, the medium is the **Troposphere**; the lowest layer of the Earth's atmosphere.

The troposphere starts at the Earth's surface. The boundary with the layer above the troposphere, the stratosphere, is called the tropopause and its height varies with both temperature and latitude. The tropopause is the highest at the equator, where it can be as high as 20 km (66,000 ft). It is much lower at the North and South Poles, where it may be only 7 km (23,000 ft) high in winter. The height of the tropopause at mid-latitudes, over North America, Central Europe and the United Kingdom, is typically 11 km (35,000 ft) [19].

The troposphere is the heaviest layer of the atmosphere accounting for approximately 75% of the total mass because it contains approximately 99% of the moisture in the atmosphere. Other important properties of the troposphere are its pressure and temperature. Air pressure is created by the mass of the air pressing down on it from above. Therefore, air pressure decreases as height increases (this was discussed in Chapter 3 when the problems of measuring aircraft height were considered). The pressure gradient is a relatively stable quantity: if it was not, it would not be a safe means for measuring aircraft height. The pressure decreases by nominally 1 h Pa for every 30 ft increase in height. By comparison, the temperature, which usually decreases with height, 6.5°Kelvin per km, is less reliable and sometimes the temperature gradient reverses; a condition known as a **temperature inversion**.

5.6.2 Refraction

Taken together the temperature, pressure and moisture content determine the refractive index of the atmosphere in accordance with the following equations:

$$\text{Refractive index } n = c/v \tag{5.40}$$

where c is the velocity of light in a vacuum and v is the velocity of the electromagnetic wave in the medium (in this case the atmosphere).

The numerical value of n is very small. A typical value at the Earth's surface is 1.0003. It is usual to use another variable to describe the Refractive Index, N; defined thus:

$$N = (n - 1) \times 10^6 \tag{5.41}$$

N is calculated using the following formula:

$$N = 77.6\frac{P}{T} + 3.673 \times \frac{10^5 e}{T^2} \tag{5.42}$$

where P is the pressure measured in mBar; T is the temperature in degrees Kelvin; e is the partial pressure of water vapour.

The above formula has two components. The first variables in the formula (pressure and temperature) are commonly referred to as the **dry terms** which account for the dry elements of the atmosphere, principally nitrogen and oxygen. The second set of variables is referred to as the **wet terms**. Strictly, partial pressure is associated with each component of the atmosphere but its effect on water vapour dominates [20].

An important point to note about (5.39) is that none of the terms in the equation refer to wavelength. In other words, the refractive index is independent of frequency and all wavelengths are refracted (bent) by the same amount[l] [21].

5.6.2.1 The practical effects of refraction

Consideration of the above formulae provides valuable insights about the effect of the troposphere. The refractive index is a function of pressure, temperature and moisture content, but none of these is constant as a function of height above the ground. It follows that the refractive index cannot be constant as a function of height. If the refractive index was constant, then electromagnetic waves would travel in a straight line through the troposphere. As it is not constant then **electromagnetic waves must refract (bend) as they travel through the troposphere**. That bending effect is in accordance with Snell's Law:

$$n_1 \sin \theta_1 = n_2 \sin \theta_2 \tag{5.43}$$

Hence, if the refractive index n_2 is less than n_1, then for the equality to hold θ_2, the angle with the local vertical must be greater than θ_1. This concept is illustrated in Figure 5.13 applied to the troposphere. As an electromagnetic wave passes from a more dense, higher refractive index, to a less dense, lower refractive index medium, then the angle to the local vertical increases and the wave bends downwards.

Although some analytical treatments in the literature use the idea of a stratified troposphere, in reality the process is continuous and the ray describes a curve. The effect of this is to make the radio horizon appear farther away than the physical horizon, or that the Earth appears larger to a radio wave than it really is. It was

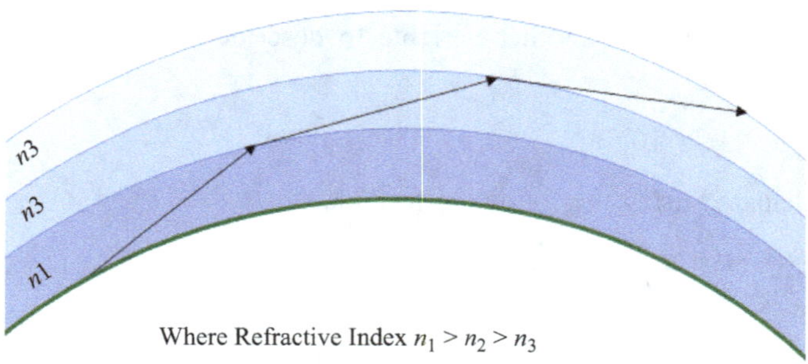

Figure 5.13 Snell's law and the troposphere

[l]There are several bands of frequencies at which the general rules of refraction do not apply. Oxygen and water vapour molecules can resonate at a number of different frequencies. Quantum mechanical molecular resonance effects attenuate transmissions at these frequencies. The frequencies include 22 GHz for water vapour and 57–60 GHz and 119 GHz for oxygen molecules. For a more detailed list of frequencies and values for the attenuation, refer to ITU-R P.676.

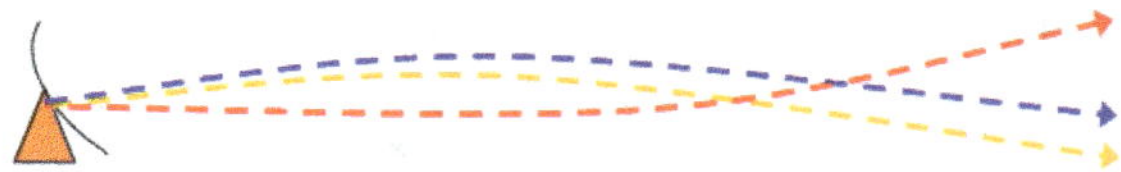

Figure 5.14 Refraction categories

PHOTO 1: The chimney on a normal day (at the foot of the right-hand wind turbine).

PHOTO 2: Super refraction makes the chimney appear much taller.

Figure 5.15 Example of super refraction (after Wisher reproduced with permission and courtesy of the RSGB)

pointed out in Chapter 3 that a good approximation to this modification is 4/3 Earth radius. This is important when considering the geometry of radio wave propagation, a point that is picked up in the discussion below on mapping.

Another important conclusion that can be made from the formula for calculating refraction is that it is a function of temperature and moisture content. In other words, refraction is weather dependent and varies from day to day, being less on some days than the norm and greater on some days than the norm. There are three broad categories of refraction as illustrated in Figure 5.14.

It is possible to observe this variation with the eye. Figure 5.15, reproduced by kind permission of the Radio Society of Great Britain (RSGB), shows the effect of super refraction at optical frequencies [22].

5.6.2.2 Does refraction make a difference?

When a terrain profile plot is being created, the question might occur, does allowing for the refraction of radio waves make a material difference? To address this question an aerodrome was chosen at random, a wind turbine location was invented exactly 100 km away from the aerodrome radar. This distance was chosen because this is close to the maximum range of an ASR and any effects would be made clearer. The implementation of refractivity in the model used was based on the Earth radius. Three values of Earth radius were modelled:

- The mean Earth radius, 6,371 km.
- The 4/3 Earth radius model value, 8,500 km (actually 4/3 $\times$ 6,371 is 8,495 but 8,500 is the widely accepted figure).

- A 4/3 refraction correction of the true Earth radius at the aerodrome chosen is 8,741 km.

Terrain profile plots were made for each value and are shown in Figures 5.16–5.18. As an additional metric, the heights of the line of sight from the radar to the location of the imaginary wind turbine were compared for the different cases. The results are presented in Table 5.6. The difference between the standard refraction correction and making no correction results in a line-of-sight difference that is greater than the height of many wind turbines. The difference between the true local Earth radius (refraction corrected) is of less concern but if a piecewise linear model was used to represent the wind turbine the difference is greater than the typical dimension of a piece.

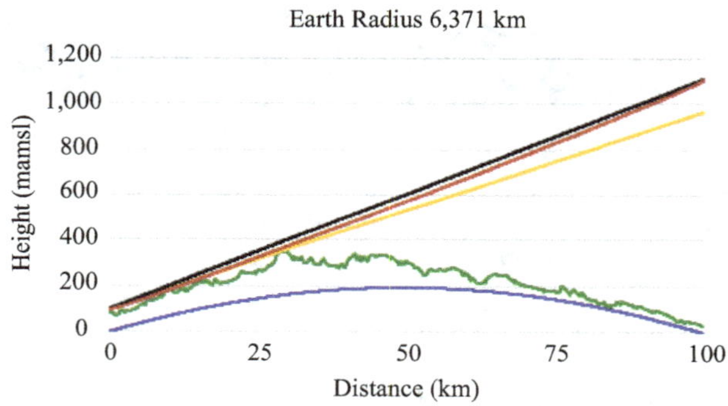

Figure 5.16 No correction for refraction

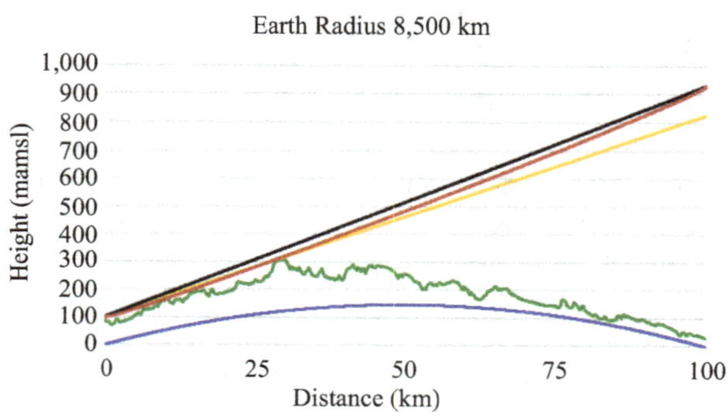

Figure 5.17 Standard refraction correction

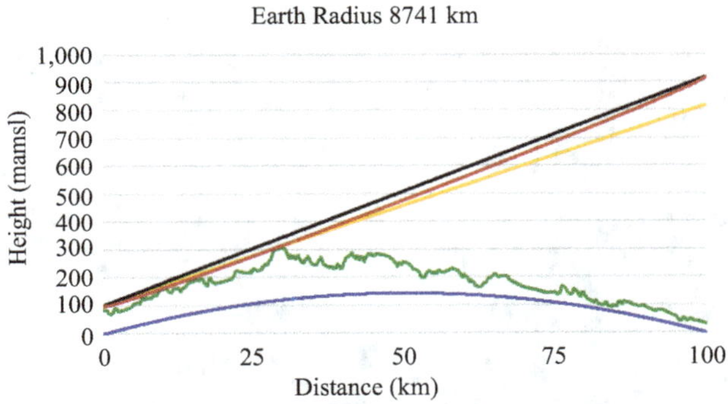

Figure 5.18 Local Earth radius correction

Table 5.6 Refraction correction comparison

Case	Height of line of sight (mamsl)
No refraction correction	1,076.7
Standard refraction correction	898.7
Local refraction correction	885.6

The next major topic of propagation to consider is diffraction. However, there are three important concepts that make it easier to understand diffraction and these are discussed in preparation. All three of these concepts rely on the wave-like behaviour of radio. In this regard the physical behaviour of radio waves is not unique to electromagnetic waves, the same behaviour can be observed in other waves, as can be seen in Figure 5.19.

The three preparatory concepts are: Huygens' construction, Fresnel Zones and the third is the Cornu spiral. The first to be considered in Huygen's construction.

5.6.3 Huygens' construction

Christiaan Huygens (1629–1695) was the first person to develop a description of how electromagnetic waves propagated between two points that were able to address all the observed behaviour. Contemporary thought, led by scientists like Newton, was that light was corpuscular in nature but this did not provide an adequate explanation of phenomena such as refraction. In 1690, Huygens published a wave model of the behaviour of light that was able to account for observations like diffraction and the model remains important today.

The two principal elements of the Huygens model, illustrated in Figure 5.20, are **wavelets** and **re-radiation**.

Each wavelet consists of an electromagnetic wave with a sinusoidal form and having the property of phase, see Figure 5.21. This shows one complete cycle of a waveform with the progression of the cycle measured in degrees from 0° to 360°.

Figure 5.19 Lulworth Cove, Dorset (image reproduced from Shutterstock under license. Photo Shaun Jacobs, reproduced by kind permission of Shaun Jacobs, Original Concept Dr David Bacon)

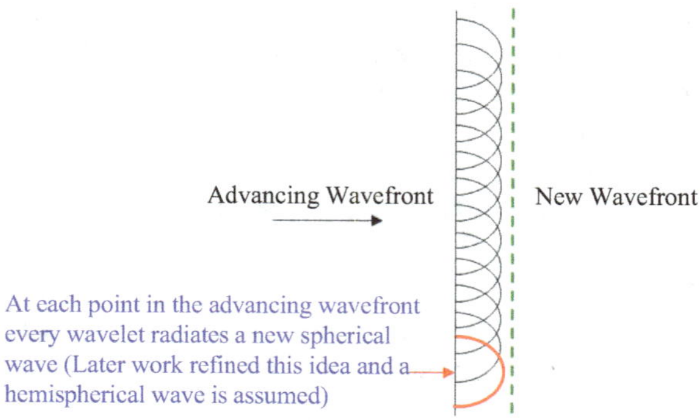

Figure 5.20 Huygens construction

The amplitude of the wave at any point is the vector sum of all the component wavelets. The power of this construction is that if the wavefront is partially obscured by an obstruction, the resulting amplitude remains the vector sum of the remaining un-obscured wavelets.

5.6.4 Fresnel zones

The work of Huygens was extended by the French scientist, Augustin Jean Fresnel (1788–1827). Fresnel investigated the behaviour of waves in space obeying Huygens' principles of propagation.

Figure 5.22 shows two points A and B. A transmitting device at A radiates a signal (a collection of wavelets) and the transmission is received at location B. The signal amplitude detected at B is the vector sum of all the wavelets from A. Some of those wavelets will have arrived travelling via wavelets re-radiated at points C and C'. Note that it is not necessary for wavelets at C and C' to be reflected in any way, as Huygens' construction tells us that the wavelets at these points will re-radiate.

Points C and C' can be chosen in such a way that:

$$\text{Paths ACB} = \text{AB} + \lambda/2 \text{ and paths AC'B} = \text{AB} + \lambda/2 \tag{5.44}$$

where λ is the wavelength of the transmission; C and C' can move around and the relationship is still maintained. The locus as C and C' move describes a **prolate ellipsoid** and the space within the ellipsoid is called the **First Fresnel Zone**. There are an infinite number of Fresnel zones, each additional zone having an additional half wavelength longer circumference than its predecessor.

5.6.5 Plotting the Fresnel zones

For reasons to be explained later, it is sometimes helpful to be able to visualise the Fresnel zone and this is relatively simple to carry out.

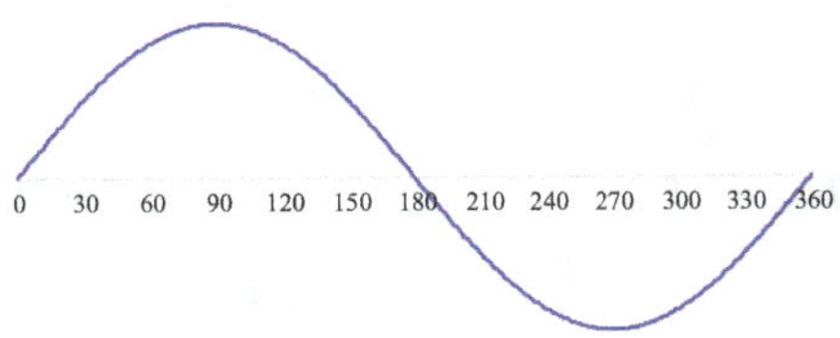

Figure 5.21 A sinusoid waveform

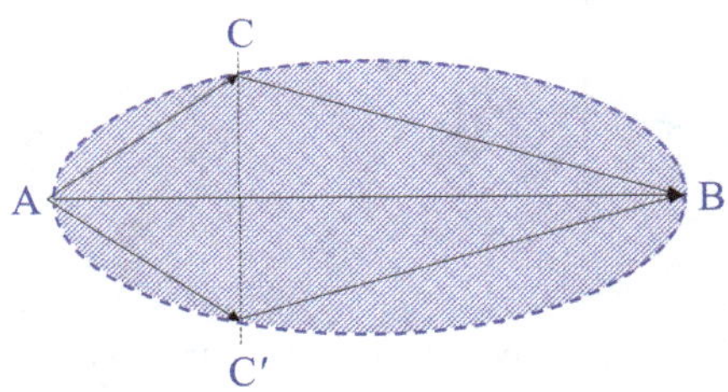

Figure 5.22 Fresnel zones

Using the geometry illustrated in Figure 5.23, the radius of the Fresnel zone at a point on the axis between A and B such that $d_1 + d_2$ is the distance between A and B, then the radius of the Fresnel zone for the *n*th Fresnel zone will be:

$$R_{FN} = \sqrt{((N\lambda d_1 d_2)/(d_1 + d_2))} \tag{5.45}$$

Repetition of this formula for different values of d_1 and d_2 allows the Fresnel zone to be plotted as shown in Figures 5.24 and 5.25.

5.6.6 The Cornu spiral

If a wavefront in the form of Huygens' construction is approaching a point, what is the signal strength at this point? For example, that point might be a radar antenna or a wind turbine blade. The problem is set out in Figure 5.26.

The means of solving the problem was worked out by the distinguished French Physicist Marie Alfred Cornu (1841–1902). The solution is a problem in vector algebra. As shown in Figure 5.26, one wavelet will arrive at the measurement point in advance of all the others. An adjacent wavelet arrives after a phase delay, etc. It is convenient to consider all the wavelets on one side of the wavefront first and then repeat the exercise for the other side.

As successive wavelets arrive at the measurement point, the phase lag increases as shown in Figure 5.27. The wavelets from the other side of the

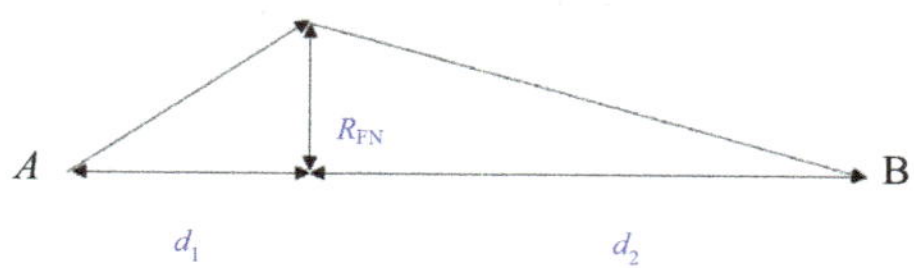

Figure 5.23 Plotting the Fresnel zone

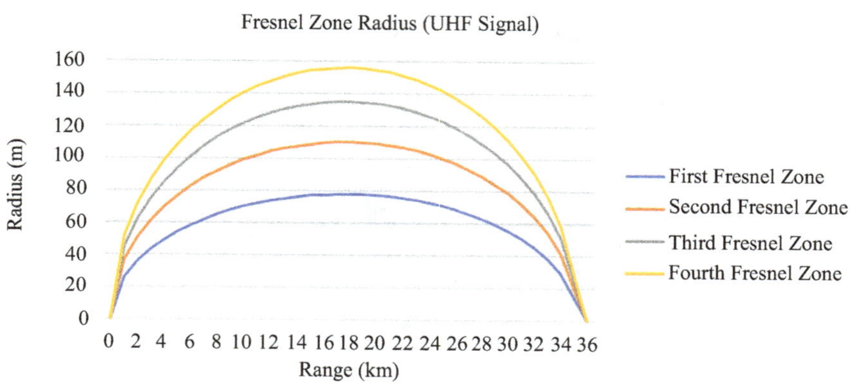

Figure 5.24 The first four Fresnel zones

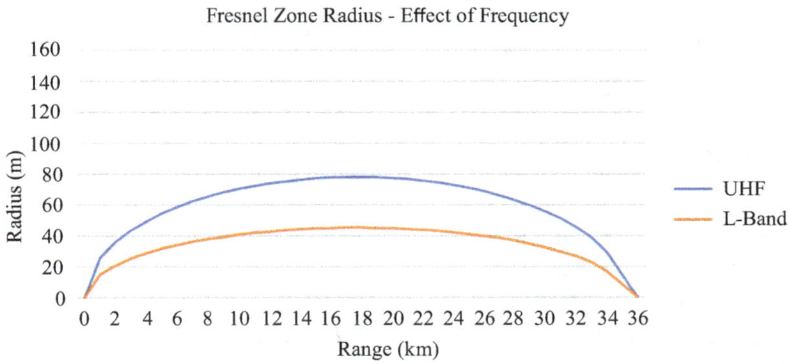

Figure 5.25 Fresnel zones at different frequencies

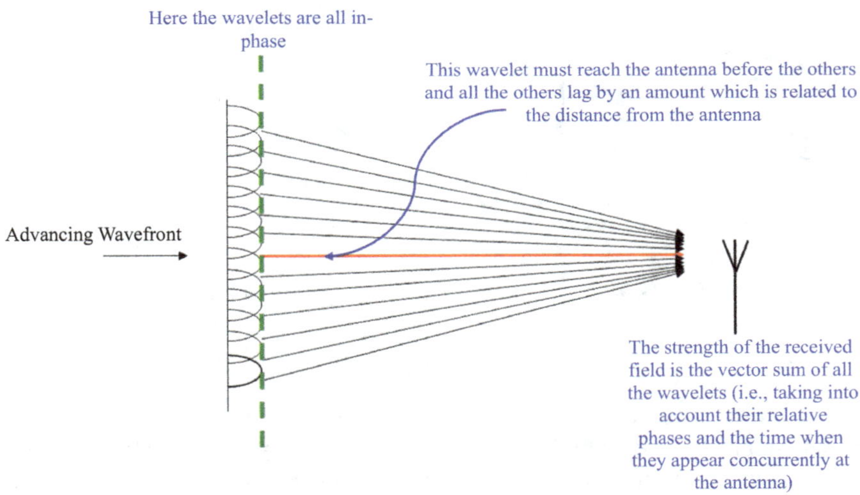

Figure 5.26 Cornu's problem

wavefront will behave in exactly the same way. The vectors are subject to the same delays. Conventionally, the vectors would lie in the same position on the phasor diagram obscuring the first set. However, custom and practice is to show the vectors from the other side of the wavefront radiating from the initial vector on the other side of the diagram. This is illustrated in Figure 5.28. If this process is extended to infinity in each direction, the resulting phasor diagram is shown in Figure 5.29. The shape formed has become known as the **Cornu spiral**. The spiral converges on the point $+0.5 + 0.5j$.

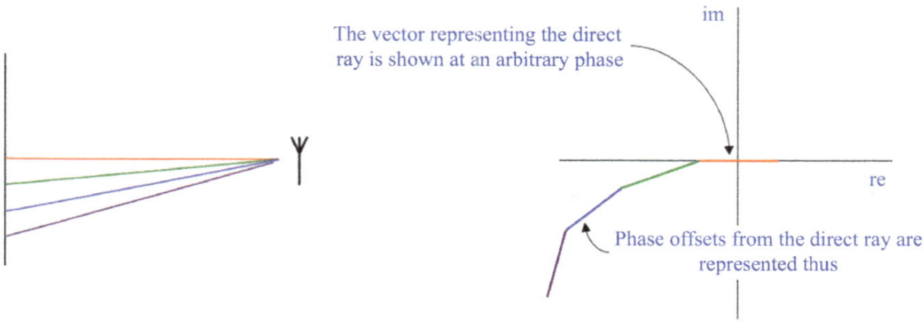

Figure 5.27 Summing one side of the wavefront after Cornu

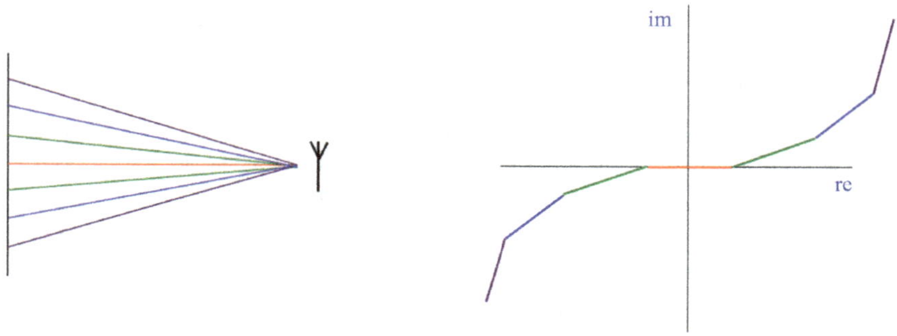

Figure 5.28 Adding the effect of the other side of the wavefront

The utility of the Cornu method is that the phasor diagram represents all the wavelets summed. If some of the wavelets are obscured by an obstruction, then this can be accounted for in the construct. A simple example illustrates the point. As the summation is extended to infinity the vector sum of the whole spiral is shown as the red vector. This vector is the field strength of the whole wavefront. Figure 5.30 shows an obstruction obscuring half the wavefront and the effect on the Cornu spiral shown on the right. The length of the vector is halved. This can be expressed as a loss in dB: $20 \times \log10\ (0.5) = 6$ dB. This may seem counterintuitive as only half the field is obscured but its capacity to pass power and energy goes down by 6 dB. However, it is the field strength that has been halved and the relationship to power is the square of the field strength [23].

The general form of solving the Cornu spiral is as an integral (i.e., the summing of the wavelets) carried out between the limits determined by the geometry of any obscuration. How that geometry is worked out is discussed below along with a simpler form of solution. Bacon provides a complete discussion of the form of the integral [24].

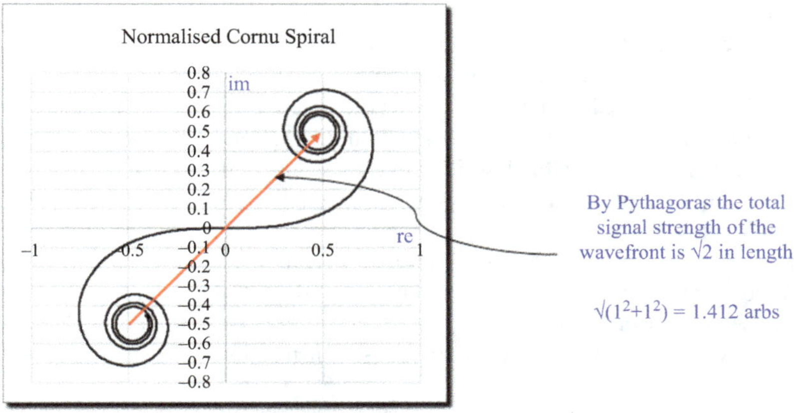

Figure 5.29 The Cornu spiral

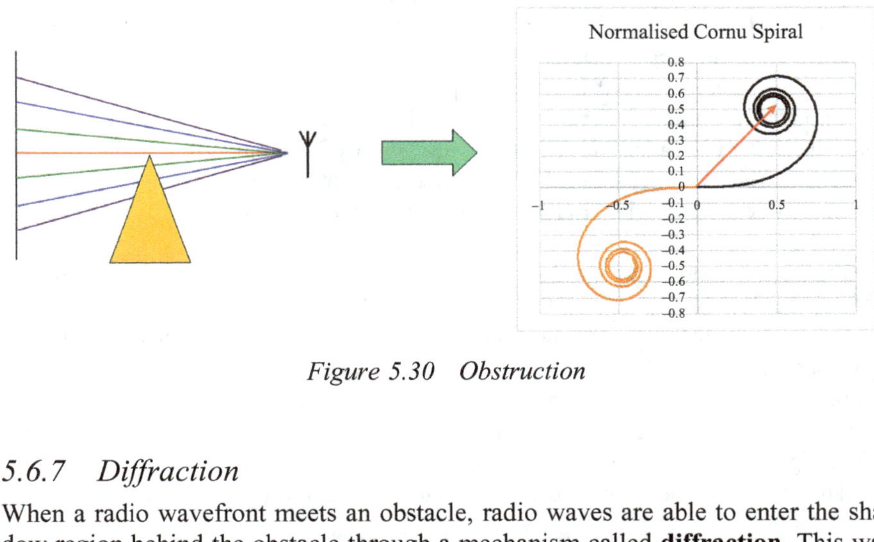

Figure 5.30 Obstruction

5.6.7 Diffraction

When a radio wavefront meets an obstacle, radio waves are able to enter the shadow region behind the obstacle through a mechanism called **diffraction**. This was described in Chapter 4 when shadow regions were discussed. The extent to which this is possible is a function of the geometry and the frequency of the radio waves. It is possible to compute the proportion of the waves that enter the shadow compared with a free space path; in other words, the **diffraction loss** can be calculated. It is important to remember that the diffraction loss is in addition to the normal losses that occur in free space. Diffraction analysis is applicable to radio transmissions involving obstacles from all types of radio systems including aviation systems such as radar, ILS and AGA communications.

Diffraction is a complex subject and it is recommended that detailed and accurate analysis should be carried out by specialists in the field. What follows is an introduction to the subject that will be useful in discussing problems with experts as well as some techniques that should be useful for quick look analyses to determine if a detailed investigation is merited.

The benefits of understanding the power loss caused by diffraction are obvious but it is equally important to understand when diffraction is not a factor and when the path of radio waves can be described as a **Free Space Path**. In turn, this informs the definition of another important concept in aviation; that is the **Base of Solid Radar Cover**.

This section leads to the descriptions of an analysis of multiple terrain obstructions which has become the industry standard for complex path analysis; namely the **Delta Bullington** method. But first it is necessary to define a number of introductory concepts, starting with Huygens Construction.

5.6.8 The knife-edge diffraction problem and the Fresnel–Kirchhoff parameter, a simplified method of calculating diffraction loss

Integrating the Cornu spiral to determine diffraction loss is computationally complex. Methods have been developed which are more limited in scope than the Cornu method but they are much simpler to compute and may provide the analyst with a quick check to see if a more complex approach is warranted.

Consider the following example of a radar and a wind turbine. A radio wavefront, created by a radar, travels in the direction of a wind turbine. There is an obstacle in the terrain **(a breakpoint)** that prevents a line of sight from being formed between the radar and the wind turbine; this geometry is illustrated in Figure 5.31.

Figure 5.31 shows a radar located at position A and a wind turbine located at position B. The breakpoint in the terrain is a distance d_1 from the radar and a distance d_2 from the wind turbine. The breakpoint is a distance h above the line of sight that would have been formed between the radar and the wind turbine in the absence of the obstruction. In the figure, h is shown above the line of sight and is positive. But the method works equally well if h is negative, that is, it is below the line of sight.

The method to be described is only valid if certain criteria are satisfied:

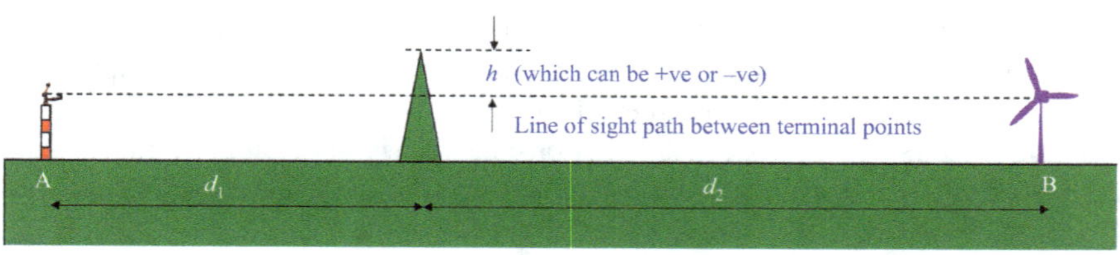

Figure 5.31 Fresnel–Kirchhoff knife edge diffraction geometry

- The distance d_1 must be very much greater than value h. This criterion ensures that the wavefront at h approximates being coherent** (all the wavelets comprising the wavefront must be parallel and in-phase); some authors refer to this condition as a **uniform plane wave**. This concept was discussed earlier in this chapter.
- The distance d_2 must also be very much greater than h, for the same reason, that when the wavefront reaches the wind turbine, it should approximate to being coherent.
- The wavelength of the transmission, λ, must also be very small compared with both d_1 and d_2. In all practical situations, this is likely to be the case.

Another criterion that must be satisfied is that the obstacle should behave like a **Knife Edge** (hence **Knife Edge Diffraction**).

- The first quality of a knife-edge is that it must be perfectly absorbing. If this requirement is met, the portion of the wavefront that is obscured by the knife edge plays no further part in the process, and there is no residual energy from that part of the wavefront to consider.
- The second requirement is that the knife edge be infinitely long. In the context of Figure 5.31, this means the two-dimensional representation of the knife edge in the figure is a section through the knife edge and the object being represented continues out of the page in both directions to infinity (a condition that prevents leakage around the edges). Although the criterion appears to impose an impossible constraint, many practical situations are good approximations. However, with real terrain, and more especially with the built environment, this criterion should not be taken for granted. It is important to remember that although it is convenient to show terrain profiles as being two-dimensional, they are three-dimensional and if the breakpoint is narrow (in the out-of-page direction), then energy can 'leak' around the obstacle and reduce the diffraction loss. If it is important that a certain level of loss or greater is needed to render a wind turbine invisible, then if leakage is possible the result may not be safe. A simple check can provide greater confidence. The breakpoint will be on a particular bearing from the radar, it is prudent to investigate the path on either side of that bearing, say, the azimuthal beamwidth of the radar, to make sure the breakpoint is not on the edge of a change in terrain, say for example, on a cliff edge.

5.6.9 Analysing the geometry

There is an infinite number of possible arrangements of radar, breakpoint and wind turbine but a single parameter capturing the essential elements of the geometry called the **Fresnel Parameter** or the **Fresnel–Kirchhoff Parameter**, can be

**Recall it can only be truly coherent if the distance d_1 is infinite.

calculated. Using the terms shown in Figure 5.31, the Fresnel–Kirchhoff Parameter is defined thus:

$$\text{Fresnel} - \text{Kirchhoff parameter, } \nu = h\sqrt{(2/\lambda((1/d_1) + (1/d_2))} \tag{5.46}$$

In the case of a radar, it is important to know the two-way loss to account for the losses in the return path experienced by the echo. If the path is reversed, then d_1 would become d_2 and vice versa. Note that swapping these terms makes no difference to the calculation of the Fresnel–Kirchhoff Parameter; that is, the path geometry is treated as though it is reciprocal [25].

5.6.10 Approximating the diffraction loss

The following formula can be found in recommendation propagation (*P*) −526 published by the International Telecommunications Union (ITU) Radiocommunications Sector (ITU-R) Fresnel–Kirchhoff Parameter, ν, greater than −0.78:

$$\text{Diffraction loss} = 6.9 + 20\log_{10}(\sqrt{((\nu - 0.1)2 + 1 + \nu - 0.1))} \tag{5.47}$$

The formula describes a curve that replicates the solution of the Fresnel Integral within the prescribed limits. Note the formula only predicts the one-way diffraction loss.

Figure 5.32 illustrates some important features of diffraction. The grazing case where the breakpoint touches the line-of-sight ray gives rise to a 6 dB diffraction loss. This was discussed above, being arrived at using the Cornu spiral.

Yet another artefact that the equation (correctly) predicts is that the loss increases as the square root of the reciprocal of the wavelength. That is the square root of frequency. Thus, as frequency increases, the losses increase. This can be

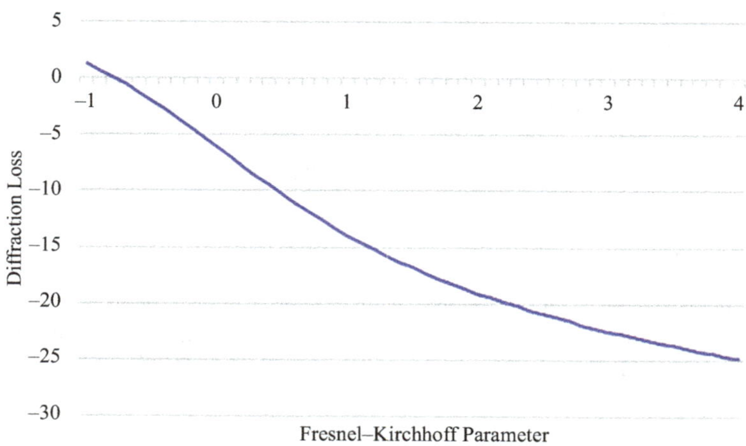

Figure 5.32 Diffraction loss versus Fresnel–Kirchhoff Parameter

regarded in two ways, lower frequencies penetrate shadows better than high frequencies or, alternatively, as frequencies increase the shadow gets deeper and the behaviour becomes more like a (hard) optical shadow. This was discussed in Chapter 4 in the context of the shadow behind wind turbine blades.

5.6.11 Free space path losses

A second feature that can be seen in Figure 5.32 is that the loss does not fall to zero until the Fresnel–Kirchhoff Parameter value reaches a value of 0.75. This value corresponds to a geometry where the obstruction is 0.6 of the first Fresnel zone from the direct ray, that is, the Free Space Path is established. In other words, if any obstruction intrudes into the space around a radio ray corresponding to the 0.6 of the first Fresnel zone, there must be some degree of diffraction loss. Provided this criterion is met, then there will be no diffraction loss and the radio is only subject to **Free Space Path Losses**.

Free space path losses, sometimes referred to as Basic Transmission Losses, are incurred because the signal expands as it moves farther from the transmitter, sharing its energy over a larger area. This is accounted for by a loss in signal strength calculated as follows [26]:

$$\text{Basic transmission loss, } L_b = 32.4 + 20 \times \log10\,(f) + 20 \times \log10\,(d) \tag{5.48}$$

where f is the frequency in MHz; d is the path length in km; 32.4 is a constant that corrects for the use of non-SI units and the physical constants required by the equation[††].

The 0.6 of the first Fresnel zone criterion is used, inter alia, for ensuring safeguarding of line-of-sight paths, when it is referred to as the Bacon method [27]. Terrain profile plots such as Figures 5.16–5.18 include the radio ray formed by the half ellipse of the 0.6 of the first Fresnel zone to mark the point where the losses are due to Free Space Path.

5.6.12 Case study one

The following exemplar is a genuine case that arose where it was necessary to predict if a wind farm would be visible to a radar. The radar in question is real. It is located in the United Kingdom. The example is a single turbine from a group of turbines making up a potential farm. The location of the turbine has been changed to protect the location of the proposed farm. Figure 5.33 shows the terrain profile plot between the radar and the turbine. The ordinate (y) axis of the plot is the vertical height referenced to Mean Sea Level and measured in metres, hence mamsl, metres above mean sea level. The abscissa (x) axis is the distance from the

[††]The formula is presented by the ITU in a number of their recommendations. It is custom and practice for them to present such material using units that are most commonly encountered for the subject in question; in this case in kilometres and MHz. This can be quite confusing for readers more used to System International (SI) units for equations.

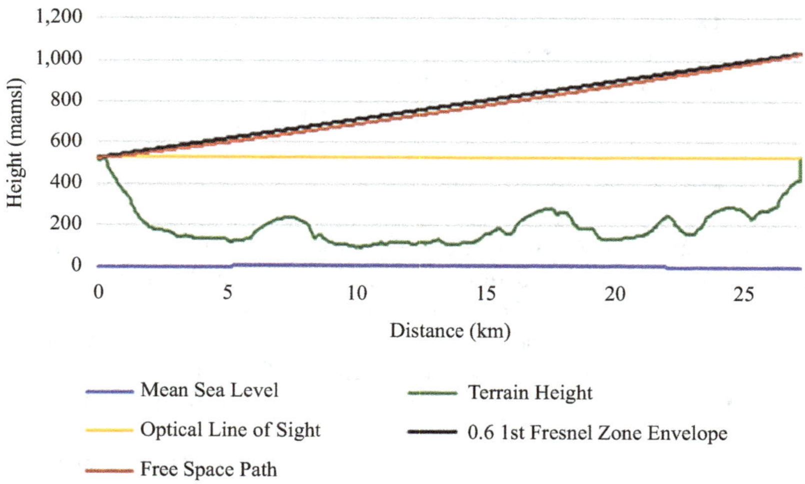

Figure 5.33 Terrain profile plot turbine A

radar measured in kilometres. The blue line in the plot is the mean sea level. Notice the line is curved, from which it would be correctly deduced that the plot takes account of Earth curvature. Not only is the Earth's curvature accounted for in such plots it also takes into account the effective Earth radius to correct for the refraction of radio waves. In this case, the plot was created based on the assumption of a 50% atmosphere. The green plot shows the terrain height above mean sea level. Note that at the right-hand end of the line, the vertical feature is the wind turbine. A tip height of 100 m was chosen to simplify this example by making the terminal points the same height above ground level, that is, to make it consistent with the simple method of calculating diffraction loss without having to perform any additional calculations. The yellow line is the optical line-of-sight formed in the direction of the turbine from the radar. The black line is the radio ray from the radar that follows a free space path (one not interfered with by terrain) and the associated brown line shows the locus of the 0.6 times the First Fresnel zone.

Looking at the plot it would be correct to assume that the radar is located on a hill overlooking the surrounding terrain. A feature of terrain profile plots is that the scale of the ordinate is always far greater than that of the abscissa; which has the effect of grossly exaggerating the features in the terrain. Note in the case of Figure 5.33, the ordinate axis represents 1,200 m and the abscissa over 25 km (27.28 km to be exact). Not only are the physical height of features exaggerated but also angles look much larger than they are in practice. The angle formed by the optical line of sight and the free space path ray is actually slightly less than 1°. Note also that free space path ray and the optical line of sight are both presented on the plot as straight lines. But the reader may ask why are they straight when both should be curved because of refraction. It is common for terrain plots to represent the curved lines of the electromagnetic wave as though they were straight lines and to accommodate the curvature by distorting the other features of the plot. However, this places an onus on the analyst to ensure refraction is accounted for.

In terms of deciding if the wind turbine will be visible to the radar, the principal factor that stands out is the height of the free space path ray above the tip of the turbine. This height is approximately 500 m, a large value and in other circumstances, without further analysis, it might be assumed that the turbine would not be visible to the radar. However, on further inspection, the optical line of sight is in fact formed at the tip of the wind turbine, 100 m above the local terrain.

The Fresnel–Kirchhoff for this path was calculated using the parameters: d_1 = 251 m, d_2 = 27.3 km, and h = 0.1 m. Initially, it will be assumed the wavelength of the radar is 23 cm (the actual wavelength of the radar in this case). Substituting these values into the above equation yields a value for the Fresnel–Kirchhoff parameter of 0.02. This value was then substituted into the equation for calculating diffraction loss and this yields a one-way loss value of 6.6 dB. This process was repeated for higher frequencies (shorter wavelengths) and the results are shown in Table 5.7.

Table 5.7 also shows that the diffraction loss is little changed by frequency and that the loss is small. It would have a very limited impact on the visibility of the turbine (despite the relatively large difference between the height of the tip of the turbine and the free space path ray.

5.6.12.1 Base of solid radar cover

The example above is a good vehicle to discuss the concept of the base of a solid radar cover. The diffraction losses caused by the obstruction are relatively small (approximately 6.6 dB). The wind turbine location is just over 27 km distant from the radar. Even without calculating whether an echo from the wind turbine would be detectable, with such low losses that is not in question. So, is the base of solid radar cover the height of the turbine tip (or even lower)? Even though the losses are small there are still losses. The diffraction losses do not return to zero above the wind turbine location until the beam is 1,040 m above mean sea level (mamsl) and approximately 600 m above ground level (magl), see Figure 5.33. It would be possible to detect a standard target size at a lower altitude but below the ray formed by 0.6 of the first Fresnel Zone it will not be not possible to detect everything that is detectable in the absence of an obstruction. Thus, the height of the ray formed by

Table 5.7 Wavelength versus one-way diffraction loss case study one

Wavelength (cm)	One-way diffraction loss (dB)
23	6.56
10	6.60
3	6.70

the 0.6 of the First Fresnel Zone at any point on the ground is a conservative Base of Solid Radar Cover.

5.6.13 Case study two

This exemplar is another genuine case but the turbine tip height has been reduced to illustrate better the principles of diffraction. The radar in the example is no longer operational but was an airfield radar in the United Kingdom operating in the 2.7–2.9 GHz band (S-Band). The wind turbine tip height, 18 m, might be consistent with a large domestic premise or small holding (Figure 5.34).

The radar is located on relatively flat ground consistent with an aerodrome and approximately 13 km distant there is a small hill that screens the turbine from the radar. The Fresnel–Kirchhoff for this path was calculated using the parameters: $d_1 = 13.5$ km, $d_2 = 2.6$ km and $h = 19.5$ m, yielding a value of 3.4. Substituting this value into the equation for the one-way diffraction loss yields a value of 12.2 dB. This calculation was repeated for a 23-cm and a 3-cm radar and the results are presented in Table 5.8.

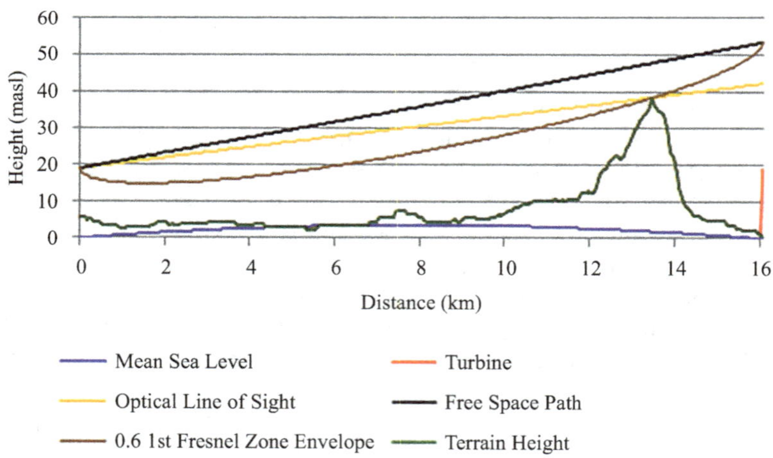

Figure 5.34 Terrain profile plot turbine B

Table 5.8 Wavelength versus one-way diffraction loss case study two

Wavelength (cm)	One-way diffraction loss (dB)
23	12.2
10	14.6
3	18.7

Table 5.8 illustrates the much greater diffraction loss associated with a higher value of Fresnel–Kirchhoff Parameter but, in addition, it also shows that the losses are lower with the longer wavelength radar demonstrating that longer wavelengths are able to penetrate shadows better than shorter wavelength signals.

Thus, methods have been discussed for the calculation of diffraction losses based on a single obstruction.

5.6.14 Diffraction loss and multiple obstructions

5.6.14.1 General observation – the spherical Earth

Although even short stretches of terrain can be complex, more complex terrain is often associated with longer paths over higher ground. When diffraction losses were calculated for a single obstruction, it was pointed out that the total pass loss also has to take into account the losses associated with a radio wave travelling the distance between the terminal points. In that sense, the diffraction loss is an additional loss. Therefore, the total path loss is the sum of the Free Space Path loss and the diffraction loss caused by the obstruction.

As the path length increases, there comes a point where the far end of the path for the radio wave passes beyond the horizon. Therefore, the surface of the Earth becomes an obstruction and the radio wave experiences diffraction losses. There are a number of ways to calculate this loss. ITU-R-P.526 provides a comprehensive treatment. Bretton points out that 'little physical insight' is associated with the formulae for calculating the loss [28]. The various formulae are empirical, providing a good fit to match what is observed in practice.

The loss due to diffraction over a spherical Earth after Ho and Tan is calculated as follows [29]:

$$L_{DS} = F(X) - (G_1 + G_2) \tag{5.49}$$

where $F = 17.6X - 10\log(X) - 11$; $X = 22f^{1/3}\,a_e^{-2/3}\,d$; A_e is the effective Earth radius usually assumed to be 8,500 km; f is the frequency in GHz; d is the path length in km; G_1 and G_2 are the antenna height gains in dB. The gains are available from ITU-R-P-526 (Equations (5.11) and (5.11A)).

This equation takes into account the heights of the terminal points and the distance between those points. The latter informs the amount by which the path is affected by the Earth's curvature, the longer the path the greater the effect. However, the higher the terminal points, the more the path rises above the Earth's surface.

5.6.14.2 Multiple obstruction methods

A number of different methods have been devised to estimate the path losses when multiple obstructions affect the radio path. As with the losses due to the spherical Earth, the methods attempt to provide a best-fit to the observed data rather than to any differences in the basic physics. The diffraction of radio waves over multiple obstructions is an active area of research. Techniques used include the Epstein Peterson, Deygout and Bullington methods. The current preferred method, at the time of writing, is a variation of the Bullington method called Delta Bullington, but in the future, this may change.

5.6.14.3 The Bullington method

The method was originally designed by the US electrical engineer, Kenneth Bullington (1913–1984). The core Bullington method is illustrated in Figure 5.35. The method requires the terrain to be searched to find the points in the terrain which create the horizon as seen from each terminal point. The lines forming the horizons are projected until they meet creating, in effect, a synthetic point in the terrain. The point where this imaginary feature meets the ground is called the Bullington Point. Unlike other methods, the Bullington method ignores all the other features in the terrain.

The first step in the method is to find out if the path between (in this case) the radar and the wind turbine is line-of-sight.

If the path is line-of-sight, the Fresnel–Kirchhoff parameter, v, is calculated for the path and an un-corrected path loss, L_{UC}, is calculated as it was for a single obstruction. This accounts for any intrusions into the 0.6 of the first Fresnel zone.

If the path is as shown in Figure 5.35, it is said to be a '**transhorizon**' path. The Fresnel–Kirchhoff parameter is worked out using the geometry defined by the imaginary point above the Bullington point and L_{UC} is calculated as it was for the line-of-sight case.

The final step of the method is to calculate the actual Bullington loss using the following empirical formula:

$$\text{Bullington diffraction loss}, L_B = L_{UC} + [1 - \exp(-L_{UC}/6)](10 + 0.02\,\text{d}) \tag{5.50}$$

where d is the path length.

The path loss calculated using the basic Bullington method is known to under-predict path losses when compared with measured values. This is attributed to ignoring some of the features in the landscape by generating an artificial feature above the Bullington point. To improve the performance of the algorithm a variation called Delta Bullington method was devised.

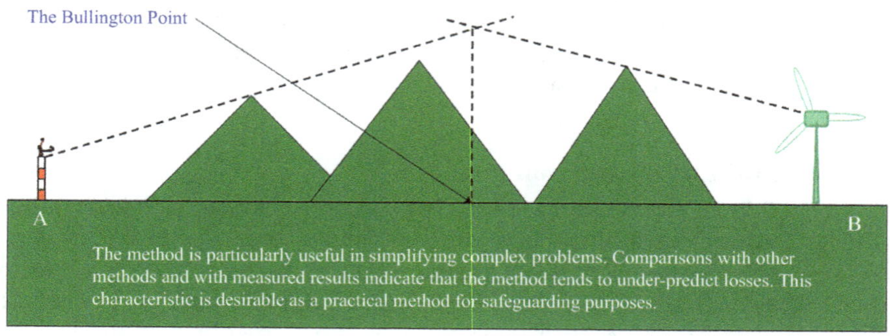

Figure 5.35 The Bullington method

5.6.14.4 The Delta Bullington method

The first step in the modified algorithm is to use the original Bullington method to create the Bullington diffraction loss, now called the actual Bullington loss, L_{BA}.

Step two in the algorithm is to calculate the path loss between the terminal points as though the Earth was smooth. This loss is usually called, L_{BS}.

Step three calculates the spherical Earth loss, described above, L_{DS}.

The final step is to calculate the Delta Bullington *L*oss, L_{DB} which is the sum of L_{BA} and whichever is the greater of L_{DS}–L_{BS} and 0.

5.7 Mapping

5.7.1 Good practice

One of the most contentious subjects surrounding the prediction of the effects of wind turbines and wind farms is working out where the wind turbine is planned to go and its location with respect to the systems described in Chapter 3. This should not be the case, but it is. There are many reasons.

From the initial assessment of a site, a wind turbine layout will be designed. However, Chapter 4 shows there are many stages to assessing the feasibility of a site. As the process of analysing and reviewing feasibility progresses, it is often necessary to move or even drop turbine locations to account for new findings that must be dealt with. It is also easy to work unintentionally with an out-of-date map of a site. ***The analyst must be assured by the developer that any site maps and coordinate lists are current.***

Ideally, turbine locations would be provided in a format that can be used directly by whatever tool the analyst must use. Frequently, this is not the case. An element of manual transcription of data will take place. It is easy to make mistakes. Some of the more extreme mistakes are easy to detect, a longitude and latitude that are transposed is one example: at least it is easy to detect provided there is a process looking for errors. If using grid locations, it is also easy to transpose Eastings and Northings. This can be harder to detect if the Easting and Northing values are similar. The hardest errors to detect are those where individual numbers are transposed. ***The analyst should validate wind turbine locations before carrying out any work.*** Useful practice is to have a third party check all the locations. A further useful practical technique is to plot locations on a map and compare maps with those provided by another source. The analyst must remember that following best practice, being careful and doing good analysis will be of no value to a client if the wind turbine locations are incorrect. It will appear shoddy and will call into question the whole analysis. ***When reporting on analysis, it is good practice to report to the client the sources of all the information used and the date of receipt, even if the client was the source.***

It is easy to think of identifying the correct location as being a wind farm problem. It is equally important to validate the locations of aviation radio and radar systems. One of the reasons Chapter 3 included images of system antennas was to allow the analyst to use tools such as Google Earth to verify locations. ***The analyst***

should verify the location of all radio and radar system antennas. Such systems have long lives but they eventually must be replaced (usually when maintenance becomes too expensive). ***The analyst should verify that radio and radar systems are current.*** Related to this problem, suppose a system is scheduled for replacement before a prospective wind farm is built. If the system in question is important, then it may not be acceptable to have any loss of service. This situation means that it cannot be assumed that a replacement system will be in the same location as an existing system. It may also be the case that systems will be discontinued. The client will not thank the analyst for wasting resources on a system that is to be withdrawn in a year's time. ***The analyst should check if there are plans for new aviation radio and radar systems or plans for withdrawing existing systems from service.***

Having confirmed the location of a wind turbine, perhaps to the nearest metre, it must be remembered that when the wind farm is built, micro-siting changes may move the planned location by as much as 50 m. Rarely, but sometimes, moving a turbine a few metres may make a significant difference to its detectability. Tools for performing terrain profile analysis will usually provide information on the path between the radio or radar system and a wind turbine. They do not tell the analyst what is happening beyond or on either side of the wind turbine. It is prudent to investigate terrain beyond and on either side of a wind turbine[‡‡]. ***The analyst should investigate not just the locations of wind turbines but also understand the environment around the wind turbines.***

Wind turbine locations and radio and radar systems are not the only features that can be dynamic. The built environment changes continuously. Domestic settings are unlikely to have a material effect on terrain path analysis but industrial development may. Sometimes the land itself changes. Figure 5.36 shows the land-art form of Northumberlandia, sometimes referred to as the 'Lady of the North'. This artwork is 34 m high and 400 m long. It was built in less than 2 years, starting in 2011 as part of an opencast coal mine development but it was a number of years before it appeared on maps[§§]. The possibility of construction is an obvious thing to be alert to. But occasionally things can happen to even ancient woodland. Loss of woodland may not be deliberate but sometimes natural damage can occur. Blights such as Dutch Elm disease and Ash dieback can decimate woodland as well as fire and wind. The trees in commercial forestation will also be harvested from time to time. The analyst cannot foresee every change that might take place in the environment during the life of a wind farm. However, ***if a screening by trees is a critical part of an analysis alerting a client is prudent.***

5.7.2 The terrain profile mapping

To assess whether wind turbines will have an effect on an aviation radio or radar system, it will be necessary to have some form of software, terrain profile mapping,

[‡‡]This has been useful for the author and his clients on multiple occasions.

[§§]This image was included because it was material in a wind farm-radar path analysis.

Figure 5.36 Northumberlandia

tool. These tools may prefilter sites, deciding whether or not greater screening is needed or they may provide advice on different propagation models. Care must be exercised when selecting and using models.

5.7.2.1 What does a terrain profile provide?

A terrain profile map is created by using a database of the area of interest that has, at some regular distances apart, a value for the height above a datum for the terrain at that point. A term that is used in the literature is a **post-spacing**. Such maps of course must be updated on a regular basis: for the reasons set out above, the terrain changes.

NASA's Shuttle Radar Topography Mission (SRTM) was, arguably, one of the most comprehensive terrain profile data collection exercises but it was carried out over 20 years ago. The data is still useful as a cross-check against other data sources but the core data set is now too old to be relied upon as an independent source. This mission measured the height from an orbiting spacecraft of the radio-reflecting surface of terrain underneath the craft as it orbited the Earth. This was very useful for a while providing guidance to the nature of the surface when compared with an alternative ground-based measurement. For example, if a ground-based measurement gave one measurement and SRTM data did not agree it may, for example, have been because the surface at that point was forested [30]. But it is over 20 years since this was updated. By comparison, in the United Kingdom, Ordnance Survey terrain profile databases are updated annually.

Haslett, recommends that the granularity of the map should offer a post spacing of 50 m and ideally 20 m. This recommendation was within the context of the safeguarding of fixed links but the recommendation might be used for the safeguarding of radar and radio systems too.

Finally and importantly, as a minimum, terrain profile maps should provide corrections for the refraction of radio waves. This is a very basic level of capability.

References

[1] Restrepo, J. (2016), *Radio Regulations*, World Radiocommunication Seminar 2016, https://www.itu.int/dms_pub/itu-r/md/15/wrs16/sp/R15-WRS16-SP-0003!!PDF-E.pdf. Retrieved 5 March 2023.

[2] Ofcom (2107), *The UK Frequency Allocation Table*, Issue 18 v1.1, Ofcom, 12 January 2017.

[3] FCC (2023), *Radio Spectrum Allocation*, FCC. https://www.fcc.gov/engineering-technology/policy-and-rules-division/general/radio-spectrum-allocation. Retrieved 5 March 2023.

[4] Skolnik, M.I. (2001), *Introduction to Radar Systems*, 3rd ed., McGraw Hill.

[5] For a discussion on the Power Equation see, for example, Nahvi, M. and Edminster, J.A. (1979), *Electromagnetics*, 5th ed., McGraw Hill Schaum's Outlines.

[6] Johnson, J.B. (1928), 'Thermal agitation of electricity in conductors', *American Physical Society, Physics Review*, 32:97 and Nyquist, H. (1928), 'Thermal agitation of electric charge in conductors", pp110.

[7] Skolnik, M.I. (2008), *Radar Handbook*, 3rd ed., McGraw Hill. Also available as an E-book.

[8] Blake, L.V. (1980), *Radar Range-Performance Analysis*, Lexington Books.

[9] Douthwaite, D.M., Newsome, K.D., Edwards, M.B., *et al.* (2013), 'Receiver protection in S-Band radars for mitigation of 4G signal interference', *Microwave Journal*.

[10] Skolnik, M.I. (1980), *Introduction to Radar Systems*, McGraw Hill International.

[11] Blake, L.V. (1980), *Radar Range-Performance Analysis*, Lexington Books.

[12] Norton, K.A. and Omberg, A.C. (1943), 'Maximum range of a radar set', *Proceedings of the IRE*, 35(1):4–24.

[13] Blake, L.V. (1980), *Radar Range-Performance Analysis*, Lexington Books.

[14] Barton, D. (2012), *Radar Equations for Modern Radar*, Artech House.

[15] ITU (2022), *Radio Noise*, ITU Recommendation ITU-R P.372-7.

[16] Blake, L.V. (1972), *Radar/Radio Tropospheric Absorption and Noise Temperature*, Naval Research Laboratory Report AD-753-197, 30 October 1972.

[17] Toomay, J.C. (2012), *Radar Principles for the Non-Specialist*, Scitech Publishing Inc.

[18] Gerlock, R.A. (1962), *Study of Interference Aspects of Fresnel Region Phenomena*, Report RADC-TDR-62-496 Vol 1, 1 August 1962.

[19] CFS (2017, *Meteorology*, The Central Flying School Manual of Flying, vol. 10, AP3456. https://assets.publishing.service.gov.uk/government/uploads/

system/uploads/attachment_data/file/857308/Volume_10_Meteorology__CFS_version_.pdf. Consulted January 2023.

[20] For a derivation of the equations see Bean and Dutton (1955), *Radio Meteorology*, US National Bureau of Standards, 1966 (Chapter 1).

[21] ITU (2022), Attenuation by Atmospheric Gases and Related Effects, ITU Recommendation P.676.

[22] Wisher, S. (2012), 'Super-refraction at visible and radio wavelengths', *Radio Communication*, June 2012, Image reproduced by kind permission of the Radio Society of Great Britain (RSGB).

[23] Haslett, C. (2008), *Radio Wave Propagation*, Cambridge University Press.

[24] Bacon, D.F. (2003), 'Introduction to diffraction', Chapter 8 in Barclay (2003), *Propagation of Radiowaves*, 2nd ed., Les Barclay (ed.), IET.

[25] Haslett, C. (2008), *Radio Wave Propagation*, Cambridge University Press.

[26] ITU-R PN 525-2, Annex 2.

[27] Bacon, D.F. (2002), A Proposed Method for Establishing an Exclusion Zone Around a Terrestrial Fixed Radio Link Outside of Which a Wind Turbine Will Cause Negligible Degradation of the Radio Link Performance, https://www.ofcom.org.uk/__data/assets/pdf_file/0031/68827/windfarmdavidbacon.pdf. Retrieved February 2023.

[28] Bretton, D.J. (2022), A Study on the Delta-Bullington Irregular Terrain Radiofrequency Propagation Model Assessing Model Suitability for Use in Decision Support Tools, Engineer Research and Development Center, US Army Corps of Engineers.

[29] Ho and Tan (2000), 'Interference Effects of Deep Space Network Transmitters on IMT2000/UMTS Receivers at S-Band'.

[30] Haslett, C. (2008), *Radio Wave Propagation*, Cambridge University Press.

Chapter 6
Mitigation

6.1 Definition and challenges

Dictionary definitions of the term ‘mitigation’ take the form:

> ‘To cause to become less harsh or hostile or to lessen the seriousness [of a situation]’ [1]

In the context of this book, to such definitions might be added ‘to a level that allows the task in hand to be carried out safely’. The task in hand, the capability to provide an Air Traffic Service, must be carried out safely and mitigation is concerned with retaining or maintaining this capability. But if the capability is provided by a new or modified system, how might it be determined if it will be safe under all the circumstances that will be encountered during its life? Suggestions taken from the wider field of electronic systems are introduced in this chapter.

However, before discussing these issues, this chapter will address the three broad categories of mitigation options, that is modification of:

- The wind farm proposal.
- The aviation service being delivered.

  or

- Modification or replacement of the system that is predicted to be affected.

And/or a combination of these three possibilities. These options are discussed in turn.

6.2 Modification of the wind farm proposal

6.2.1 Removal of wind turbines

If some or all of the turbines in a wind farm are visible to a radar, the simplest mitigation that might be considered by a wind farm developer is to modify the turbine layout so that they all become invisible. It is worth making the point that the aim must be to render the turbines invisible not just to reduce their visibility. Partial visibility means that some of the time a turbine(s) would be visible and distracting to air traffic controllers.

If only a few turbines in a wind farm are visible, then it may be economically viable to remove the visible turbines from the plans.

6.2.2 Reduced height of wind turbines

Another possible option might be to reduce the height of turbine tips. Earlier, this option might have been more practical than it may be today. In Chapter 4, it was shown that the problems of wind turbulence in some sites can be addressed by making turbine heights higher as increasing height lifts the blades out of turbulent air. If turbine height was increased to make them operate in the local wind environment, then reducing the height to make them invisible to radar may simply not be viable. Furthermore, the visual appearance of a wind farm is less acceptable if the the tip heights vary across the farm. Reducing the height of some turbines in a development may eliminate one problem but cause another.

There are also other trends. Turbine sizes are increasing to make them harvest more wind, to produce more electricity. Reducing the turbine height runs contrary to this trend.

Moreover, in some terrains, the turbine size could be drastically reduced in height and the turbines would still remain visible. A good example of this problem is illustrated off-shore. As the ability to build wind farms further off-shore becomes a realistic option, there may be some situations where reducing the height of the turbine might render them invisible. However, the reduced wind turbulence over water might mean the turbines could already be relatively low compared to an on-shore turbine of the same productivity.

6.2.3 Special coatings

Another mitigation possibility is to reduce the visibility of turbines by the use of special coatings for the tower, nacelle and blades. These coatings reduce the RCS of turbines. Coupled with changes to the radar, this may be efficacious. However, in this section, the consideration is changes to the wind farm alone. Reducing the turbine RCS might be practicable in some circumstances where there is already some screening available from the terrain. However, on its own, the problem would remain that the turbine must be rendered invisible not just reduced in visibility.

In summary, while reduction in turbine size might have been an option in some situations, today it is likely to be a less attractive option and more likely to render a plan unaffordable.

6.2.4 Turbine curtailment

There is one option for rendering a turbine invisible while retaining height, that is, to stop the blades from rotating. This was discussed at some length in Chapter 4. There are some mitigations in place that rely on this option but, as were explained, there are many reasons why this is undesirable. If the curtailment is driven by the reduction in demand, the wind turbine operator may benefit from some payment; however, as a means of aviation mitigation, it could reduce the operator's revenue. Curtailment is also highly undesirable for the electricity distributor, especially if unplanned because the electricity must be sourced from elsewhere. The problems are exacerbated for all concerned if curtailment is used frequently.

6.3 Modification of the aviation service being delivered

Before responding to a proposal to build a wind farm that has been submitted for planning permission the Air Navigation Service Provider (ANSP) will consider whether they need to object or whether an operational workaround would be more appropriate. Regulators recognise that this is a valid approach to the problem. In the United Kingdom, for example, CAP 764 and 670 address whether it is appropriate to modify air traffic control procedures to accommodate wind farms. These procedural modifications include, for example, re-routing aircraft and changing the way airspace is used.

The following discussion is intended to help a layperson understand discussions of this complex topic. For authoritative information on these operational matters, the appropriate national regulators' rules and regulations should be consulted.

6.3.1 Operational workarounds

This approach accepts that there is clutter present and the operator has to work around it for example by rerouting aircraft. Figure 6.1 is a sketch of an ATC screen that might be seen at an aerodrome. There are three wind farms shown in the sketch each one creating a region of clutter. Wind turbine clutter can appear like an aircraft flying over a wind farm but Air Traffic Controllers cannot be certain that the returns observed are caused by clutter or by, say, a GA (light) aircraft. Thus, avoiding action will be necessary to ensure separation rules are maintained as illustrated in Figure 6.2.

This approach may be acceptable if the likelihood of needing to reroute aircraft is low. However, if the airspace is used a lot, then it is likely that this will lead to an objection. Moreover, the possibility of having to reroute aircraft often is in no one's interests: time will be added to the flight of the diverted aircraft, extra fuel consumed and the tempo of the controller's job is affected calling for extra attention to the situation. Nor, in the planning stages, is it in the wind farm developer's interests. If planning permission is being sought, the risk of reduced benefits from the wind farm by such diversions is unwelcome. Above all, safety must not be compromised. However, when a wind farm is being commissioned, if local circumstances permit, then for short periods of time, the risk might be managed until a permanent mitigation is made operational. The risk *may* be accepted only after a proper evaluation of the circumstances as a workaround.

6.3.2 Changes to airspace

Changing the use of airspace, such as changing the class of airspace, might be considered with a view to improving the conspicuity of GA aircraft. In effect, such a change would either exclude GA from some volumes of airspace or mandate them to improve their equipage [2]. In practical terms that would mean requiring aircraft to carry SSR or

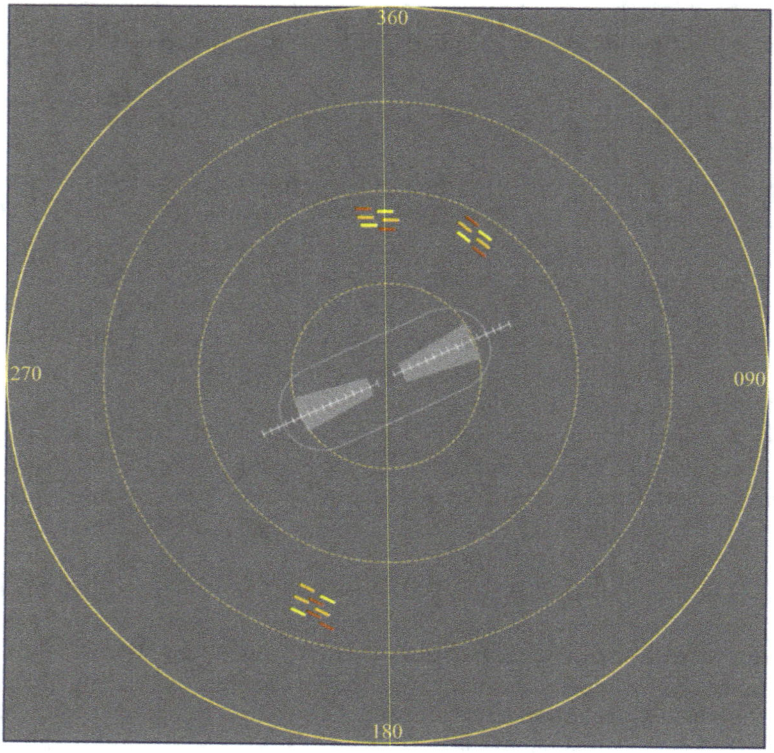

Figure 6.1 ATC picture with clutter

ADSB transponders. However, such changes are contrary to the ICAO presumption of using the least amount of regulation of airspace (discussed in Chapter 3).

There are also practical concerns with such an approach. It can take a long time to change airspace. There are exhaustive procedures to follow, which can be time consuming [3]. One change to airspace usage that has been used for mitigating the impact of wind farms is the imposition of the transponder mandatory zone (TMZ).

6.3.2.1 Transponder mandatory zones

There are some volumes of airspace where special measures have to be put in place to maintain safety by improving the 'conspicuity'* of aircraft. One way of achieving this is to create a TMZ, also referred to as 'SSR only zones' [4]. As the name implies, aircraft entering a TMZ must be equipped with an SSR transponder which must be capable of reporting the height of the aircraft using a pressure-derived altimeter. In addition, the transponder should be capable of operating in Mode S. The transponder must also be operating although, if the equipment fails or if the transponder is only capable of operating in Mode A/C then, ATC can grant access as an exception. It will be obvious

*Making it easier for the controllers to see aircraft.

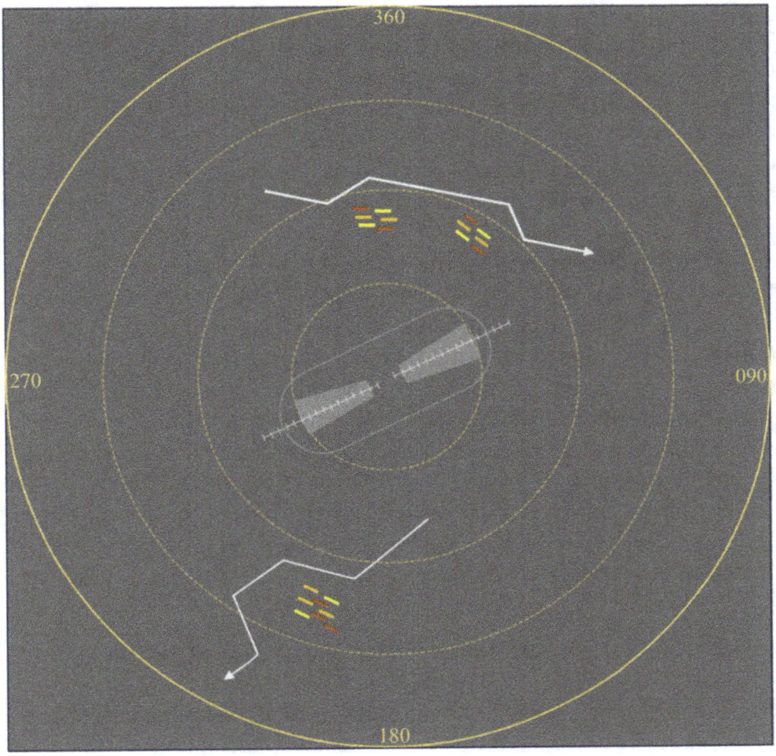

Figure 6.2 Rerouting aircraft around clutter

that a TMZ will have a lateral extent but it may also have vertical limits. For example, in the United Kingdom, there are two TMZs in the airspace over Stansted Airport; one at either end of the control zone and they extend from the surface to 1,500 ft. The TMZs were put in place in 2009 after a high volume of serious airspace incursions [5]. Other TMZs are associated with airspace used for military exercises; for example, in the Moray Firth. The United Kingdom is far from unique in using TMZs. The TMZ at Memmingen in Germany has recently been in the news because it has been extended [6], and almost all of Dutch Airspace is covered by TMZs [7].

A TMZ can be used to provide wind farm mitigation. So that, as the aircraft transits the airspace, it can always be distinguished from wind turbine returns on primary radar and if there is more than one aircraft in the vicinity it will be possible for controllers to maintain separation.

As all aircraft using such airspace must carry SSR, the possibility of primary returns adding clutter to the picture can be eliminated by using one of the forms of blanking in the region; for example, RAG mapping described later in this chapter. Blanking eliminates clutter from the picture but quite aside from the obvious advantage of removing the source of clutter, desirable in its own right, it also removes the possibility that the clutter may obscure part of the SSR Track Data Block.

A TMZ may not be appropriate for some airspace because its dependence on cooperative surveillance and the associated removal of the primary radar returns would not be acceptable on defence grounds.

TMZs can be used for the purpose of windfarm mitigation; for example, in UK waters, the London Array and the Thanet windfarms both have TMZ in place and both have vertical limits as well as lateral limits. For example, the TMZ for Thanet Windfarm extends from the surface to Flight Level 105. All the airspace associated with these TMZs are Class G [8].

One of the disadvantages of a TMZ is that it departs from the general principle, described in Chapter 3, that airspace should have a minimum of regulation. The introduction of a TMZ may be opposed by the General Aviation community; whose aircraft may not be equipped with transponders. TMZs are more attractive off-shore because there is less use of the airspace offshore by GA. In the future, TMZs may become more acceptable because other forms of cooperative surveillance are becoming available which may be more affordable for the GA community.

Implementation of a TMZ is likely to require a period of consultation after which an Airspace Change Process must be followed in accordance with the local regulator. For example, in the United Kingdom, the CAA process for changing the use of airspace is set out in CAP 725.

6.4 Modification or replacement of affected systems

In this section, mitigations requiring modifications to the radar are considered. Radar mitigation options are evolving: the process is not complete. It is useful to consider the form that evolution is taking and also consider what still needs to be done.

- **Clutter removal**: This method prevents the operator from observing and being distracted by wind farm echoes by removing them from the display (or the radar receiver). Examples include Range Azimuth Gating and Antenna Tilting.
- **Clutter removal and augmentation**: This method removes the distraction of clutter but provides a means of providing aircraft returns in the airspace over the wind farm,
    - Using Secondary Surveillance Radar (SSR). This method is associated with a change to the airspace to make it a TMZ. The method has the advantages and disadvantages of implementing a TMZ; the additional regulation of airspace and the presumption that all aircraft entering the airspace are cooperating using an SSR/IFF transponder.
    - Using Primary Surveillance Radar (PSR). This method requires the use of an in-fill radar to provide primary returns from aircraft over the wind farm area.
    - Using a-priori information about aircraft.
- **Clutter discrimination**: The radar in this method is capable of discriminating wind farm returns from aircraft returns so the wind farm returns can be discarded leaving only the aircraft to be tracked.

The discussion starts with the first method, that is, clutter removal: techniques to prevent the radar ‘seeing’ the wind farm.

6.4.1 Clutter removal

There are a number of ways to remove clutter from the picture seen by the operator these are discussed in turn.

6.4.1.1 Clutter removal – range azimuth gating maps

Range gating has been a design feature used in generations of radars for different purposes depending on the type of radar. The feature is usually implemented as part of the initial stage of the signal processing part of the radar and the principle is simple. The 'gate' may be opened or closed, allowing targets through or preventing them from being processed further, see Figure 6.3. In some radars, they were used to simplify tracking of targets closing the gate most of the time and only opening it to allow returns through when the target being tracked was expected. In the context of surveillance radars, the gate is used in the opposite sense. Most of the time the gate is open, allowing returns through to be presented to the operator on the display, but it can be selectively closed to eliminate returns from specific ranges and azimuth bearings, hence **Range Azimuth Gating**. For example, a common source of clutter on radar displays is vehicular traffic: cars, lorries, etc. The traffic is moving and therefore is not removed by moving target filters and the echoes come from predictable locations like roads. Filtering out returns in limited areas of the radar coverage is helpful because it removes a source of distraction for the operator. The areas where targets are gated out are maintained in a map, hence, Range Azimuth Gating Maps or **RAG Mapping**.

Multiple gated regions can be used together in a single radar, as illustrated in Figure 6.4. However, as more regions are blanked the cumulative impact of the blanked regions will themselves become a distraction for the operator.

Adding turbine locations to the RAG map is a simple means of eliminating returns from an area being presented to the operator. However, this method is not able to distinguish clutter from genuine targets and not only will clutter be removed from the picture but aircraft in the RAG mapped region will also disappear as illustrated in Figure 6.5.

Whether this is tolerable depends on the operational assessment, taking into account factors such as how distracting the clutter is and the relative importance of having the airspace edited out of the display. The overriding consideration is

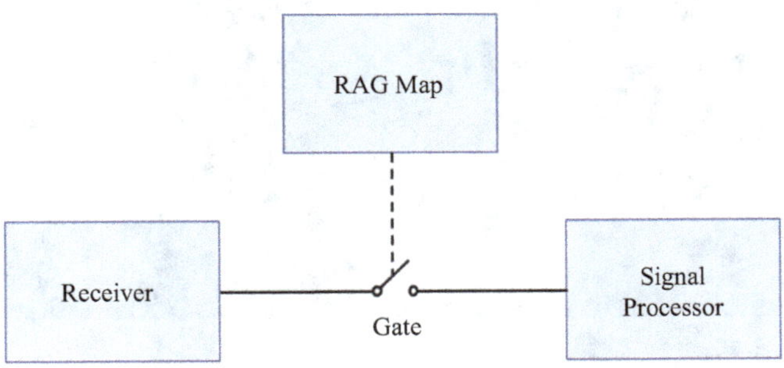

Figure 6.3 RAG mapping concept

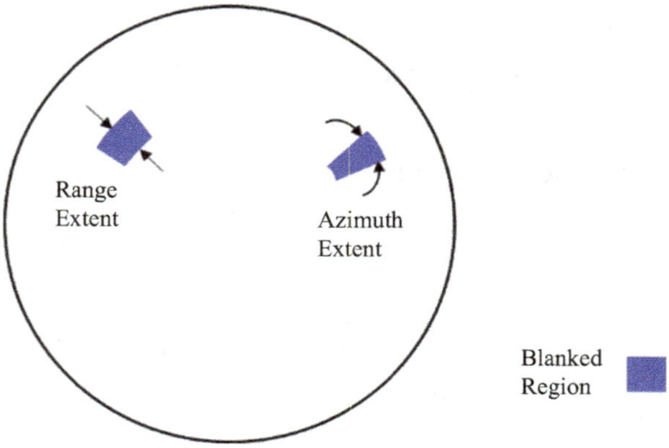

Figure 6.4 RAG-mapped display

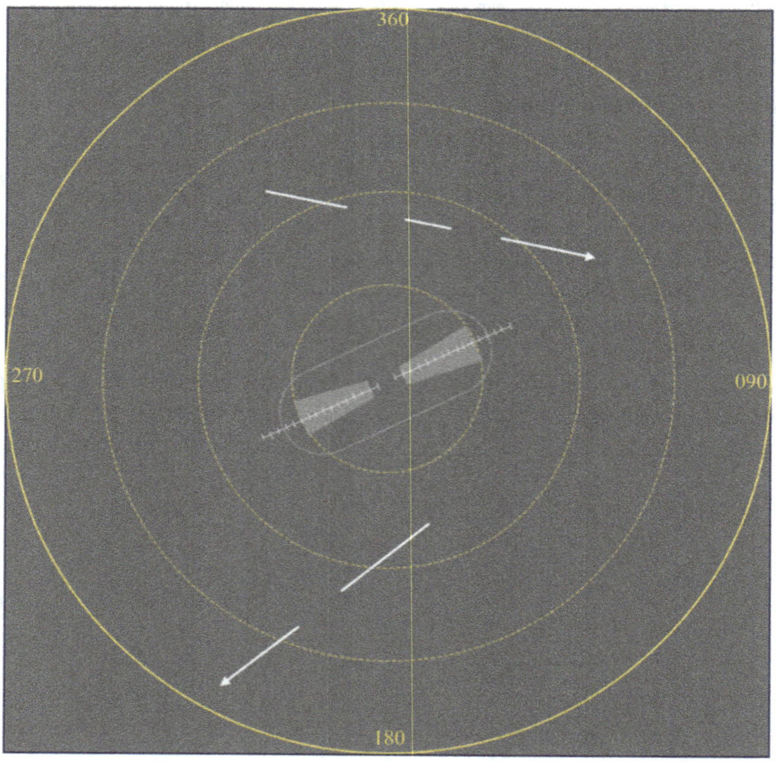

Figure 6.5 Clutter removal using RAG mapping

whether a robust safety case can be made taking into account the loss of the data. A concern would be if an air traffic service was being provided in the airspace where the data is being lost.

In some radars, the gating mechanism is more sophisticated; for example, the ASR-11 [9]. In this and similar radars when an aircraft is being tracked and it moves into the volume of airspace that is gated out, the radar is capable of overriding the gating. This will retain the target data and allow it to be displayed. This information should be considered with the discussion below on Non-Auto Initiation Zones (NAIZ).

Air defence radars have had much finer-grained control over the processing of returns for a long time, which provides greater flexibility when dealing with jamming. The processing in small regions of coverage (cells) can be modified and this technology is now available in some Air Traffic Control radars. For example, the gain applied to cells can be modified providing what is effectively a very granular form of RAG mapping. All the usual constraints apply with this technology: the processing is not able to distinguish between genuine targets and clutter so if the gain of a cell is reduced to zero both will be lost.

6.4.1.2 Weather RAG mapping

A further sophistication of simple RAG mapping is the ability of some radars to RAG map weather information. It may also be possible to edit into the RAG-mapped region the mean weather conditions in adjacent areas. This concept may be a fruitful area for research for its application to wind turbine clutter removal.

6.4.1.3 Clutter removal – sector blanking or PSR blanking

Another technique for manipulating the radar behaviour and the radar picture is **Sector Blanking**. Sector blanking can be used to control clutter but other applications include management of interference. In general, the term Sector, in this context, refers to performing an action on selected azimuths, as illustrated in Figure 6.6. Different radars may perform sector blanking either by stopping

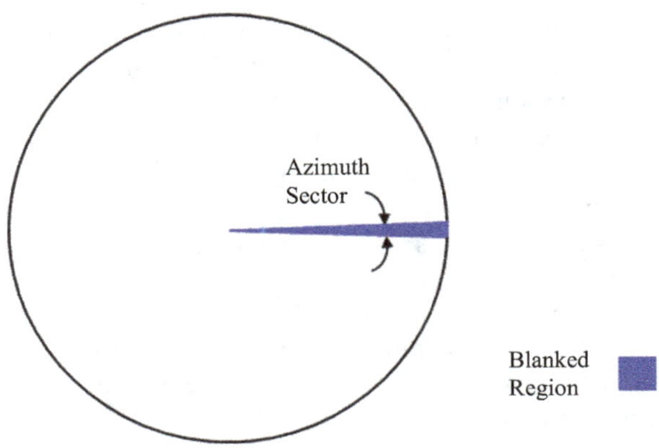

Figure 6.6 Sector blanking

transmitting or stopping receiving on the selected azimuths: **transmit blanking** and **receive blanking**, respectively. If transmit blanking is used, then no transmissions are made within the blanked sector, it follows therefore that there can be no echoes received from that sector at all ranges.

If received blanking is used, it would be possible to sub-divide the range and allow echoes to be detected within the blanked sector. This arrangement can be used to facilitate those more sophisticated radars that allow targets that are already in track to be tracked through regions of blanking but this is not the norm.

A further corollary of transmit blanking is that because dual beam antennas only transmit using the main beam, both the main beam and auxiliary beam are blanked. In the case of receive blanking then the main beam processing and auxiliary beam processing can be independent.

Just as is the case with RAG mapping, whether this level of editing of the picture would be acceptable would depend on the balance between reducing the workload on the operator and the penalty of removing genuine aircraft echoes from the display. Ultimately it depends on whether it is considered safe to operate with a permanently blanked volume of airspace.

6.4.1.4 Clutter removal – beam tilting

The de facto standard for an aerodrome radar antenna is the Cosecant Squared antenna described in Chapter 3. The peak gain of the main beam of such an antenna is usually at an elevation between 3° and 4° and the auxiliary beam between 6° and 7°. However, this angle can be adjusted. The tilt mechanism is mounted at the base of the antenna, on the rotating part of the antenna. The mechanism is like a jack that can lift the whole antenna. Tilting affects both the main and the auxiliary beams, which is to say, whatever tilt is applied to the main beam, then the auxiliary beam is increased or decreased by the same amount. There are limits to prevent extreme tilts and the tilting is typically limited to 4° or 5° greater than the nominal value and 2° or 3° below the nominal value. The adjustment is not continuous and will only be permitted in increments of either a quarter or a half of a degree.

The purpose of being able to adjust the antenna tilt is to optimise the balance between the detection of aircraft at low elevation angles and the level of ground clutter. A part of the commissioning process of a new radar is to tailor the coverage at individual aerodromes and set the tilt value. For example, a new radar has just been installed at a UK RAF base and the antenna tilt angle has been set to −1°.

CAP 764 suggests that one means of mitigating a wind farm might be to adjust the mechanical tilt angle.

There may be some situations where a wind farm could appear at a low elevation in the radar coverage and a measure of mitigation might be afforded by an upward tilt of the antenna. A feature of the cosecant squared antenna is the sharp cut-off at low elevations which minimises ground clutter. The greatest rate of change of gain occurs between 0° degrees and −2° elevation where the gain may fall by 10 dB (each way). Potentially, this could make a difference in the detectability of a wind farm.

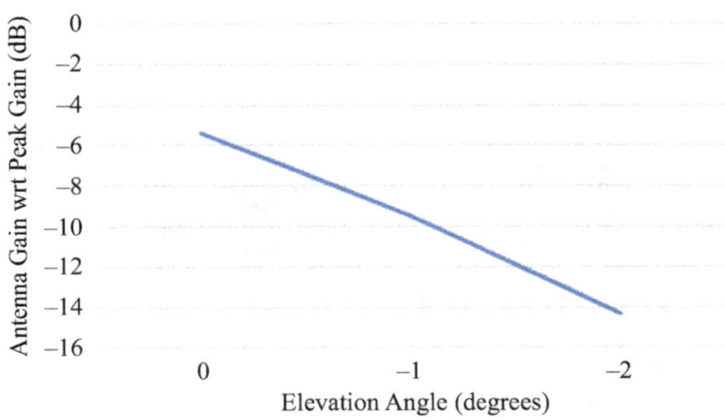

Figure 6.7 Typical Cosecant squared one-way antenna low elevation characteristics

However, note:

- Figure 6.7 is the gain relative to the peak gain, usually approximately 34 dB_i in the case of a cosecant squared antenna. Therefore, for example, the gain at $-1°$ is -10 dB with respect to the peak gain but this is still a gain of 24 dB_i.
- Also note that the penalty of tilting the antenna upwards would be a commensurate loss in aircraft detectability close to the horizon.
- Moreover, if the radar is currently in service, then it will already have been optimised for its role. Any change will compromise this optimisation.
- Because this is mechanical tilting, the same tilt is applied to the whole 360° coverage of the radar.

6.4.1.5 Clutter removal – electronic beam tilting

Electronically scanned arrays of the type used by air defence radars steer by phase or frequency control and these usually synthesise a cosecant squared-like pattern by rapidly switching the beam position while rotating in azimuth by mechanical means. For the purposes of tilting the beam, this may be carried out by using a combination of both frequency or phase control [10,11].

Electronic-tilt (E-tilt) allows the beam pattern being formed to be shifted at chosen azimuths. Figure 6.8 illustrates the concept applied to a frequency-steered array. The array uses a serpentine feed to drive the array. At different frequencies, the array will steer to different elevation angles. The drive to the array itself passes through a set of ferrite phase shifters. When desired, an additional steering command can be applied and this has the effect of lifting the whole pattern. The example showed how this might be implemented in a frequency scanned array but it is, if anything, easier to implement with a phase steered array.

The benefit of e-tilt compared with mechanical tilting is that the beam may be shifted over limited azimuths. Changes in gain are similar to a mechanical tilt.

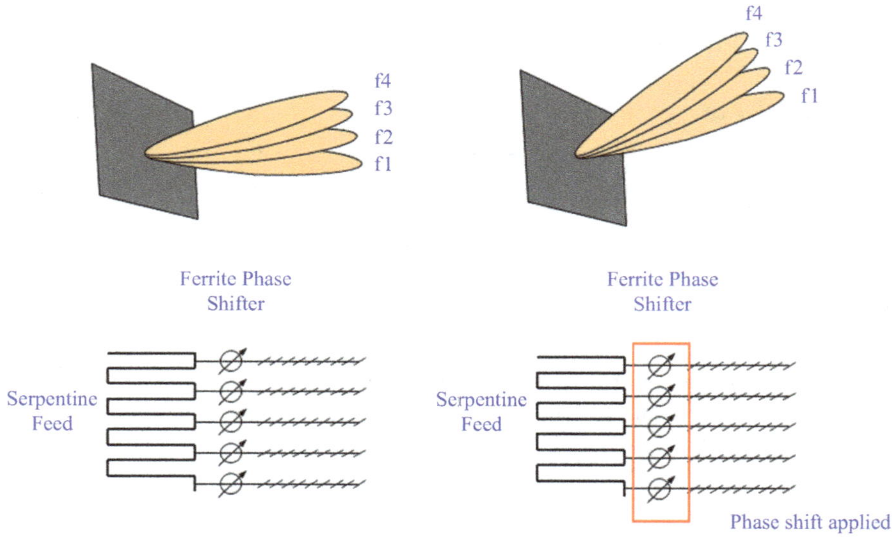

Figure 6.8 Electronic-tilt in a frequency-steered array

6.4.1.6 Clutter removal – amplitude thresholding

In more modern radars much greater control and more fine-grained control can be exercised over the radar's surveillance allowing control of the amount of gain applied at different parts, called **cells**, of the radar's coverage. The technique is referred to as **Amplitude Thresholding**. Applications of amplitude thresholding include preventing overload. In general, in a plot-extracted radar there will be some limit on the number of plots that can be handled in any sector of the radar coverage. The limit will be chosen to be much higher than is expected in normal operations. However, there may be circumstances where the processing could be overloaded such as handling flocks of birds or, in some circumstances, hydrometeors (rain, snow and sleet). Setting an amplitude threshold can thus remove a large number of targets that do not reach the threshold. In some radars, the threshold can be applied at either the input to or the output from the plot processor.

In principle, amplitude thresholding can be very helpful in allowing low-level target returns from clutter to be removed without affecting greater magnitude returns from aircraft. However, there is an obvious limitation: if the clutter being controlled presents a larger signal than an aircraft this can lead to the same disadvantage as sector blanking an RAG mapping; the aircraft returns will be edited out as well as the clutter.

6.4.2 Clutter removal with augmentation

As set out in the introduction, having removed wind turbine clutter from the picture, the picture can then be augmented using either secondary or primary information from other sensors or using a-priori information. The use of SSR data is associated with the use of a TMZ, a change of airspace, and this was covered above. Therefore, this discussion starts with the considerations of using a PSR in-fill radar.

6.4.2.1 Clutter removal with augmentation – in-fill radars

An in-fill radar provides coverage of the airspace over a wind farm without itself suffering the effects of that wind farm. The in-fill radar may exploit the terrain to block its view or be a wind farm-tolerant radar. The former situation is depicted in Figure 6.9.

The slant range problem

A disadvantage of using an in-fill radar is that the two radars produce separate 'pictures' which must be integrated into a single picture to be used by ANSP. The integration problem is complicated because ASR and ARSR are 2-D. That is, they can report the range and azimuth of a target from the radar but they cannot, without other systems such as SSR/IFF, determine the height of the aircraft. The situation is illustrated in Figure 6.10. An extreme case illustrates the problem best. Suppose an aircraft is located at a range of 42,000 ft. A 2-D radar has no way of knowing if the aircraft is directly overhead (for the sake of this argument, ignoring the fact that if it was directly overhead it would be in the cone of silence and would be unlikely to be detected) or 42,000 ft (12.8 km) away on the ground. More generally, there is always a measurement uncertainty associated with the range measurement from a 2-D radar. This is illustrated in the right-hand panel of Figure 6.10.

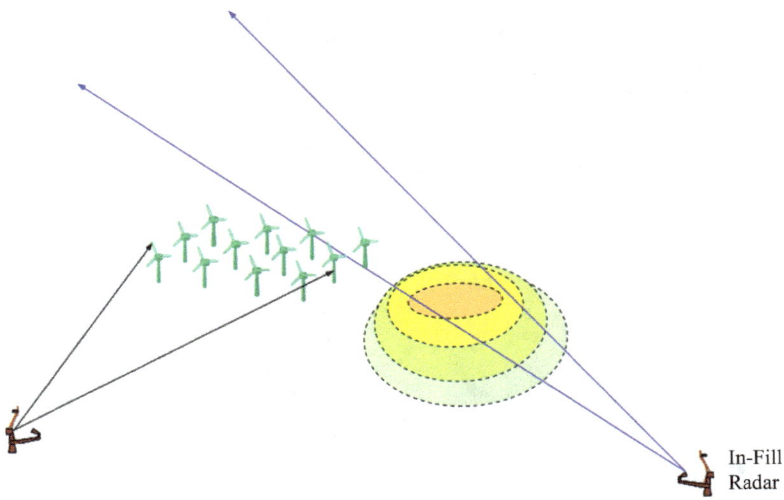

Figure 6.9 Terrain screening for in-fill

The picture integration problem

Now consider the problem that arises when two 2-D radars must contribute to the same displayed picture. The problem is depicted in Figure 6.11. Note that a term has been used in the figure that has been adopted from the wider topic of interference. A radar (or any receiver) suffering interference is called the 'victim'. This has been used here because it is a term that might be encountered in the literature.

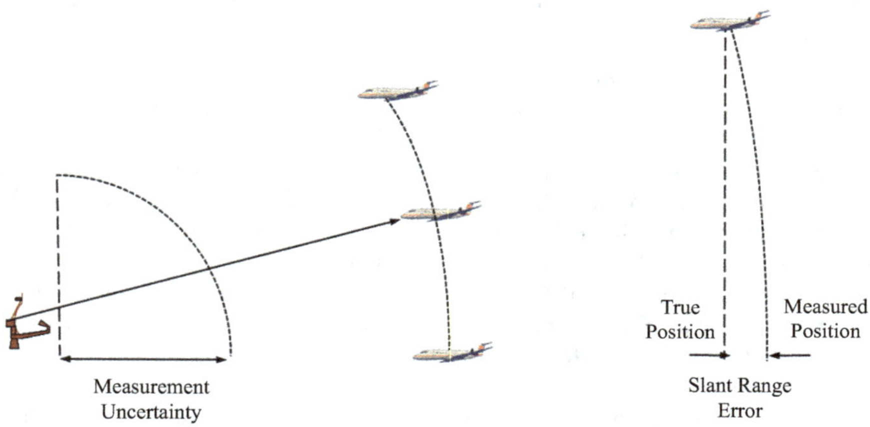

Figure 6.10 Measurement uncertainty and slant range error

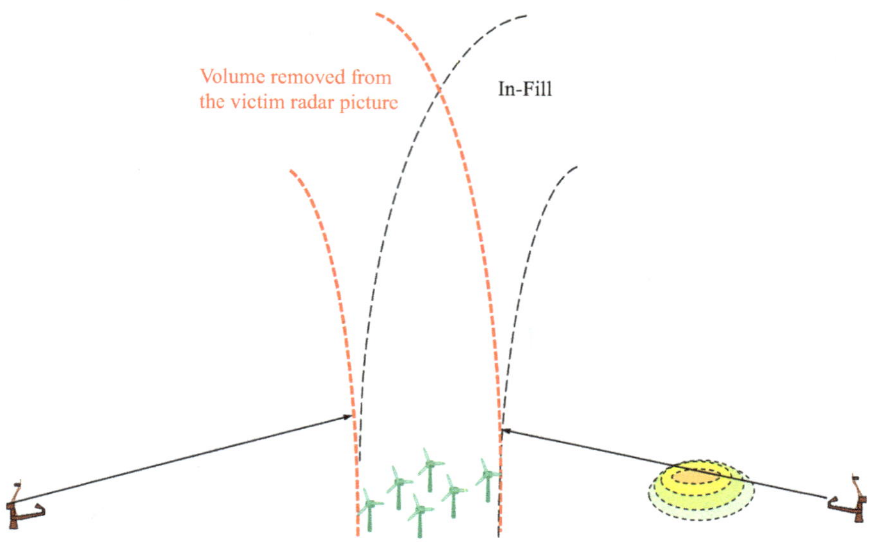

Figure 6.11 In-fill

At first, the problem seems simple. All that must be done is to remove an area from the picture being provided by the aerodrome ASR, or in principle any 2-D radar, and replace it with the picture from the in-fill radar's view of the same airspace. The technique is sometimes referred to as **mosaicking**; as though the radar picture were being made from mosaic tiles.

Hard boundary mosaics

In a hard boundary mosaic, the volume of airspace over a wind farm is removed from the display and is replaced by the coverage from the in-fill radar.

Consider an aircraft flying over the wind farm as shown in Figure 6.12. First consider what happens when the aircraft flies at Flight Level X. When the aircraft is in the region A, it is visible to the principal radar and appears in that radar's part of the picture. As the aircraft flies on it passes into region B. Now the radar is still in the coverage of the first radar, it has not yet reached the coverage of the in-fill radar but it is in part of the picture that is being filled in by the in-fill radar. In this region, the operator will not be able to see the primary returns from the aircraft, they will have disappeared off the screen. If secondary surveillance is available that will be unaffected. Flying on further, the aircraft will reach region C. The aircraft is now in the coverage of the in-fill radar and appears in the 'tile' that is being inserted into the radar picture. When the aircraft reaches the region D, then the aircraft returns to airspace in the coverage of the main radar and the part of the picture that is being

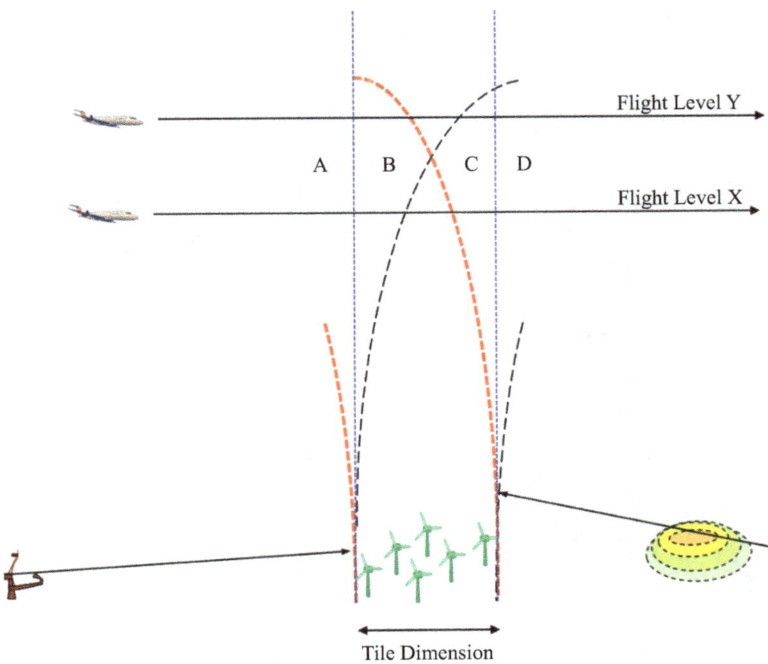

Figure 6.12 Hard boundary in-fill

provided by the main radar. The concern is that there is a region, B, where the aircraft will disappear from view. The length of time, and the number of primary returns that are missed depends on its altitude. An aircraft flying at Flight Level Y would be absent from the picture for a longer period of time. This method of integration is called a **hard border**. The cartoons exaggerate the problem to explain it but the problem is real and integration of an in-fill radar in this fashion can result in aircraft not being observed as they cross the boundary into the in-fill coverage. Losing an aircraft from coverage is undesirable, although secondary radar data will still be available. However, there is an alternative; the soft boundary mosaic.

Soft boundary mosaic

In a soft boundary mosaic, the region over the wind farm viewed from the main radar must be removed from the operator's display. However, instead of replacing just this volume with the picture from the in-fill radar, the in-fill radar is allowed to be displayed over a wider region. Therefore, around the wind farm tile part of the picture is being provided by two radars. This situation is illustrated in Figure 6.13.

Consider an aircraft flying from left to right in the illustration. In region A, the aircraft is reported by the main radar. As it flies into region B the display will be provided with information from both radars. In region C, only the in-fill radar will provide information to the display. In region D both radars will provide information. And, finally, in region E only the main radar will provide information to the display.

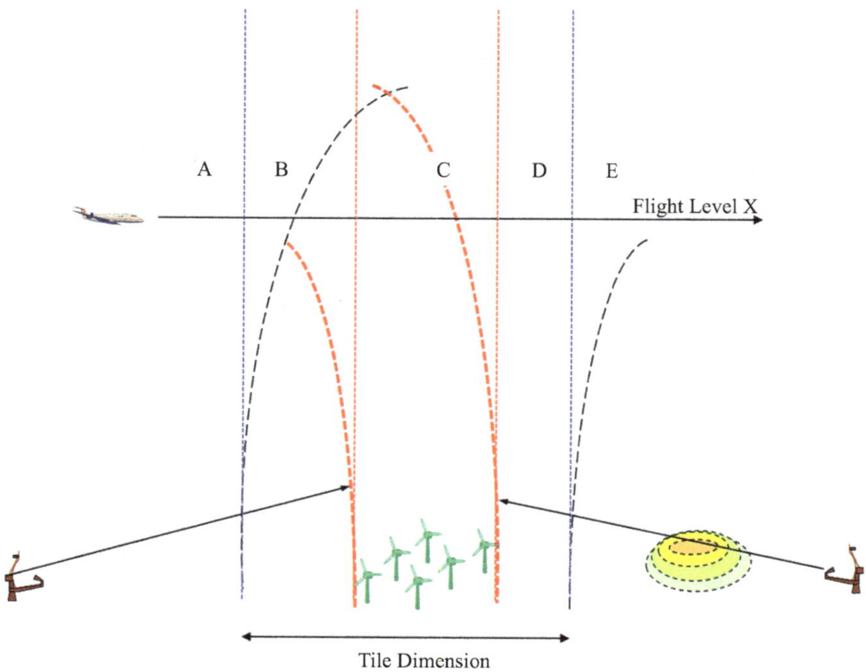

Figure 6.13 Soft boundary in-fill

This approach appears to solve the problems of the hard border. However, unless the contributions from both radars are perfectly registered, then at points in the aircraft transit, it may be reported twice in slightly different positions and because of the measurement uncertainty the two pictures cannot be perfectly registered at all altitudes. A common approach is to register the two radars at an altitude which is operationally important.

If the main radar and the in-fill radar are both 2-D, there is no perfect way of integrating the picture. This does not mean that it is not possible to arrive at a satisfactory integration.

Other observations about in-fill and mosaicking

A prerequisite for an in-fill radar is that it is not able to see the wind farm which is to be mitigated. Therefore, the radar must be in a different place from the first. This carries with it a number of problems:

- A suitable location must be found for the radar. This may not be simple on-shore and may be far more problematic off-shore.
- An additional radar requires a spectrum in which to operate. Spectrum is becoming increasingly congested and identifying a suitable spectrum may be difficult. Furthermore, this will pose an additional cost to the project caused by increases in licence fees.
- When the in-fill radar is not located on the aerodrome, it needs to be protected separately from the other equipment on the aerodrome.
- The maintenance of a remote radar may be problematic.
- During the lifetime of the wind farm, there are now two radars that need to have life cycle management, that is, mid-life upgrades and potential replacement.
- The data feed from the in-fill needs to be assured.
- Fall-back scenarios must be identified in case of equipment failure. For example, given that part of the display over the wind farm is going to be edited from the display (possibly by RAG mapping), then if there is a failure of the remote radar the display must fail safe.
- However, there is less chance of the two radars interfering with each other.

Thus far, only a single wind farm has been considered. If there are multiple wind farms, then there may need to be multiple in-fill tiles to be added to the mosaic. For the reasons set out above, no registration is perfect and the more parts of the display with imperfect registration, the solution to the clutter problem caused by wind turbines becomes a distraction.

A means of mitigating the registration problem is if the in-fill radar is 3-D. Adding height to the information from the in-fill can resolve some of the uncertainty in the position assessment.

Figure 6.14 illustrates the two different forms of boundary definition in an ATC picture. The top two wind farms are using hard boundaries. Hard boundaries carry with them the potential for small gaps in coverage which are exaggerated in this sketch to illustrate the point. The wind farm at the bottom of the display is using a soft

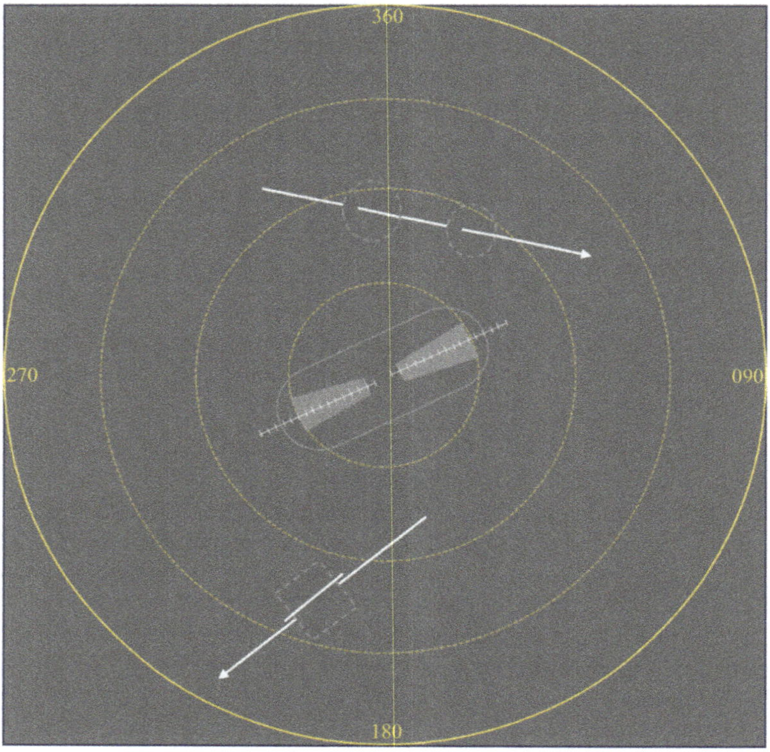

Figure 6.14 ATC picture with in-fill

boundary. There are two points to note. The first is a slight mis-registration of the in-fill and main radar picture. Mis-registration can occur with both hard and soft boundaries but it is convenient to show it here because the second feature of the hard fill is the double target reporting; note how the tracks overlap.

Clutter removal with augmentation – the NAIZ

A non-auto (or automatic) initiation zone (NAIZ) or a non-auto initiation area (NAIA) is a zone where the radar processor does not allow new tracks to be initiated. Not every radar has the ability to implement a NAIZ. When they do, the geographic limits of the area where the NAIZ is to be implemented are calculated and are provided to the radar, where they are usually stored in a computer file that is read each time the radar software is re-started. NAIZ are usually available in Air Defence radars and Dixon identifies them as a useful mitigation technique [12].

Under normal circumstances, tracks will start anywhere in the radar's field of regard, provided the criteria for track initiation are met (discussed in Chapter 3). However, the clutter returns from a wind farm could inadvertently meet these criteria and start a track which potentially gives the impression that a genuine aircraft is present in the airspace above the wind farm. The disadvantage of using a NAIZ is when say a track is terminated because track seduction has caused it to be corrupted and in whatever new direction the track is headed it receives no updates. A new track cannot be started until the aircraft has left the NAIZ. This is unlikely to be a serious problem unless the NAIZ is large, but some wind farms are large.

Clutter removal with augmentation – using a-priori information
The radar operator may wish to tailor the display to his or her preference, for example, zooming in on a particular part of the area of interest. Making such changes does not require the radar to change how it produces data. These sorts of changes are handled external to the radar, in the radar data processor or even in the display driver. It follows that these systems have to be able to correlate an area of the display with a volume of airspace (strictly the 2-dimensional representation of the volume). It is possible to define areas of the display where processing can be modified and new radar returns originating within the designated area can be set to a background display colour. This has the effect of making the plots disappear. If the sole intention is to prevent the operator from being distracted by seeing these clutter returns, then this approach may be satisfactory and it imposes no processing overhead in the radar. The returns cannot be used for any other purpose. For example, it is not possible to retrack them.

Clutter removal with augmentation – Further work
Brookner points out that track initiation and clutter removal are 'intertwined'. Clutter can give rise to false track initiation but one way of dealing with clutter is to allow tracks to be formed and have the software characterise the returns as clutter. The basis of this argument is that tracking algorithms, for aviation applications, are based on models of behaviour of aircraft. Clutter does not behave in the same way as aircraft and it is possible to configure trackers to discard groups of clutter returns. In effect, it is argued that tracking algorithms can act as clutter filters.

6.4.3 Clutter discrimination (wind farm tolerance)

Discrimination is the process of recognising and being able to tell the difference between one thing and another; in this case, a radar return caused by a wind turbine and a radar return caused by an aircraft. It does not matter what sort of discrimination a radar is attempting to perform; it might be discriminating a warhead from a rocket body or an aircraft from a wind turbine: there are a set of processes that have to take place:

- Radar data is collected.
- Features are extracted from the data. A **feature** is a recognisable characteristic. For example, in the case of a wind turbine, the return may be observed to come from a known location or it may exhibit a spread of Doppler frequencies characteristic of wind turbine blades. These features may be determined from a single radar return. Alternatively, a feature may be built up over a number of returns, for example, the way the signal level has changed over a number of observations may be a feature. This latter process is similar to the background averaging processes used to discriminate other forms of clutter.
- A classifier determines the class to which the data belongs.

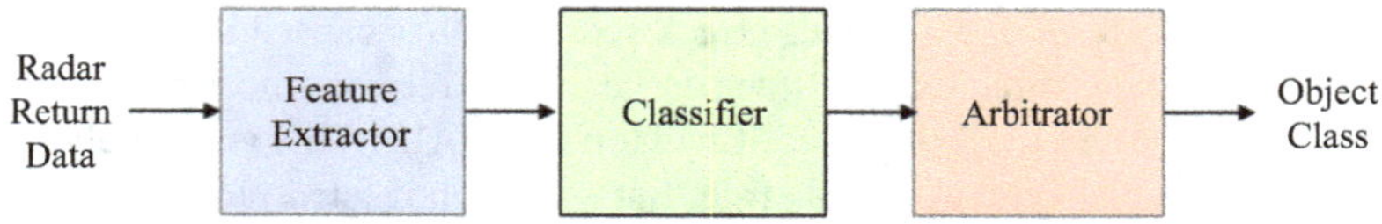

Figure 6.15 The discrimination process flow

- And, if the foregoing process does not reach a conclusive result, then an arbitration mechanism must choose the class of the return.

These processes are essentially sequential as shown in Figure 6.15.

6.4.3.1 The radar data collection problem

For the classification to work, the characterising features must be accessible in the data collected. In legacy radars, that is radars that have been in service for a long time, this may not always be the case.

Consider the Marconi S-511 radar: a highly successful and widely used radar introduced into service in 1980. The azimuth beamwidth of this radar's antenna was approximately 1.5°. It will be assumed that a wind farm is located at an arbitrarily chosen range of 30 km from the antenna. Further assume, that returns from wind turbines only appear in the half-power beamwidth of the antenna. This will not be the case. So large are the returns that they could be observed outside of this angle but the point is adequately made in making this assumption. At a range of 30 km, the half-power azimuthal beamwidth of the antenna corresponds to a linear width of 790 m. The range cell size of the S-511 is 240 m. Therefore, each sample measurement is drawn from a resolution cell size of 240 m × 790 m. The radar is 2-D and the vertical dimension can be ignored.

Now consider the wind farm. Assume the rotor diameter is 80 m and the wind turbines are set out in a 3 × 5 rotor diameter grid, then the turbine spacing in one dimension would be 240 m and 400 m in the other dimension.

Figure 6.16 shows the turbine layout and the resolution cell size of the radar on a single drawing.

The radar can also resolve targets in another dimension: the frequency, that is, Doppler, the dimension that is not shown in the diagram.

A radar return from the resolution cell shown in Figure 6.16 would comprise the vector sum of components from four wind turbines and the aircraft in amplitude and a Doppler spread from the aircraft and 12 wind turbine blades. In the S-511, or any contemporary, radar receiver, the composite signals would have no granularity. The signals from these different elements would all be convolved together and from a single return it would not be possible to extract features to perform discrimination. In practice, the situation is worse than not just being able to deconvolve the individual components, the CFAR arrangements would preclude observations of the aircraft even when it was not close to turbines.

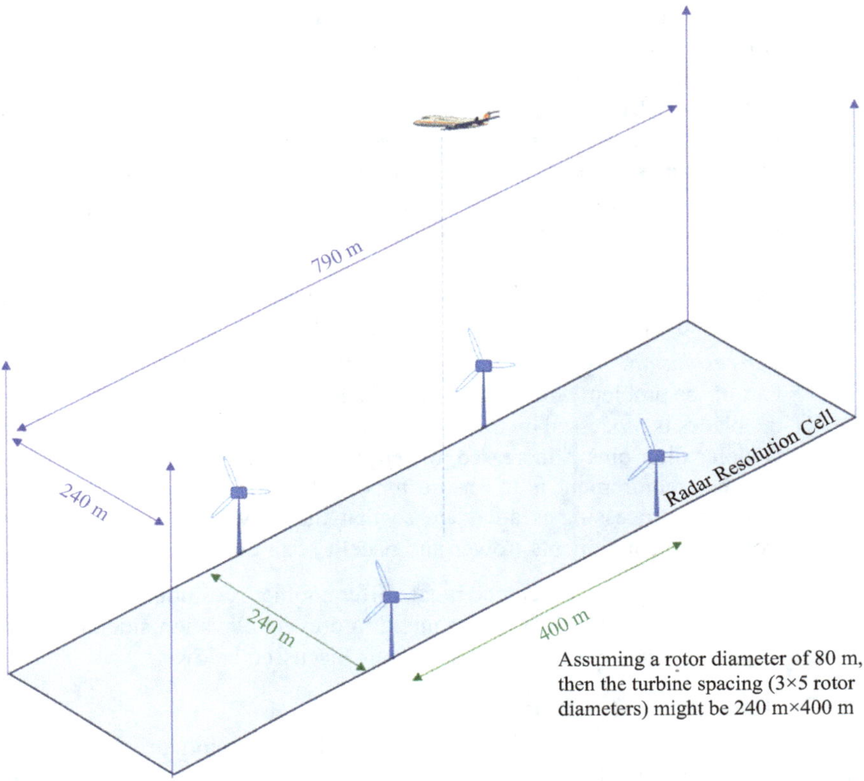

Figure 6.16 *Range cell size versus turbine spacing*

A number of methods of improving performance to provide better feature extraction are listed below. There is not a one-to-one correspondence between techniques, some serve multiple purposes.

6.4.3.2 Reducing the resolution cell size

If useful features are to be extracted, then the resolution cells must be reduced in size. The following techniques might be employed to achieve enhanced resolution:

- *Azimuth resolution*: A larger antenna reduces the azimuthal beamwidth, and, therefore, the azimuthal resolution of the radar. Thomsen reports a beamwidth of 0.32° for one radar [13]. This was achieved by using an X-Band radar. Thomsen also makes the point that the narrower the beamwidth the greater the number of 'looks' required to surveil the airspace and this also affects the maximum range that can be achieved.
- *Range resolution*: The following techniques all enhance range resolution:
  - Sufficient bandwidth is required to obtain a range resolution to resolve wind turbines and the nearby environment. Thomsen *et al.* suggest that

range resolution does not need to be so great that it provides resolution within a wind turbine and, in 2011, they proposed that range resolution need be no greater than 10–20 m [13].
  - *Enhanced ADC*: The speed of digitization must be increased to reduce the range cell and provide resolution between wind turbines and as close to the wind turbines as possible to increase the opportunities to observe aircraft returns inside the margins of the wind farm. As inter-turbine spacing increases in new wind farms, the problem is eased but there are still a lot of closely spaced wind farms in service. The more opportunities to observe aircraft the greater the chance of forming a good track on the aircraft. C-Speed reports their wind farm-tolerant radar as having a cell size <10 m [14].
- *Doppler resolution*: Doppler processing enhancement:
  - Part of the problem with legacy radars is that all of the Doppler energy from the blades is processed by the same, simple, Doppler filter. If the number of Doppler filter bins is increased, the energy in each bin is reduced.
  - Another requirement is to ensure moving target indicator/moving target detection processing is adequate so that the very large signal strengths from static components (tower and nacelle) can be eliminated.
- *Elevation resolution*: Air Defence radars offer another resolution dimension of elevation. Good taper selection is required to prevent elevation sidelobes from reducing performance. This consideration is discussed below.

6.4.3.3 Reducing saturation

Smaller range cell sizes are not effective if the signal from a wind turbine can bleed from adjacent range cells because of saturation. Techniques to minimise saturation include:

- The dynamic range of the radar receiver must be modified to be sufficient so that saturation does not occur when maximum signal strengths from turbines are encountered.
- Reduced antenna beamwidth and reduced antenna sidelobes potentially decreases the number of wind turbines observed in any one 'look' and, consequentially, reduces the signal strengths occurring in any single range cell. There are practical considerations to take into account. Antenna sizes have to be practical sizes. Higher frequencies (X-Band) are able to achieve this more readily than lower frequencies (L-Band).
- *Enhanced analogue to digital conversion*: The number of bits must be increased to be commensurate with sustaining the radar receiver's dynamic range. Terma describes their wind farm-tolerant radar as having 14-bit ADC [15].
- *Pulse compression must be enhanced*: The range sidelobes must be reduced to prevent regions close to the turbines from being obfuscated by pulse compression artefacts. This effect was discussed in Chapter 4.

Taken together an enhanced resolution cell radar might, for comparison with Figure 6.16, produce a situation as depicted in Figure 6.17.

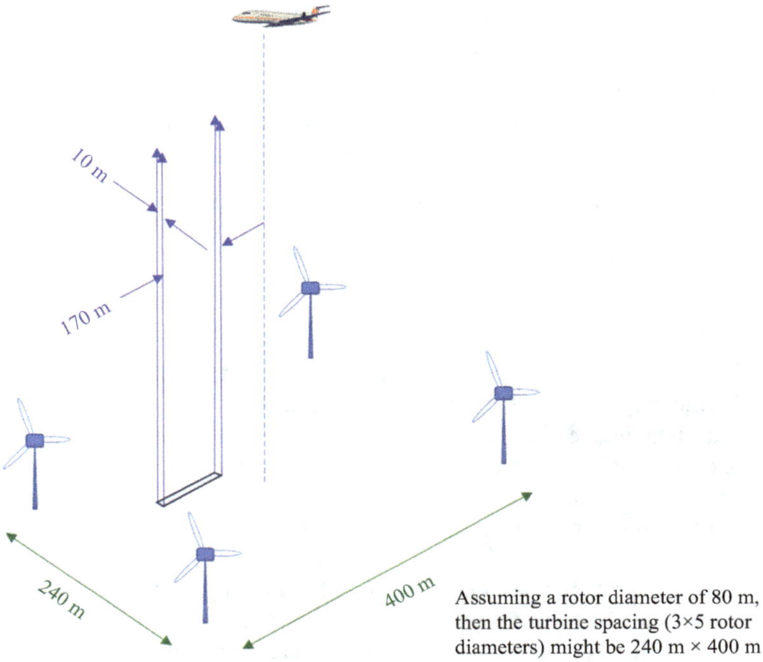

Figure 6.17 Reduced resolution cell size

6.4.3.4 Detection, tracking and processing considerations

Assuming that measures have been taken to ensure radar returns can be created free of the effects of the proximity of wind turbines, a number of processing considerations must also be addressed:

- Obscuration: CFAR algorithms must be chosen to prevent the high RCS returns from wind turbines masking/obscuring genuine targets. This is discussed below and an example of a suitable CFAR algorithm is described.
- Track seduction: Prevent wind turbine returns from updating aircraft tracks using either a priori position data (such as, a return comes from a known turbine location) or by recognising wind turbine signatures and editing out the associated returns.
- Processing overload: Ensuring the radar's throughput and/or storage cannot be exceeded by large numbers of returns from a wind farm and running the risk of genuine aircraft returns being lost. This is unlikely to be a problem with ASR, ARSR or AD radars which are designed to handle high volumes of data but could be problematic if the expected throughput is low such as with PAR.

6.4.4 Performance metrics

Metrics are needed to measure the performance of wind turbine-tolerant radars. Measurements using these metrics can then provide comparisons of performance that will inform decision making. Before suggesting some useful metrics for this purpose, what types of information are needed:

- What is the performance of the radar without the new capability?
  - This measurement provides a baseline for all comparisons.
  - Ideally, measurements in regions with and without wind farms.
- What is the performance of the radar with the new capability added?
  - Has the addition of the wind farm tolerance degraded radar performance in the absence of a wind farm.
  - Has the addition of the wind farm tolerance improved radar performance in the presence of a wind farm.
- Do performance changes degrade the capability of the radar?
  - Some might argue the need to maintain radar performance when wind turbine tolerance is added; others the need to maintain capability. A simple example illustrates the debate. Suppose that the required capability is to detect an aircraft with an RCS of 2 m^2, with Swerling I characteristics, and with an 80% probability of detection. The radar is tested in the absence of a wind farm and any wind farm mitigation. The measured radar performance is a 90% probability of detection. Wind farm mitigation is now added to the radar and the tests are repeated. In the presence of a wind farm the new probability of detection is 85%. There has been a degradation of performance. But has there been a degradation of capability?

6.4.4.1 What measures are necessary

Some of these measures are common to all types of radar. It does not matter whether the radar is for civil ATC or defence, a certain level of probability of detection will be required. Whereas a civilian radar which may be supported by secondary surveillance may be less concerned about the preservation of track identification than a military radar with the role of defending against enemy attack.

Some common measures that may be encountered are as follows:

- **Detection-based metrics**
  - **Probability of detection**: The calculation is based on the ratio aircraft detections and the number of opportunities to make a detection (it may be simpler to count the missed detections). The results may be contentious because of the arguments discussed above. An important consideration is that P_D tests are usually associated with Swerling I behaviour: the aircraft flying directly towards the radar. This detail is a factor because aircraft flying across the radar's Field of View (FOV) are associated with a lower P_D. A simple rule should be observed always compare like with like.
  - **Base of solid radar cover**: This measure would be important if the mitigation implementation relied upon an in-fill radar which used terrain to screen the in-fill from the wind turbines. However, it may also be important in other circumstances. Even with good wind farm mitigation there may be a point when an aircraft is flying at a low altitude that

performance is degraded. As a general rule, operators need to be confident of the envelope of performance of the radar. The measure is a form of Probability of Detection but in this context, it is the airspace above a wind farm where it is important to gauge what has been lost.

- **False alarm rate** (FAR): One of the key problems encountered in the presence of wind farms is clutter. A common metric is to assess the number of false alarms that occur over a wind farm. The method is to count the number of plots that are generated by wind turbines. However, caution needs to be exercised in interpreting false alarm rate results:
  - Tests are typically performed in adverse weather conditions:
    For obvious reasons it is necessary to understand how well mitigation works in as many wind states as possible. However, in such adverse weather there can be a general uplift in false alarm rates caused by rain or even something as simple as trees blowing in the wind.
  - More generally, if there are persistent returns from a single area close to a wind farm that might be indicative that the FAR is caused by a turbine. If, however, the FAR is low and distributed over a wider area, then the result is possibly less conclusive.
- **Maintenance of track identification**: For each track of an aircraft that enters the airspace over a wind farm this metric measures the success rate of the track having the same identification when the aircraft leaves. A variation on this measure is to have the aircraft perform a manoeuvre over the wind farm area.
- **Track seduction measure**: As one of the known problems with the aircraft over-flying wind farms is track seduction, then a metric giving insight is useful. This metric also provides a measure of effectiveness of a NAIZ. A feature to be aware of when considering the interpretation of the results concerns the desire to conduct tests when it is windy. This is not a suggestion that tests should be conducted when it is not windy. However, when windy, and particularly when flying at low levels where it may be turbulent, the pilot may have problems maintaining a straight and level course. For the test to be fair, it is useful to have a baseline for comparison that might be available from other parts of the sortie.

6.4.5 Integration of wind farm-tolerant radars

The functions carried out to make a radar able to discriminate between a wind turbine and an aircraft return are processing intensive. Radars with the qualities identified above tended to have quite low maximum ranges and would require integration in a similar fashion to in-fill radars discussed above.

However, the range of these radars is increasing. This is beneficial because:

- the radars do not necessarily have to be located remote from the aerodrome. This makes it easier to secure, maintain and simpler to convey data from the radar to the point where it is used;

- depending on the operational requirement at a given aerodrome, it might be possible to replace the existing ASR; or
- if the range is sufficient to reach the airspace above wind farms but not sufficient to meet the operational requirement of the ASR, the integration of the two pictures is simplified because of the relative proximity of the two radars. This is illustrated in Figure 6.18.

6.4.6 Elevation sidelobe control

Although the dimensions vary from radar to radar, most Air Defence radars have planar antennas. That is, they are two-dimensional, flat, surfaces, such as the antenna shown in Figure 6.19.

Although it is not possible to see it in the figure, the surface is populated with antenna elements. There are a number of different antenna element types but that is not important to this discussion and can be ignored. Suppose the antenna elements are equally spaced. Further, suppose the antenna is transmitting and the same amplitude signal is provided to each antenna element. Under these circumstances, the antenna is said to have a **uniform illumination function**. A similar argument applies in receive.

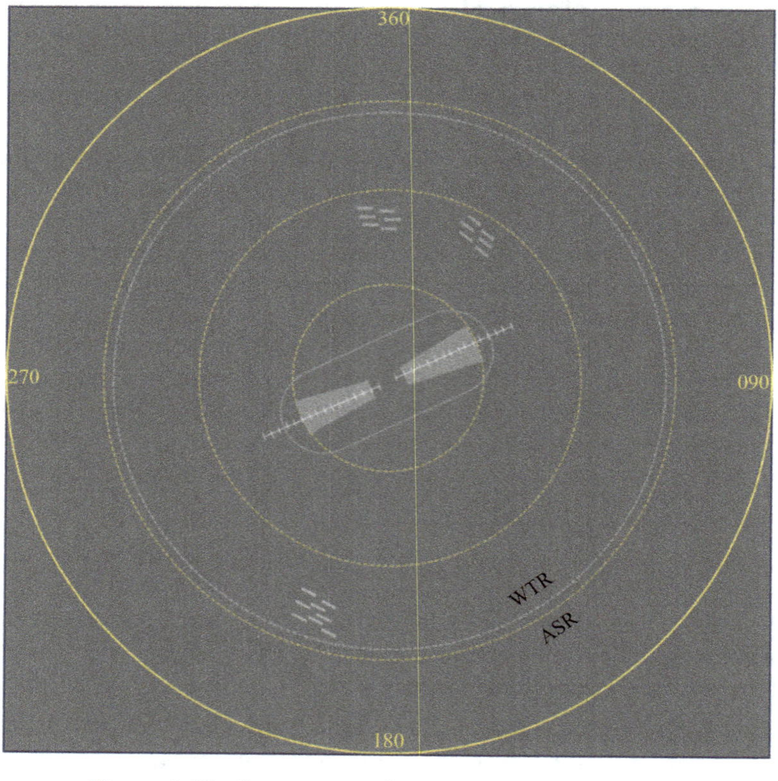

Figure 6.18 Integration of a wind farm-tolerant radar

Figure 6.19 A planar array (courtesy RAF Boulmer)

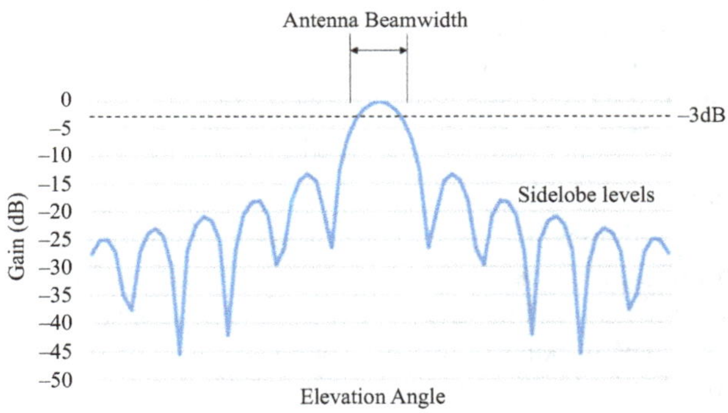

Figure 6.20 Uniform illumination far-field antenna pattern

If the same signal amplitude is taken from each element, then uniform illumination is preserved. Uniform illumination creates a particular far-field antenna pattern illustrated in Figure 6.20. This shape is the familiar (sin x)/x pattern.

One of the features of a phased array antenna is that it is possible to change the illumination function. There are a number of ways this could be done. The spacing between the antenna elements can be changed or the signal applied to each element in transmit and the signal taken from each element in receive can be altered. This is called **aperture control**. In the case of the antenna shown, the latter method is used. The antenna elements in the middle of the array have the maximum signal levels, in transmit and receive, and the level is reduced towards the edge of the antenna. This is called applying a **taper**. There are many different algorithms for calculating the taper values. Brookner [16] has identified 23 different antenna tapers and lists their characteristics. This list is not exhaustive. Doerry reports 50 different types [17].

The use of aperture control, that is, applying a taper, controls the width of the main beam and the level of sidelobes. A common taper used by many radars calculates the proportion of signal applied to and extracted from using a Taylor series. This is a polynomial expression. The precise details are not important here, it is just a way of describing a curve. Each different taper algorithm produces a different main lobe width and overall antenna gain and different sidelobe levels.

To mitigate the effects of wind turbines, it is desirable to have very low sidelobe levels. The intention is to make sure that when the main lobe is pointing to the space above a wind farm the sidelobes that will be illuminating the turbines provide a minimum of power to create echoes and minimise the echo from wind turbines. This approach could mitigate the effects of wind turbines on the upper elevation beams.

Dolph Chebyshev and Gaussian tapers give good sidelobe performance with maximum sidelobes better than 40 dB lower that the antenna's main lobe. Thomsen identifies the Nutall Window Taper as a candidate taper for use in wind farm mitigation [13]. The Nutall window taper is actually a family of related taper functions and the whole set is a member of the greater Blackman Harris family of tapers. Doerry provides characteristics for each of the family members that are useful when considering options for antenna design [17].

6.4.7 Modified CFAR

In Chapter 4, it was shown that the CFAR processor can lead to the obscuration of targets in the presence of large returns such as wind turbines. The presumption in a standard CFAR processor is that there are few genuine targets and the majority of range cells are samples of noise or clutter. Therefore, averaging the numeric values in cells adjacent, that is, both earlier and later, to a test cell provides a good measure of the average noise/clutter level in the receiver at the time when the echo signal containing a target is tested. The technique is called a Cell-Averaging CFAR (CA-CFAR) or a Mean Level Detector (MLD).

Various refinements were made to the CA-CFAR. All exploit additional information about the problem domain. Greatest Of GO-CFAR is an example. GO-CFAR addresses the clutter edge problem. This arises when there is a step function in the clutter level, for example, at the edges of clouds. Outside the cloud, in free space, the backscatter is low. But inside a cloud full of water vapour, the backscatter increases with a step function at the edge of the cloud. By understanding the environment, the basic CFAR algorithm can be enhanced. However, the early refinements still assumed that the CFAR averaging cells were occupied by noise or clutter not targets and certainly not targets creating large signals.

However, the presence of a large return in the averaging cells, such as the echo from a wind turbine, violates the assumption of the cells only containing low signal levels. Now, when the average is calculated, it is excessively high and the threshold

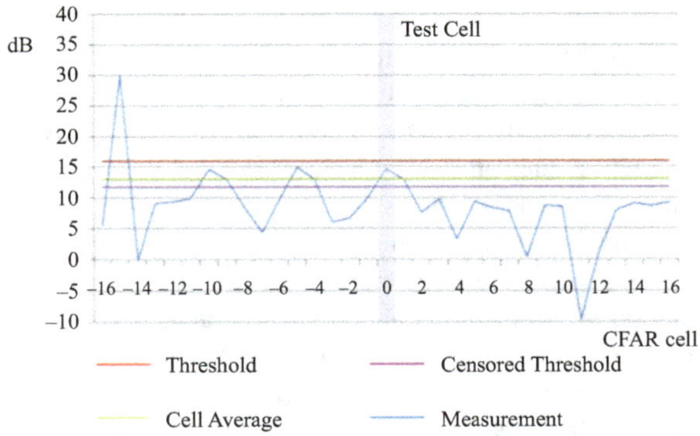

Figure 6.21 CMLD CFAR

for target detection is set too high to detect anything except possibly another wind turbine. In 1977, Rickard and Dillard reported a modified version of an MLD, called the Censored MLD (CMLD). Prior to averaging, the ensemble of returns in the averaging cells (the leading and lagging) are rank ordered. The highest values are discarded and the values remaining are averaged to calculate the threshold value. Figure 6.21 illustrates the process. A wind turbine return has just entered the cells to be averaged (that is at the extreme left). If this return was left in the ensemble, it would create an excessively high threshold (shown in red) and the simulated target present in the test cell would be assumed to be noise as well and no target would be declared. If the wind turbine return was discarded and the average recomputed, the lower (green) threshold would be created and the targets detected [18].

The CMLD offers a means of detecting smaller targets in the presence of larger returns. An implementation detail for the system designer is, if a range cell has been censored to ensure smaller targets can be detected, should that cell be subjected to a test for the presence of a target?

A further consideration that must be taken into account concerns the number of guard spaces used in the CFAR processor. Chapter 4 points out that rotor diameter sizes in coming years might reach 180 m. To put this in context, if the range cell size is, say, 10 m long. A situation might arise where a cell is being tested and there are 17 adjacent cells with wind turbine returns present. Modelling and simulation using the specific CFAR algorithm to be used will identify an optimum setting.

6.4.8 Feature extraction and classification

A simple measure of feature extraction is to identify whether a cell or cells are consistent with the known location of a wind turbine. The precise location of a wind turbine will only be known after construction because of the possibility of micro-siting. Even though this might only move the location of the wind turbine, say 50 m, this must also be considered against the possibility of a range cell being only 10 m. Moreover, the location of the wind turbine and its influence on the airspace around it must account for position uncertainty caused by slant range error. Notwithstanding these matters, accounting for the location of a wind turbine offers a means of censoring returns and enhancing aircraft track accuracy and reduction of the possibility of track seduction.

One of the most successful forms of clutter removal processes is the background averaging method. In effect, this process allows the system to adapt to the environment and it has some useful properties for a wind turbine feature extraction/ classification process. Wind turbines alter their characteristics depending on wind direction and strength. On occasions, they may present limited, if any, clutter to a radar. On these occasions, it might be undesirable to edit returns because there will be times when an aircraft appears in the same range cell as the nominal position of a wind turbine. The question that arises is how best to characterise the signature presented by a wind turbine or a wind farm. More work is needed in this area. In the future, AI may be able to recognise turbine signatures but currently even a simple high-resolution clutter map could be used to excise the turbine returns. This topic is taken up in Chapter 7.

6.5 What still needs to be done?

There is still much to be done. Chapter 7 identifies a number of items of basic research that should be followed up. Much of that is basic science. For example, understanding better the interactions of radio waves and wind turbines. One of the most challenging environments for a wind farm-tolerant radar to address is off-shore and the principal concern is defence.

6.5.1 The defence challenge

With the current generation of AD radars, there is a concern that they may not be able to detect, track and identify threat aircraft reliably approaching from off-shore. If these processes are accomplished beyond the wind farms there is a concern that tracks and target identity may be lost because of the effects of the wind turbines on the threat radar. Moreover, the problem is getting worse:

- The number of off-shore wind farms is increasing.
- The wind turbines themselves are getting bigger, increasing their 'radio footprint'.
- The wind farms are increasing in range from the shore.

6.5.2 Technical challenges

The technical challenges off-shore are also worse than on-shore:

- Because this is a defence problem, it is not possible to rely upon SSR/IFF, or its derivatives to assist with detecting objects.
- Sea clutter is one of the most challenging forms of clutter.
- Depending on the location, the radar may actually be looking down on the wind farms increasing the challenges of clutter.

- The distances off-shore and the beamwidth of antennas, even allowing for the greater inter-turbine spacing, mean that each 'look' can take in multiple wind turbines.
- The task has previously benefitted from the use of lower frequency radars. This poses a challenge to create a narrow beamwidth to counter the aforementioned problem.
- To reduce beamwidths an option might be to use X-Band which has proved effective on-shore but the distance and damp atmospheric conditions are not conducive.
- Propagation challenges are significant with the possibility of encountering strong ducting creating the potential for even more clutter from wind farms that would normally be beyond line of sight.
- Radar update rates are relatively low compared with aerodrome radars, which exacerbates problems with tracking.
- There are limits to what might be achieved by reducing the turbine RCS. A reduction in RCS mitigates the problems of saturation but if reduced too much would become problematic for marine radars.

6.5.3 Options and initiatives

The United Kingdom is actively currently investigating the problems of radar and off-shore wind turbines led by the Off-shore Wind Industry Council [19]. The work has been hampered by the COVID-19 pandemic. The studies are focusing on the following areas:

- The layout of wind turbines.
- Technologies to reduce turbine RCS.
- Machine learning to recognise wind turbine clutter.
- In-fill radar technologies.

6.6 Technology readiness

6.6.1 The problem

Part of the debate about when is good, good enough is deciding if a technology is sufficiently mature to be used in service. This question is particularly important in a safety-critical environment and begs the question; how can confidence be assured in a new technology?

There are other domains where safety is paramount and this question must also be addressed. Spaceflight for example has to be assured that technology will work and, in some cases, it is not possible to try out the technology in advance of flight. There are simulation capabilities but nothing is quite the same as flying a vehicle in space. A similar situation arises with the fielding of wind farm mitigation, there must be confidence that this mitigation will work in advance of construction. Wind farms are too expensive to build if mitigation is not going to be successful.

Some criteria for judging readiness for service are obvious. The highest level of confidence will be assigned to mitigations that have been shown to be successful in other projects. However, even if a mitigation has been shown to be successful in an earlier project, there must also be confidence that the nature of the new project is sufficiently similar to the previous one that the mitigation is still valid. But, in at least some situations, new innovative technology is required to mitigate problems. This poses a conundrum. Useful ideas would never get into service if a prerequisite was that it had been in service before. What is required is a means of assessing the technical maturity of a system. The problem of assessing technical maturity is far wider than just aviation safety mitigation and techniques have been developed for other domains which can be helpful.

Assessing the maturity of a technology is fundamental but, also, non-trivial. Perhaps in the future, decisions about technical maturity can be supported by other measures, for example, by the use of modelling and simulation. But at the time of writing, assessments to support decision-making need to be as objective as possible without the use of such supporting technology[†].

This section addresses the question: what might be learned from looking at how other technology areas deal with maturity assessment? Before answering that question, the additional question to be addressed is who needs to know the answers?

6.6.2 Stakeholders

A wide community needs to have such an understanding of technical maturity as the following examples illustrate.

In the earliest stages of a project, organisation providing finance will wish to be assured that his or her investment is sound with no unexpected risks to either the cost or timescale of an investment because the technology is insufficiently mature.

The financiers' interest is driven by the Air Navigation Service Provider (ANSP) who will wish to be assured that technology, or a novel procedure, is capable of being deployed safely as part of their existing suite of navigation systems and procedures.

Then, if the plans are contentious, legal counsel must ensure their cases are sound and the mitigations are possible within the timescale of planning permission.

6.6.3 Assessment techniques

A widely accepted method of judging technology readiness is to rank its maturity using a scale of technology readiness level (TRL) as described in this chapter.

In some instances, completely new equipment will be needed to facilitate operations in the presence of wind turbines but, it is far more likely that new technology must be incorporated into an existing suite of systems: a process often

[†]The use of modelling and simulation is undergoing a revolution and the technology is being extended into concepts such as digital twinning, in which authentic representations of operational systems can be used to predict behavioural changes in the presence of various stimuli.

called, 'technology insertion'. Even if a system has been operational in one application for a long time using a particular approach to mitigation, if it is moved to a new application, perhaps because of different geography or different numbers of turbines, it may not perform as well or even not work at all. To cater to such situations, assessments of technology readiness may still be required but it is becoming increasingly common to define a second measure of readiness, called the system readiness level (SRL) or the integration readiness level (IRL).

It is useful that everyone associated with the life cycle of a project understands TRL, IRL and SRL to minimise risk, and to have a better understanding of timescales. This chapter will also introduce these measures.

6.7 Technology readiness level

6.7.1 Background

The concept of TRL is rooted in a more general concept that products have a 'life cycle'. Life cycle is defined by the international standard as [20]:

> 'A system progresses through its life as a result of actions, performed and managed by people in organizations, using processes for execution of these actions'.

It is generally accepted that there are six stages in the life cycle of a system. The following list of stages is taken from the international standard. The life cycle for military systems is composed slightly differently and is referred to as the concept, assessment, development, in-service and disposal or termination (CADMID/CADMIT) cycle. The terminology used for military systems is shown in brackets after the civilian term. This is often introduced pictorially as in Figure 6.22.

In the United Kingdom, for reference, the end of the concept phase is when the decision is made about whether a project can move to the assessment phase, that is the initial gate decision. At the end of the Assessment phase, the decision is made about whether the project can move on to development: the Main Gate decision.

The life cycle of a system starts with a ***concept (concept and assessment)***, which must then go into ***development (development)***. The developed concept then goes into ***production (manufacture)***, and once produced the system enters the ***utilization (In-service)*** stage. To sustain the system, a ***support*** stage is required. The final stage of a system life cycle is ***retirement (disposal)***.

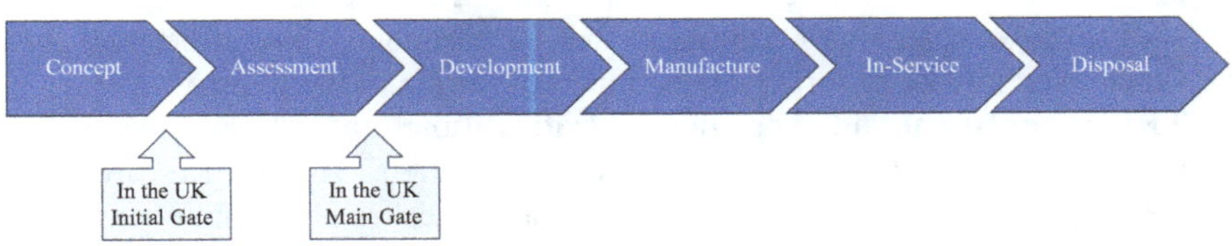

Figure 6.22 CADMID cycle

6.7.2 NASA TRL

The TRL concept was first developed by NASA in 1974 by Stan Sadin, a senior scientist working at the time at NASA Headquarters. The first assessments of TRL were based on a seven-layer model and were used for the development of a spacecraft to orbit the planet Jupiter. The process was deemed successful and it was formally adopted in 1989 for other work [21]. In the 1990s, the model was extended to include the nine-stage model that is in use today. TRL assessments using the NASA model have now been widely accepted by other organisations such as the European Space Agency (ESA), the European Commission, the UK Ministry of Defence (MOD) and more widely by industry [22].

The descriptions of TRL are as follows. TRL are numbered: TRL 1 is the least developed and the maximum readiness level is TRL 9. Some examples of evidence that might support any given level of development have been added by the author.

- **TRL 1: basic principles observed and reported:** Transition from scientific research to applied research. Essential characteristics and behaviours of systems and architectures. Descriptive tools are mathematical formulations or algorithms.

  Evidence might include reference to the technique or observations in scientific papers perhaps released in the proceedings of conferences. Reports of observations in scientific and engineering journals are particularly valuable because these types of publications are usually peer reviewed which lends weight to the evidence. Even though evidence of later stages of development might be commercially protected, it is often the case that basic observations are more readily available.
- **TRL 2: technology concept and/or application formulated:** Applied research. Theory and scientific principles are focussed on specific application areas to define the concept. Characteristics of the application are described. Analytical tools are developed for simulation or analysis of the application.

  Evidence might still include references in scientific papers and conference proceedings. But, by its very nature, evolution to this stage of development implies a level of investment by an organisation and this may make these organisations more reluctant to provide evidence. The counter-argument is that this is still a very low level of maturity and organisations are sometimes more keen to demonstrate progress to secure funding for further development. For someone assessing TRL for a particular application it may also prove useful, even at this low level of maturity, to review how it is envisaged that the technology may be used. For example, was it envisaged that the technology would be used for a single wind turbine or for a large commercial farm of wind turbines?
- **TRL 3: analytical and experimental critical function and/or characteristic proof of concept:** Proof of concept validation. Active Research and Development (R&D) is initiated with analytical and laboratory studies. Demonstration of technical feasibility using breadboard or brassboard implementations that are exercised with representative data.

A breadboard and brassboard are tools used in a design laboratory to build prototype circuits. These circuits may only consist of a few components and they will, generally, be working in isolation and not part of a larger system or configuration. As with TRL 2, to have moved to this stage of maturity further investment will have been made. Even if the investment in hardware may be small at this stage, the investment of time by specialist personnel will be growing. Also, as with TRL 2, there may be a reluctance to release proprietary information unless it is intended to seek additional funding from outside sources. It is also worth noting that the higher the TRL the greater the value of the information garnered. Even negative results may have commercial value as maturity increases.

- **TRL 4: component/subsystem validation in laboratory environment:** Standalone prototyping implementation and test. Integration of technology elements. Experiments with full-scale problems or data sets.

  An important distinction that would be indicative that this stage had been reached would be by using the system or device with real problem data. Successful handling of real data is an important step in the evolution of a product or device and may be indicative of its eventual success. It may be the case that if governments wish to foster development in a particular problem area, they may offer grants to support companies achieve higher levels of readiness. A government would be unlikely to help a commercial organisation get a product to the marketplace or make the system or device operational (TRL 9), but a grant might be awarded to elevate maturity to say TRL 7 or 8. Evidence transacted in such arrangements might be very useful in readiness assessment by a third party and additional confidence might be gained from the knowledge that government agents have already reviewed evidence. Once again, an important factor to be noted is the scope of the product or device. To repeat the previous example, have tests been conducted against individual turbines or against a commercial wind farm?

- **TRL 5: system/subsystem/component validation in relevant environment:** Thorough testing of prototyping in a representative environment. Basic technology elements integrated with reasonably realistic supporting elements. Prototyping implementations conform to the target environment and interfaces.

  Arguably, the distinction between TRL 4 and 6 is the most subtle for each of the levels. The difference between TRL 4 and TRL 5 hinges on 'thorough' testing. Evidence that might support the evolution to TRL 5 might, for example, be that while in stage 4, it has been shown that a mitigation technique might be successful in mitigating the effect of some number of wind turbines. A demonstration that that level of capability is achievable with more representative levels of genuine aircraft targets and other forms of clutter (described in more detail in Chapter 4) would be valuable. Evidence that might support this level of maturity might take the form of internal company reports. The progression to what is deemed to be a relevant environment will be useful in assessing whether a specific application might be out of the normal scope of a system or device.

- **TRL 6: system/subsystem model or prototyping demonstration in a relevant end-to-end environment (ground or space):** Prototyping implementations on full-scale realistic problems. Partially integrated with existing systems. Limited documentation is available. Engineering feasibility fully demonstrated in the actual system application.

  The evidence base for this stage of maturity might be found in engineering reports. All ANSP-related systems will need to demonstrate that they are safe in operation. Part of the process of building a safety case is being able to demonstrate that safety has been 'tracked' from the earliest stages of a project. At this level of maturity, evidence should be starting to be gathered for these purposes.
- **TRL 7: system prototyping demonstration in an operational environment (ground or space):** System prototyping demonstration in an operational environment. The system is at or near scale of the operational system, with most functions available for demonstration and test. Well integrated with collateral and ancillary systems. Limited documentation is available.

  It is usually the case that the integration process will create documentation that provides evidence of the level of maturity. Such documentation might consist of, for example, integration test plans and their associated reports. Planning for such tests may include evidence of what is expected from the system or device and this will be useful in assessing the scope in which it is intended to operate in.
- **TRL 8: actual system completed and 'mission qualified' through test and demonstration in an operational environment (ground or space):** End of system development. Fully integrated with operational hardware and software systems. Most user documentation, training documentation and maintenance documentation are completed. All functionalities are tested in simulated and operational scenarios. Verification and Validation (V&V) completed.

  As set out in this definition of TRL 8, a large amount of documentation is created when a system achieves this level of maturity. This evidence will illustrate the scope of the operational usage of the system.
- **TRL 9: actual system 'mission proven' through successful mission operations (ground or space):** Fully integrated with operational hardware/software systems. The actual system has been thoroughly demonstrated and tested in its operational environment. All documentation is completed. Successful operational experience. Sustaining engineering support in place.

The highest technical readiness level is associated with a system entering operational service. This transition may also mark the transfer of the system from a contractor who has developed the system to the operating authority, a process called Transfer and Turnover. Although it would be unusual for a system to enter service with a mission-limiting defect, sometimes minor defects may still be present that will be rectified while in service. If the system is also being considered for a different application, having knowledge of any limitations in an existing application may be informative.

6.7.3 Assessing the level

In practice, assigning a specific level to a system can be difficult and it may be easier to identify that a technology's maturity falls within bands of TRL, say between levels 3 and 4 or 7 and 8. Only when a system has entered service can it be categorically argued that level 9 has been reached.

6.7.4 Examples of TRL

In Chapter 1, it was shown that three different technologies were brought together to produce wind turbines. The development of those technologies illustrates very well how they matured through the different stages of readiness.

In the case of radar, the observations of natural phenomena, corresponding to TRL 1, were carried out by Christian Oersted and Michael Faraday in the 1820s and 1830s. The work of James Clerk Maxwell in the 1850s to develop a theoretical model of electromagnetic waves raised the TRL to level 2. Illustrating the point that assignment of a unique level is sometimes difficult, it was the work of Heinrich Hertz in the 1980s that validated Maxwell's model increasing the maturity to somewhere in level 3 or 4. Even at this stage of maturity, key properties of electromagnetic waves were still not understood and it took the work of Guglielmo Marconi to extend the understanding (of propagation) to increase the maturity to TRL 5. Note that at this stage, the technologies might be applied to a number of different applications. The first proof of concept demonstration of a radar device was carried out by Christian Hülsmeyer elevating the readiness level to 6. The development of a radar system was, for example, carried out by a number of organisations in the same period of the 1930s like the work of Robert Watson Watt thus maturing radar to TRL 9.

The development of wind turbines also illustrates the readiness levels but, in addition, it shows that the maturation process is not always a chronological progression. The initial observations of being able to extract power from wind were made in ancient times. Blyth and Brush also built systems that showed it was possible to create electricity from wind power in the late 1880s. The development of a theoretical model of a wind power extraction device was carried out by John Smeaton in 1759 and was significantly extended by Poul La Cour in the early 1890s. Taken together this work elevated the readiness to TRL 2. La Cour's turbines at Askov validated the theory, maturing the concept to TRL 3/4. Betz added further extensions to the theoretical model in 1926 arguably achieving TRL 5 and this work was validated by the construction of the Yalta turbine in 1931. The first prototypical high-powered, grid-supplying turbine, corresponding to TRL 6, was the Smith Putnam turbine built in 1941. TRL 7-9 was achieved with the construction of the Plum Brook and Crotched Mountain turbines in 1975 and 1980, respectively.

6.8 Observations on maturity

6.8.1 Increasing maturity

Repeating the point made earlier, it is imperative to have confidence in a system or device before planning permission is granted. However, this poses another conundrum, since useful ideas would never get into service if a prerequisite was that it had been in service before.

Thus, while it is useful to judge the level of technology readiness, it is also helpful to consider what is involved in increasing the maturity of a system so that it might progress to becoming operational. Moreover, if a system or device is on the verge of becoming operational might it be practicable for it to become operational within the life of planning permission? There are no simple answers to these questions because there are so many factors involved; including, but not limited to, the system complexity and the scale of the problem being addressed. The aim here is to review the factors that contribute to increasing the maturity of a system or device.

Obviously, the higher the TRL, the more likely it is to be able to achieve TRL 9 in a limited timescale and systems with a very low TRL would be highly unlikely to be viable [21]. NASA point out in their System Engineering Handbook that a minimum TRL of 6 is required for a system to enter their Systems Engineering processes (preparing a system for launch). But launching a spacecraft is a very complex process that can take many years[‡]. It was observed in the discussion of TRL definitions that a lot of work has to be accomplished to move from TRL 7 to TRL 8. For example, a great deal of documentation has to be prepared. Taking these factors into account, in practice, it is likely that a system would need to be a minimum of TRL level 8 before being considered appropriate for flight.

6.8.2 System/integration readiness level

Thirty years after the introduction of TRL, another metric was introduced: integration readiness levels.

6.8.3 Definition of integration readiness levels

Sauser *et al.* defined these levels as follows [23]:

IRL 1: An interface between technologies has been identified with sufficient detail to allow characterization of the relationship. This is the lowest level of integration readiness and describes the selection of a medium for integration.

IRL 2: There is some level of specificity to characterize the interaction (i.e. ability to influence) between technologies through their interface. Once a medium has been defined, a 'signalling' method must be selected such that two integrating technologies are able to influence each other over that medium. Since IRL 2 represents the ability of two technologies to influence each other over a given medium, this represents the integration proof-of-concept.

IRL 3: There is compatibility (i.e. common language) between technologies so they may integrate in an orderly and efficient manner and interact together. IRL 3 represents the minimum required level to provide successful integration. This means that the two technologies are able to not only influence

[‡]The time it takes to plan a spacecraft launch is a recognised problem within the space industry. New initiatives are aimed at reducing this time.

each other but also communicate interpretable data. IRL 3 represents the first tangible step in the maturity process.

IRL 4: There is sufficient detail in the quality and assurance of the integration between technologies. Many technology integration failures never progress past IRL 3, due to the assumption that if two technologies can exchange information successfully, then they are fully integrated. IRL 4 goes beyond simple data exchange and requires that the data sent is the data received and that there is a mechanism for checking it.

IRL 5: There is sufficient control between technologies necessary to establish, manage and terminate the integration. IRL 5 simply denotes the ability of one or more of the integrating technologies to control the integration itself; this includes establishing, maintaining and terminating.

IRL 6: The integrating technologies can accept, translate and structure information for its intended application. IRL 6 is the highest technical level to be achieved, it includes the ability not only to control integration, but to specify what information to exchange, labels to specify what the information is, and the ability to translate from a foreign data structure to a local one.

IRL 7: The integration of technologies has been verified and validated with sufficient detail to be actionable. IRL 7 represents a significant step beyond IRL 6; the integration has to work from a technical perspective, but also from a requirements perspective. IRL 7 represents the integration is meeting requirements such as performance, throughput and reliability.

IRL 8: Actual integration completed and mission qualified through test and demonstration, in the system environment. IRL 8 represents not only that the integration is meeting requirements but also a system-level demonstration in the relevant environment. This will reveal any unknown bugs/defects that could not be discovered until the interaction of the two integrating technologies was observed in the system environment.

IRL 9: Integration is mission proven through successful mission operations. IRL 9 represents the integrated technologies being used in the system environment successfully. In order for a technology to move to TRL 9, it must first be integrated into the system, and then proven in the relevant environment. So, attempting to move to IRL 9 also implies maturing the component technology to TRL 9.

6.8.4 TRL and IRL application to wind turbine mitigation

Other industries need to be assured of the readiness of systems (hardware and software) for safety-critical applications. A system of TRLs has been developed and has worked successfully for many years. However, this system has been improved upon addition of another metric, the integration readiness level.

For a system to be capable of being widely accepted as a means of wind farm mitigation, it would be appropriate for TRL and IRL metrics to be applied to indicate the maturity of designs.

6.9 Implementing mitigation

6.9.1 Introduction

In some instances, to provide mitigation of the predicted effects of a wind farm, it will be necessary for a new system to be purchased either to replace or augment what is already in place. This might be an in-fill radar to provide radar coverage of the airspace above the wind farm or a supplementary, wind farm-tolerant radar that could provide coverage over a wider operational area.

This section will discuss the factors to be considered when procuring, integrating, and satisfying a Grampian condition, by putting the new system to work. This last phrase, 'putting the new system to work' is important. Providing a radar system is not just about providing a new piece of equipment, it is providing a new *capability*. Berry puts this succinctly, '[It is] about people and technology' [24].

Every project of this nature is more successful if there is a system in place to plan and organise it. Beasley, Sillitto and Squires point out that this is a systems engineering activity and at the heart of the matter is risk management [25]. There are a number of civilian organisations that provide helpful guidance on such systematic approaches, for example, the International Council On Systems Engineering (INCOSE), the US Institute of Electrical and Electronic Engineers (IEEE) Systems Council and the US National Aeronautics and Space Administration (NASA). However, the vehicle that is going to be used to discuss matters here was developed by the military.

The military of all nations of the North Atlantic Treaties Organisation (NATO) have useful strategies in place when new equipment is to be brought into service which ensure they meet the goals for the equipment and that it enters service successfully. The generic name for these very similar strategies is the Defence Lines of Development (DLOD). Berry refers to them simply as Lines of Development for application to civilian projects [24]. The distinction is not important here because both military and civilian aviation radar systems need mitigation.

What are the DLOD? Collectively, they are a systematic approach for making sure every aspect of the project is considered and is in place when required. In the United Kingdom, the DLOD are listed using a memorable acronym, TEPIDOIL:

- Training
- Equipment
- Personnel
- Infrastructure
- Doctrine
- Organisation
- Information
- Logistics

In the United States and in NATO, the same concept is used but the less memorable acronym DOTMLPF is used, standing for Doctrine, Organization, Training, Material, Leadership and Education, Personnel and Facilities. Within NATO, remembering that it is a military alliance of many nations, there is an additional DLOD that must always be considered, that is Interoperability. That is the ability to make materials from different nations communicate together electronically and be usable in a multinational force.

The DLOD will be discussed in turn using the UK form and, where there is a difference with other nations, this will be discussed.

6.9.2 Training

6.9.2.1 Training requirement

Within the topic of training, there are a number of factors that must be taken into account. The personnel in place at the Air Navigation Service Provider (ANSP) will have to operate and maintain the new equipment. It might reasonably be assumed that these personnel have suitable qualifications and experience to carry out the necessary types of work required but it cannot be assumed that they will have enough experience with the new equipment being provided to mitigate a wind farm(s).

To enable the personnel to take over responsibility for new equipment, they must be trained. Equipment suppliers will provide training on their equipment. If the equipment is being bought from a reputable company, they will have lots of experience in providing such training and how best to tailor it for an individual establishment. This training will have to be paid for and this will be a direct cost to the project and cost may be expected to fall on the project sponsor, the wind farm developer.

6.9.2.2 Scheduling

However, there are some hidden problems with training. The aerodrome or service provider will have personnel to man the facility but they will not have 'spare' staff waiting to be trained and then rotated into the regular shift patterns at the facility. Ensuring personnel get trained and that there is adequate cover for provision of the service takes careful planning. This problem may not seem like the business of the wind farm developer, but it has a direct impact on project timescales. The equipment cannot enter service until personnel have been trained but it is even more complicated than that. It may also be necessary to have the equipment in place and ready to operate for hands-on training and the qualification of personnel. Part of the training effort required is to provide personnel with the opportunity to practice on new equipment. This again is a factor that will need to be accounted for in the project planning stage.

Modelling and simulation are increasingly being used to support training. This approach can be implemented in a number of ways. Equipment may be supplied with a training mode so that personnel can practice on the actual equipment. Alternatively, there may be a remote suite of equipment where training and practice can be carried out. This situation has both advantages and disadvantages. Removing personnel from the working environment allows them to focus on the task at hand but scheduling the training becomes more complex. Increasingly, however, it is possible to provide comprehensive training using personal computer equipment.

6.9.2.3 Life-cycle considerations

Training will have to be sustained throughout the life of the project to ensure new staff can operate and maintain the equipment. The point will come where training is part of the routine concerns for the establishment receiving the equipment. However, the initial tranche of training is unusual because every member of the personnel team with a need to be trained will have to be trained in preparation for the new equipment entering service.

This will be covered in the discussion about equipment but it must be remembered that during the life of the wind farm, it is highly likely that equipment will have to be replaced and additional training will be required.

6.9.3 Equipment or material[§]

6.9.3.1 Definition stage

To state the obvious, critical to the success of the project is choosing the right equipment to do the job. The decision about which equipment is appropriate has to be made by the ANSP and their technical advisors but the developer will wish to be part of that process for their own internal governance. Furthermore, expenditure on equipment will be in advance of revenue from the wind farm, the finance must come from somewhere. The developer's backers, usually their banks or venture capitalists, will wish to have oversight of large capital expenditures to ensure costs are under control and the financial forecast is on track.

Currently, the number of manufacturers who can provide the equipment to meet this specialised requirement is limited and, therefore, the shortlist of suppliers may be very limited. Even if the supplier has been chosen in advance and the type of equipment chosen, there will be a stage in the project where the precise equipment configuration is to be selected. Illustrations of the decisions that must be made include, for any given radar type:

- What power level is required for this specific site? Usually nowadays this is a question of how many transmitter modules are required?
- How many receiver channels are required?
- How must the signal processor be configured? etc.

This detailed planning will take time. It may be undertaken as part of a feasibility study, or sometimes these are referred to as a trade study, which investigates the trade-offs between different options, in advance of the main contract to purchase. One of the principal outcomes of such as study will be a detailed schedule that will form the basis for the procurement contract. Even large manufacturers are unlikely to have the resources for such an investigation straight away. It must be assumed that from the commissioning of such an investigation to having the results will take a number of months.

[§]Materiel is the military term for equipment.

6.9.3.2 Construction phase

Equipment such as radars cost a lot of money. At the time of writing, an ASR will cost in the order of £2 Million to £3 Million (€2.3M to €3.5M, or $2.5M to $3.8M). This figure is approximate because no two procurements are the same. This will be discussed more in the line of development considering infrastructure. The significance of such a high capital expenditure to this discussion is that the company from whom the equipment is being purchased is unlikely to hold a whole radar in stock. There will be a lead time for the manufacturer to procure and build the equipment. It is probable that the lead time for equipment procurement alone from placing an order will be a major part of the overall project; perhaps as long as a year.

6.9.3.3 Secondary equipment

The basic equipment schedule is required to perform the ATC role including the mitigation of the wind farm(s). However, depending on the maintenance policy being adopted there may also be a need for specialised test equipment. If the equipment comes with a service plan, test equipment may not be needed but that depends on the custom and practice at the aerodrome.

6.9.3.4 Life-cycle considerations

The wind turbine interference will endure for the life of the wind farm which is driven by a range of complicated factors, but eventually turbines wear out and will need to be replaced. The life of the whole wind farm is likely to be 20 or 30 years. For those currently in the planning cycle, once they are installed the lifetime might be as high as 35 years and, if repowered, even longer. The lifetime of a radar is very similar to that of a wind farm. Currently, in the United Kingdom, many aerodromes are replacing, the very good, Watchman radars installed in the 1980s. It follows that, in the lifetime of the wind farm, it may be necessary to replace the wind farm mitigation radar. Whether any replacement would be like-for-like is debatable. Technology changes such as multilateration and electronic conspicuity, may alter the basic requirements in the next 20 years. However, it would be prudent to assume that at least one replacement system will have to be procured. Moreover, in some regards, replacing a system is more complex than adding a new capability. The service from an existing facility cannot be suspended while it is being replaced. Projects can be a lot more complex therefore when an existing capability is being replaced At the time of writing, whether these life cycle considerations are the responsibility of the ANSP or the developer is a matter of great, contentious, debate.

6.9.3.5 Personnel (including leadership and training)

One definition of the scope of the personnel line of development is:

> 'The timely provision of sufficient, capable and motivated personnel to deliver [Defence] outputs, now and in the future'[¶].

[¶]Based on a UK MOD definition.

Taking a wider view, who are the stakeholders in a project of this nature and what are their personnel requirements to execute a project and deliver a capability:

- The wind farm developers need the capability to be installed and are sponsoring the activity. The personnel within this organisation will be experienced in developing wind farms. Some of the larger developers may have dedicated staff with experience in aviation and radar, but not all do. All developers will have had access to aviation and radar expertise to have been able to prepare and submit an Environmental Impact Assessment. Where there may be a potential shortfall in experience and expertise is in the radar capability delivery.
- The ANSP will have people expert and experienced in operating and managing an ATC capability and they may also have personnel with expertise and experience in maintaining equipment. All these personnel will need some form of training on the new equipment. It is likely that they will have some experience of replacing or installing equipment but this cannot be assumed and they may have a shortfall of experienced personnel in capability delivery.
- The ANSP may sub-contract ATC services to a third party. If this is the case, it is possible that that agency will have a wider experience base and may have access to personnel with capability delivery.
- Safety case support. If a new capability is being added to an aerodrome (or any ANSP facility), then the new capability must be demonstrated to be safe to operate. Although nominally responsible for the safety case, as the licensee of the facility, it is likely that the ANSP will have some form of independent agency to advise on and prepare the safety case for submission to the regulator. These agency personnel may not require experience in capability delivery but they will need to be aware of the scope of what is to be delivered and all matters related to the safe provision of service.

Returning to the suggested definition of personnel, the requirement for the managers of the project will be to ensure that all the personnel are in place and have the necessary resources in a timely manner.

6.9.4 Infrastructure

Figure 6.23 illustrates a modern ASR where the most obvious structure in the image is the tower. Whether this is classed as part of the radar equipment or the infrastructure is less important than remembering that the radar needs a tower. If the tower is new, then in some circumstances it may need planning permission. Planning permission is on the basis of aviation safety, so while planning permission cannot be taken for granted, neither can it be assumed. Irrespective of the outcome, there is a time consideration that must be taken into account. Plans will have to be submitted and the outcome awaited.

Infrastructure considerations include, but are not limited to:

- The radar tower and cabin are on a concrete pad. The pad had to be built. The tower is substantial but it is free-standing (there are no guy wires). The pad has to be capable of dealing with the load.

Figure 6.23 A modern air surveillance radar (image courtesy RAF Brize Norton)

- The tower has to have lightning protection. The pad must have a good earth.
- The pad must have access to prime power to drive the radar.
- The pad must have communications to the control tower to pass the radar output to where it is needed. Ducting and cabling must be in place. Note that may be non-trivial to provide. Long distances may be involved and the task is made more complex if working runways must be crossed.
- The communications must be in accordance with the correct protocols and standards that can be used by the integration facility at the tower. It is likely that the output from the radar will be in a standard form. However, relying on 'it is likely' is not a good integration strategy. This consideration is what would be the interoperability DLOD in NATO.
- Some standards vary from country to country, it is worth checking if standards are nationally compatible.
- An important infrastructure consideration is defining system boundaries and who is responsible for each side of that boundary.

- There may be a need for the radar to be controlled remotely which requires more cabling to be installed.
- If the radar is new and there is an existing radar, which remains in place until the new one is fully commissioned, then attention must be paid to obtaining permission to use more spectrum and ensure that the choice of operating frequencies does not cause interference with any other nearby radar.
- The radar cabin must have some way to drain rainwater. A sump or drain must be planned and incorporated.
- There will be motor vehicles next to the radar. In other words, there is vehicular access to be organised.
- The output from the radar must be taken to where it can be used in the control tower. There is ducting and cabling to organise.
- The tower and cabin must be made secure.
- Safety lighting may be required.
- The cabin will require Heating, Ventilation and Air Conditioning (HVAC).

All of the above factors have to be planned and sequenced.

In the US DLOD, one of the lines of development was Facilities. The US way of classifying equipment comprises two classes: Prime Mission Equipment (PME) and Real Property Installed Equipment (RPIE). A facility is an alternative form of RPIE. Some schools of thought no longer support this concept. For example, PME may only operate correctly if the equipment is installed in a room in which the temperature should not be outside certain bounds. The air conditioning in the room that maintains the temperature would be classed as RPIE (or Facility). If the air conditioning fails, the PME might not be able to function. The boundary between PME and RPIE is becoming less clearcut today.

6.9.5 Doctrine

Doctrine is a contraction of the term concept and doctrine. In military circles, the term is described as what is known and believed to provide the context for operations. Military aviation systems can also be affected by wind turbines and as such the systems are utilised within the context of Air Power Doctrine. For example, in the United Kingdom, this is set out in UK Air and Space Power [26]. This document sets out how things like Air Traffic Control are part of the overall concept of Air and Space Power as well as the relationship with the corresponding civilian organisations which collectively provide a national capability. US Air Power Doctrine is very similar in concept, identifying ATC as one of the 'Functional Communities' forming part of the US national capability [27]. There is a corresponding civilian doctrinal function. For example, it is known and believed that to maintain air safety minimum separation standards must be observed. Thus, the infrastructure of ATC must be designed, deployed and operated in such a way to maintain safety.

It may be questioned what this has to do with buying a new radar for wind farm mitigation? Setting the context for operations, the top-level concepts, form the basis for defining the operational requirement for systems.

6.9.5.1 Life-cycle considerations

Concepts and doctrine evolve. In the future, the way Air Traffic Control and Air Defence is carried out may change. Equally it may not. For example, it seems rational to imagine moving towards a more cooperative means of carrying out ATC using systems like ADSB, M-Lat and WAM. However, it may be that Cyber threats mean these technologies will have inherent risks. Concepts and doctrine will respond to the changing environment and the operational requirements for future systems will change with them.

6.9.6 Organisation

This DLOD (or LOD) is concerned with the functional organisation of how to ensure the capability is delivered. Making sure the organisation is understood and who reports to whom is important. The following non-exhaustive list of questions may have to be addressed:

- Is the mitigation capability being funded by an individual developer or by a community of developers?
- Is there a lead developer who can make decisions?
- If there is a community, how is the financial share worked out? Is the cost per MW or per turbine? Etc.
- Who is responsible for communicating with the other stakeholders?
- If a contentious matter arises, who is the arbitration authority?
- Who is the integration authority?
- Who will provide flight trials?
- Who has the authority to control flight trials?
- If the integration authority decides a change is required, who has the financial authority to allow agree or deny the change?
- Who is responsible for permitting access to the aerodrome?
- Who can speak for each developer, the integration authority, the aerodrome, etc.?
- Who 'owns' the site configuration control/who is the site Installation Design Authority?
- Who is the radar design authority?
- Who pays the bills for resources used prior to handover?
- Who can sign off that the Grampian Condition is satisfied?
- Who is responsible for safety?
- If the supplier nation is a foreign entity, which nations, laws and safety regulations apply?
- What happens if there is an accident?
- What happens if there is a fire?
- Is there a publicity strategy?
- Who is responsible for quality assurance?
- What working hours are allowed?
- If equipment is being imported, who pays the duty and has the authority to recover the equipment from customs officials?
- Who owns the equipment on completion of delivering the equipment?

- How are residual warranty issues resolved?
- Who 'owns' the project schedule and has the authority to change it?
- Who is responsible for applying for radio frequency licensing?
- How often are progress review meetings?
- Who calls meetings and compiles the minutes?
- Who is responsible for trenching and ducting and other types of site infrastructure?
- Who is responsible for cyber security?
- Who will pay for maintenance?

6.9.7 Information

The Information DLOD is a schedule of everything that is required for the capability to be delivered. The information includes but is not limited to:

- Drawings
- System boundaries schedule (who has responsibility for equipment on each side of the boundary)
- Design documents/system descriptions
- Technical/maintenance manuals
- Operating procedures
- Statements of conformity
- Warranty
- Training manuals
- Training certificates
- Critical spares lists
- Spares lists

6.9.8 Logistics

In the military, world logistics is a very broad topic covering everything from how to evacuate injured troops to the replacement of ammunition. In the present context, one of the key matters that is part of the Logistics DLOD/LOD is system maintenance. Today, complex systems such as radars usually provide Built-In-Tests (BIT) as part of their capability. The system tests itself periodically to determine if it is fully operable and reports faults. The results of these tests will be presented on a maintenance display. There may be a software equivalent of BIT which will present errors detected in an error log. It is commonplace too for systems to provide historical data on performance. Another common feature today is a remote monitoring capability so that manufacturers can monitor equipment remotely. Remote monitoring does not just identify faults and errors: the data collected can be analysed to identify trends which allows the manufacturers to predict when actions will be needed. These practices allow a compilation of what is called a sustainment plan.

There are two broad classes of maintenance Preventative Maintenance (PM) and Corrective Maintenance (CM), both names are indicative of the action carried out.

PMs might consist, for example, of checking gearbox oil levels in the antenna and topping up with the appropriate lubricant if required. Vacuum cleaning cabinets are another PM that is sometimes identified. The benefits of this level of cleanliness may not seem obvious but electronic components overheat and making sure there is a free flow of air in the cabinet is important. On a similar basis checking that nothing is blocking air vents is a helpful PM procedure. Airfield engineers should routinely visit all the equipment on an aerodrome to ensure everything is well with the equipment. Such inspections may be fortnightly or monthly.

CM of complex systems is categorised into 'lines of maintenance'. **First-line** maintenance consists of identifying error messages if something goes wrong and looking up the error message to determine the course of action. Replacing fuses and replacing ink cartridges in printers would fall into this category. **Second-line** maintenance consists of replacing faulty modules, as indicated by BIT. The sorts of modules that can be replaced in this fashion are called Line Replaceable Units (LRU). **Third-line** maintenance would consist of repairing LRU. This latter activity used to be undertaken at aerodromes but now is exclusively the domain of manufacturers.

One of the key decisions that must be made concerning equipment is the Maintenance Policy Statement (MPS). Simply, who will be responsible for maintaining the equipment and to which line of maintenance?

If second-line maintenance is to be carried out locally, part of the logistics challenge is arranging for spares holding and the management of spares. Questions to be dealt with include:

- What level of spares should be held to be immediately available for use (what traditionally was called bench stock)?
- At what point do spares need replenishing?
- Where are spares sourced?

And,

- Who pays for spares?

6.10 Putting to work

The DLOD/LOD above show the material and the other things that must be put in place to deliver a capability. However, it says little about what is required to make the considerations a successful project.

Figure 6.24 illustrates the features that would be present in a project to deliver a capability. No timescales are included in this sketch, which would be project specific. This is only an overview of the project.

The project would start with the developer receiving planning permission to proceed with the construction of a wind farm. Assuming a radar system provider has been identified during the planning application process, one of the first milestones is to place an order with that company for the development of a detailed specification. On completion and agreement of the specification the project proper can start. In parallel with the construction of the radar wind farm development and preparatory

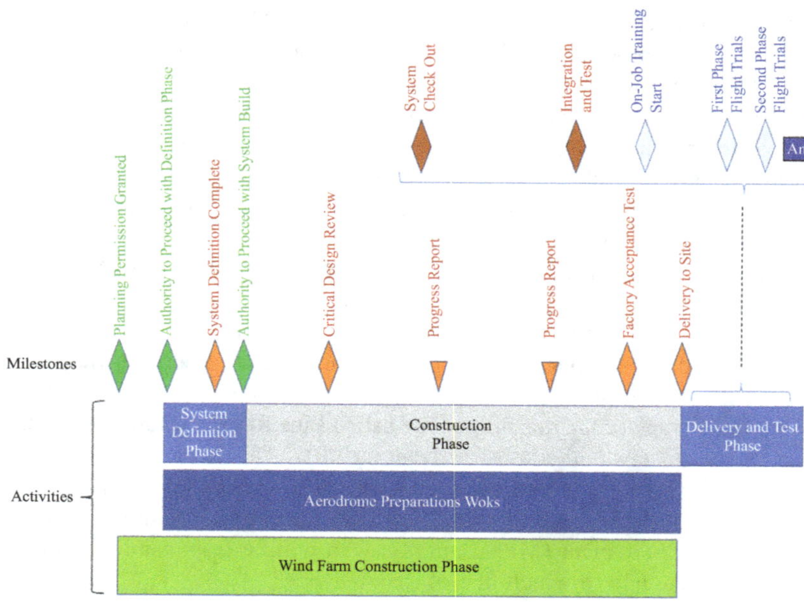

Figure 6.24 Project overview

works can be carried out at the wind farm and radar sites. Early in the project, it would be normal to conduct a detailed design review, often called a Critical Design Review (CDR). This is the radar designer/manufacturer agreeing with the customer about key design features that will be delivered and receiving agreement that that is what is required. Assuming that the milestone is successfully met, manufacturing will proceed. Regular progress reviews should take place following that until a Factory Acceptance Test (FAT). When this stage is completed successfully, the radar can be shipped to the chosen radar site and installation can begin.

Once the radar system is delivered to the site many activities can proceed; integration, training and preparation for flight trials can go ahead. There are two phases to flight trials. The first stage must demonstrate that the radar can detect a target of a given size at a specific distance. Depending on aircraft availability, the radar may need to be modified for this first stage. If the trials aircraft is larger than the specification requirement, then modifications may be required. An example might make the requirement clearer. It is common for a radar to be able to detect a 2 square metre target (3 dBsm) RCS target at a distance of, say 100 km with, say an 80% probability of detection. What happens if the aircraft has an RCS of 4 square metres? The sensitivity of the radar would be adjusted to make 4 square metre targets appear as though they are 2 square metres. The aircraft can be flown to a distance 100 km from the radar and data is collected to see if the aircraft is detected 80% of the time. The radar can then be restored to its normal configuration and additional tests can be carried out to demonstrate its efficacy in detecting aircraft in close proximity to wind farms. If all proceeds well and the regulator is convinced the capability is acceptable, then Initial Operating Capability (IOC) can be declared. Full Operating Capability can be declared when all the other DLOD/LOD deliverables are shown to be in place.

6.11 Wind farm construction

The construction of the wind farm(s) that is being mitigated would take place in parallel with the delivery of the mitigation capability. Many of the factors that must be considered are discussed in Chapter 4. In summary, the key milestones are:

- Ordering the turbines.
- Ordering cranage and other specialised equipment. It is important that this activity starts early because ideally, the construction of wind turbines is undertaken when the wind is at a minimum; usually during the Summer. With limited numbers of machines capable of the construction of large wind turbines, booking early is advised.
- Confirming grid access.
- Making ready road access.
- Making ready on-site road access.
- Building site infrastructure.
- Digging turbine foundations.
- Trenching and ducting for electricity, control and monitoring.
- Laying turbine foundations and allowing them to cure.
- Delivering turbines to the site.
- Erecting towers.
- Hanging blades.
- Commissioning the turbines.
- Connecting turbines to a site electricity substation.
- Making grid connection.

References

[1] Longman Dictionary.

[2] See for example, Dunstane, G. (2005), *Aircraft Equipage Update*, Air Services Australia, The Third Meeting of Automatic Dependent Surveillance – Broadcast (ADS-B) Study and Implementation Task Force (ADS-B TF/3), 23–25 March 2005.

[3] CAA (2021), *Airspace Change: Guidance on the Regulatory Process for Changing the Notified Airspace Design and Planned and Permanent Redistribution of Air Traffic, and on Providing Airspace Information*, CAP1616, UK Civil Aviation Authority, March 2021.

[4] See for example, CAA (2022), *Policy for Radio Mandatory Zones and Transponder Mandatory Zones*, Safety and Airspace Regulation Group (SARG) of the Civil Aviation Authority, 13 January 2022.

[5] ASI (2023), *Stansted TMZ.* https://airspacesafety.com/wp-content/uploads/2022/10/StanstedTMZ_ASI1022.pdf. Retrieved 13 February 2023.

[6] Fliegergruppe Leutkirch (2021), *Update on the New ICAO chart 2021: TMZ Memmingen Extended, Marking of IFR Final Approach Areas*. Available from,https://www.fliegergruppe.de/update-aenderung-der-tmz-um-memmingen-edja/ Retrieved 13 February 2023.

[7] ATC Netherlands (2014), *Transponder Mandatory Zones*. https://www.zweefvliegopleiding.nl/images/navigatie/EH-ENR-6-2-6.pdf. Retrieved 13 February 2023.

[8] NATS (2011), *Introduction of London Array and Thanet Offshore Windfarm Transponder Mandatory Zones*, National Air Traffic Services. http://www.aopa.co.uk/images/stories/Documents/aic%20yellow%2082_2011.pdf. Retrieved 13 February 2023.

[9] Gilman, P. (2020), *Offshore Wind Turbine Radar Interference Mitigation (WTRIM) Series Air Traffic Control Terminal Radar*, US Department of Energy. OSW Turbine Radar Interference Mitigation Webinar Series, 26 October 2020. https://www.energy.gov/sites/prod/files/2020/11/f80/offshore-wind-turbine-radar-interference-mitigation-webinar-10-26-2020.pdf. Retrieved 31 January 2023.

[10] Brookner, E. (1987), 'Array radars: an update, Part 1', *Microwave Journal*, 25:213.

[11] Jackson, C. and Butler, M. (2007), 'Options for mitigation of the effects of wind farms on radar systems', in: *IEEE Radar Conference* 2007.

[12] Dixon, M. (2018), 'Wind turbine impact on military radars a growing perfect storm – or a perfect opportunity?', *Presentation to Global Wind Summit*, Hamburg, 2018.

[13] Thomsen, A.C.K., Riis, M.A., Marqversen, O., and Pedersen, M.Ø. (2011), 'Air traffic control at wind farms with TERMA SCANTER 4000/5000', in *IEEE Proceedings*.

[14] C-Speed (2022), *Lightwave Radar*, https://www.cspeed.com/_files/ugd/624505_3ca4d79a49104139b5394727df5c94b5.pdf?index=true. Retrieved June 2023.

[15] Terma (2023), *Terma 4002 Radar Specification*, https://www.terma.com/media/dogb0jqt/termascanter4002_a4_dec2021.pdf. Retrieved June 2023.

[16] Brookner, E. (1991), *Practical Aspects of Phased Array Antenna Systems*, Artech House.

[17] Doerry, A. (2017), *Catalog of Window Taper Functions for Sidelobe Control*, Sandia National Laboratories, SANDIA REPORT SAND2017-4042. Unlimited Release Printed April 2017.

[18] Presley, J.A. (1984), *The Censored Mean Level Detector for Multiple Targets Environments*, Naval Ocean Systems Center, Report 228, March 1984.

[19] OWIC (2022), *Mitigating the Adverse Effects of Offshore Wind Farms on Air Defence Radar: Concept Demonstrations*, Off-shore Wind Industry Council, May 2022.

[20] International Standards Organisation (ISO), International Electrotechnical Commission (IEC), Institute of Electrical and Electronics Engineers (IEEE) 15288: Systems and Software Engineering – System Life Cycle Processes.

[21] Hirshorn, S.R. (2007), *NASA Systems Engineering Handbook*, NASA SP-2016-6105, Rev 1, December 2007.

[22] https://esto.nasa.gov/files/trl_definitions.pdf.

[23] Sauser, B.J., Forbes, E., Long, M., McGrory, L., and McGrory, S.E. (2009), 'Defining an integration readiness level for defense acquisition', in *Proceedings of International Symposium of the International Council on Systems Engineering*, Singapore, Japan, July 20–23.

[24] Berry, J. (2022), *TEPID OIL Makes Desired Capability a Reality*, Blog Post, 25 March 2022. Timeless Time Consultants.

[25] Beasley, R., Sillitto, H., and Squires, A. with contributions from Heidi Davidz (2023), 'Developing systems engineering capabilities within businesses and enterprises', in: *Art Pyster SEBok Guide to Systems Engineering Body of Knowledge*. https://sebokwiki.org/wiki/ Developing_Systems_Engineering_Capabilities_within_Businesses_and_Enterprises. Retrieved June 2023.

[26] MOD (2017), *UK Air and Space Power*, Joint Doctrine Publication 0-30, 2nd ed., December 2017.

[27] USAF (2023), 'US Air Power Doctrine', https://www.doctrine.af.mil/. Retrieved June 2023.

Chapter 7
Future work

7.1 Introduction

In addition to the application of known techniques and technologies, there is also a need for research to address a range of problems. Some of that research is basic science; for example, the interaction of radio waves and wind turbines. Other research is more applied in nature; for example, the standardisation of conspicuity techniques. This chapter provides some research suggestions for industry and academia.

The following topics are discussed:

- Standardisation of turbine conspicuity methods.
- Interactions between wind turbines and radio waves:
  - An uncertainty model.
  - Electromagnetic compatibility of turbines and radar systems.
  - Submarine communications.
  - Shadow effects.
  - 3-D radar signatures.
  - Thresholds for fast fading.
  - Effects on modulation depth.
- Performance prediction of the interactions between wind turbines and sensors.
- Investigation of new and enhanced signal processing techniques to mitigate the effects of wind turbines. For the purposes of this chapter, signal processing includes detection, tracking and discrimination.
- Some of the problems identified might benefit from the application of Artificial Intelligence techniques. This approach to solving a problem poses some challenges in this domain, principally safety related, that would benefit from research.

7.2 Standardisation of turbine conspicuity

In conversation with wind farm developers, when research was being carried out for this book, they expressed concerns about the lack of standardisation for wind turbine 'conspicuity'. Conspicuity refers to techniques used to make the presence of wind turbines more noticeable to an observer or, in planning parlance, a receptor; in other words, to make wind turbines more conspicuous.

Conspicuity is a very broad and complex topic. For example, the observer might be the pilot of a low-flying aircraft, birds or bats at risk of flying into blades or, electronic systems surveilling airspace. A simple example illustrates the meaning of the term well, the conspicuity of wind turbines can be enhanced by lighting them up when it is dark. The topic of lighting also serves to illustrate the complexity of conspicuity. It is important to understand what is legally required and lighting may or may not be a statutory requirement. For example, in the United Kingdom, obstruction lighting is only required on structures taller than 150 m, and at the time of writing, there are many wind turbines that have tip heights of 147 m to avoid the requirement for lighting [1]. However, if lighting is required, there are environmental concerns about light pollution and it is desirable that lighting is only used as and when needed. Furthermore, legal requirements may change. In the United States, it was once mandatory that wind turbine towers were illuminated at night with a flashing red light. However, in some states of the United States, for example Colorado, Kansas and Washington, the law has changed because of safety concerns about drivers of motor vehicles being distracted by the flashing red lights. Now the red lights must not be turned on unless an aircraft is approaching wind turbines. The situation is complex. One way of satisfying the new requirement in these states is, for example, by having radar automatically turn on the lighting if an aircraft is detected approaching the wind turbines. This is one of a number of methods called **electronic conspicuity** [2]. Another example of electronic conspicuity is to equip the wind turbines with Automatic Dependent Surveillance-Broadcast (ADS-B). Then, the lighting can be turned on when the turbines are alerted to the presence of aircraft and/or the aircraft will be aware of the proximity of the wind turbines via ADS-B.

If wind turbines do carry obstruction lighting, the lights are mounted on the turbine nacelles; hence, they have to operate in extreme conditions and must be effectively maintenance free. In some countries, there are several, sometimes seemingly conflicting, requirements for lighting. If lighting must be used, then it must not be so bright that pilots could be dazzled. One approach is to use infrared lighting which is readily detectable by military pilots flying low and using night vision goggles. Lighting that is not visible to the naked eye solves one problem but then it becomes problematic knowing if the lighting is working. Operators, where this approach is practiced, must carry out routine maintenance checks to make sure the infrared 'lighting' is still operating.

It is not surprising, given the complexity of the requirements, that different nations have arrived at different solutions. However, it is desirable for further work to be carried out to harmonise technical capability and airspace regulations to the benefit of all parties [3,4].

7.3 Interactions between wind turbines and radio

7.3.1 An uncertainty model

The standard method used for determining if a proposed wind turbine will be visible to a radar is to make certain assumptions about the radar, estimate the radar cross-section of the wind turbine, estimate the path losses between the radar and the wind turbine and decide if the signal that would be returned from the turbine exceeds the minimum detectable signal (MDS) of the radar.

Typically, the results of these calculations are presented as the predicted signal strength in dB, more usually dBm, from each wind turbine and an indication of whether the signal level is above or below the MDS. For example, if the MDS of the subject radar is −107 dBm, if the predicted return from a wind turbine is −105 dBm, it will be highlighted in some way, say, **−105 dBm** or −105 dBm. However, little attention is paid to the uncertainty present in this information.

Considering the radio propagation element in this calculation alone, there are a range of factors including temperature inversions and moisture content, etc. that affect the day-to-day propagation losses. Moreover, there are other factors that occur with lower frequency such as ducting (which enhances propagation reducing losses) and fading (which degrades propagation and increases losses). These latter conditions might be characterised as anomalous propagation and they may only occur a few days each year. However, even in what might be considered normal propagation, there is variation. A widely accepted construct is the idea of mean propagation, sometimes referred to as a 50% atmosphere; that is, from the average value of propagation loss, half the year propagation losses are higher and half the year they are lower. Without taking into account anomalous propagation, the 'normal' variations due to refraction may be ±3 dB [5].

This concern is not intended to be alarmist and call into question whether the recognised techniques are valid. Other sources of uncertainty, such as estimates of loss in the radar system, almost certainly add to the robustness of the safeguarding provided by the method, tending to over-predict the signal levels that would be encountered. Furthermore, the reference used, the MDS must be understood. A possible misconception, given that a radar is usually specified to achieve a certain level of performance, say an 80% probability of detection of 2 m^2 target, is that the MDS is the point at which that level of performance is met: it is not. The MDS is the point at which detection is only just starting to be a non-zero percentage. The signal levels would need to be perhaps 4 or 5 dB higher to meet the specification. The point of accounting for uncertainty is to enhance the information available to the potentially affected Air Navigation Service Provider so they are better informed of what to expect.

Two useful topics for research arise from investigation uncertainty:

- Development of an uncertainty model for predicting the effect of wind turbines on radar systems.
- Once a wind turbine or wind farm has been built, investigate whether the predicted effects are realised in practice. In addition to enhancing the models of uncertainty, gathering post-construction results would be useful in enhancing the prediction process itself.

7.3.2 Electromagnetic compatibility

The principal focus of this book is on the negative effects that wind turbines can have on radio and radar systems used for aviation. Conversely, little attention has been given by industry or academia to the potential negative effects that radio systems and, in particular, because of their high power, radar systems might have on wind turbines. It is an industry concern that there is a need for a better understanding of this aspect of electromagnetic compatibility (EMC).

Wind turbines have complex electrical and electronic systems to manage the generation system and these must comply with EMC directives. Karpat and Karpat have identified the following national compliance requirements: USA Federal Communications Commission (FCC) Part 15; Japan VCCI; Canada ICES; and the European Union Directive 2004/108 IEC [6]. As yet, there is little understanding of the susceptibility of these systems themselves to high-powered radio sources such as radar transmitters. This situation will become more problematic as consideration is given to the use of radars within and around off-shore wind farms for aviation and ecological purposes.

7.3.3 Submarine interactions

During discussions with stakeholders when this book was being prepared, it was reported that German Defence Forces have expressed concerns about the possible interactions between off-shore windfarms and submarine detection and communications. Detection and communications use VLF frequencies (3–30 kHz) because of their ability to penetrate sea water.

A literature search identified a body of work on the effects of electromagnetic waves on human health [7] and there have been investigations of general propagation mechanisms of VLF frequencies [8]. No reports were found in open-source literature on the interactions of wind turbines and such low-frequency electromagnetic waves.

7.3.4 Shadow effects

Anecdotally, at least some classes of radar system experience shadow effects that may go beyond what might be predicted by signal blockage described in Chapter 4, that is, there is a volume of airspace behind a wind turbine where there is a reduced Probability of Detection (P_d). As discussed in Chapter 4, blockage does occur but, even at the highest frequencies of interest, any blockage effects are reduced by diffraction, that is, the propagation mechanism that fills in the shadow behind obstructions.

There is little published evidence for a shadow effect over and above the effects of diffraction. The CAA has reported trial data which indicates shadowing is limited to ‘a few hundred metres’, consistent with the findings reported in Chapter 4 [9]. Despite this, rumours still persist.

Interference investigations have also shown that there are regions behind wind turbines where signals are depolarised. It is postulated that a shadow-like effect could be caused by depolarisation of the forward scattered wave behind wind turbines. The signal returning from a target beyond the wind turbine will be further depolarised on the return path to the radar. The nett effect could be a reduction in the probability of detection by the radar. If depolarisation does reduce probability of detection in the manner suggested, the effect could be mitigated by using circular polarisation. Circular polarisation is commonly used to improve the visibility of targets in the rain (and other hydrometeors, snow, sleet, etc.) and this may have a bearing on why there is no evidence of the effect. However, this discussion is largely speculative. To gain a better understanding and either confirm or refute the shadow and depolarisation hypothesis either flight trial measurement or theoretical modelling, or ideally both, would be helpful.

7.3.4.1 Flight trial measurements

A practical measurement campaign to collect empirical data would be helpful. Figure 7.1 illustrates a genuine terrain profile (terrain profiles are described in Chapter 5) from a radar on the left-hand side of the graph to a point 60 km away. A wind turbine is located on the brow of a hill approximately 27 km from the radar. Beyond the turbine, there is further hilly terrain to the terminal point on the graph 60 km from the radar. The wind turbine is visible to the radar at the ground level. Beyond the turbine, there are volumes of airspace where it might be possible for an aircraft to fly, provided it is deemed safe. If an aircraft was to fly in this region, data might be collected to investigate the phenomenon. For example, in the terrain profile plot in Figure 7.1, there is a region beyond approximately 43 km from the radar, which is in the radio far field behind the turbines that might be particularly suitable. Ideally, data might be collected with the radar being operated in linear polarisation and circular polarisation modes and when an aircraft is in the shadow of turbines and, by flying higher, where an aircraft would be out of the shadow.

7.3.4.2 Ground measurements

As desirable as they are, flight trials are expensive. Empirical data could also be gathered on the ground without the use of aircraft. It is possible to measure the

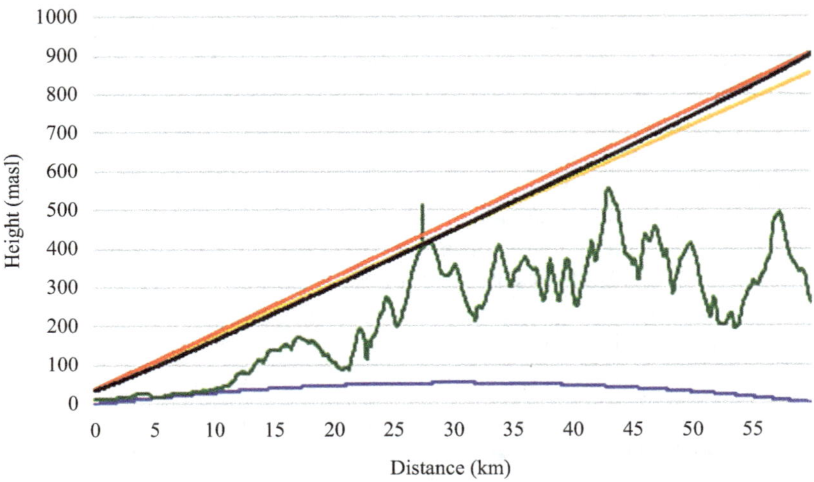

Figure 7.1 Example terrain profile

signal levels from different types of radars using the type of equipment that might be used to make radio interference measurements; for example, spectrum analysers and real-time spectrum analysers. If pursuing this, care must be taken when making radar measurements to follow best practices as set out in both ITU and US National Telecommunications and Information Administration (NTIA) guidance. Critical factors to take into account are:

- Selection of the optimum measurement bandwidth.
- Measurement in the far field.
- Adequate measurement dynamic range.
- Adequate measurement integration time [10–13]*.

Ideally, data would be collected behind and in front of wind turbines and with, if possible, different polarisation modes employed.

7.3.4.3 Modelling

To provide a better understanding of the depolarising effects of wind turbines blades and the impact of depolarisation on radar performance, a desk-based investigation would be useful. Such a study should focus on the differences in performance delivered by linearly and circularly polarised radar.

7.3.5 3-D radar signatures

Another cause for concern is the incomplete understanding of the 3-D RCS signature of wind turbines and their aggregation as wind farms. This concern is relevant to the accurate prediction of fading, when reflections from wind turbines interfere with signals received directly from the ground transmitter which was discussed in Chapter 4. The phenomenon of fast fading is relevant to communications and navigation systems (in the latter case, in a lot of literature, this phenomenon is referred to as scalloping).

Standard guidance on interference with AGA communications is contained in documents such as the CAA's Air Traffic Services Safety Manual, CAP 670. The method specified there for the computation of Carrier to Interference (C:I) ratio† relies upon the reflection of radio waves from turbines as set out in ITU-R BT-805 [14]. This recommendation introduces a model that distinguishes between forward scatter and general scatter regions. A formula is provided to calculate the value of the signal in the forward scatter region with a correction factor for the general scatter region. The correction factor is −10 dB; although another ITU recommendation, ITU-R BT 2142-2 [15] states that it is not clear how this figure is arrived at. The method was put forward in the context of interference effects on Broadcast Television (BT) signals from wind turbines‡. A feature implicit in the scenario set

*The reader should not be put off by the reference in the ITU publications to out-of-band emissions. The focus of the measurement regimen is interference but it is intended to ensure that out-of-band emissions are measured accurately against the reference of the in-band measurement.

†Sometimes referred to as the wanted to unwanted (W:U) ratio.

‡Hence ITU-R-BT-805.

out in ITU-R BT-805 is that the television transmitter, the wind turbine(s) and the receiver are all effectively co-planar; or at least the angular separations in the vertical plane are small, see Figure 7.2. However, in the case of aviation, the receiver/transmitter is on-board an aircraft and the sources of the interfering signal are not co-planar and can have high angular separations in the vertical plane.

It may still be entirely appropriate to use the ITU-R-BT805 model for this application but it would be helpful to confirm this and provide greater resolution of the RCS profile of different turbines and turbine classes. To illustrate why it may be, beneficial consider AGA communication (described in Chapter 3). The communications take place using vertical polarisation at VHF or UHF frequencies. Major contributors to the RCS come from the tower, the blades and resonance effects between the two when the blades are in specific orientations. Figures 7.3 and 7.4 illustrate, courtesy of Vattenfall, the differences in the optical area projected by a turbine in the coplanar and vertical planes, respectively. The photographs suggest that the projected area of the tower and the vertical turbine blades offer low optical projected areas and, in particular the tower, which if treated using the canonical form of a long cylinder has little reradiation along the axis of the cylinder.

The availability of drone technology may offer a means of carrying out an investigation of RCS at different viewing angles. Moreover, the ITU now publish guidance on the use of commercial drones for making spectrum monitoring tasks [16].

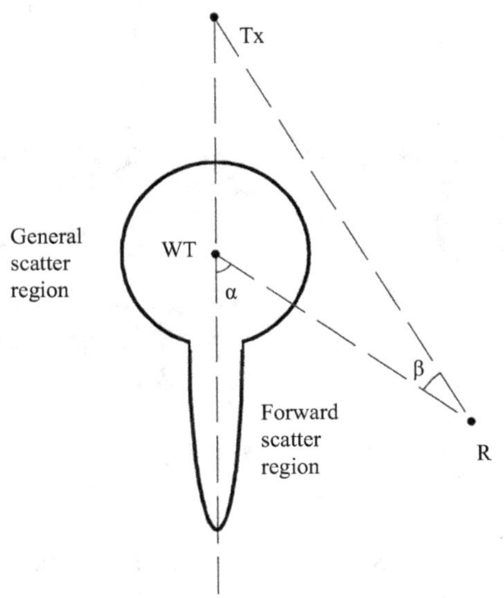

Figure 7.2 Scatter regions from ITU-R-BT-805. Tx is the television transmitter, WT is the wind turbine and R is the television receiver (reproduced by the kind permission of the ITU)

Figure 7.3 *Approximately coplanar view of wind turbine (image courtesy Vattenfall PYC)*

Figure 7.4 *Vertical plane view of wind turbine from below (image courtesy Vattenfall PYC)*

7.3.6 A threshold for fast fading interference and cumulative effect

7.3.6.1 Introduction

Chapter 4 described the problem of fast fading caused by the interference of direct waves from ground transmitters and reflection off turbine structures. The receivers may be for the purposes of either communications or navigation. The safeguarding test requires the calculation of the effect of the wind turbine that creates the worst-case carrier-to-interference ratio, and also the carrier to interference ratio for the sum of all the wind turbines in the wind farm on the assumption that all the turbines sum coherently. The C:I ratio is then compared with a threshold.

There would be a benefit in researching an optimum value for a threshold setting for this comparison for the different receiver types and then developing a method for generating a rationale for a cumulative impact statistic.

7.3.6.2 Background publications

NASA has investigated the cumulative effect of portable electronic devices (PED) on aircraft communications and navigation receivers [17]. This work formed the basis for providing telephones, computers, etc. with an airplane mode to disable their rf communications systems.

The ITU has considered methods for calculating interference between terrestrial television services and satellites, including cumulative effects and threshold values [18].

7.3.7 Effects on modulation depth and thresholds

A standard approach to determining the impact of wind turbines on ILS is to address fast-fading analysis in a fashion similar to that used for AGA communications. Concerns have been expressed about whether this takes sufficient account of modulation depth [19]. A description of ILS was provided in Chapter 3. In summary, the bearing of an aircraft with respect to the glideslope is determined by the proportion of two modulating signals. The signals are transmitted at different frequencies. When the two modulating signals are the same amplitude, the aircraft is on the glideslope and on the extended centreline of the runway. If one modulation dominates, then depending on the frequency, the aircraft is off the desired location in either or both the horizontal and vertical dimensions. The concern is that a wind farm may reflect and cause interference in such a way that one modulation depth is favoured over another and that this may lead to misleading navigational guidance. Say for example, if the wind farm, or any other strong reflector of radio waves, was on one side of the extended centreline of the runway and the reflected signals could influence the signals on the other side of the centreline.

A detailed investigation of whether it is possible to corrupt modulation depth would provide greater confidence in the analysis of these problems.

A starting point for an investigation might be provided by the work of Gierhart and Johnson [20] who, in 1965, investigated the effects of co-channel interference on ILS. Although the work was concerned with additive, and not multiplicative, interference, and no distinction is made on modulation depth, interference levels are set out along with the effects at different heights.

7.4 Performance prediction

When this book was being researched, it became clear that there is a debate about the role of modelling and simulation in the prediction of the impact of wind farms on aviation interests. It is widely accepted that models of radio propagation (described in Chapter 5) are useful tools for predicting whether turbines will be detectable by radars. But, in some circles, there is a reluctance to use modelling for a wider assessment of the impact of prospective windfarms on aviation systems in case the models are not rich enough to represent the interactions with adequate fidelity. Moreover, if resources are limited it may be more effective to improve solutions than attempt to model imperfect situations. However, as set out in Chapter 4, it remains the case that, currently, performance must be predicted in advance of planning permission being granted for a wind farm. Furthermore, modelling and simulation offer the opportunity to exercise mitigation in a wider range of circumstances than is practical in, what are relatively short tests, such as flight trials. For example, even in a relatively long flight trial the weather conditions that will be encountered are unlikely to be representative of all the weather conditions that will be experienced during the life of a wind farm.

7.4.1 Modelling granularity and scope

The current methods for modelling wind farm effects on radar systems are intended to provide confidence that the wind farm will have no effect, that is, the signal level from echoes from the wind turbines do not exceed the minimum signal level detectable by the radar. If that is the case, then there is no concern, and development can proceed without an objection on radar grounds. If echoes from the wind farm might be detectable, then some form of mitigation is required.

However, as mitigations are integrated into radar systems, until they are sufficiently mature to be considered reliable under all circumstances, there will be a need to be able to predict their efficacy against a wind farm that has not yet been built. Therefore, it becomes important to know if the proposed wind farm can be modelled adequately to have confidence in the results. The problem has parallels in the world of Artificial Intelligence. An AI system can provide advice on problems provided the problem is within the scope of its training set. The question becomes: what needs to be incorporated into the model of performance to have confidence predictions will be accurate? Using the AI metaphor, what factors need to be included in the training set? Some of the algorithms of interest may be proprietary, further complicating the problem. However, many algorithms start life in academia and it is important that modelling and algorithm development proceed in parallel. Furthermore, there is also a need to recognise important features to allow independent assessment, and possibly comparative assessment of different systems by other communities. For example, how can an Air Navigation Service Provider (ANSP) have confidence that a particular mitigation technique will provide what they will need?

The elements of modelling will need to cover:

- The performance of the wind turbine. For reasons pointed out in Chapter 4, there are good practical reasons why the wind turbine type may not be known and a general model is required. Representative RCS models will be required and a clear understanding of why they are representative.

- The environment including the terrain and a range of wind strengths and directions.
- The effect(s) on the radar resulting from the nett effects of the wind turbines and their situation in the environment.

Recognising the features that are important is itself important for other reasons. It will also provide guidance on what needs to be included in flight trials and, equally importantly, what is not necessary. For example, suppose the radar mitigation is sensitive to the distance between wind turbines, then it is important to incorporate inter-turbine spacing into the modelling.

The goal of research should be to determine those factors that are critical to assessing performance.

A technology that might be a useful starting point for investigations is how these concepts are dealt with in the emerging field of digital twinning.

It is possible that this topic might exploit AI as discussed below.

7.5 Novel and enhanced signal processing techniques

7.5.1 Non-traditional detection

The current methods of detection of targets by radar systems are described in Chapter 3. In summary, the presence of a target is declared when a Constant False Alarm Rate (CFAR) processor indicates that the measured signal level exceeds a threshold. The threshold is adaptive, attempting, as the name indicates, to maintain the rate of false alarms at a constant level that is sustainable through the subsequent processing in the radar. The optimum, most reliable, setting for the threshold, based on Neyman–Pearson principles, is that the threshold should be a value where the probability of a miss (P_{miss}) and the probability of a false alarm (P_{fa}) are closely matched.

It was shown in Chapter 4 that the basic CFAR process has a number of failure modes which include (but is not limited to) the presence of wind turbines which cause the condition called 'masking'. This condition is where sample cells contain large signal measurements (from the wind turbine) which so load the assessment of average noise level as to make genuine targets appear to be below the noise level.

In Chapter 6, one method of preventing desensitisation of the detector (masking) by wind turbines is to adopt a Censored Mean Level Detector (CMLD) algorithm. This type of algorithm edits out statistically 'large' returns from what should be a sample of just the noise environment.

In summary, CFAR detection has evolved as new problems have been exposed and alternative algorithms developed and adopted. If a common theme might be identified it is perhaps that more and more empirical knowledge of the target/clutter environment has been accounted for by new and modified algorithms.

The question arises, can anything else be done to enhance or replace detection algorithms to make them more effective in an environment in which there are wind turbines and wind farms? A starting point is to examine, and question, the basic assumptions about detection.

7.5.1.1 Signal-to-noise ratio

A fundamental assumption about target detection is that a high signal-to-noise ratio can be reliably assumed to indicate the presence of a target (an aircraft). But it was shown in Chapter 4 that this leads to problems with detections because the clutter returns from wind turbines exceed the signal-to-noise ratios of genuine targets. Moreover, whereas volume clutter may give rise to high signal-to-noise ratios which CFAR can handle, wind turbines are point sources of clutter (hence they can be handled using CMLD). Simply put, the assumption that a high value of signal-to-noise ratio is caused by a genuine target is no longer valid.

Moreover, as wind turbines get larger, their RCS increases, and, hence, the amplitude of their returns increase, whereas genuine targets of interest are decreasing in RCS. For example, the use of composite materials in aircraft manufacture is giving rise to reductions in RCS. Precise figures for the reductions tend not to be in the public domain because of the security implications for defence and the figures that are published tend to be for higher frequencies than for those of principal interest. However, the indicative figures may be at least a 10 dB reduction in RCS compared with non-composite materials [21,22]. Another category of target that is of concern is Unmanned Air (or aircraft) Systems (UAS) or, Unmanned Air Vehicles (UAV), or in some literature, simply, drones. UAS have become a serious problem because of their incursions into Control Zones and Controlled Airspace and the consequent disruption to commercial flights [23]. UAS can be very small and have small RCS values. Oswald and Baker state the RCS of a typical hobby-scale UAS at L-Band (1.2 GHz) is between −20 dBsm and −30 dBsm [24]. A good summary of RCS and RCS characteristics is published by the IET [25]. One of the most complex UAS problems to deal with is tracking swarms of UAS, Easthope and Collinson (2018) introduce these problems [26].

Therefore, the question arises whether there are alternatives to using SNR as an indicator of the presence of a target. A subsequent suggestion relates to the use of cross-polarisation data products to discriminate between wind turbine clutter and different forms of volume clutter. One possible alternative might be the use of cross-polarisation.

7.5.2 Tracking and data fusion

A feature of wind turbine signatures is that the RCS is dependent, inter alia, on the yaw angle of the radar to the wind turbine. It follows that two radars observing a wind turbine from different viewing angles will be presented with different signatures. It is also the case that an aircraft flying above a windfarm may be viewed by multiple radars.

7.5.2.1 Multiple sensors

Tracking using multiple sensors might be used to enhance the detection of targets in high levels of clutter from wind farms. Research is, therefore, required to characterise the tracking requirements in a data fusion situation where more than one sensor is available.

The legacy radar networks have been designed around communications protocols designed for highly reliable but relatively low throughput capacity. Today much higher throughputs are possible without sacrificing assuredness. Lower-level data, for example, returns or plot-level data can be remotely fused without them being tracked before transmission. As more measurable data on targets and the environment can be moved to a coordination centre the easier, it becomes to exploit data fusions realising:

- Higher update rates.
- Enhanced ability to detect low observables.
- Enhancing accuracy.
- Enhanced sensor management.
- Better options for discrimination.

These benefits of data fusion set out above are well understood but there is considerable scope for investigating how they could be applied to the domain of interest here. Hall and McMullen (2004) provide a good introduction to newcomers to this field [27].

7.5.2.2 Improved return to track association and classification

Another consideration worthy of investigation is Return To Track Association (RTTA) algorithms which were discussed in Chapter 4. RTTA is an intrinsic part of tracking and is usually only concerned with spatial and kinematic factors; simply put, where is the aircraft now and where might it be in some number of seconds time. Huizing *et al.* [28] argued that RTTA was also an intrinsic element of discrimination, referred to in their paper as classification, as well as tracking. They made the point that if a track is corrupted by associating a wrongly classified return, then this leads to further confusion when attempting to classify the track. Implicit in this observation is the need for a reassess the association rules taken into account in RTTA algorithms. Huizing *et al.* used a helicopter track and identified different classifications of returns using different algorithms. This paper may be a useful starting point for consideration of wind turbine returns.

7.5.3 Discrimination using cross-polarisation

Although meteorological radars are outside the scope of this book, there have been recent innovations introduced into the processing of their radar data that with further research might be useful.

A good introduction to the subject is provided in the UK Met Office's Factsheet 15 [29]. Meteorological radars usually operate in the S-Band (2.7–3.1 GHz) or C-Band (4–8 GHz). At these frequencies, hydrometeors, that is, rain, snow or hail, present as volume clutter and the backscatter can be readily detected. However, wind turbines have presented significant problems for meteorological radars. Harrison provides a description of the UK network of meteorological radars and a useful summary of the effects caused by interactions with wind turbines [30]. The Operational Programme for the Exchange of Weather Radar Information (OPERA), a programme of the European Meteorological Network (EUMETNET) also provides a discussion on the effects of these interactions and an analysis of radar data [31].

In summary, meteorological radars detect the primary echoes from wind turbines but they also detect echoes created as a result of reflections off the turbines hitting the terrain around the wind turbine. There may even be multiple reflections between the ground and the turbine. The latter detections are received later than the direct echoes from the turbine and thus appear to be at a longer range than the turbine. The returns depend on wind speed and direction and in some circumstances they appear to have the same characteristics as persistent rain. This can be problematic because when rain falls for a long period of time it can trigger flood alerts. One way of discriminating the clutter created by wind turbines from genuine rain is to compare radar data with satellite imagery. If there are no clouds visible near the wind turbines then the returns are not from rainfall. The absence of a cloud is an effective discriminant but it may not be available in real time.

The benefits of dual polarisation as a means of characterising rainfall have been known for a long time; Rutherford Appleton Laboratory (RAL) described using dual polarisation in 1984 based on data collected using the Chilbolton radar [32]. The French Weather Service started to experiment with dual-polarised radars in 2004 [33]. Snow, sleet, different forms of graupel (snow crystals) and rain can each be identified using dual polarisation as well as some non-hydrometeors such as burning chaff and ground clutter echoes which may appear in the sidelobes of the radar antenna. There is, however, a disadvantage of using dual polarisation created by having to transmit signals in both polarisations; R L Ice found additional losses in dual polarisation [34].

7.5.3.1 Recommended work

Meteorological radar cells are much larger than aviation radar resolution cells and aviation radars do not employ cross-polarisation but it would be valuable to investigate whether any of the concepts from the meteorological radar work can be used in aviation radars.

7.5.3.2 Background publications

The benefits of dual polarisation were judged to outweigh the disadvantages and weather networks started to convert their radars to dual polarisation: the US Weather Service started to modify their NEXRAD radars in 2011 [35] and the UK's Met Office started to modify their network in 2013 and the programme was completed in 2018 [36].

As more meteorological radars have been modified to provide dual polarisation, more data about the interactions of this class of radar and the ability to discriminate wind turbine clutter from hydrometeors and other forms of clutter has become available. The domain performance metric is called the Heidke Skill Score; this is a measure of how well a weather forecast performs. In 2016, Beauchamp and Chandrasekar of Colorado State University proposed cyclostationary[§] properties in

[§]Stationary in the statistical sense but composed of a repetitive function caused by the rotation of the blades.

the RCS time intensity[¶] might be used as a discriminant [37]. The benefit of the method proposed and similar methods are that the discrimination process can be carried out on the data from individual resolution cells. In 2017, Hall *et al.* of Bristol University have analysed data from two UK meteorological radars and identified a Fuzzy Logic Membership function that is able to discriminate wind turbine clutter present in the radar resolution cells that result in an increase in the Heidke Skill Score from 67.4% to 97.8% [38]. In 2022, Darlington believed the Met Office was close to being able to discriminate successfully wind turbine returns from other forms of ground clutter [39].

7.6 Artificial Intelligence

Artificial Intelligence (AI) is already used to carry out a number of complex tasks supporting wind-powered electricity generation; the following two tasks are examples. The amount of power generated by a wind turbine is dependent on the complex behaviour of the wind. In still conditions, energy must be obtained from other sources. AI is proving useful to predict the amount of energy that will be produced at least a day in advance. This information helps to optimise the management of the supply of electricity to the grid [40]. Another complex task caused by remoteness and difficulty in accessing wind turbines and their components is planning preventative maintenance to ensure that it is carried out before components fail. AI is proving effective at determining corrosion levels of turbine blades and predicting in advance when blade failure might occur [41].

AI is also starting to be used in some specialist types of radar system called cognitive radars. Cognitive radar systems monitor the spectral environment, identifying and countering the impact of unintentional interference and malicious jamming by adapting the transmitter and antenna performance. For example, if an interference source is detected on a particular bearing, then the antenna performance can be modified to reduce the gain in the direction of the interference or to exploit a different polarisation [42].

The question arises, might AI be useful to detect and help mitigate the effect of wind turbine clutter on radar systems?

7.6.1 Background publications

The foregoing examples were examples of machine learning systems (MLS). There are many different forms of MLS such as deep learning algorithms and artificial neural networks (ANN). The topics are covered in textbooks such as Russel and Norvig's 'Artificial Intelligence: A Modern Approach' [43] and the key issues are addressed in the IET's fact file 'Artificial Intelligence and Functional Safety' [44].

[¶]The variation of RCS versus time.

7.6.2 Desirable functionality

With a sufficiently large learning set and appropriate control data, MLS technology allied with Big Data may be able to detect signatures that are not obvious to humans. This could be useful going forward. However, the problem is not static, it was pointed out in Chapter 1 that as wind turbines are increasing in size, their radio signatures will change accordingly. Therefore, machine learning will need to advance with changes in the technology. This points to two useful avenues of research: identification of those factors that are representative of turbine behaviour and, if the system is to learn in real time, what is required to ensure the system can be safety accredited; in other words what is necessary to establish confidence in the role of the AI system?

7.6.3 Critical behavioural characteristics

One of the greatest challenges to be resolved is determining those factors that would build confidence in a method of filtering wind turbine returns. This problem was introduced in Chapter 4 but is summarised here for convenience.

As organisations develop mitigations for wind turbines, algorithms have to be tested to prove that they work. Implicit in this requirement is that the algorithms have to work in all the circumstances that a radar would meet in service such as different:

- Wind turbine sizes.
- Combinations of turbine sizes.
- Radar-turbine geometry.
- Blade speeds.
- Wind speeds, etc.

There is a long list of factors to be considered. But it may be that relatively few of these are important to a given mitigation design. For example, perhaps the critical factor may not be turbine height but the inter-turbine spacing. AI coupled with the allied technology of Big Data may be able to provide a shortlist of critical factors that are essential to be assessed.

7.6.4 System confidence

An AI system must be capable of certain levels of functionality in order for it to be useful; these include:

- Consistency – presented with similar situations will the system provide similar answers?
- Completeness – how well does the system cope when new situations are encountered?
- Soundness – are the opinions provided, correct?
- Precision – If judgements have to be qualified, are the qualifications precise?
- Practicality – Can the system be used effectively, for example, can it be provided with all the information it needs? [45].

Without satisfying these requirements, the value of a system will be limited. However, for the user to have confidence in the system another quality is required; the system must be able to demonstrate how conclusions were reached, that is it must be transparent.

Research is therefore needed to establish whether AI can assist in identifying critical parameters for inclusion in modelling.

References

[1] CAA (2016), *Air Navigation Order (ANO) 2016*, Article 222 of the Civil Aviation Authority ANO 2016.

[2] Flatt, C. (2023), *No More Blinking Lights at Night, Washington Wind Turbines to Go Dark*, Reporter for News Station KUOW, Report 11 May 2023. https://www.kuow.org/stories/washington-pushes-back-on-light-pollution-from-wind-farms. Retrieved 18 May 2023.

[3] See for example, https://landuse.co.uk/visual-effects-wind-farm-lighting/. Retrieved 23 March 2023.

[4] See for example, https://orga.nl/obstruction-lighting/. Retrieved 23 March 2023.

[5] Personal conversations between the author and Dr David Bacon.

[6] Karpat, E. and Karpat, F. (2016), 'Electromagnetic compatibility (EMC) in wind turbines', In *the Proceedings of 68th IIER International Conference*, Barcelona, Spain.

[7] Knopper, L.D., Ollson, C.A., McCallum, L.C., *et al.* (2014), *Wind Turbines and Human Health*, Frontiers in Human Health, 19 June 2014. https://www.frontiersin.org/articles/10.3389/fpubh.2014.00063/full. Retrieved 4 April 2023.

[8] Zhao, S., Shen, X., Zhou, C., and Xia, Z. (2023), 'Space- and ground-based observations of ELF (extremely low frequency)/VLF (very low frequency) electromagnetic waves and their propagation mechanisms', *Research topic in Frontiers in Astronomy and Space Sciences*. https://www.frontiersin.org/research-topics/46453/space–and-ground-based-observations-of-elfextremely-low-frequencyvlfvery-low-frequency-electromagnetic-waves-and-their-propagation-mechanisms#overview. Retrieved 4 April 2023.

[9] CAA (2019), *Air Traffic Services Safety Requirements*, CAP-670, the [UK] Civil Aviation Authority, Issue 3 Amendment 1, 7 June 2019.

[10] ITU (2003), *Measurement of Unwanted Emissions*, ITU-R-SM-1177, June 2003.

[11] ITU (2015), *Unwanted Emissions in the Out-of-Band Domain*, ITU-R-SM-1541, August 2015.

[12] Sanders, F., Hinkle, R.L., and Ramsey, B.J. (2005), *Measurement Procedures for the Radar Spectrum Engineering Criteria (RSEC)*, TR-05-420.

[13] Sanders, F. and Dalke, R. (2012), *Relationships Between Measured Power and Measurement Bandwidth for Frequency-Modulated (Chirped) Pulses*, TR-12-488.

[14] ITU (1992), *Assessment of Impairment Caused to Analogue Television Reception by a Wind Turbine*, Recommendation ITU-R BT.805, March 1992.

[15] ITU (2015), *The Effect of Scattering of Digital Television Signals from Wind Turbines*, Report ITU-R BT.2142-2, Dated 2015.

[16] ITU (2021), *Use of Commercial Drones for ITU-R Spectrum Monitoring Tasks*, Report ITU-R SM.2486-0, June 2021.

[17] Nguyen, T.X. (2015), *Cumulative Interference to Aircraft Radio from Multiple Portable Electronic Devices*, NASA Langley Research Center, Hampton, VA. https://ntrs.nasa.gov/api/citations/20150009465/downloads/20150009465.pdf. Retrieved 23 February 2023.

[18] ITU (2019), *Interference Calculation Methods*, Report ITU-R-BO. 2019-1.

[19] Private communications between the author and Glasgow Prestwick International Airport.

[20] Gierhart, G.D. and Johnson, M.E. (1965), *Interference Predictions for the Instrument Landing System*, US National Bureau of Standards, Technical Note 324, 25 October 1965.

[21] Mitrano, *et al.* (2008), 'Characterization of radar cross section of carbon fiber composite materials', In *Radar Conference*, 2008. RADAR '08. IEEE, 26–30 May 2008.

[22] Marro, F., Lecini, J., Tamburrano, A., Pisu, L., and Sarto, M.S. (2018), 'Electromagnetic wave absorption and structural properties of wide-band absorber made of graphene-printed glass-fibre composite', Science Report, 8:12029. https://www.ncbi.nlm.nih.gov/pmc/articles/PMC6089982/. Retrieved 22 February 2023.

[23] SESAR (2022), *Making Headway in Protecting Airports from Drone Incursions*, SESAR. https://www.sesarju.eu/news/making-headway-protecting-airports-drone-incursions. Retrieved 22 February 2023.

[24] Oswald, G. and Baker, C. (2021), *Holographic Staring Radar*, IET, Section 4.1.2.

[25] Patel, J.S., Fioranelli, F., and Anderson, D. (2018), *Review of Radar Classification and RCS Characterisation Techniques for Small UAVs or Drones*, IET. https://ietresearch.onlinelibrary.wiley.com/doi/10.1049/iet-rsn.2018.0020. Retrieved 22 February 2023.

[26] Easthope, P.F. and Collinson, A. (2018), 'Tracking swarms of, un-manned aerial systems", in *Automatyka i informatyka przemysłowa (Journal of Automation and Industrial Informatics)*, vol. 43, pp. 47-63, 2018.

[27] Hall, D.L. and McMullen, S.A.H. (2004), *Mathematical Techniques in Multisensor Data Fusion*, Artech House.

[28] Huizing, A.G., Theil, A., and Groen, F.C.A. (1997), 'A comparison of a possibilistic and a probabilistic classifier in a multitarget tracking environment', in *IEE Radar Conference*, Edinburgh, 1997, pp. 614–619.

[29] Met Office (2019), *Fact Sheet 15, Weather Radar*, National Meteorological Library and Archive. https://www.metoffice.gov.uk/binaries/content/assets/metofficegovuk/pdf/research/library-and-archive/library/publications/factsheets/factsheet_15-weather-radar-2020_temp.pdf. Retrieved 20 February 2023.

[30] Harrison, D. (2012), 'Impact of operational wind farms in the United Kingdom on weather radar products', in *The Seventh European Conference on Radar Meteorology and Hydrology*.

[31] OPERA (2006), *Impact of Wind Turbines on Weather Radars*, OPERA II Report WP WD_2006_18.

[32] Martin, P.M., Goddard, J.W.F., and Cherry, S.M. (1984), 'Identification of hydrometeors and other targets by dual-polarization radar', *Radio Science*, 19(1):132–140.

[33] Chatelet, J., Testud, J., Andrieu, H., and Tabary, P. (2004), 'A first dual polarization weather radar in the French network: test of its usefulness for hydrology', in *Proceedings of European Conference on Radar in Meteorology and Hydrology (ERAD)*, 2004, pp. 421–424.

[34] Ice, R.L. (2010), *Sensitivity Analysis Comparing the WSR-88D Baseline and L3 Baron Dual Polarization Modification*, WSR-88D Radar Operations Center, Revision 1.

[35] Cate, G. (2010), 'Dual polarization is coming to NEXRAD!', in *NEXRAD*, Issue 20, December 2010.

[36] *Fact Sheet 15, Weather Radar*, National Meteorological Library and Archive.https://www.metoffice.gov.uk/binaries/content/assets/metoffice-govuk/pdf/research/library-and-archive/library/publications/factsheets/factsheet_15-weather-radar-2020_temp.pdf. Retrieved 20 February 2023.

[37] Beauchamp, R.M. and Chandrasekar, V. (2016), 'Dual-polarization radar characteristics of wind turbines with ground clutter and precipitation', *IEEE Transactions on Geoscience and Remote Sensing*, 54(8):4833–4846.

[38] Hall, W., Rico-Ramirez, M.A., and Kramer, S. (2016), 'Offshore wind turbine clutter characteristics and identification in operational C-band weather radar measurements', *Quarterly Journal of the Royal Meteorological Society*, 143(703):720–730.

[39] Darlington, T. (2022), 'Case study from MET Office – use radar in weather forecasting and what are the factors to consider', Radar 2022, Edinburgh October 2022. In response to a question from Matthew Keal, Vattenfall, indicated he was optimistic that discrimination of wind turbine clutter using dual polarisation could be carried out successfully, Author's contemporaneous note.

[40] Clifton, A., Kilcher, L., Lundquist. J.K., and Fleming, P (2016), 'Using machine learning to predict wind turbine power output, *Environmental Research Letter*, 8:024009.

[41] Martinez, C.R., Yeboah, F.A., Herford, J.S., Brzezinski, M., and Puttagunta, V. (2019), 'Predicting wind turbine blade erosion using machine learning', *Data Science Review*, 2(2):Article 17.

[42] Gini, F. (2021), *Grand Challenges in Radar Signal Processing*. https://www.frontiersin.org/articles/10.3389/frsip.2021.664232/full#B34. Retrieved 22 February 2023.

[43] Russel, S. and Norvig, P. (2022), *Artificial Intelligence: A Modern Approach*, Pearson Education.

[44] IET (2023), *Artificial Intelligence and Functional Safety*. https://www.theiet.org/impact-society/factfiles/engineering-safety-factfiles/artificial-intelligence-and-functional-safety/. Retrieved 21 February 2023.
[45] Sell, P.S. (1985), *Expert Systems – A Practical Introduction*, Cambridge University Press.

Bibliography

Introduction

There are many references throughout this book but if the reader wanted to investigate some of the subjects further, the following textbooks are regarded as some of the best in their subject areas.

Chapter 1: Introduction

A very good introduction to the subject of wind turbines can be found in Anderson's:

"Wind Turbines: Theory and Practice", Colin G. Anderson, Publisher: Cambridge University Press, 2020, ISBN: 978-1-108-47832-8.

Chapter 2: A brief history of windmills, electricity generation and radar

On the history of windmills

The origins of extracting energy from the wind in the Middle East, and many other forms of technology is covered in Al-Hassan and Hill's work:

"Islamic Technology: An Illustrated History", Ahmad, Y. al-Hassan and Donald Hill, Publisher: Cambridge University Press, 1996, ISBN: 0-521-42239-6.

The late Rex Wailes was a recognised authority on windmills in the United Kingdom. His beautifully illustrated work on windmills in England covers their introduction and development.

"Windmills in England: A Study of Their Origin, Development and Future", Rex Wailes, Publisher: Charles Skilton Ltd., 1975.

Hills' scholarly work on the rise and fall of the traditional windmill is an interesting account that concludes with a chapter dedicated to the use of wind power for the generation of electricity.

"Power from Wind: A History of Windmill Technology", Richard L. Hills, Publisher: Cambridge University Press, 1996, ISBN: 0-521-41398-2.

On the history of electricity generation

The archives of Michael Faraday are held by the UK Institution of Engineering and Technology (IET), and they can be viewed by appointment. The Hathi Trust has digitised Faraday's Experimental Researches in Electricity and these are available on line.*

A useful consolidated version of Michael Faraday's memoires is contained in:

"The Discovery of Induced Electric Currents: Memoirs by Michael Faraday", Joseph Henry and Michael Faraday, Publisher: Franklin Classics, 2022, ISBN: 9-780342-331437.

The library of the IET also holds a number of valuable documents including Hertz's memoires:

"Electric Waves: Being Researches on the Propagation of Electric Action with Finite Velocity Through Space", Dr Heinrich Hertz. Authorised English translation by D.E. Jones. With a preface by Lord Kelvin, 1893.

This document is particularly interesting because it was once owned by Sir Oliver Lodge who made many insightful hand-written notes in pencil in the margins of the book.

A comprehensive and very readable biographical work on Faraday and Maxwell was written by Forbes and Mahon:

"Faraday, Maxwell and the Electromagnetic Field: How Two Men Revolutionized Physics", Nancy Forbes and Basil Mahon, Publisher: Prometheus Books, 2014, ISBN: 978-1-61614-942-0.

On the more specific topic of the use of wind power to generate electricity, the work by the late Edward Golding provides an interesting historical introduction as well as, a well-developed technical introduction to the subject.

"The Generation of Electricity by Wind Power", E. W. Golding, Publisher: E and F N Spon Ltd, 1976, ISBN: 0-419-11070-4.

On the history of radar

A summary of the early history of radar, with very good illustrations, can be found in:

"Radar Origins Worldwide: History of Its Evolution in 13 Nations Through World War II", Dr Raymond C. Watson Jr, Publisher: Trafford Publishing, 2009, ISBN: 978-1-4269-2110-0.

*Information provided courtesy of Anne Locker, the IET.

A summary of early British radar history which has excellent maps, diagrams and images:

"Watching the Skies: The History of Ground Radar in the Air Defence of the United Kingdom", Jack Gough, Publisher: Her Majesty's Stationery Office (HMSO), 1993, ISBN: 0-11-772723-7.

Chapter 3: Aviation systems

On aviation regulation

A number of important documents on the interaction of wind turbines and aviation radio and radar systems are identified in Chapter 3.

On aviation

Binns overview of Aviation systems is highly recommended.

"Aircraft Systems: Instruments, Communications, Navigation, and Control", Chris Binns, Publisher: Wiley IEEE Press, 2019, ISBN: 978-1119259541.

To provide a context for aviation systems, there are a number of works that introduce ATC. A good example is:

"Fundamentals of Air Traffic Control", 3rd ed., Michael S. Nolan, Publisher: Brooks/Cole Publishing Company, 1999, ISBN: 0-534-56795-9.

On air–ground–communications

Binns 2019 includes a useful discussion of radio systems as well as aviation electronic systems in general. The following book is dedicated to the topic of AGA communications.

"Aircraft Radio Systems", James Powell, Publisher: Jeppesen Sanderson, 1990, ISBN: 978-0-89-100356-4.

There are also many fine books on general Communications Theory. The following is a good example:

"Communication Systems", 2nd ed., Marcel S. Alencar and Valdemar C. da Rocha Jnr, Publisher: Springer, 2005, ISBN: 978-3-030-25461-1.

On ILS

The following work also makes extensive use of diagrams and images and covers all the essential concepts of ILS.

"ILS Fundamentals", Revision 8.2, Roger Holm, Publisher: CreateSpace Independent Publishing Platform; Edition 8.2, 2017, ISBN: 978-1-97-919924-7.

On navigation systems

Binns (2019) provides a good introduction to the subject of navigation. A more comprehensive treatment of the siting issues and propagation factors relating to VOR, DME and VORTAC can be found in:

"VOR, VOR/DME and VORTAC Siting Criteria", [US] Federal Aviation Administration, [US] Department of Transportation, Publisher: Scholar's Choice, Original Work 1986, ISBN: 978-1-29-604361-2.'

On primary radar

There are numerous good text books on the subject of radar in general. The two best known are written and edited, respectively, by Merrill Skolnik.

"Introduction to Radar Systems", 3rd ed., Merrill I. Skolnik, Publisher: McGraw-Hill, 2002, ISBN: 0-07-290980-3.

"Radar Handbook", 3rd ed., Merrill I. Skolnik, Publisher: McGraw-Hill, 2008, ISBN: 978-0-07-148547-0.

The following book covers similar ground to that covered by the Skolnik works but, as the title indicates, Nathanson's work spends more time covering how radar interacts with the environment. In particular, a strong feature in this book is his coverage of Radar Cross Section of aircraft.

"Radar Design Principles: Signal Processing and the Environment", 2nd ed., Fred E. Nathanson, Publisher: Scitech, 1991, ISBN: 1-891121-09-X.

Despite the title, the following work has many chapters devoted to general radar principles. The unique selling point of this book is its emphasis on the use of diagrams and visual representation to explain the most complex of concepts.

"Introduction to Airborne Radar", 2nd ed., George W. Stimson, Publisher: Scitech, 1998, ISBN: 0-85296-942-2.'

On radar tracking

To date, one of the most comprehensive treatments of tracking using radar is by the late Eli Brookner.

"Tracking and Kalman Filtering Made Easy", Eli Brookner, Publisher: Wiley Interscience, 1988, ISBN: 0-471-18407-1.'

On secondary radar

Not just a good text on secondary surveillance radar but also a good model for how to write technical works:

"Secondary Surveillance Radar", Michael C. Stevens, Publisher: Artech House, ISBN: 0-89006-292-7.

Chapter 4: Wind turbines and their effects

There are many academic papers and numerous reports that deal with the effects of wind turbines on systems but currently there are very few books. There are, however, a number of books which provide comprehensive treatments of the production of wind energy and these provide references to the problems of site planning. Thus, the following has reference to some of the problems of compatibility with electronic systems in general.

"Wind Energy Handbook", 3rd ed., Tony Burton, Nick Jenkins, Ervin Bossanyi, David Sharpe and Michael Graham, Publisher: John Wiley and Sons Ltd, 2021, ISBN: 978-1-119-45109-9.

Chapter 5: Analysis

On the subject of radio propagation

The following work by the late Professor Les Barclay provides a comprehensive treatment of the propagation of radio waves. In the context of this book, Chapter 8 by David Bacon which discusses diffraction is particularly relevant.

"Propagation of Radiowaves", 2nd ed., Edited by Les Barclay, Publisher: IET, 2003, ISBN: 978-0-85296-102-5.

On the subject of mapping

The US Geological Survey publishes a lot of material on mapping; the problems of providing a two-dimensional representation of a three-dimensional world are illustrated by one of their publications:

"An Album of Map Projections", John P. Snyder and Philip M. Voxland, US Geological Survey, Professional Paper 1453, Publisher: US Government Printing Office, 1989.

On the subject of radar performance

Many books deal with radar performance: for example, all the books mentioned above in the list of Chapter 3 radar textbooks discuss performance. There is, however, one particularly valuable book dedicated to the subject by Blake whose name has become synonymous with radar performance through his eponymous chart for calculation of radar range.

"Radar Range-Performance Analysis", Lamont V. Blake, Publisher: Lexington Books, 1980, ISBN: 0-669-00781-1.

Another useful text for a wide range of radar performance issues is by Barton. Chapter 1 is given over to the calculation of radar range performance.

"Modern Radar System Analysis", David Knox Barton, Publisher: Artech House, ISBN: 0-89006-170-X.

On the general subject of radar cross section

For detailed treatments of RCS, any text written by Eugene Knott is recommended. Knott was also published in

"Aspects of Modern Radar", Edited by Eli Brookner, Publisher: Artech House, ISBN: 1-58053-128-8.

and

"Radar Handbook", 3rd ed., Merrill I. Skolnik, Publisher: McGraw-Hill, 2008, ISBN: 978-0-07-148547-0.

A further useful reference on the general subject of RCS is:

"Radar Cross Section Lectures", Allen E. Fuhs, Publisher: American Institute of Aeronautics and Astronautics, ISBN: 0-915028-88-4.

On the subject of the RCS of wind turbines

"Wind Turbine Effects on Radar Performance: Effect Analyses and Mitigation Approaches", Osman Karabayır, Publisher: LAP LAMBERT Academic Publishing, 2017, ISBN-13: 978-3-330-03508-9.

"Wind Farms Impact on Radar Aviation Interests – Final Report", Gavin Poupart, 2003, FES W/14/00614/00/REP, DTI PUB URN 03/1294.

"Feasibility of Mitigating the Effects of Windfarms on Primary Radar", M.M. Butler and D.A. Johnson, 2003, ETSU W/14/00623/REP, DTI PUB URN No. 03/976.

Chapter 6: Mitigation

To date there are relatively few books given over to mitigation. The most comprehensive treatment of wind turbine mitigation is contained in Oswald and Baker's book which describes the Holographic Staring Radar; one of the prime objectives for designing the radar was to mitigate wind turbine effects.

"Holographic Staring Radar", Gordon Oswald and Chris Baker, Publisher: IET Scitech, 2022, ISBN-13: 978-1-78561-389-0.

The following work provides a more generic approach to the subject dealing with a diverse range of radar topics:

"Wind Turbine Effects on Radar Performance: Effect Analyses and Mitigation Approaches", Osman Karabayır, Publisher: LAP LAMBERT Academic Publishing, 2017, ISBN-13: 978-3-330-03508-9.

The title of the following book implies that it may be outside of the scope of the material covered by this book. However, some of the concepts and principles described have a wider context; for example, the book considers the effects of ducting and this has direct relevance to the off-shore environment. The book has some very graphic examples of the problems that arise off-shore and it also serves to illustrate that systems must address the specific operational requirements of a domain and must also be able to operate in specific environments; in this case the marine environment.

"Wind Turbine Generator Impacts to Marine Vessel Radar", The Ocean Studies Board Division of Earth and Life Studies, National Academies of Sciences, Engineering and Medicine, Publisher: National Academies Press, 2022, ISBN: 978-0-309-27548-4.

Glossary

AC:	alternating current
ACAS:	airborne collision avoidance system (an alternative name for traffic alert and collision avoidance system)
ACP:	Airspace Change Process
AD:	air defence
ADC:	analogue to digital converter
ADF:	automatic direction finder
ADSB:	automatic dependent surveillance – broadcast
AEEC:	Airlines Electronic Engineering Committee
AFB:	[US] Air Force Base
AFS:	aeronautical fixed service
AFTN:	aeronautical fixed telecommunications network
AGA:	air-ground-air [communications]
agl:	above ground level
AI:	altitude indicator
AI:	artificial intelligence
Air Cdre:	Air Commodore
AM:	amplitude modulation
AMS:	aeronautical mobile service
amsl:	above mean sea level
ANN:	artificial neural networks
ANO:	air navigation order
ANSP:	air navigation service provider
AOR:	area of responsibility
ARCEP:	Autorité de Régulation des Communications Électroniques et des Postes
ARINC:	Aeronautical Radio Incorporated
ARP:	airport reference point
ARSR:	air route surveillance radar
ASDE:	airport surface detection equipment
asl:	above sea level

ASR: airfield surveillance radar
ATC: air traffic control
ATCO: air traffic control officer
ATCRBS: air traffic control radar beacon system
ATIS: automatic terminal information service
ATM: air traffic management
ATZ: aerodrome traffic zone
BCAS: beacon collision avoidance system
BCE: before the current era
BDS: [Comm] B data select (a subset of mode S messages)
BGS: British Geological Survey
BIT: built in test
BLOS: beyond line of sight
BRA: buildings in a restricted area
BT: [ITU] Broadcast Television [Recommendations]
BVOR: Broadcast VHF omnidirectional range
C: centigrade
CAA: [UK] Civil Aviation Authority
CACFAR: cell averaging constant false alarm rate [processor]
CADMID: concept, assessment, development, manufacture, in-service, development
CADMIT: concept, assessment, development, manufacture, in-service, termination
CAP: [UK] Civil Aviation Publication
Cdr: [Naval] commander
CDR: critical design review
CE: current era
CFAR: constant false alarm rate
CM: corrective maintenance
CMLD: censored mean level detector (a CFAR algorithm)
CNS: communications, navigation and surveillance
CPDLC: controller pilot data link communications
CRC: cyclic redundancy check
CSB: carrier and sideband signal (ILS signal)
CTZ: control zone
CVOR: conventional VHF omnidirectional range
CW: continuous wave
C:I: carrier to interference ratio
DABF: digital adaptive beamforming

DABS: discrete address beacon system (alternative name for Mode S)
dB: decibel
dBi: decibel values for antennas which are referenced to the isotropic antenna
dBm: decibel values for power referenced to 1 mW
dBsm: decibel values for radar cross sections which are referenced to a one square metre RCS object
dBW: decibel values for power referenced to 1 W
DC: direct current
DC: [Washington] District of Columbia
DC: Dolph Chebyshev (a phased array antenna taper)
DF: downlink format (of SSR Mode S messages)
DFAG: doubly fed asynchronous generator
DFIG: doubly fed induction generator
DFS: Deutsche Flugsicherung
DLOD: defence lines of development
DME: distance measuring equipment
DNO: distribution network operator
DNS: developments of national significance
DOD: [US] Department of Defense
DP: data processor
DPSK: differential phase shift keying (modulation used for Mode S interrogations)
DSB: double sideband
DSNA: Direction des Services de la navigation aérienne
DSP: digital signal processor
DTI: Department of Trade and Industry
DVOR: Doppler VHF omnidirectional range
EASA: European [Union] Aviation Safety Agency
EEZ: exclusive economic zone
EIA: Environmental Impact Assessment
EIS: Environmental Impact Statement
EMC: electromagnetic compatibility
ERAD: European Conference on Radar in Meteorology and Hydrology
ESA: electronically scanned array
EU: European Union
EUMETNET: European Meteorological Network
FAA: Federal Aviation Administration
FAF: final approach fix

FAR: false alarm rate
FAT: factory acceptance test
FCC: Federal Communications Commission
FIR: flight information regions
FIS: flight information service
FM: frequency modulation
F_N: noise factor
FOC: full operational capability
FOWT: floating offshore wind turbine
GA: general aviation (light aircraft)
GCA: ground controlled approach (an alternative name for a precision approach radar)
GCI: ground controlled interception (of aircraft)
GE: general electric
GHG: greenhouse gases
GIS: geographical information system
GOCFAR: greatest of constant false alarm rate [processor]
Gp Capt: Group Captain
GPIA: Glasgow Prestwick International Airport
GPS: global positioning system
G/S: ILS glide slope
HAWT: horizontal axis wind turbine
HF: high frequencies
HP: horse power
hPA: hecto Pascal
HPDME: high power distance measuring equipment
IAA: Irish Aviation Authority
IATA: International Air Transport Association
ICAO: International Civil Aviation Authority
IEA: International Energy Agency
IEC: International Electrotechnical Commission
IEE: Institution of Electrical Engineering
IEEE: Institute of Electrical and Electronics Engineers
IET: [UK] Institution of Engineering and Technology
IFF: identify friend or foe (military term for secondary surveillance radar)
IFR: instrument flight rules
ILS: instrument landing system
INCOSE: International Council On Systems Engineering

IOC: initial operational capability
IRENA: International Renewable Energy Agency
IRL: integration readiness level
ISA: International Seabed Authority
ISBN: International Standard Book Number
ISO: International Standards Organisation
ITU: International Telecommunications Union
I&Q: in-phase and quadrature (pair of signals)
JPDF: joint probability density function
KPP: key performance parameter
KRA: key resource areas
LDA: landing distance available
LEE: [blade] leading edge erosion
LEP: [blade] leading edge protection
LFM: linear frequency modulation (one of the forms of chirp used in pulse compression)
LHCP: left-hand circularly polarized
LIDAR: light detection and ranging
LLZ: instrument landing system localiser
LO: local oscillator
log: logarithm
LORAN: long-range navigation
LOS: line of sight
LPA: Local Planning Authority
LPDME: low power distance measuring equipment
LPS: [turbine blade] lightning protection system
LRU: line replaceable unit
LVA: large vertical aperture (SSR antenna)
MAA: [UK] Military Aviation Authority
magl: metres above ground level
mamsl: metres above mean sea level
masl: metres above sea level
MATZ: military aerodrome traffic zone
mbar: millibars [unit of pressure]
MCA: marine conservation areas
MCS: [UK] Microgeneration Certification Scheme
MDS: minimum detectable signal
MIP: mode interlace pattern

M-Lat: multilateration
MLD: mean level detector (an alternative name for cell averaging constant false alarm rate processing)
MLS: machine learning systems
MOD: Ministry of Defence
MON: [VHF omnidirectional range] minimum operating network
MPS: maintenance policy statement
MRP: [UK] Military Aviation Authority Regulatory Publications
msl: mean sea level
MSSR: monopulse secondary surveillance radar
MTI: moving target indication
MUF: maximum usable frequency
mV: millivolt (a thousandth of a volt)
NAIA: non-auto initiation area (alternative name for an non-auto initiation zone)
NAIZ: non-auto initiation zone
NASA: National Aeronautics and Space Administration
NATO: North Atlantic Treaty Organisation [OTAN]
NATS: [UK] National Air Traffic Services
NAVAID: navigational aid
NDB: non-directional beacon
NERL: [UK] NATS En-Route Limited
NF: noise figure
NLFM: non-linear frequency modulation (one of the forms of chirp used in pulse compression)
Nm: nautical miles
NOTAM: notice to airmen/notice to air missions
NRL: [US] Naval Research Laboratory
NTIA: National Telecommunications and Information Administration
NTSB: [US] National Transportation Safety Board
OHL: overhead lines (for electricity distribution)
OPEC: Organisation of Petroleum Exporting Countries
OPERA: operational programme for the exchange of weather radar information
OTHR: over the horizon radar
PAR: precision approach radar
PBN: performance-based navigation (new paradigm for flexible navigation methods)
PDF: probability density function
PED: portable electronic devices

P_{fa}:	probability of false alarm
PGES:	[UK] Parliamentary Group for Energy Studies
PLM:	piecewise linear model (RCS model of wind turbines)
PM:	preventative maintenance
PME:	prime mission equipment
PPM:	pulse position modulation (Mode S replies)
PRBS:	pseudo-random binary sequences
PRF:	pulse repetition frequency
PSR:	primary surveillance radar
P&L:	Penrhyddlan and Llidiartywaun [windfarms]
radar:	radio detection and ranging
RAE:	Royal Aircraft Establishment
RAF:	[UK] Royal Air Force
RAG:	range azimuth gating
RAL:	Rutherford Appleton Laboratory
RAP:	recognised air picture
RCA:	Radio Corporation of America
RCS:	radar cross-section
RDP:	radar data processor
Retd.:	retired
RFC:	Royal Flying Corps
RHCP:	right-hand circularly polarized
RN:	Royal Navy
RPIE:	real property installed equipment
rpm:	revolutions per minute
RSEC:	radar spectrum engineering criteria
RSGB:	Radio Society of Great Britain
RTTA:	returns to track association (part of a radar tracking software)
RVR:	runway visual range
RX:	receiver
R&D:	research and development
R/T:	radio telephony
SAC:	special areas of conservation
SAGE:	semi-automatic ground environment
SARG:	[UK CAA] Safety and Aerospace Regulation Group
SAM:	surface to air missiles
SAW:	surface acoustic wave
SBO:	side band only (ILS signal)

SCIG:	squirrel cage induction generator (type 1 generator)
SERA:	Standardised European Rules of the Air
SESAR:	Single European Sky ATM Research
SI:	International System of Units
SIF:	[SSR/IFF] selective identification feature
SNR:	signal-to-noise ratio
SOA:	state of the art
SOLAS:	safety of life at sea
SPA:	special protection areas
SQEP:	suitably qualified and experienced person
SRL:	system readiness level
SRTM:	[space] shuttle radar topography mission
SS:	steam ship
SSB:	single sideband
SSR:	secondary surveillance radar
SSV:	standard service volumes (of VOR)
SSSI:	site of special scientific interest
STC:	sensitivity time control
TACAN:	tactical air navigation
TDB:	track data block
TDOA:	[multilateration] time distance of arrival
TEM:	transverse electromagnetic wave
TEPIDOIL:	training, equipment, personnel, infrastructure, doctrine, organisation, information, logistics
TLP:	tension leg platform (floating wind turbine)
TMZ:	transponder mandatory zone
TRL:	technology readiness level
TSV:	terminal service volume [of a DME]
TVOR:	terminal VHF omnidirectional range
TX:	transmitter
T_0:	reference temperature
UAR:	upper air routes
UAS:	unmanned air (or aircraft) system (a drone)
UAV:	unmanned aerial vehicle (an alternative name for a drone)
UHF:	ultra high frequency
UIR:	upper information region
UKADGE:	United Kingdom Air Defence Ground Environment
UN:	United Nations

UNCLOS:	United Nations Convention on the Law Of the Sea
UNESCO:	United Nations Educational, Scientific and Cultural Organization
URATS:	user requirements for air traffic services
US:	United States of America
USSR:	Union of Soviet Socialist Republics
UV:	ultraviolet
V:	voltage
VAR:	visual aural range
VAWT:	vertical axis wind turbine
VFM:	visual flight minima
VFR:	visual flight rules
VHF:	very high frequency
VMC:	visual meteorological conditions
VOR:	VHF omnidirectional range
VORTAC:	VHF omnidirectional range/tactical air navigation
VOT:	VOR test
VTS:	vessel traffic service
WAM:	wide area multilateration
WGS:	world geodetic system
WO:	Warrant Officer
WPAC:	Wind Power Aviation Consultants
WRC:	World Radio Conference
WTG:	wind turbine generator
W:U:	wanted to unwanted [signal] ratio
ZTV:	zone of theoretical visibility
μV:	microvolt (a millionth of a volt)

Index

www.ingramcontent.com/pod-product-compliance
Lightning Source LLC
LaVergne TN
LVHW011548200926
848484LV00001B/297

* 9 7 8 1 8 3 9 5 3 8 4 5 2 *